Cannabinoid Modulation of Emotion, Memory, and Motivation

Patrizia Campolongo • Liana Fattore
Editors

Cannabinoid Modulation of Emotion, Memory, and Motivation

Springer

Editors
Patrizia Campolongo
Dept. of Physiology and Pharmacology
Sapienza University
Rome
Italy

Liana Fattore
Institute of Neuroscience CNR
University of Cagliari
Cagliari
Italy

ISBN 978-1-4939-4875-8 ISBN 978-1-4939-2294-9 (eBook)
DOI 10.1007/978-1-4939-2294-9

Springer New York Heidelberg Dordrecht London

Softcover reprint of the hardcover 1st edition 2016

Printed on acid-free paper

Springer is part of Springer Science+Business Media (www.springer.com)

To Damiano, Matteo, Marco and Nicola, for their patience in conceding us to pursue our passion for science.

Preface

Memory, emotions, reward, motivation, dependence, appetite, sociability. These are only some of the multiple domains in which the (endo)cannabinoid system is involved. We usually think of marijuana and cannabis derivatives as recreational compounds. But what do we know about their actions on the brain, about how the endocannabinoid system modulates such numerous and important aspects of our life? Are we really aware of the uniqueness of this, until recently unknown but, ubiquitous neuromodulatory system? What can we learn from recent progress in research?

These are all questions that recent research allowed us to start addressing. After the finding of specific receptors that are activated by smoking marijuana, it was the time of the discovery of a number of endogenous marijuana-like substances called endocannabinoids followed by the identification of metabolic enzymes for such ligands. Other groundbreaking advances in the field then paved the way for an enthusiastic research activity on this fascinating regulatory system. And the more we know the more we want to know. This book is intended to offer an all-embracing overview of the most recent discoveries on the role played by the endocannabinoid system in the modulation of memory, emotions, reward and motivation, and how it interferes with the actions of other drugs of abuse and underlying neurotransmission systems.

In *Cannabinoid Modulation of Emotion, Memory, and Motivation* leading experts in the field critically illustrate and discuss in dedicated chapters recent breakthroughs on the effects of cannabinoids on memory, learning and cognition, fear-coping strategies and emotional processing, motivation and reward. A particular emphasis is given to the delicate issues of cannabis use by adolescents and the emerging role of gender and sexual hormones in the frequency and consequences of its use, the problem of poly-substance abuse, and the diffusion of potent synthetic cannabinoids on the internet.

The book is organized into three distinct sections. Part I focuses on the modulation of memory and emotions by cannabinoids, featuring the underlying neurobiology and emphasizing their effects on fear, anxiety and depression. Part II is centered on reward and motivation; it discusses subjective, cognitive, and social effects of cannabinoids and their impact of the motivational brain system with a particular attention on age and sex effects. Finally, interactions of cannabinoids with other drugs of abuse such as nicotine, alcohol, opioids and methamphetamine are illustrated in Part III, with a special focus on their interaction with the dopaminergic neurotransmission system.

This book will stimulate curiosity toward research on (endo)cannabinoids from molecular neurobiology to behavior to therapeutic implications and will be of help to students, scientists and clinicians for better appreciating this captivating brain endogenous system and its powerful modulatory action.

Patrizia Campolongo
Liana Fattore

Contents

Contributors

Irit Akirav Department of Psychology, University of Haifa, Haifa, Israel

Mano Aliczki Department of Behavioural Neurobiology, Institute of Experimental Medicine, Budapest, Hungary

Alessia Auber Preclinical Pharmacology Section, Behavioral Neuroscience Research Branch, Department of Health and Human Services, National Institute on Drug Abuse, IRP, Baltimore, MD, USA

Marco Bortolato Department of Pharmacology and Toxicology, School of Pharmacy, University of Kansas, Lawrence, Kansas, USA

Patrizia Campolongo Department Physiology and Pharmacology, Sapienza University of Rome, Rome, Italy

Benjamin Chadwick Departments of Neuroscience and Psychiatry, Friedman Brain Institute, Icahn School of Medicine at Mount Sinai, New York, NY, USA

Joseph F. Cheer University of Maryland School of Medicine, Baltimore, MD, USA

Giancarlo Colombo Institute of Neuroscience-Cagliari, National Research Council (CNR), Monserrato, Italy

Caroline Davis Department of Kinesiology and Health Sciences, York University, Toronto, Ontario, Canada

Filippo Drago Department of Clinical and Molecular Biomedicine, Section of Pharmacology and Biochemistry, Medical School, University of Catania, Catania, Italy

Marie R. Ehrler Department of Psychiatry, University of Utah School of Medicine, Salt Lake City, UT, USA

Liana Fattore Institute of Neuroscience-Cagliari, National Research Council (CNR), Monserrato, Italy

CNR Institute of Neuroscience-Cagliari, National Research Council-Italy @ Department of Biomedical Sciences, University of Cagliari, Cittadella Universitaria di Monserrato, Monserrato, Cagliari, Italy

François Georges IINS-UMR5297, CNRS, Bordeaux, France

Steven R. Goldberg Preclinical Pharmacology Section, Behavioral Neuroscience Research Branch, Department of Health and Human Services, National Institute on Drug Abuse, IRP, Baltimore, MD, USA

Jozsef Haller Department of Behavioural Neurobiology, Institute of Experimental Medicine, Budapest, Hungary

Institute of Experimental Medicine, Hungarian Academy of Sciences, Budapest, Hungary

Yasmin L. Hurd Departments of Neuroscience and Psychiatry, Friedman Brain Institute, Icahn School of Medicine at Mount Sinai, New York, NY, USA

Icahn School of Medicine at Mount Sinai, 1470 Madison Avenue, New York, NY, USA

Zuzana Justinova Preclinical Pharmacology Section, Behavioral Neuroscience Research Branch, Department of Health and Human Services, National Institute on Drug Abuse, IRP, Baltimore, MD, USA

Jana Kucerova Department of Pharmacology, CEITEC (Central European Institute of Technology) Masaryk University, Brno, Czech Republic

Paola Maccioni Institute of Neuroscience-Cagliari, National Research Council (CNR), Monserrato, Italy

Antonia Manduca Department of Science, Roma Tre University, Rome, Italy

Eva María Marco Departamento de Fisiología (Fisiología Animal II), Facultad de Biología, Universidad Complutense de Madrid (UCM), Madrid, Spain

Giovanni Marsicano University of Bordeaux, INSERM U862 NeuroCentre Magendie, Bordeaux, France

Erin C. McGlade Department of Psychiatry, University of Utah School of Medicine, Salt Lake City, UT, USA

Miriam Melis Department of Biomedical Sciences, University of Cagliari, Cagliari, Italy

Mathilde Metna-Laurent University of Bordeaux, INSERM U862 NeuroCentre Magendie, Bordeaux, France

Vincenzo Micale Department of Pharmacology, CEITEC (Central European Institute of Technology) Masaryk University, Brno, Czech Republic

Michael Michaelides Departments of Neuroscience and Psychiatry, Friedman Brain Institute, Icahn School of Medicine at Mount Sinai, New York, NY, USA

Michael L. Miller Departments of Neuroscience and Psychiatry, Friedman Brain Institute, Icahn School of Medicine at Mount Sinai, New York, NY, USA

Jeffery H. Moran Arkansas Department of Health, Arkansas Public Health Laboratory, Little Rock, AR, USA

Maria Morena Department Physiology and Pharmacology, Sapienza University of Rome, Rome, Italy

Claudia V. Morris Departments of Neuroscience and Psychiatry, Friedman Brain Institute, Icahn School of Medicine at Mount Sinai, New York, NY, USA

Luis A. Natividad Research Associate, Committee on the Neurobiology of Addictive Disorders, The Scripps Research Institute, La Jolla, CA, USA

Erik B. Oleson University of Maryland School of Medicine, Baltimore, MD, USA

Lucas de Oliveira Alvares Psychobiology and Neurocomputing Lab - LPBNC Dept. of Biophysics,

IB & Neurosciences Graduate Program, ICBS, Universidade Federal do Rio Grande do Sul (UFRGS), Porto Alegre, Brazil

Leigh V. Panlilio Preclinical Pharmacology Section, Behavioral Neuroscience Research Branch, Intramural Research Program, Department of Health and Human Services, National Institute on Drug Abuse, IRP, National Institutes of Health, Baltimore, MD, USA

Loren H. Parsons Committee on the Neurobiology of Addictive Disorders, The Scripps Research Institute, La Jolla, CA, USA

Jorge Alberto Quillfeldt Psychobiology and Neurocomputing Lab - LPBNC Dept. of Biophysics,

IB & Neurosciences Graduate Program, ICBS, Universidade Federal do Rio Grande do Sul (UFRGS), Porto Alegre, Brazil

Maria Scherma Department of Biomedical Science, Section of Neuroscience and Clinical Pharmacology, University of Cagliari, Cagliari, Italy

Kathryn A. Seely Arkansas Department of Health, Arkansas Public Health Laboratory, Little Rock, AR, USA

Edgar Soria-Gómez University of Bordeaux, INSERM U862 NeuroCentre Magendie, Bordeaux, France

Katarina Tabiova Department of Pharmacology, CEITEC (Central European Institute of Technology) Masaryk University, Brno, Czech Republic

Simone Tambaro Department of Pharmacology and Toxicology, School of Pharmacy, University of Kansas, Lawrence, Kansas, USA

Viviana Trezza Department of Science, Roma Tre University, Rome, Italy

Louk J. M. J. Vanderschuren Department of Animals in Science and Society, Division of Behavioural Neuroscience, Utrecht University, Utrecht, The Netherlands

María-Paz Viveros Departamento de Fisiología (Fisiología Animal II), Facultad de Biología, Universidad Complutense de Madrid (UCM), Madrid, Spain

William D. Wessinger Department of Pharmacology and Toxicology, College of Medicine, University of Arkansas for Medical Sciences, Little Rock, AR, USA

Deborah A. Yurgelun-Todd Department of Psychiatry, University of Utah School of Medicine, Salt Lake City, UT, USA

Part I
Cannabinoid Modulation of Memory and Emotions

Chapter 1
Endocannabinoid Modulation of Memory for Emotionally Arousing Experiences

Maria Morena and Patrizia Campolongo

Abstract There is extensive evidence that the endocannabinoid system is a key modulator of memory for emotionally arousing experiences. We have demonstrated that endocannabinoids play an essential role in regulating glucocorticoid effects on different memory processes.

In this chapter we will summarize findings describing cannabinoid effects on emotional memory acquisition, consolidation, retrieval and extinction. Then, we will present evidence indicating a critical involvement of the endocannabinoid system in mediating stress effects on memory. Finally, we will describe how endocannabinoids bidirectionally modulate memory processes depending on the level of stress at the time of drug administration, raising the possibility that endocannabinoids may act as an emotional buffer modulating stress effects on memory for emotional experiences.

Keywords Basolateral amygdala · Emotional arousal · Endocannabinoids · Glucocorticoids · Memory consolidation

Introduction

A large body of evidence indicates that the endocannabinoid system is crucially involved in the modulation of memory consolidation for stressful experiences [1–5]. However both impairing and enhancing effects have been reported with respect to cannabinoid effects on memory. Nowadays, the hypothesis is emerging that cannabinoid drugs can induce distinct, and often opposite, effects on behavior depending on the level of stress and/or aversiveness induced by the environmental context [6–9]. In this chapter we will first summarize the different effects induced by endocannabinoid system modulation on learning and memory functions, comparing the results obtained with different methodological approaches. Then we will focus on the fact that cannabinoid effects on cognitive processes are often biphasic,

P. Campolongo (✉) · M. Morena
Department of Physiology and Pharmacology, Sapienza University of Rome,
P.le A. Moro 5, 00185 Rome, Italy
e-mail: patrizia.campolongo@uniroma1.it

P. Campolongo, L. Fattore (eds.), *Cannabinoid Modulation of Emotion, Memory, and Motivation,* DOI 10.1007/978-1-4939-2294-9_1

providing evidence that such effects are strongly dependent on the aversiveness of environmental context and on the level of stress at the time of drug administration. Finally, we will provide hypotheses that may help to explain cannabinoid dual effects on emotional memory functions.

The Endocannabinoid System in the Brain

The endocannabinoid system is a neuromodulatory lipid system which consists of the cannabinoid receptor type 1 and type 2 (CB1 and CB2, respectively) [10–12] and the two major endogenous ligands for these receptors, the *N*-arachidonoyl ethanolamine (anandamide, AEA) [10] and the 2-arachidonoyl glycerol (2-AG) [13]. Endocannabinoids are synthesized on demand from phospholipid precursors on the postsynaptic membrane by Ca^{2+}-dependent and independent mechanisms [3]. These lipophilic molecules are released directly into the synaptic cleft and act in a retrograde manner on the presynaptic neuron where the cannabinoid receptors are expressed. Activation of the CB1 receptor modulates intracellular transduction pathways through activation/inhibition of several ion channels and kinases, thus inducing the inhibition of further neurotransmitter release [3, 14]. AEA and 2-AG are subsequently taken back into the cell by a still poorly defined uptake process mediated by a transporter mechanism [15, 16] and mainly degraded by the fatty acid amide hydrolase (FAAH) and monoacylglycerol lipase (MAGL) [3], respectively.

CB receptors couple to $G_{i/o}$ proteins which function to inhibit adenylyl cyclase activity, activate potassium channels and inhibit voltage-gated calcium channels [17]. CB1 receptors represent the most abundant class of G-protein-coupled receptors in the central nervous system, but are also present in a variety of peripheral tissues [17], while CB2 receptors are mostly peripherally located on immunological tissues and, only recently, detected in neuronal and glial cells in diverse rat brain areas, including the cerebellum and hippocampus [18, 19]. Within the cortico-limbic system, the most prominent expression of CB1 receptors has been detected in the hippocampus, the basolateral complex of the amygdala (BLA), and the prefrontal cortex (PFC) [20, 21]. Only recently, CB1 mRNA expression has been clearly detected at low levels in the central nucleus of the amygdala (CeA) [22]. Similar to the CB1 receptor, FAAH and MAGL are found in high levels in the BLA, whereas only low levels can be found in the CeA [23]. Several compounds, able to inhibit endocannabinoid transport or degradation, have been characterized and are currently used as pharmacological tools to increase endocannabinoid tone. Among them, the most well characterized compounds are the transport inhibitor AM404 [24], the FAAH inhibitor URB597 [25], the MAGL inhibitor JZL184 [26] and the dual FAAH and MAGL inhibitor JZL195 [27]. Moreover, several extensively used drugs such as acetaminophen [28] and propofol [29, 30] are able to indirectly increase endocannabinoid tone trough inhibition of FAAH.

Within the limbic regions, CB1 receptor is expressed at very high levels in cholecystokinin-positive GABAergic interneurons [31–33] and at moderate to low levels in glutamatergic terminals [3, 34, 35]. Additionally, this receptor has also been detected on serotonergic, noradrenergic and dopaminergic terminals [36–38]. As endocannabinoids and CB1 receptors differentially mediate homeostatic, short- and long-term synaptic plasticity processes [39–41], and neuronal firing [42] throughout the brain, the endocannabinoid system has been reported to be crucially involved in learning and memory processes [1–5].

Cannabinoid Effects on Different Memory Stages

Discrepant findings have been described concerning the role of the endocannabinoid system in the modulation of cognitive processes. Some studies report that cannabis users present short-term memory deficits [43, 44] and impairments in various aspects of executive functioning such as planning, working memory, and mental flexibility [45] after cannabis consumption. However, other studies did not detect any cannabis-related deficit in executive function [46]. Due to the different participant selection strategies used, in terms of poly-drug abuse, pre-existing cognitive and emotional criteria and widely differing methodologies (i.e. chronic vs acute use) it is difficult to draw any clear conclusion from human studies [43, 47].

In this context, basic research is of critical importance to elucidate the neural mechanisms of cannabinoid effects on cognition. However, also preclinical studies in this field are producing some apparent controversial results.

Early studies, examining the effects of systemic pretraining administration of the cannabinoid agonists Δ9-Tetrahydrocannabinol (Δ9-THC) and WIN55,212-2 (WIN) on memory acquisition reported impairing effects on several behavioral tasks in rodents [48–50], whereas other studies reported that intraperitoneal administration of the cannabinoid antagonist rimonabant induces similar effects to those induced by the agonists [51]. Similarly, the exogenous amplification of the endocannabinoid tone induces impairments in memory acquisition in a recognition memory task [6] and in the inhibitory avoidance task [52]. Local infusions into distinct brain regions have given clearer results in this regard. Pretraining administration of a CB1 receptor agonist into the hippocampus has consistently been shown to impair spatial learning [48, 53–55], whereas bilateral blockade of BLA CB1 receptor transmission has been reported to prevent the acquisition of associative fear memory in an olfactory fear conditioning paradigm [56].

Concerning cannabinoid effect on memory consolidation, systemic post-training administration of cannabinoid receptor agonists or of the FAAH inhibitor URB597 have been shown to impair memory consolidation [57–59]. Consistently, systemic post-training injection of cannabinoid receptor antagonists improves the same memory process [60, 61]. Central effects of cannabinoid compound on memory consolidation appear more controversial. Post-training intra-hippocampal administration of the synthetic cannabinoid receptor agonist WIN has been reported to impair memory consolidation [58, 62, 63]. However, other authors reported enhancing effects of anandamide when infused into the hippocampus [64] and of WIN when infused into the BLA [2].

Evidence regarding cannabinoid effects on memory retrieval is still scarce but very consistent. Cannabinoid agonists seem to impair memory retrieval when administered either systemically [65, 66] or in discrete brain areas [67–69].

With regard to memory extinction, literature data have abundantly demonstrated that the cannabinoid system facilitates this memory process. Using a fear conditioning procedure, Marsicano et al. (2002) and subsequent investigators demonstrated that inhibition of endocannabinoid transmission robustly inhibits fear extinction [55, 70–72]. Conversely, stimulation of the endocannabinoid system accelerates fear extinction [71, 73, 74]. Interestingly, Niyuhire and coworkers (2007) reported that rimonabant administration significantly disrupted extinction in two different aversively motivated behavioral tasks (e.g., conditioned freezing and inhibitory avoidance) but failed to affect extinction in an appetitively motivated operant conditioning task [75].

Cannabinoid heterogeneous effects on learning and memory may be due to differences on the dose and route of drug administration, the nature of the task used, the kind of memory (emotional vs non-emotional) [1], the brain areas involved and the memory stage under investigation (acquisition, consolidation, retrieval, and extinction).

However, variations in the stressful conditions employed in the different studies are implicated as well. Recent evidence suggests that the neural processes underlying emotional memory formation seem to be differently sensitive to cannabinoids depending on the levels of emotional arousal associated to the experimental context [9, 76]. In the next paragraph we will briefly describe the interaction between glucocorticoids and arousal-induced norepinephrine in modulating emotional memory, providing evidence which demonstrates how endocannabinoids are crucial mediators of stress effects on memory.

Stress and Endocannabinoids in the Regulation of Memory Function

Memories for emotional events are more persistent and vivid than other memories [77]. Studies examining emotional memory have focused on the highly arousing nature of emotional stimuli or experimental contexts as the key component contributing to enhanced memory [77–80]. The brain regions mainly involved in emotionality are represented by cortico-limbic structures, such as the amygdala, hippocampus, ventral striatum, and medial and orbital regions of the PFC [81]. Among these brain regions, the amygdala represents a key structure for assigning emotional salience to external stimuli and for orchestrating the use of various memory systems in different brain regions, for fear and anxiety responses but also for processing of positive emotions, during periods of emotional arousal [82–86].

Emotional and stressful experiences, *via* the activation of specific hormonal and brain systems, modulate brain function and regulate memory storage. The response to stress involves the release of epinephrine and glucocorticoids (cortisol in humans; corticosterone in rodents) from the adrenal gland into the bloodstream. Consequently, the peripheral stimulation of the vagal nerve and the activation of the nucleus tractus solitarius (NTS) induce a strong noradrenergic input from the locus coeruleus (LC) to several limbic structures including the amygdala shortly after stress [87]. The same neurons also receive high levels of corticosterone which binds with higher affinity to mineralcorticoid receptors (MRs) and lower affinity to glucocorticoid receptors (GRs) [88], so that during basal conditions only MRs are occupied while during stress conditions both MRs and GRs are occupied and mediate genomic and rapid non-genomic actions [89, 90]. Typically, stress hormones mediate the selective enhancement of consolidation of memory for emotionally significant experiences [89, 91–94]. Conversely, glucocorticoids impair memory retrieval and working memory during emotionally arousing tests [93, 95–98].

Considerable evidence indicates that emotional memory modulation requires activation of the BLA specifically. Lesions of the BLA block the memory enhancing effects of systemic injections of GR agonists on inhibitory avoidance retention, whereas lesions of the CeA are ineffective [99]. During emotionally arousing training, norepinephrine is also released into the amygdala to enhance memory consolidation [77, 100–102], whereas β-adrenoceptor antagonists infused into the BLA, but not into the CeA, block the memory enhancement induced by a glucocorticoid administered either systemically or directly into the BLA [103, 104]. Considerable evidence indicates that glucocorticoids interact with this training-induced noradrenergic activation within the amygdala in enhancing the consolidation of memory of emotionally arousing training experiences [105]. The selective influence of glucocorticoids in modulating memory for emotionally arousing information [106] indicates that glucocorticoid effects on memory processes require concurrent noradrenergic activation in the amygdala [93, 107].

Compelling evidence has been reported in the literature demonstrating a strong bidirectional interaction between endocannabinoids and stress-activated hormones such as glucocorticoids and norepinephrine. For instance, stressful experiences significantly alter endocannabinoid content in limbic brain regions resulting in opposing actions that can both increase and terminate the stress response [108]. Conversely, in the hypothalamus, glucocorticoids induce endocannabinoid signaling [109], which suppresses hypothalamic-pituitary-adrenal (HPA) axis activity by inhibiting the release of glutamate in the paraventricular nucleus [109, 110]. Furthermore, the existence of CB1 receptors in the LC and NTS suggests that cannabinoids may modulate noradrenergic activity [111–115]. Intravenous injection of cannabinoid agonists dose-dependently increases the firing rate of LC noradrenergic neurons *via* the activation of CB1 receptors [116] as well as noradrenergic release in cortical and limbic brain regions [117, 118]. In concert with this glucocorticoid-noradrenergic interaction, the endocannabinoid system has emerged as a crucial key mediator of stress effects on memory function. Indeed, we and others have shown that the cannabinoid receptor antagonist AM251 is able to block the ability of systemically

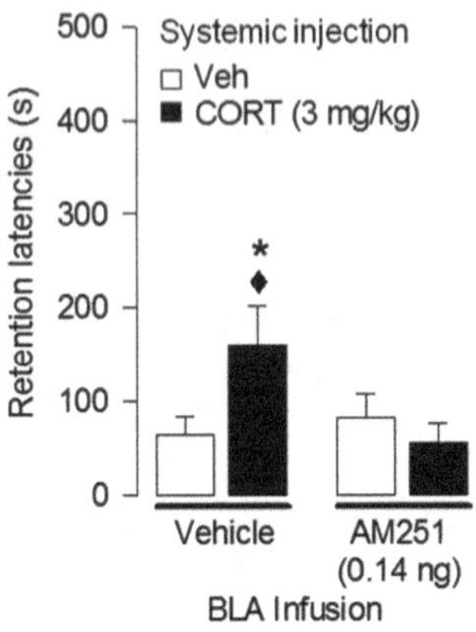

Fig. 1.1 Endocannabinoids in the BLA enable glucocorticoid modulation of memory. Immediate post-training bilateral infusions of the CB1 receptor antagonist AM251 (0.14 ng/0.2 μ(micro)l) into the BLA block retention enhancement induced by systemic injection of corticosterone (CORT; 3.0 mg/kg, subcutaneously). Data represent step-through latencies (mean ± SEM) in seconds on the 48-h inhibitory avoidance retention test. $*P<0.05$ vs the corresponding vehicle group; ♦ $P<0.05$ vs the corresponding AM251 group. (Adapted from [2]; used with permission)

injected corticosterone (or the synthetic analogue dexamethasone) to enhance memory consolidation of inhibitory avoidance training when directly infused into the BLA (Fig. 1.1) [2] or into the hippocampus [119]. These findings provided the first *in vivo* evidence in mammals of the existence of this pathway [120]. Additionally, we reported that also the endocannabinoid oleoylethanolamide (OEA) enhances memory consolidation *via* a norepinephrine-dependent mechanism by demonstrating that the β-adrenoceptor antagonist propranolol, infused into the BLA, blocks the memory enhancing effects induced by systemic administration of OEA [121]. Besides the enhancing effects of glucocorticoids on memory consolidation, many studies demonstrated that such hormones typically impair memory retrieval and working memory during emotionally arousing tests [93, 96–98].

Recently, we investigated the interaction between glucocorticoids and the endocannabinoid system in modulating contextual fear memory retrieval. The cannabinoid antagonist AM251 infused into the dorsal hippocampus blocked the impairing effects on memory retrieval of systemic administered corticosterone; such impairing effects were mediated by elevation of hippocampal 2-AG. Moreover, the β-adrenoceptor antagonist propranolol blocked the impairing effect of WIN on memory retrieval and, conversely, the CB1 receptor antagonist AM251 infused into hippocampus together with an impairing dose of norepinephrine failed to abolish the impairing effect of norepinephrine, thus indicating that norepinephrine is functionally located downstream from the endocannabinoid system [67]. Research from clinical studies demonstrated that exposure to a psychosocial stressor impaired the retrieval of emotional, but not neutral, words learned 24-h before [122]. Other evidence also shows that cannabinoid drugs preferentially modulate memory for emotionally arousing, and not mundane, experiences [123]. Collectively, these findings indicate that endocannabinoids interact with glucocorticoids and, depending on the availability of arousal-induced activation of noradrenergic system, they might differentially modulate memory functions.

In the next paragraphs we report evidence demonstrating how cannabinoid effects on memory can strongly depend on: (i) the level of emotional arousal associated to the experimental context which is originated by elements strikingly related to the cognitive tasks (i.e. footshock intensity on an inhibitory avoidance task); and on (ii) previous stress experiences completely unrelated to the cognitive task; or (iii) to the combination of both factors.

Influence of Emotional Arousal Associated to the Experimental Context

The evidence that CB1 receptor is highly expressed in limbic structures [112, 124] suggests that endocannabinoid signaling has a key role in the control of neuronal responses induced by environmental challenges involving an emotional dimension. We recently demonstrated that exogenously-induced enhancement of endocannabinoid signaling before the training trial, impaired the acquisition of a novel object recognition task in rats tested under high arousal (HA) condition, but had no effect in rats tested under low arousal (LA) condition [6]. Rats under the HA condition were not handled and tested under bright light in an empty arena, whereas animals in the LA condition, were daily handled, habituated to the experimental arena for 1 week and tested under dim red light in an arena with a familiar bedding [6]. Our study demonstrated that cannabinoid effects on memory are dependent on the arousal state at the time of testing. In a subsequent study we investigated the importance of emotional arousal in influencing cannabinoid effects on short- and long-term object recognition memory retention [125]. By following a previous described procedure [106], in order to induce in rats two different levels of emotional arousal, one group of rats received extensive prior habituation to the training apparatus (in the absence of any objects), while a second group was never exposed to the experimental apparatus until the training trial. Unlike the previous study, rats were administered with the CB receptor agonist WIN immediately after the training trial. As shown in Fig. 1.2, WIN induced different effects on short- and long-term memory depending on the level of emotional arousal at the time of training and drug injection. The cannabinoid agonist impaired short-term memory retention in rats not habituated to the experimental arena, while enhanced it in well habituated rats (Fig. 1.2a, b). In contrast, the effects of post-training WIN administration on long-term memory of the object recognition training were different: WIN enhanced long-term retention of object recognition memory in non-habituated rats, but had no effect on long-term memory of extensively habituated animals (Fig. 1.2c, d). WIN effects on memory in not habituated rats were highly comparable to those induced by glucocorticoids in the same experimental protocol [106, 126]. This evidence, together with the fact that cannabinoids closely interact with glucocorticoids, prompted us to explore the possibility that the divergent effects of systemic WIN administration on object recognition memory could be related to differential effects

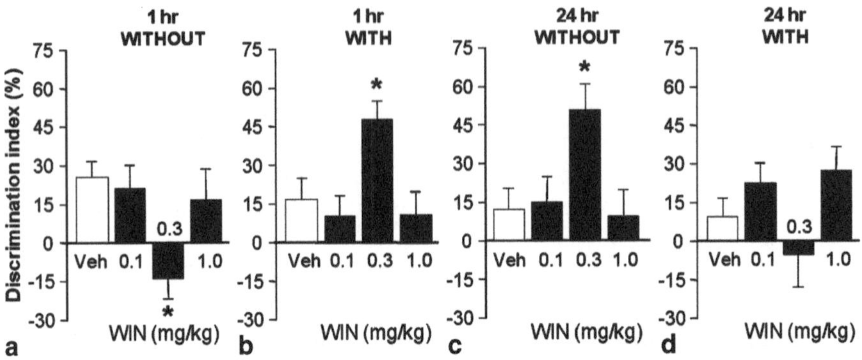

Fig. 1.2 Effects of the CB receptor agonist WIN on short- and long-term retention of object recognition training are influenced by the level of training-associated emotional arousal. Rats were either habituated for 7 days (WITH) or not habituated (WITHOUT) to the training context. On day 8, they were given a 3-min training trial during which they could freely explore two identical objects, and training was followed by an intraperitoneal (i.p.) administration of WIN (0.1, 0.3 or 1 mg/kg,). Retention was tested 1 h or 24 h later. Data represent discrimination index (%) at the retention trial, expressed as mean ± SEM. The discrimination index was calculated as the difference in the time spent exploring the novel and the familiar object, expressed as a ratio of the total time spent exploring both objects. Post-training WIN dose-dependently impaired 1-h object recognition performance of non-habituated rats (a) but enhanced performance of habituated rats (**b**). In contrast, post-training administration of WIN, at a dose that impaired 1-h retention, enhanced 24-h object recognition performance of non-habituated rats (**c**) but not of habituated rats (**d**). *$P<0.05$ vs. vehicle. (Adapted from [125]; used with permission)

of WIN on training-induced glucocorticoid levels in rats in these two habituation conditions. Confirming our hypothesis, WIN elevated plasma corticosterone levels in non-habituated rats whereas it decreased corticosterone levels in habituated rats. Furthermore, we also demonstrated that adrenocortical suppression with the corticosterone-synthesis inhibitor metyrapone in non-habituated rats altered the effect of post-training WIN administration on both short- and long-term recognition memory in such a way that their cognitive performance became similar to that seen in habituated animals (Fig. 1.3) [125]. It is likely that post-training WIN administration on short-term memory could have influenced memory retrieval, while in the long-term, WIN could have specifically affected memory consolidation. Similarly, de Oliveira Alvares et al. (2010) reported that hippocampal endocannabinoid system is recruited to enhance memory consolidation in a contextual fear conditioning paradigm only under high arousal condition. In this study the cannabinoid antagonist AM251 infused into the dorsal hippocampus impaired the consolidation of a strong conditioning training (0.7 mA) while it did not induce any effect on a weak paradigm (0.3 mA) [119]. Clearly, these findings indicate that some degree of training-associated emotional arousal is essential for enabling glucocorticoid effects on memory acquisition, consolidation and retrieval, supporting the idea that the origin of the altered sensitivity to cannabinoids results from a differential activation of the noradrenergic system during arousing versus low-arousing conditions [117, 118, 127, 128].

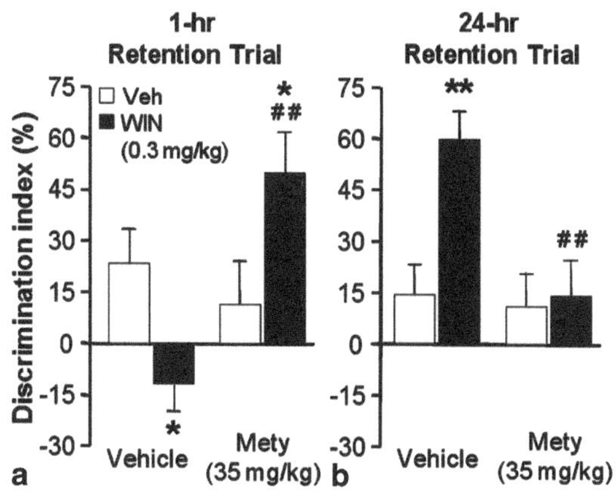

Fig. 1.3 Effects of the CB receptor agonist WIN on short- and long-term retention of object recognition in rats trained under high arousal conditions and pretreated with the corticosterone synthesis inhibitor metyrapone (Mety). Metyrapone (35 mg/kg, i.p.) administered to non-habituated rats 40 min before training reverted the impairing effect of post-training WIN (0.3 mg/kg, i.p.) on 1-h retention performance (**a**) and the enhancing effect of WIN (0.3 mg/kg, i.p.) on 24-h retention performance (**b**) in such a way that their performances became similar to that seen in habituated animals. Data are expressed as means ± SEM. *$P<0.05$; **$P<0.01$ vs. the corresponding vehicle group; ##$P<0.01$ vs. WIN alone group. (Adapted from [125]; used with permission)

Influence of Stress Unrelated to the Behavioral Task

Few studies have been reported in the literature investigating the effects of stress experience completely unrelated to the experimental context and its interaction with endocannabinoid system in modulating cognitive functions. De Oliveira Alvares and coworkers (2010) showed that intra-hippocampal infusions of the cannabinoid antagonist AM251 had no effect *per se* on memory consolidation of a weak contextual fear conditioning paradigm (i.e. 0.3 mA footshock intensity) but reverted the memory enhancing effects of a stressor (i.e. two 0.1 mA footshocks in a different context) administered immediately before conditioning [119]. It has been reported that exposure to an out-of-context stressor (i.e. elevated platform) after an object recognition training enhances long-term memory only in rats that were not previously habituated to the experimental apparatus [129]. The same group has recently shown that intra-BLA infusions of the cannabinoid agonist WIN inhibit the increase in plasma corticosterone levels in rats exposed for 30 min to a stressor (i.e. elevated platform) [130]. Moreover, they demonstrated that WIN infusion did not have any effect by itself but prevented the memory enhancing effects on the acquisition and the impairment on the extinction of an inhibitory avoidance task induced by the elevated platform stress exposure [130]. In a separate study, they also reported that WIN (5 μ(micro)g/side) infused into the BLA did not show any effect on memory consolidation by itself. Conversely, when administered before stress exposure (i.e. elevated platform for 30 min), it blocked the enhancing effects on memory consolidation induced by stress [131]. This evidence appears to be at odd with our finding that intra-BLA infusions of WIN (50 ng/side) immediately after the training trial of an inhibitory avoidance task enhanced memory consolidation [2]. It is likely that differences in doses or/and in the behavioral task used may account for such discrepancy. In a very recent study Segev et al. (2014) reported that 3 days of WIN administration during a 21-day exposure to a chronic mild stress (a common paradigm for stress-induced depression) prevents the stress-induced alterations in memory extinction *via* CB1 receptor activation [132].

Taken together, these findings give evidence that factors related to arousal, stress, and emotional state at the time of training may differentially influence cannabinoid effects on memory.

Assumption to Explain Biphasic Cannabinoid Effects on Memory: A Putative Model

To summarize the findings reported above, the role of the endocannabinoid system on memory modulation is strictly dependent on the aversiveness of the environmental condition and on the level of emotional arousal at the time of training. Therefore, the interaction with glucocorticoids and norepinephrine is of crucial importance in determining the impairing or enhancing effects of cannabinoids on memory. Stress effects on both consolidation and retrieval of emotionally arousing experiences require concurrent glucocorticoid and noradrenergic activity [107, 133]. Corticosterone rapidly elevates endocannabinoid levels in the amygdala [134], conversely, cannabinoid administration can both activate and inhibit the HPA axis [130, 135] and a blockade of CB1 receptor activity in the BLA prevents corticosterone-induced memory enhancement [2]. Extensive evidence indicates that the BLA preferentially modulates memory of emotionally arousing training experiences [77]. In the BLA, CB1 receptors are expressed in GABAergic cells, thus, an activation of CB1 receptors can suppress the release of GABA [136–138]. It is well established that the amygdala GABAergic transmission is involved in memory modulation [133] and that inhibition of GABAergic activity within the BLA enhances memory consolidation by increasing the release of norepinephrine [139]. Thus, in view of this evidence, we and others previously proposed a model: glucocorticoids *via* a rapid, non-genomic effect [140], bind to a membrane bound receptor in the BLA that activates a G-protein signaling cascade to stimulate the synthesis of endocannabinoids. Therefore, endocannabinoids might increase BLA neuronal activity by decreasing GABAergic neurotransmission, leading to increased noradrenergic activity within the same brain region (Fig. 1.4). Nevertheless, it is possible that glucocorticoid-induced memory effects might be also a result of the endocannabinoid-mediated changes in glutamatergic signaling [141].

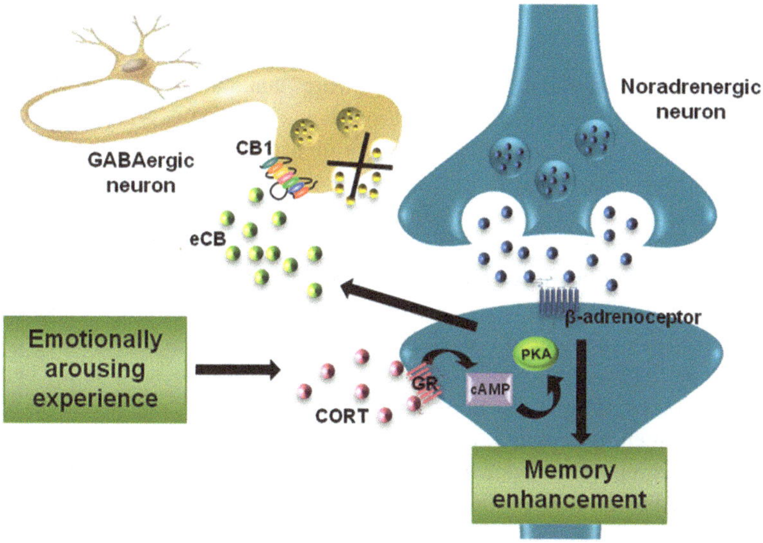

Fig. 1.4 Putative model of role of the endocannabinoid system in the modulation of memory consolidation within the BLA. Stress hormones (i.e. CORT and epinephrine) are released into the bloodstream during training. CORT binds metabotropic GRs within the BLA, activating the Gs–cAMP/PKA pathway to induce endocannabinoid (eCB) synthesis. Endocannabinoids are released into the synaptic cleft where they bind CB1 receptors on GABAergic terminals, thereby inhibiting GABA release. Suppression of GABAergic transmission results in the disinhibition of noradrenergic neurons and increases noradrenergic activation of postsynaptic β-adrenoceptors, enhancing the consolidation of emotionally arousing memories. (Adapted from [76]; used with permission)

Several characteristics of the endocannabinoid system may also account to the opposing effects of cannabinoids on memory functions. Endocannabinoids are synthesized on demand and, as a result, they are released only in those brain regions where and when there is an active endocannabinoid signaling. As a consequence, depending on the pharmacological tool used for a certain experiment, it is possible to appreciate different effects. For instance, direct agonists can bind all cannabinoid receptors both in the brain and in the periphery regardless of their specific involvement in a particular process. In contrast, drugs inducing an amplification of endocannabinoid response act only in those brain areas where and when the signaling is already active. Furthermore, CB1 receptors can suppress the release of neurotransmitters such as GABA and glutamate [3, 31, 32, 34] which often act in an opposite way in the control of several neurophysiological processes related to memory and emotional responses [39, 142–144]. Besides CB1-dependent effects, endocannabinoids also activate CB2 receptors, the peroxisome proliferator-activated nuclear receptor (PPAR) and the transient receptor potential vanilloid type 1 (TRPV1) which have been shown to modulate both emotional responses [145] and aversive memory processes [19, 52, 121].

Conclusions

The evidence summarized in this chapter indicates that the endocannabinoid system plays a key role in mediating emotional arousal and stress effects on memory, shedding light on the neurobiological mechanism involved in the differential impact of stress on memory processes.

The endocannabinoid system modulates cognitive function in a manner strictly dependent on the aversiveness of the environmental condition and on the level of emotional arousal at the time of testing, thus making it possible to hypothesize that the interaction with stress hormones is of crucial importance in determining modulatory effects of cannabinoid compounds on memory processes. It is likely that, depending on the availability of stress hormones, the subsequent interplay between endocannabinoids and glucocorticoids and/or norepinephrine might result in opposing effects on memory processes.

Further research is warranted to disentangle the complex neurobiological mechanisms involved in the unique endocannabinoid modulatory action on memory.

Role of Funding Source P.C.'s research is supported by grants from Human Frontier Science Program (RGY0077/2012), the Italian Ministry of Education MIUR [FIRB Futuro in Ricerca (RBFR10XKHS_003), and PRIN 2010–2011 (2010BN3MXM_003].

References

1. Akirav I. The role of cannabinoids in modulating emotional and non-emotional memory processes in the hippocampus. Front Behav Neurosci. 2011;5:34.
2. Campolongo P, Roozendaal B, Trezza V, Hauer D, Schelling G, McGaugh JL, Cuomo V. Endocannabinoids in the rat basolateral amygdala enhance memory consolidation and enable glucocorticoid modulation of memory. Proc Natl Acad Sci U S A. 2009;106(12):4888–93.
3. Kano M, Ohno-Shosaku T, Hashimotodani Y, Uchigashima M, Watanabe M. Endocannabinoid-mediated control of synaptic transmission. Physiol Rev. 2009;89(1):309–80.
4. Marsicano G, Lafenetre P. Roles of the endocannabinoid system in learning and memory. Curr Top Behav Neurosci. 2009;1:201–30.
5. Wotjak CT. Role of endogenous cannabinoids in cognition and emotionality. Mini Rev Med Chem. 2005;5(7):659–70.
6. Campolongo P, Ratano P, Manduca A, Scattoni ML, Palmery M, Trezza V, Cuomo V. The endocannabinoid transport inhibitor AM404 differentially modulates recognition memory in rats depending on environmental aversiveness. Front Behav Neurosci. 2012;6:11.
7. Haller J, Barna I, Barsvari B, Gyimesi Pelczer K, Yasar S, Panlilio LV, Goldberg S. Interactions between environmental aversiveness and the anxiolytic effects of enhanced cannabinoid signaling by FAAH inhibition in rats. Psychopharmacology (Berl). 2009;204(4):607–16.
8. Szuster RR, Pontius EB, Campos PE. Marijuana sensitivity and panic anxiety. J Clin Psychiatry. 1988;49(11):427–9.
9. Zanettini C, Panlilio LV, Alicki M, Goldberg SR, Haller J, Yasar S. Effects of endocannabinoid system modulation on cognitive and emotional behavior. Front Behav Neurosci. 2011;5:57.

10. Devane WA, Hanus L, Breuer A, Pertwee RG, Stevenson LA, Griffin G, Gibson D, Mandelbaum A, Etinger A, Mechoulam R. Isolation and structure of a brain constituent that binds to the cannabinoid receptor. Science. 1992;258(5090):1946–9.
11. Herkenham M, Lynn AB, Little MD, Johnson MR, Melvin LS, de Costa BR, Rice KC. Cannabinoid receptor localization in brain. Proc Natl Acad Sci U S A. 1990;87(5):1932–6.
12. Matsuda LA, Lolait SJ, Brownstein MJ, Young AC, Bonner TI. Structure of a cannabinoid receptor and functional expression of the cloned cDNA. Nature. 1990;346(6284):561–4.
13. Sugiura T, Kondo S, Sukagawa A, Nakane S, Shinoda A, Itoh K, Yamashita A, Waku K. 2-Arachidonoylglycerol: a possible endogenous cannabinoid receptor ligand in brain. Biochem Biophys Res Commun. 1995;215(1):89–97.
14. Turu G, Hunyady L. Signal transduction of the CB1 cannabinoid receptor. J Mol Endocrinol. 2010;44(2):75–85.
15. Fu J, Bottegoni G, Sasso O, Bertorelli R, Rocchia W, Masetti M, Guijarro A, Lodola A, Armirotti A, Garau G, Bandiera T, Reggiani A, Mor M, Cavalli A, Piomelli D. A catalytically silent FAAH-1 variant drives anandamide transport in neurons. Nat Neurosci. 2011;15(1):64–9.
16. Hillard CJ, Edgemond WS, Jarrahian A, Campbell WB. Accumulation of N-arachidonoylethanolamine (anandamide) into cerebellar granule cells occurs via facilitated diffusion. J Neurochem. 1997;69(2):631–8.
17. Howlett AC, Barth F, Bonner TI, Cabral G, Casellas P, Devane WA, Felder CC, Herkenham M, Mackie K, Martin BR, Mechoulam R, Pertwee RG. International Union of Pharmacology. XXVII. Classification of cannabinoid receptors. Pharmacol Rev. 2002;54(2):161–202.
18. Van Sickle MD, Duncan M, Kingsley PJ, Mouihate A, Urbani P, Mackie K, Stella N, Makriyannis A, Piomelli D, Davison JS, Marnett LJ, Di Marzo V, Pittman QJ, Patel KD, Sharkey KA. Identification and functional characterization of brainstem cannabinoid CB2 receptors. Science. 2005;310(5746):329–32.
19. Onaivi ES, Ishiguro H, Gong JP, Patel S, Perchuk A, Meozzi PA, Myers L, Mora Z, Tagliaferro P, Gardner E, Brusco A, Akinshola BE, Liu QR, Hope B, Iwasaki S, Arinami T, Teasenfitz L, Uhl GR. Discovery of the presence and functional expression of cannabinoid CB2 receptors in brain. Ann N Y Acad Sci. 2006;1074:514–36.
20. Marsicano G, Kuner R. Anatomical distribution of receptors, ligands and enzymes in the brain and the spinal cord: circuitries and neurochemistry. In: Kofalvi A, editor. Cannabinoids and the brain. New York: Springer; 2008. pp. 161–202.
21. McPartland JM, Glass M, Pertwee RG. Meta-analysis of cannabinoid ligand binding affinity and receptor distribution: interspecies differences. Br J Pharmacol. 2007;152(5):583–93.
22. Kamprath K, Romo-Parra H, Haring M, Gaburro S, Doengi M, Lutz B, Pape HC. Short-term adaptation of conditioned fear responses through endocannabinoid signaling in the central amygdala. Neuropsychopharmacology. 2010;36(3):652–63.
23. Ramikie TS, Patel S. Endocannabinoid signaling in the amygdala: anatomy, synaptic signaling, behavior, and adaptations to stress. Neuroscience. 2012;204:38–52.
24. Beltramo M, Stella N, Calignano A, Lin SY, Makriyannis A, Piomelli D. Functional role of high-affinity anandamide transport, as revealed by selective inhibition. Science. 1997;277(5329):1094–7.
25. Mor M, Rivara S, Lodola A, Plazzi PV, Tarzia G, Duranti A, Tontini A, Piersanti G, Kathuria S, Piomelli D. Cyclohexylcarbamic acid 3'- or 4'-substituted biphenyl-3-yl esters as fatty acid amide hydrolase inhibitors: synthesis, quantitative structure-activity relationships, and molecular modeling studies. J Med Chem. 2004;47(21):4998–5008.
26. Long JZ, Li W, Booker L, Burston JJ, Kinsey SG, Schlosburg JE, Pavón FJ, Serrano AM, Selley DE, Parsons LH, Lichtman AH, Cravatt BF. Selective blockade of 2-arachidonoylglycerol hydrolysis produces cannabinoid behavioral effects. Nat Chem Biol. 2009;5(1):36–44.
27. Long JZ, Nomura DK, Vann RE, Walentiny DM, Booker L, Jin X, Burston JJ, Sim-Selley LJ, Lichtman AH, Wiley JL, Cravatt BF. Dual blockade of FAAH and MAGL identifies behavioral processes regulated by endocannabinoid crosstalk in vivo. Proc Natl Acad Sci U S A. 2009;106(48):20270–5.

28. Hogestatt ED, Jonsson BA, Ermund A, Andersson DA, Bjork H, Alexander JP, Cravatt BF, Basbaum AI, Zygmunt PM. Conversion of acetaminophen to the bioactive N-acylphenolamine AM404 via fatty acid amide hydrolase-dependent arachidonic acid conjugation in the nervous system. J Biol Chem. 2005;280(36):31405–12.
29. Patel S, Wohlfeil ER, Rademacher DJ, Carrier EJ, Perry LJ, Kundu A, Falck JR, Nithipatikom K, Campbell WB, Hillard CJ. The general anesthetic propofol increases brain N-arachidonylethanolamine (anandamide) content and inhibits fatty acid amide hydrolase. Br J Pharmacol. 2003;139(5):1005–13.
30. Hauer D, Ratano P, Morena M, Scaccianoce S, Briegel I, Palmery M, Cuomo V, Roozendaal B, Schelling G, Campolongo P. Propofol enhances memory formation via an interaction with the endocannabinoid system. Anesthesiology. 2011;114(6):1380–8.
31. Marsicano G, Lutz B. Expression of the cannabinoid receptor CB1 in distinct neuronal subpopulations in the adult mouse forebrain. Eur J Neurosci. 1999;11(12):4213–25.
32. Azad SC, Kurz J, Marsicano G, Lutz B, Zieglgansberger W, Rammes G. Activation of CB1 specifically located on GABAergic interneurons inhibits LTD in the lateral amygdala. Learn Mem. 2008;15(3):143–52.
33. Morozov YM, Torii M, Rakic P. Origin, early commitment, migratory routes, and destination of cannabinoid type 1 receptor-containing interneurons. Cereb Cortex. 2009;19(Suppl 1):i78–89.
34. Monory K, Massa F, Egertova M, Eder M, Blaudzun H, Westenbroek R, Kelsch W, Jacob W, Marsch R, Ekker M, Long J, Rubenstein JL, Goebbels S, Nave KA, During M, Klugmann M, Wolfel B, Dodt HU, Zieglgansberger W, Wotjak CT, Mackie K, Elphick MR, Marsicano G, Lutz B. The endocannabinoid system controls key epileptogenic circuits in the hippocampus. Neuron. 2006;51(4):455–66.
35. Kawamura Y, Fukaya M, Maejima T, Yoshida T, Miura E, Watanabe M, Ohno-Shosaku T, Kano M. The CB1 cannabinoid receptor is the major cannabinoid receptor at excitatory presynaptic sites in the hippocampus and cerebellum. J Neurosci. 2006;26(11):2991–3001.
36. Haring M, Marsicano G, Lutz B, Monory K. Identification of the cannabinoid receptor type 1 in serotonergic cells of raphe nuclei in mice. Neuroscience. 2007;146(3):1212–9.
37. Oropeza VC, Mackie K, Van Bockstaele EJ. Cannabinoid receptors are localized to noradrenergic axon terminals in the rat frontal cortex. Brain Res. 2007;1127(1):36–44.
38. Hermann H, Marsicano G, Lutz B. Coexpression of the cannabinoid receptor type 1 with dopamine and serotonin receptors in distinct neuronal subpopulations of the adult mouse forebrain. Neuroscience. 2002;109(3):451–60.
39. Chevaleyre V, Takahashi KA, Castillo PE. Endocannabinoid-mediated synaptic plasticity in the CNS. Annu Rev Neurosci. 2006;29:37–76.
40. Katona I, Freund TF. Multiple functions of endocannabinoid signaling in the brain. Annu Rev Neurosci. 2012;35:529–58.
41. Marsicano G, Lutz B. Neuromodulatory functions of the endocannabinoid system. J Endocrinol Invest. 2006;29(3 Suppl):27–46.
42. Pistis M, Perra S, Pillolla G, Melis M, Gessa GL, Muntoni AL. Cannabinoids modulate neuronal firing in the rat basolateral amygdala: evidence for CB1- and non-CB1-mediated actions. Neuropharmacology. 2004;46(1):115–25.
43. Ranganathan M, D'Souza DC. The acute effects of cannabinoids on memory in humans: a review. Psychopharmacology (Berl). 2006;188(4):425–44.
44. Riedel G, Davies SN. Cannabinoid function in learning, memory and plasticity. Handb Exp Pharmacol. 2005;168:445–77.
45. Verdejo-Garcia A, Bechara A, Recknor EC, Perez-Garcia M. Executive dysfunction in substance dependent individuals during drug use and abstinence: an examination of the behavioral, cognitive and emotional correlates of addiction. J Int Neuropsychol Soc. 2006;12(3):405–15.
46. Fisk JE, Montgomery C. Real-world memory and executive processes in cannabis users and non-users. J Psychopharmacol. 2008;22(7):727–36.

47. Karila L, Roux P, Rolland B, Benyamina A, Reynaud M, Aubin HJ, Lancon C. Acute and long-term effects of cannabis use: a review. Curr Pharm Des. 2014;20(25):4112–8.
48. Lichtman AH, Dimen KR, Martin BR. Systemic or intrahippocampal cannabinoid administration impairs spatial memory in rats. Psychopharmacology (Berl). 1995;119(3):282–90.
49. Da S, Takahashi RN. SR 141716A prevents delta9-tetrahydrocannabinol-induced spatial learning deficit in a Morris-type water maze in mice. Prog Neuropsychopharmacol Biol Psychiatry. 2002;26(2):321–5.
50. Pamplona FA, Takahashi RN. WIN 55212–2 impairs contextual fear conditioning through the activation of CB1 cannabinoid receptors. Neurosci Lett. 2006;397(1–2):88–92.
51. Robinson L, McKillop-Smith S, Ross NL, Pertwee RG, Hampson RE, Platt B, Riedel G. Hippocampal endocannabinoids inhibit spatial learning and limit spatial memory in rats. Psychopharmacology (Berl). 2008;198(4):551–63.
52. Mazzola C, Medalie J, Scherma M, Panlilio LV, Solinas M, Tanda G, Drago F, Cadet JL, Goldberg SR, Yasar S. Fatty acid amide hydrolase (FAAH) inhibition enhances memory acquisition through activation of PPAR-alpha nuclear receptors. Learn Mem. 2009;16(5):332–7.
53. Egashira N, Mishima K, Iwasaki K, Fujiwara M. Intracerebral microinjections of delta9-tetrahydrocannabinol: search for the impairment of spatial memory in the eight-arm radial maze in rats. Brain Res. 2002;952(2):239–45.
54. Wegener N, Kuhnert S, Thuns A, Roese R, Koch M. Effects of acute systemic and intracerebral stimulation of cannabinoid receptors on sensorimotor gating, locomotion and spatial memory in rats. Psychopharmacology (Berl). 2008;198(3):375–85.
55. Abush H, Akirav I. Cannabinoids modulate hippocampal memory and plasticity. Hippocampus. 2010;20(10):1126–38.
56. Tan H, Lauzon NM, Bishop SF, Chi N, Bechard M, Laviolette SR. Cannabinoid transmission in the basolateral amygdala modulates fear memory formation via functional inputs to the prelimbic cortex. J Neurosci. 2011;31(14):5300–12.
57. Mackowiak M, Chocyk A, Dudys D, Wedzony K. Activation of CB1 cannabinoid receptors impairs memory consolidation and hippocampal polysialylated neural cell adhesion molecule expression in contextual fear conditioning. Neuroscience. 2009;158(4):1708–16.
58. Yim TT, Hong NS, Ejaredar M, McKenna JE, McDonald RJ. Post-training CB1 cannabinoid receptor agonist activation disrupts long-term consolidation of spatial memories in the hippocampus. Neuroscience. 2008;151(4):929–36.
59. Busquets-Garcia A, Puighermanal E, Pastor A, de la Torre R, Maldonado R, Ozaita A. Differential role of anandamide and 2-arachidonoylglycerol in memory and anxiety-like responses. Biol Psychiatry. 2011;70(5):479–86.
60. Wise LE, Iredale PA, Lichtman AH. The cannabinoid CB(1) receptor antagonist CE prolongs spatial memory duration in a rat delayed radial arm maze memory task. Eur J Pharmacol. 2008;590(1–3):246–9.
61. Wolff MC, Leander JD. SR141716A, a cannabinoid CB1 receptor antagonist, improves memory in a delayed radial maze task. Eur J Pharmacol. 2003;477(3):213–7.
62. Jamali-Raeufy N, Nasehi M, Zarrindast MR. Influence of N-methyl D-aspartate receptor mechanism on WIN55,212-2-induced amnesia in rat dorsal hippocampus. Behav Pharmacol. 2011;22(7):645–54.
63. Zarrindast MR, Navaeian M, Nasehi M. Influence of three-day morphine-treatment upon impairment of memory consolidation induced by cannabinoid infused into the dorsal hippocampus in rats. Neurosci Res. 2011;69(1):51–9.
64. De Oliveira Alvares L, Genro BP, Diehl F, Quillfeldt JA. Differential role of the hippocampal endocannabinoid system in the memory consolidation and retrieval mechanisms. Neurobiol Learn Mem. 2008;90(1):1–9.
65. Mishima K, Egashira N, Hirosawa N, Fujii M, Matsumoto Y, Iwasaki K, Fujiwara M. Characteristics of learning and memory impairment induced by Δ^9-tetrahydrocannabinol in rats. Jpn J Pharmacol. 2001;87:297–308.

66. Niyuhire F, Varvel SA, Martin BR, Lichtman AH. Exposure to marijuana smoke impairs memory retrieval in mice. J Pharmacol Exp Ther. 2007;322(3):1067–75.
67. Atsak P, Hauer D, Campolongo P, Schelling G, McGaugh JL, Roozendaal B. Glucocorticoids interact with the hippocampal endocannabinoid system in impairing retrieval of contextual fear memory. Proc Natl Acad Sci U S A. 2012;109(9):3504–9.
68. Piri M, Zarrindast MR. Modulation of WIN55,212-2 state-dependent memory by alpha2-adrenergic receptors of the dorsal hippocampus. Arch Iran Med. 2011;14(6):389–95.
69. Segev A, Akirav I. Differential effects of cannabinoid receptor agonist on social discrimination and contextual fear in amygdala and hippocampus. Learn Mem. 2011;18(4):254–9.
70. Marsicano G, Wotjak CT, Azad SC, Bisogno T, Rammes G, Cascio MG, Hermann H, Tang J, Hofmann C, Zieglgansberger W, Di Marzo V, Lutz B. The endogenous cannabinoid system controls extinction of aversive memories. Nature. 2002;418(6897):530–4.
71. Suzuki A, Josselyn SA, Frankland PW, Masushige S, Silva AJ, Kida S. Memory reconsolidation and extinction have distinct temporal and biochemical signatures. J Neurosci. 2004;24(20):4787–95.
72. Pamplona FA, Prediger RD, Pandolfo P, Takahashi RN. The cannabinoid receptor agonist WIN 55,212-2 facilitates the extinction of contextual fear memory and spatial memory in rats. Psychopharmacology (Berl). 2006;188(4):641–9.
73. Chhatwal JP, Davis M, Maguschak KA, Ressler KJ. Enhancing cannabinoid neurotransmission augments the extinction of conditioned fear. Neuropsychopharmacology. 2005;30(3):516–24.
74. Barad M, Gean PW, Lutz B. The role of the amygdala in the extinction of conditioned fear. Biol Psychiatry. 2006;60(4):322–8.
75. Niyuhire F, Varvel SA, Thorpe AJ, Stokes RJ, Wiley JL, Lichtman AH. The disruptive effects of the CB1 receptor antagonist rimonabant on extinction learning in mice are task-specific. Psychopharmacology (Berl). 2007;191(2):223–31.
76. Morena M, Campolongo P. The endocannabinoid system: an emotional buffer in the modulation of memory function. Neurobiol Learn Mem. 2014;112C:30–43.
77. McGaugh JL. Memory–a century of consolidation. Science. 2000;287(5451):248–51.
78. Cahill L. Neurobiological mechanisms of emotionally influenced, long-term memory. Prog Brain Res. 2000;126:29–37.
79. Cahill L, McGaugh JL, Weinberger NM. The neurobiology of learning and memory: some reminders to remember. Trends Neurosci. 2001;24(10):578–81.
80. Dolcos F, LaBar KS, Cabeza R. Remembering one year later: role of the amygdala and the medial temporal lobe memory system in retrieving emotional memories. Proc Natl Acad Sci U S A. 2005;102(7):2626–31.
81. Price JL, Drevets WC. Neurocircuitry of mood disorders. Neuropsychopharmacology. 2010;35(1):192–216.
82. Packard MG, Wingard JC. Amygdala and "emotional" modulation of the relative use of multiple memory systems. Neurobiol Learn Mem. 2004;82(3):243–52.
83. Aggleton JP. The contribution of the amygdala to normal and abnormal emotional states. Trends Neurosci. 1993;16(8):328–33.
84. Davis M, Rainnie D, Cassell M. Neurotransmission in the rat amygdala related to fear and anxiety. Trends Neurosci. 1994;17(5):208–14.
85. Baxter MG, Murray EA. The amygdala and reward. Nat Rev Neurosci. 2002;3(7):563–73.
86. Pape HC, Pare D. Plastic synaptic networks of the amygdala for the acquisition, expression, and extinction of conditioned fear. Physiol Rev. 2010;90(2):419–63.
87. Sara SJ. The locus coeruleus and noradrenergic modulation of cognition. Nat Rev Neurosci. 2009;10(3):211–23.
88. Reul JM, de Kloet ER. Two receptor systems for corticosterone in rat brain: microdistribution and differential occupation. Endocrinology. 1985;117(6):2505–11.
89. Joels M, Baram TZ. The neuro-symphony of stress. Nat Rev Neurosci. 2009;10(6):459–66.
90. Tasker JG, Di S, Malcher-Lopes R. Minireview: rapid glucocorticoid signaling via membrane-associated receptors. Endocrinology. 2006;147(12):5549–56.

91. de Kloet ER, Oitzl MS, Joels M. Stress and cognition: are corticosteroids good or bad guys? Trends Neurosci. 1999;22(10):422–6.
92. Oitzl MS, de Kloet ER. Selective corticosteroid antagonists modulate specific aspects of spatial orientation learning. Behav Neurosci. 1992;106(1):62–71.
93. Roozendaal B. 1999 Curt P. Richter award. Glucocorticoids and the regulation of memory consolidation. Psychoneuroendocrinology. 2000;25(3):213–38.
94. Sandi C, Rose SP. Corticosterone enhances long-term retention in one-day-old chicks trained in a weak passive avoidance learning paradigm. Brain Res. 1994;647(1):106–12.
95. De Boer SF Koopmans SJ, Slangen JL, Van der Gugten J. Plasma catecholamine, corticosterone and glucose responses to repeated stress in rats: effect of interstressor interval length. Physiol Behav. 1990;47(6):1117–24.
96. de Quervain DJ, Aerni A, Schelling G, Roozendaal B. Glucocorticoids and the regulation of memory in health and disease. Front Neuroendocrinol. 2009;30(3):358–70.
97. de Quervain DJ, Roozendaal B, McGaugh JL. Stress and glucocorticoids impair retrieval of long-term spatial memory. Nature. 1998;394(6695):787–90.
98. Roozendaal B, de Quervain DJ, Schelling G, McGaugh JL. A systemically administered beta-adrenoceptor antagonist blocks corticosterone-induced impairment of contextual memory retrieval in rats. Neurobiol Learn Mem. 2004;81(2):150–4.
99. Roozendaal B, McGaugh JL. Amygdaloid nuclei lesions differentially affect glucocorticoid-induced memory enhancement in an inhibitory avoidance task. Neurobiol Learn Mem. 1996;65(1):1–8.
100. Quirarte GL, Galvez R, Roozendaal B, McGaugh JL. Norepinephrine release in the amygdala in response to footshock and opioid peptidergic drugs. Brain Res. 1998;808(2):134–40.
101. Galvez R, Mesches MH, McGaugh JL. Norepinephrine release in the amygdala in response to footshock stimulation. Neurobiol Learn Mem. 1996;66(3):253–7.
102. McIntyre CK, Hatfield T, McGaugh JL. Amygdala norepinephrine levels after training predict inhibitory avoidance retention performance in rats. Eur J Neurosci. 2002;16(7):1223–6.
103. Quirarte GL, Roozendaal B, McGaugh JL. Glucocorticoid enhancement of memory storage involves noradrenergic activation in the basolateral amygdala. Proc Natl Acad Sci U S A. 1997;94(25):14048–53.
104. Roozendaal B, Quirarte GL, McGaugh JL. Glucocorticoids interact with the basolateral amygdala beta-adrenoceptor–cAMP/cAMP/PKA system in influencing memory consolidation. Eur J Neurosci. 2002;15(3):553–60.
105. Roozendaal B, McEwen BS, Chattarji S. Stress, memory and the amygdala. Nat Rev Neurosci. 2009;10(6):423–33.
106. Okuda S, Roozendaal B, McGaugh JL. Glucocorticoid effects on object recognition memory require training-associated emotional arousal. Proc Natl Acad Sci U S A. 2004;101(3):853–8.
107. Roozendaal B. Stress and memory: opposing effects of glucocorticoids on memory consolidation and memory retrieval. Neurobiol Learn Mem. 2002;78(3):578–95.
108. Hill MN, McEwen BS. Involvement of the endocannabinoid system in the neurobehavioural effects of stress and glucocorticoids. Prog Neuropsychopharmacol Biol Psychiatry. 2010;34(5):791–7.
109. Di S, Malcher-Lopes R, Halmos KC, Tasker JG. Nongenomic glucocorticoid inhibition via endocannabinoid release in the hypothalamus: a fast feedback mechanism. J Neurosci. 2003;23(12):4850–7.
110. Di S, Malcher-Lopes R, Marcheselli VL, Bazan NG, Tasker JG. Rapid glucocorticoid-mediated endocannabinoid release and opposing regulation of glutamate and gamma-aminobutyric acid inputs to hypothalamic magnocellular neurons. Endocrinology. 2005;146(10):4292–301.
111. Derbenev AV, Stuart TC, Smith BN. Cannabinoids suppress synaptic input to neurones of the rat dorsal motor nucleus of the vagus nerve. J Physiol. 2004;559(Pt 3):923–38.
112. Herkenham M, Lynn AB, Johnson MR, Melvin LS, de Costa BR, Rice KC. Characterization and localization of cannabinoid receptors in rat brain: a quantitative in vitro autoradiographic study. J Neurosci. 1991;11(2):563–83.

113. Jelsing J, Larsen PJ, Vrang N. Identification of cannabinoid type 1 receptor expressing cocaine amphetamine-regulated transcript neurons in the rat hypothalamus and brainstem using in situ hybridization and immunohistochemistry. Neuroscience. 2008;154(2):641–52.
114. Mailleux P, Vanderhaeghen JJ. Distribution of neuronal cannabinoid receptor in the adult rat brain: a comparative receptor binding radioautography and in situ hybridization histochemistry. Neuroscience. 1992;48(3):655–68.
115. Matsuda LA, Bonner TI, Lolait SJ. Localization of cannabinoid receptor mRNA in rat brain. J Comp Neurol. 1993;327(4):535–50.
116. Muntoni AL, Pillolla G, Melis M, Perra S, Gessa GL, Pistis M. Cannabinoids modulate spontaneous neuronal activity and evoked inhibition of locus coeruleus noradrenergic neurons. Eur J Neurosci. 2006;23(9):2385–94.
117. Oropeza VC, Page ME, Van Bockstaele EJ. Systemic administration of WIN 55,212–2 increases norepinephrine release in the rat frontal cortex. Brain Res. 2005;1046(1–2):45–54.
118. Page ME, Oropeza VC, Sparks SE, Qian Y, Menko AS, Van Bockstaele EJ. Repeated cannabinoid administration increases indices of noradrenergic activity in rats. Pharmacol Biochem Behav. 2007;86(1):162–8.
119. de Oliveira, Alvares L, Engelke DS, Diehl F, Scheffer-Teixeira R, Haubrich J, de Freitas Cassini L, Molina VA, Quillfeldt JA. Stress response recruits the hippocampal endocannabinoid system for the modulation of fear memory. Learn Mem. 2010;17(4):202–9.
120. Hill MN, Patel S, Campolongo P, Tasker JG, Wotjak CT, Bains JS. Functional interactions between stress and the endocannabinoid system: from synaptic signaling to behavioral output. J Neurosci. 2010;30(45):14980–6.
121. Campolongo P, Roozendaal B, Trezza V, Cuomo V, Astarita G, Fu J, McGaugh JL, Piomelli D. Fat-induced satiety factor oleoylethanolamide enhances memory consolidation. Proc Natl Acad Sci U S A. 2009;106(19):8027–31.
122. Kuhlmann S, Piel M, Wolf OT. Impaired memory retrieval after psychosocial stress in healthy young men. J Neurosci. 2005;25(11):2977–82.
123. Ballard ME, Bedi G, de Wit H. Effects of delta-9-tetrahydrocannabinol on evaluation of emotional images. J Psychopharmacol. 2012;26(10):1289–98.
124. Tsou K, Brown S, Sanudo-Pena MC, Mackie K, Walker JM. Immunohistochemical distribution of cannabinoid CB1 receptors in the rat central nervous system. Neuroscience. 1998;83(2):393–411.
125. Campolongo P, Morena M, Scaccianoce S, Trezza V, Chiarotti F, Schelling G, Cuomo V, Roozendaal B. Novelty-induced emotional arousal modulates cannabinoid effects on recognition memory and adrenocortical activity. Neuropsychopharmacology. 2013;38:1276–86.
126. Roozendaal B, Okuda S, Van der Zee EA, McGaugh JL. Glucocorticoid enhancement of memory requires arousal-induced noradrenergic activation in the basolateral amygdala. Proc Natl Acad Sci U S A. 2006;103(17):6741–6.
127. Patel S, Hillard CJ. Cannabinoid-induced Fos expression within A10 dopaminergic neurons. Brain Res. 2003;963(1–2):15–25.
128. Carvalho AF, Van Bockstaele EJ. Cannabinoid modulation of noradrenergic circuits: implications for psychiatric disorders. Prog Neuropsychopharmacol Biol Psychiatry. 2012;38(1):59–67.
129. Maroun M, Akirav I. Arousal and stress effects on consolidation and reconsolidation of recognition memory. Neuropsychopharmacology. 2008;33(2):394–405.
130. Ganon-Elazar E, Akirav I. Cannabinoid receptor activation in the basolateral amygdala blocks the effects of stress on the conditioning and extinction of inhibitory avoidance. J Neurosci. 2009;29(36):11078–88.
131. Ramot A, Akirav I. Cannabinoid receptors activation and glucocorticoid receptors deactivation in the amygdala prevent the stress-induced enhancement of a negative learning experience. Neurobiol Learn Mem. 2012;97(4):393–401.
132. Segev A, Rubin AS, Abush H, Richter-Levin G, Akirav I. Cannabinoid receptor activation prevents the effects of chronic mild stress on emotional learning and LTP in a rat model of depression. Neuropsychopharmacology. 2014;39(4):919–33.

133. McGaugh JL, Roozendaal B. Role of adrenal stress hormones in forming lasting memories in the brain. Curr Opin Neurobiol. 2002;12(2):205–10.
134. Hill MN, Karatsoreos IN, Hillard CJ, McEwen BS. Rapid elevations in limbic endocannabinoid content by glucocorticoid hormones in vivo. Psychoneuroendocrinology. 2010;35(9):1333–8.
135. Cota D, Steiner MA, Marsicano G, Cervino C, Herman JP, Grubler Y, Stalla J, Pasquali R, Lutz B, Stalla GK, Pagotto U. Requirement of cannabinoid receptor type 1 for the basal modulation of hypothalamic-pituitary-adrenal axis function. Endocrinology. 2007;148(4):1574–81.
136. Katona I, Rancz EA, Acsady L, Ledent C, Mackie K, Hajos N, Freund TF. Distribution of CB1 cannabinoid receptors in the amygdala and their role in the control of GABAergic transmission. J Neurosci. 2001;21(23):9506–18.
137. Katona I, Sperlagh B, Sik A, Kafalvi A, Vizi ES, Mackie K, Freund TF. Presynaptically located CB1 cannabinoid receptors regulate GABA release from axon terminals of specific hippocampal interneurons. J Neurosci. 1999;19(11):4544–58.
138. Ohno-Shosaku T, Maejima T, Kano M. Endogenous cannabinoids mediate retrograde signals from depolarized postsynaptic neurons to presynaptic terminals. Neuron. 2001;29(3):729–38.
139. Hatfield T, Spanis C, McGaugh JL. Response of amygdalar norepinephrine to footshock and GABAergic drugs using in vivo microdialysis and HPLC. Brain Res. 1999;835(2):340–5.
140. Hill MN, Tasker JG. Endocannabinoid signaling, glucocorticoid-mediated negative feedback, and regulation of the hypothalamic-pituitary-adrenal axis. Neuroscience. 2012;204:5–16.
141. Popoli M, Yan Z, McEwen BS, Sanacora G. The stressed synapse: the impact of stress and glucocorticoids on glutamate transmission. Nat Rev Neurosci. 2011;13(1):22–37.
142. Millan MJ. The neurobiology and control of anxious states. Prog Neurobiol. 2003;70(2):83–244.
143. Myhrer T. Neurotransmitter systems involved in learning and memory in the rat: a meta-analysis based on studies of four behavioral tasks. Brain Res Brain Res Rev. 2003;41(2–3):268–87.
144. Metna-Laurent M, Soria-Gómez E, Verrier D, Conforzi M, Jego P, Lafenetre P, Marsicano G. Bimodal control of fear-coping strategies by CB1 cannabinoid receptors. J Neurosci. 2012;32(21):7109–18.
145. Maione S, Bisogno T, de Novellis V, Palazzo E, Cristino L, Valenti M, Petrosino S, Guglielmotti V, Rossi F, Di Marzo V. Elevation of endocannabinoid levels in the ventrolateral periaqueductal grey through inhibition of fatty acid amide hydrolase affects descending nociceptive pathways via both cannabinoid receptor type 1 and transient receptor potential vanilloid type-1 receptors. J Pharmacol Exp Ther. 2006;316(3):969–82.

Chapter 2
Cannabinoids Modulation of Emotional and Non-Emotional Memory Processes After Stress

Irit Akirav

Abstract The endocannabinoid system (ECS) is involved in regulating the stress response and subsequent changes in neuroendocrine function and emotional behavior. It is also a critical neuromodulatory system that affects learning and memory. Generally systemically administered cannabinoid agonists have an impairing effect on memory processes although enhancing effects are also reported.

Stress is a potent modulator of brain function and cognition that has differential effects on memory function depending on a number of factors (such as stress duration, stress intensity, timing and the source of the stress, as well as the learning type under study). Most of the tasks to investigate learning and memory in laboratory rodents are stressful for the animals (i.e. the cognitive task includes intrinsic stress) as opposed to extrinsic stress which refers to outside stress that occurs before or after the cognitive task. Several lines of evidence suggest that cannabinoids differentially affect different memory phases (acquisition, consolidation, retrieval and extinction), and that the type of cognitive task (emotional or aversive versus non-emotional) also determines the neural substrates underlying the effects of cannabinoids on memory.

In this chapter I will describe the interaction between the effects of activating the ECS and stress exposure on emotional (i.e., aversive) and non-emotional learning and memory processes in animal models. I will argue that administering cannabinoid agonists in proximity to extrinsic stress exposure normalizes stress modulation of emotional memory. A possible model of the effects of cannabinoids on emotional memory after stress is also presented.

Keywords Endocannabinoids · Learning and memory · Emotional learning · Stress · Hypothalamic-pituitary-adrenal axis

I. Akirav (✉)
Department of Psychology, University of Haifa, Haifa 31905, Israel
e-mail: irit.akirav@gmail.com

P. Campolongo, L. Fattore (eds.), *Cannabinoid Modulation of Emotion, Memory, and Motivation,* DOI 10.1007/978-1-4939-2294-9_2

Introduction

The endocannabinoid system (ECS) plays a modulatory role in many cognitive and emotional processes. Cannabis has been used recreationally for its mood-enhancing and stress-alleviating properties for centuries. The discovery of Δ9-tetrahydrocannabinol (THC), the primary psychoactive constituent of marijuana, along with its biologically active analogs has stimulated extensive research devoted to understanding how these compounds exert their effects on emotionality within the brain.

Multiple animal models have been used to assess the effects of the ECS on various stages of memory (acquisition, consolidation, retrieval and extinction), using a wide range of behavioral paradigms [1–5]. Both acute and chronic exposure to cannabis is associated with impairments in attention, working memory, verbal learning and memory functions [5–8]. Working memory refers to a brain system that provides temporary storage and manipulation of the information necessary for complex cognitive tasks. Long-term heavy cannabis users show impairments in memory and attention that, depending on the task analyzed, might be reversible [9], although in some cases they persist beyond the period of intoxication and get worse with increasing years of regular cannabis use [10].

Most of the studies that examined the effects of the ECS on hippocampal-dependent memory focused on spatial learning [3]. Spatial learning is the acquisition of information about one's surroundings. In general, the findings are that exogenous and endogenous cannabinoid agonists impaired working memory and the acquisition of long-term memory [11], but had no effect on memory retrieval [12–13]. Other studies, usually focused on fear-related paradigms, demonstrated that activating the ECS facilitated extinction [4, 14–15]. Hence cannabinoids differentially affect the different memory phases but it seems that the type of cognitive task (emotional or aversive versus non-aversive) also determines the neural substrates underlying the effects of cannabinoids on memory [5]. The effects of inhibiting the ECS on learning and memory has been reviewed elsewhere [16].

Most of the tasks currently used to investigate learning and memory in laboratory rodents can be considered as being stressful for the animals: they are based on the application of stressful manipulations and/or stimuli to motivate animals to learn. "Intrinsic stress", refers to situations in which stress is originated by elements related to the cognitive task, and "extrinsic stress", refers to those situations in which stress is originated by conditions completely unrelated to the cognitive task and thus generally occurring temporally dissociated from such task (i.e., either before or afterwards) [17].

Intrinsic stress that is related to the cognitive task generally enhances the consolidation of memory through actions of norepinephrine and glucocorticoids on the neural circuits activated by the learning experience (see review by [17]). When the stress derives from conditions other than the cognitive task (i.e. extrinsic stress), then the effects are more varied and more specific to the type of learning involved [17]. For example, acute extrinsic stress enhanced aversive hippocampal-dependent tasks such as contextual fear conditioning and trace eye-blink conditioning (e.g., [18–19]), but impaired hippocampal-dependent spatial memory retrieval (see review by [20]). In contextual fear conditioning an animal learns the association between the shock (i.e., the unconditioned stimulus) and the context in which conditioning occurs (the conditioned stimulus) whereas in trace eye-blink conditioning the animal learns the association between a shock or an air puff (i.e., the unconditioned stimulus) and a tone (the conditioned stimulus).

Stress is known to be a potent modulator of brain function and cognition. While prolonged and/or excessive stress generally exerts negative effects on learning and memory processes, acute stress can have differential effects on memory function depending on a number of factors (such as stress duration, stress intensity, timing and the source of the stress, as well as the learning type under study).

In this chapter I will describe the modulatory effects of activating the ECS on aversive (i.e., emotional) and non-aversive learning and memory processes in animal models, with or without exposure to extrinsic environmental stress. I will argue that activating the ECS has a different effect on learning and memory processes when ECS activation occurs shortly before or after an exposure to a stressful experience. Hence, the administration of cannabinoid agonists or exposure to stress may enhance or impair memory, but when cannabinoid agonists are administered in proximity to stress exposure (i.e., before or after stress exposure), they prevent the stress-induced alterations in memory. To summarize I present a possible model of the effects of cannabinoids on memory after stress that involves the interaction between the ECS and the hypothalamic-pituitary-adrenal axis system.

The Endocannabinoid System and Emotional Memory

The ECB system, which includes cannabinoid receptors (CB1 and CB2) and endocannabinoids (*N*-arachidonylethanolamine [anandamide; AEA] and 2-arachidonoyl-glycerol [2-AG]) [21–23], has been repeatedly implicated in the effects on emotionality within the brain. The ECS has recently emerged as a promising therapeutic target for the treatment of stress-related emotional disorders [24–27]. In support, a growing literature base has collectively demonstrated that facilitation of endocannabinoid signaling promotes antidepressant- and anxiolytic-like responses in preclinical animal models [15, 28–33].

Emotional learning is extremely important for the survival of an individual. In studies of emotional behavior, the amygdala, medial prefrontal cortex, and hippocampus have received the most attention because structural and functional abnormalities within these regions are most commonly associated with mood disorders in clinical populations [34]. These three areas are a key circuit in the adaptive and maladaptive responses to stress as they undergo stress-induced structural remodeling, which alters behavioral and physiological responses, including anxiety, aggression, mental flexibility, memory and other cognitive processes [35, 36]. This is in accordance with the fact that both glucocorticoid receptors and CB1 receptors are located within this brain circuit [37–40].

The amygdala acts as an interface between sensory inputs and cortical processing, and activation of this structure is directly linked to the generation of fear and anxiety [41–42], and promotes activation of the hypothalamic-pituitary-adrenal axis [43]. The prefrontal cortex is involved in higher-order processing and is explicitly involved in the recognition of aversive stimuli and in drawing conclusions about the controllability of stimuli [44–45]. The hippocampus interacts with the prefrontal cortex to suppress the hypothalamic-pituitary-adrenal axis and promote recovery to homeostasis following a stressful expeirence [46].

The Effects of Cannabinoids on Aversive and Non-Aversive Learning Tasks

There have been reports of deficits in memory following administration of cannabis extracts or cannabinoid agonists in rodents and in humans and this has been reviewed extensively before [47–55]. Nevertheless, several reports suggest enhancing effects of cannabinoids on memory [56–57].

I will briefly describe some of these studies that together demonstrate apparent discrepancies in the effects of cannabinoid agonists on learning and memory. Several reasons could explain these effects, among them are different spatial and temporal ways of activating cannabinoid receptors by exogenous pharmacological agents; the time point of drug administration in relation to the cognitive task being examined; systemic versus local intra-cerebral administration of the drugs; the different pharmacokinetics of each drug that can affect the system for hours or days. Other reasons are related to the memory phase under investigation, the aversiveness of the cognitive task etc.

Non-Aversive Memory Tasks

As described earlier, most of the memory tasks in animals involve an aversive component. In recognition tasks, on the other hand, no rewarding or aversive stimulation is used during training, so the learning occurs under conditions of relatively low stress or arousal [58]. Object recognition memory is the ability to discriminate the familiarity of previously encountered objects and it is based on the spontaneous exploration behavior of the rat.

Systemic administration of Δ9-THC or the CB1/2 receptor agonist WIN55,212–2 impaired object recognition in rats [59–60]. Suenaga & Ichitani [61] found that intra-hippocampal WIN55,212–2 (1–2 μg/side) did not affect object recognition memory but impaired the ability to recognize an object that was moved to another location (hippocampal-dependent spatial recognition).

Recently, Campolongo et al. [62] investigated the effects of cannabinoid administration on both short- and long-term object recognition memories under two experimental conditions that differed with respect to their training-associated arousal level (i.e., by manipulating the level of habituation to the apparatus). They found

Table 2.1 The effects of cannabinoids on performance in non-aversive tasks

Species	Drug	Test	Effect on memory	References
Rat	Δ9-THC 5 mg/kg, IP	Object recognition	Impaired acquisition	[59]
Rat	WIN55,212–2 1.2 mg/kg, IP	Object recognition	Impaired acquisition	[60]
Rat	WIN55,212–2 1–2 μg, intra-hippocampal	Object recognition	No effect	[61]
Rat	WIN55,212–2 0.3 mg/kg, IP	Object recognition-habituated	No effect retrieval	[62]
Rat	WIN55,212–2 1–2 μg, intra-hippocampal	Spatial recognition	Impaired acquisition	[61]
Rat	WIN55,212–2 0.6 or 1.2 mg/kg, IP	Object recognition	Impaired acquisition	[63]
Rat	WIN55,212–2 0.6 or 1.2 mg/kg, IP	Social recognition	Impaired acquisition	[63]
Rat	WIN55,212–2 5 μg, intra-hippocampal	Social recognition	Impaired acquisition and retrieval	[64]
Rat	WIN55,212–2 5 μg, intra-BLA	Social recognition	Impaired retrieval	[64]

A general summary of the pharmacological studies examining the effects of exogenous cannabinoids on non-aversive memory paradigms. Δ9-THC, Δ9- tetrahydrocannabinol, *IP* Intraperitoneal

differential effects of cannabinoids depending on the rats' level of arousal and the memory under investigation (short- versus long-term memory). Post-training systemic administration of WIN55,212–2 (0.3 mg/kg) impaired short-term retention while enhancing long-term retention in non-habituated highly aroused rats. In habituated rats, WIN55,212–2 enhanced short-term retention with no effect on long-term retention.

The social recognition test is similar to the object recognition test but uses conspecifics instead of objects as the stimuli. WIN55,212–2 (0.6 and 1.2 mg/kg) impaired the performance of rats in the social and object recognition task [63]. We found that WIN55,212–2 (5 μg/side) impaired acquisition and retrieval of social recognition when microinjected into the hippocampus and impaired acquisition when microinjected into the medial amygdala [64]. See Table 2.1 for a general summary of the pharmacological studies examining the effects of exogenous cannabinoids on non-aversive memory paradigms.

Aversive Memory Tasks

Water maze procedures which focus on spatial memory have been extensively used to test the effects of cannabinoids on different stages of hippocampal-dependent memory. In the water maze task, the animals are trained to escape to a submerged platform in a tank filled with opaque water suggesting that this task is aversive.

In mice and rats, acute systemic administration of Δ9-THC [8 mg/kg, Intraperitoneal (IP)] or WIN55,212–2 (1 and 3 mg/kg) before the training session disrupted acquisition in the water maze test [65–66]. However, Δ9-THC in doses known to impair acquisition did not impair memory retrieval in the water maze [67–68]. Impaired place-learning in the water maze was also demonstrated in rats treated repeatedly with Δ9-THC [69] or acutely with Δ8-THC [70] or synthetic CB1 receptor agonists such as HU-210 [50], but not with the synthetic agonist nabilone [70].

In contextual fear conditioning, the agonist WIN55,212–2 (2.5 and 5.0 mg/kg), given 30 min before the conditioning phase, impaired acquisition of contextual fear conditioning [54]. When WIN55,212–2 (10 or 30 ng in 0.5 μL) was infused into the hippocampus 1 h before the retention test, it impaired retrieval of contextual fear memory [71].

In the inhibitory (passive) avoidance task, systemic injections of Δ9-THC or intra-hippocampal injections of WIN55,212–2 impaired memory acquisition, consolidation, and recall in rats and mice [67, 72]. We found that intra-basolateral amygdala or intra-CA1 WIN55,212–2 (5 μg) had no effect on inhibitory avoidance conditioning [14, 73]. In the light-dark inhibitory avoidance paradigm the animal experiences a pairing of a previously neutral stimulus, the dark context, with an aversive stimulus, footshock, a pairing which results in an increase in latency to enter the dark chamber at testing. Interestingly, several studies found that cannabinoid agonists may enhance memory consolidation. Intra- basolateral amygdala WIN55,212–2 (5–50 ng per side), infused immediately after inhibitory avoidance training, induced dose-dependent enhancement of 48-h retention [56] and propofol, which inhibits fatty acid amide hydrolase, the enzyme that degrades the endocannabinoid anandamide, administered intraperitoneally after training also significantly increased memory consolidation [57]. See Table 2.2 for a general summary of the pharmacological studies examining the effects of exogenous cannabinoids on aversive memory paradigms.

Taken together, these results suggest that the type of cognitive task can determine the neural substrates underlying the memory impairment produced by cannabinoids. The time of drug injection in relation to the learning phase under examination is also a critical factor. Finally, it should be noted that the fact that cannabinoid receptors are localized in different brain structures suggests the modulation of distinct memory process and may explain cases where microinfusion of cannabinoid compounds into specific areas can produce effects different from those seen with systemic administration.

Extinction

Fear inhibition in the form of extinction learning is also considered as an aversive or at least an emotional learning paradigm. In the majority of the studies described,

Table 2.2 The effects of cannabinoids on performance in aversive tasks

Species	Drug	Test	Effect on memory	References
Mice	Δ9-THC 6 or 10 mg/kg, IP	Water maze	Impaired acquisition	[65]
Rat	WIN55,212–2 1 or 3 mg/kg, IP	Water maze	Impaired acquisition	[66]
Rat	Δ9-THC 6 or 10 mg/kg, IP	Water maze	No effect retrieval	[67]
Mice	Δ9-THC 3, 10 or 30 mg/kg, IP	Water maze	No effect retrieval	[68]
Rat	WIN55,212–2 2.5 or 5 mg/kg, IP	Contextual fear conditioning	Impaired acquisition	[54]
Rat	WIN55,212–2 10 or 30 ng, intra-hippocampal	Contextual fear conditioning	Impaired retrieval	[71]
Rat	WIN55,212–2 5 μg, intra-amygdala	Inhibitory avoidance	No effect acquisition	[73]
Rat	WIN55,212–2 5 μg, intra-hippocampal	Inhibitory avoidance	No effect acquisition	[14]
Rat	Δ9-THC 10 mg/kg, IP	Inhibitory avoidance	Impaired acquisition	[67]
Mice	WIN55,212–2 0.25, 0.5 or 1 5 μg, intra-hippocampal	Inhibitory avoidance	Impaired retrieval	[72]
Rat	WIN55,212–2 5–50 ng intra-BLA	Inhibitory avoidance	Enhanced consolidation	[56]
Rat	WIN55,212–2 0.3 mg/kg, IP	Object recognition-not habituated	Enhanced retrieval	[56]
Rat	Propofol 300 or 350 mg/kg, IP	Inhibitory avoidance	Enhanced consolidation	[57]

A general summary of the pharmacological studies examining the effects of exogenous cannabinoids on aversive memory paradigms. Δ9-THC: Δ9-tetrahydrocannabinol, *IP* Intraperitoneal

cannabinoids impaired learning and memory with an aversive (water maze, contextual fear conditioning) or non-aversive (object recognition, spatial recognition, social recognition) nature [54, 60]. On the other hand, it was reported that the ECS has a specific role in facilitating fear associated extinction [4, 12, 14–15, 74].

The anandamide uptake inhibitor AM404 [IP: 10 mg/kg; 1 μg/μL, intracerebroventricular (ICV)] administered during extinction training facilitated the extinction of startle or freezing elicited by a shock-associated context [53, 75–76]. We found that intra-CA1 WIN55,212–2 facilitated the extinction of inhibitory avoidance whereas intra- basolateral amygdala WIN55,212–2 had no effect on extinction [14, 73].

Table 2.3 The effects of cannabinoids on extinction

Species	Drug	Test	Effect on extinction	References
Rat	WIN55,212–2 5 μg, intra-CA1	Inhibitory avoidance	Facilitated	[14]
Rat	AM404 200 ng, intra-CA1	Inhibitory avoidance	Facilitated	[14]
Rat	WIN55,212–2 0.25 mg/kg, IP	Contextual fear conditioning	Facilitated	[13]
Rat	WIN55,212–2 0.25 mg/kg, IP	Contextual fear conditioning	Facilitated	[53]
Rat	AM404 10 mg/kg, IP	Contextual fear conditioning	Facilitated	[53]
Rat	AM404 1 μg, ICV	Contextual fear conditioning	Facilitated	[75]
Rat	CBD 2 μg, ICV	Contextual fear conditioning	Facilitated	[75]
Rat	AEA 0.17 ng, intra-CA1	Contextual fear conditioning	Facilitated	[12]
Rat	WIN55,212–2 0.05 μg, intra-IL	Fear-potentiated startle	Facilitated	[74]
Rat	AM404 10 mg/kg, IP	Fear-potentiated startle	Facilitated	[76]
Rat	AM404 0.2 μg, intra-IL	Fear-potentiated startle	Facilitated	[74]
Rat	URB597 0.3 μg, intra-IL	Fear-potentiated startle	Facilitated	[74]
Rat	WIN55,212–2 5 mg/kg, IP	Fear-potentiated startle	No effect	[76]

A general summary of the pharmacological studies examining the effects of exogenous cannabinoids on extinction. *AEA* N-arachidonylethanolamine, *CBD* cannabidiol, *IP* Intraperitoneal, *ICV* intracerebroventricular, *IL* infralimbic

Lin and coworkers have shown that direct infusion of a CB1 receptor agonist, fatty acid amide hydrolase inhibitor, or uptake inhibitor into the ventromedial prefrontal cortex facilitated extinction of a cue-induced fear-potentiated startle response, while infusion of a CB1 receptor antagonist retarded this form of extinction learning [74]. Furthermore, activation of CB1 receptors within this region also reduced startle potentiation in the absence of cue presentation, suggesting that these receptors are not only involved in the extinction of conditioned fear, but also in adaptation to aversive situations in general [74]. Direct microinjection of cannabidiol, a non psychoactive cannabinoid compound, into the prelimbic prefrontal cortex reduced freezing induced by re-exposure to a context previously paired with footshocks [77]. However, in the more ventrally located infralimbic region of the prefrontal cortex, cannabidiol produced an opposite result, increasing the expression of contextual fear conditioning [77]. See Table 2.3 for a general summary of the pharmacological studies examining the effects of exogenous cannabinoids on extinction.

It should be noted that the facilitating effects on extinction were not generalized to another aversively motivated test, the water maze, in which THC did not affect extinction [78]. Furthermore, no effect on extinction was observed in tasks based on appetitive conditioning [79–81].

The Interaction Between Stress and Cannabinoids in Their Effects on Emotional Learning

Although cannabinoid agonists may have different effects on learning and memory, depending on several factors (such as the aversiveness of the task, the memory phase under investigation etc), accumulating data suggest that when cannabinoid agonists are administered in proximity to an environmental stressor, i.e., shortly before or after an exposure to a stressful experience, cannabinoids can normalize the effects of stress on learning and memory [28, 30–32, 73, 82].

We found that the agonist WIN55,212–2 (5 μg) microinjected into the basolateral amygdala had no effect on inhibitory avoidance conditioning or extinction by itself. However, microinjecting WIN55,212–2 into the basolateral amygdala before exposing the rats to an elevated platform stress reversed the enhancing effects of the stressor on inhibitory avoidance conditioning and its impairing effects on extinction [73]. Intra-basolateral amygdala WIN55,212 before elevated platform stress exposure also prevented the stress-induced enhancement of memory consolidation for reduction in reward magnitude [82]. In this negative emotional learning task we measure the decrease in the magnitude of the expected quantity of reinforcements in an alley maze. In contrast to other fear-related negative experiences, reward reduction is more associated with frustration and is assessed by measuring the latency to run the length of the alley to consume the reduced quantity of reward. These findings suggest that cannabinoid receptors in the basolateral amygdala are important modulators of stress-induced modulation of emotional memory [73, 82].

However, when we examined the effects of elevated platform stress on consolidation of memory in a non-emotional object location task, a different picture emerged. Rats were exposed to the elevated platform stress after the acquisition of a non-aversive hippocampal-dependent learning paradigm, the object location task. These rats were exposed to extensive prior habituation to the arena which reduced novelty stress/arousal level. Exposure to the elevated platform stressor impaired consolidation of the location task. The agonist WIN55,212–2 (5 μg) microinjected into the basolateral amygdala did not prevent the stress-induced impairment in consolidation [83].

Taken together, the data strongly points to the integration of endocannabinoids in the stress response and their role in normalizing emotional memory processes, suggesting that the effects of endocannabinoids become evident only in highly aversive situations.

Indeed, using a much more intensive stressor, the single-prolonged stress (SPS) (i.e., restraint for 2 h, forced swim for 20 min, and anesthesia) we found that intra-basolateral amygdala WIN55,212–2 (5 μg) prevented the SPS-induced enhanced conditioned avoidance and the SPS-induced impaired extinction [30]. SPS also impaired contextual fear extinction tested one week after stress exposure and intra-basolateral amygdala or intra-hippocampal WIN55,212–2 (5 μg) prevented the SPS-induced impairment in extinction [31].

Chronic WIN55,212–2 administration also prevented the effects of chronic stress on memory. WIN55,212–2 prevented the impairing effects of chronic stress on memory, even in the case in which WIN55,212–2 by itself, with no stress exposure, impaired memory. Chronic restraint-stress (2 weeks, 1 h per day) impaired hippocampal short-term memory in the spatial location task, when measured 30 days after stress termination [28]. Chronic WIN55,212–2 administration with no stress exposure (1.2 mg/kg IP, 2 weeks) also impaired performance, consistent with our previous findings of impaired short-term spatial location memory even after 75 days of withdrawal [84]. However, a combination of chronic WIN55,212–2 administration with chronic stress resulted in an intact spatial location memory [28].

It is interesting that WIN55,212–2 by itself impaired short-term location memory, but when administered in proximity to stress exposure, it attenuated the impairing effects of the stressor on memory. It seems that the effects of WIN55,212–2 on memory are dependent on the emotional state of the animal; this is supported by our previous findings that chronic WIN55,212–2 impaired short-term memory in the non-aversive spatial location task, but had no effect on spatial short-term memory in the aversive water-maze task [14]. In general, the cannabinoid system and the stress system are highly interconnected [25, 85–88] and it has been suggested that the ECS might become activated specifically in highly aversive situations, but not in non-aversive situations [79–81]. See Table 2.4 for the effects of exposure to stress and WIN55,212–2 on memory. Taken together, the data indicate that the agonist WIN55,212–2 alters the behavioral effects of environmental conditions, such as stressful experiences, on learning.

In a recent study, we found that exposure to chronic mild stress (CMS) for 3 weeks, which is considered as a rat model for depression, impaired inhibitory avoidance extinction tested 2 and 7 days after CMS ended. Importantly, 3 days administration of WIN55,212–2 (0.5 mg/kg, IP) prevented the CMS-induced impairment of extinction tested 2 and 7 days after the last stressor [32]. Figure 2.1 shows the effects of WIN55,212–2 and the CB1 receptor antagonist AM251, with or without CMS exposure, on fear retrieval and extinction tested 2 and 7 days after the last drug injection.

Rats were exposed to CMS or handled on days 1–21. The agonist WIN55,212–2 or vehicle were administered on days 19–21 (IP; 0.5 mg/kg) and rats were tested for conditioned avoidance and extinction on days 23 and 28 (Fig. 2.1a, b). When no CMS was administered, rats were injected with Vehicle, WIN55,212–2 or AM251 on days 1–3 and tested for inhibitory conditioning on day 5 (Fig. 2.1c; equivalent to testing conditioning on day 23 in Fig. 2.1a, after the drugs were injected on days 19–21) or day 10 (Fig. 2.1d; equivalent to testing conditioning on day 28 in Fig. 2.1b, after the drugs were injected on days 19–21) [32].

Table 2.4 WIN55,212–2 prevents the effects of stress on learning

Drug	Stressor	Behavioral test	Time of testing	Effects of drug	References
WIN55,212–2 5 μg intra-BLA	EPM	Inhibitory avoidance acquisition	Immediately after stress exposure	Prevented the stress-induced enhancement	[73]
WIN55,212–2 5 μg intra-BLA	EPM	Inhibitory avoidance extinction	Immediately after stress exposure	Prevented the stress-induced impairment	[73]
WIN55,212–2 5 μg intra-BLA	EPM	Inhibitory avoidance extinction	24 h after stress exposure	Prevented the stress-induced impairment	[73]
WIN55,212–2 5 μg intra-BLA	EPM	Reduction in reward magnitude consolidation	24 h after stress exposure	Prevented the stress-induced enhancement	[82]
WIN55,212–2 5 μg intra-BLA	EPM	Object location task	24 h after stress exposure	Did not prevent the stress-induced impairment	[83]
WIN55,212–2 5 μg intra-BLA	SPS	Inhibitory avoidance extinction	1 W after stress exposure	Prevented the stress-induced impairment	[30]
WIN55,212–2 5 μg intra-BLA or intra-vSub	SPS	Contextual fear extinction	1 W after stress exposure	Prevented the stress-induced impairment	[31]
WIN55,212–2 1.2 mg/kg IP	Chronic restraint	Object recognition	1 month after stress exposure	Prevented the stress-induced impairment	[28]
WIN55,212–2 1.2 mg/kg IP	Chronic restraint	Object location task	1 month after stress exposure	Prevented the stress-induced impairment	[28]
WIN55,212–2 0.5 mg/kg IP 3 days	CMS	Inhibitory avoidance extinction	48 h after stress exposure	Prevented the stress-induced impairment	[32]
WIN55,212–2 0.5 mg/kg IP 3 days	CMS	Inhibitory avoidance extinction	1 W after stress exposure	Prevented the stress-induced impairment	[32]

A general summary of the effects of exposure to stress and WIN55,212–2 on memory. *CMS* Chronic mild stress, Chronic restraint (1 h per day, 2 weeks), *EPM* elevated platform stress (30 min), *SPS* single prolonged stress (2 h restraint, 15 min of forced swim and sedation), *IP* Intraperitoneal, *BLA* basolateral amygdala

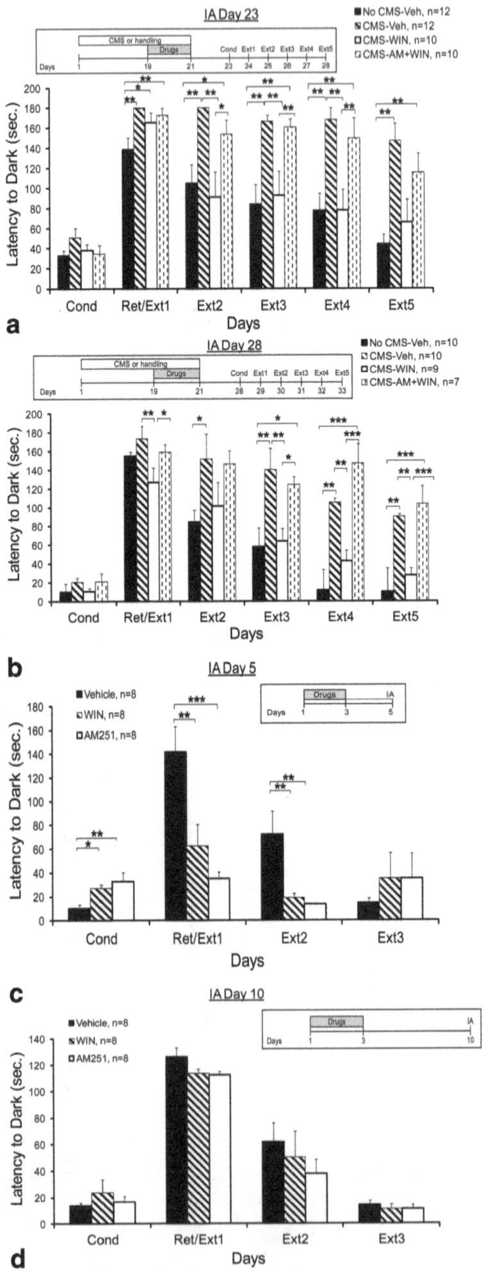

Fig. 2.1 WIN55,212–2 prevents the CMS-induced impairment in extinction **a** Vehicle (Veh), WIN55,212–2 (WIN) or AM251 (AM)+WIN IP on days 19–21. Conditioning (Cond) was tested on day 23. When tested on day 23, the no CMS-Veh group demonstrated decreased latency compared to all groups on Ext1. On Ext2, Ext3 and Ext4, the no CMS-Veh and the CMS-WIN groups demonstrated decreased latency compared to CMS-Veh and CMS-WIN+AM groups. On Ext5, the

We found that when WIN55,212–2 (and AM251) were injected without stress exposure 2 days before avoidance conditioning, rats showed impaired fear retrieval. A possible explanation for the different effects of WIN55,212–2 and AM251 on avoidance when administered 2 days versus a week before training could be that cannabinoids have delayed effects on acquisition of the avoidance memory. In any case, although the drugs impaired fear retrieval when administered 2 days, but not 7 days, before conditioning, WIN55,212–2 prevented the CMS-induced impairment in extinction on both occasions suggesting again that WIN55,212–2 has a different effect on learning and memory that is dependent on the presence (or absence) of an environmental stressor [32].

A Possible Model for the Effects of WIN55,212–2 on Memory After Stress

One of the effects of stress exposure is to activate the hypothalamic-pituitary-adrenal axis and the consequent release of glucocorticoids. It has been suggested that glucocorticoids recruit endocannabinoids signaling in the basolateral amygdala and the hippocampus to modulate aversive memory consolidation [69, 56, 89]. According to this model, corticosterone, which is released after a stressful experience, binds to a yet-uncharacterized membrane-bound glucocorticoid receptor that activates the G_s–cAMP/PKA pathway to induce endocannabinoids synthesis. Endocannabinoids are released into the synapse where they bind to CB1 receptors on GABAergic terminals inhibiting GABA release. This inhibition of GABA release disinhibits norepinephrine release and increases norepinephrine activation of postsynaptic β-adrenoreceptors, increasing the consolidation of emotionally-aversive memories [69, 56, 89].

no CMS+Veh group demonstrated decreased latency compared to CMS+Veh and CMS+WIN+AM groups (*$p<0.05$; **$p<0.01$). Hence, CMS impaired extinction and WIN55,212–2 reverted this effect. **b** Vehicle (Veh), WIN or AM+WIN IP on days 19–21. Conditioning was tested on day 28. When tested on day 28, the CMS-WIN group demonstrated decreased latency compared with the CMS-Veh and CMS+WIN+AM group on Ext1. On Ext2, the no CMS+Veh group demonstrated decreased latency compared with the CMS-Veh. On Ext3, Ext4 and Ext5, the no CMS+Veh and the CMS+WIN groups demonstrated decreased latency compared to the CMS-Veh and CMS+WIN+AM groups (*$p<0.05$; **$p<0.01$; ***$p<0.001$). Hence, CMS impaired extinction and WIN55,212–2 reverted this effect. **c** To examine the effects of the drugs on inhibitory avoidance without exposure to CMS, rats were injected with Vehicle, WIN, or AM on days 1–3 and tested for conditioning on day 5. When the drugs were injected with no stress exposure two days before conditioning, the Vehicle group demonstrated decreased latency compared with the other groups on Cond and increased latency on Ext1 and Ext2 (*$p<0.05$; **$p<0.01$; ***$p<0.001$). Hence, WIN55,212–2 and AM251 impaired retrieval. **d** To examine the effects of the drugs on inhibitory avoidance without exposure to CMS, rats were injected with Vehicle, WIN, or AM on days 1–3 and tested for conditioning on day 10. When WIN55,212–2 or AM251 were injected without stress exposure a week before conditioning, conditioned avoidance and extinction levels were not significantly different from vehicle treated rats. Hence, WIN55,212–2 and AM251 had no effect on retrieval. (Data was published by [32] in Neuropsychopharmacology)

De Bitencourt et al. [90] suggested a mechanism for fear memory extinction involving the interaction between the glucocorticoid and endocannabinoid systems. On the basis of the data suggesting that endocannabinoids [14–15] and glucocorticoids [91–92] mediate the extinction of aversive memories and that glucocorticoids can increase endocannabinoids levels [93–95], they suggested that endocannabinoids are recruited by glucocorticoids in the process of extinction of aversive memories. As endocannabinoids are released in the basolateral amygdala during fear extinction [15], and glucocorticoids can trigger endocannabinoids release at this location, they hypothesized that an endocannabinoid-mediated feedback of the hypothalamic-pituitary-adrenal axis, which is dependent on glucocorticoid release, is an important component of the mechanism of fear memory extinction. In support, we have recently found that the ameliorating effects of WIN55,212–2 on the extinction of contextual fear after exposure to SPS are mediated by glucocorticoid receptors in the basolateral amygdala and hippocampus [31].

I have recently proposed a model suggesting that the ameliorating effects of exogenously administered cannabinoids on emotional learning after *acute* stress are mediated by the decrease in the activity of the hypothalamic-pituitary-adrenal axis via GABAergic mechanisms in the amygdala [96]. The model is presented in Fig. 2.2. The mechanisms underlying the ameliorating effects of exogenously administered cannabinoids on emotional learning after *chronic* stress requires further research [28, 32].

Evidence suggest that enhancing cannabinoids prevented the stress-induced increase in glucocorticoid levels [73, 87, 97] and the stress-induced modulation of emotional memory [30–31, 73, 82]. Stressful events and the release of stress hormones can either impair or enhance emotional memory through the modulation of the basolateral amygdala [98–103].

It has been hypothesized that the anxiolytic effects of cannabinoids are mediated via CB1 receptor activation of GABAergic [40] or glucocorticoid [104] mechanisms within the amygdala. CB1 receptors are densely localized to a distinct population of GABAergic interneurons in the lateral and basal (BLA) nuclei of the amygdala [40]. The basolateral amygdala has a majority of glutamatergic neurons and a minority of GABAergic interneurons and the central amygdala (CeA) has mainly medium spiny-type GABAergic neurons [105]. The basolateral amygdala is thought to process aversive sensory stimuli via afferent inputs to CeA [106]. It is also believed that GABAergic neurons in the intercalated nuclei serve as an intermediate relay station to generate feedforward inhibition of CeA after activation by basolateral amygdala [107].

Intra- basolateral amygdala CB1 receptor agonist (e.g., WIN55,212–2) administered before or after stress exposure, reduces GABA release in basolateral amygdala interneurons, thereby reducing their inhibition of the GABAergic neurons of the intercalated nuclei, which, in turn, increases their inhibition of the pyramidal neurons of the CeA [40]. Hence, the reduction in inhibitory tone may in turn indirectly reduce the effects of stress on memory by enhancing the activity of intercalated GABAergic cells that inhibit activation of the CeA. The end result of the reduction in inhibitory tone may be reduced hypothalamic-pituitary-adrenal axis activity and

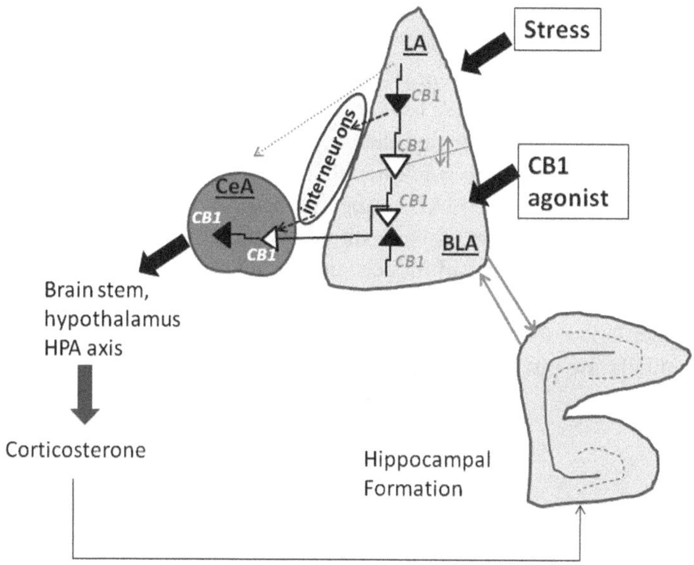

Fig. 2.2 Intra-basolateral amygdala CB1 receptor agonist immediately after stress exposure and hypothalamic-pituitary-adrenal axis activation reduces the stress response via GABAergic mechanism. The lateral amygdala (LA) is connected to basolateral amygdala (BLA) and central amygdala (CeA). A sub-population of LA neurons innervates inhibitory interneurons, which in turn are connected to CeA by inhibitory synapses. The CeA represents a main output station of the amygdala to the brain stem and hypothalamus (and the hypothalamic-pituitary-adrenal axis). A most dominant distribution of CB1 receptors is found in GABAergic (full arrow) and glutamatergic (empty arrow) neurons in the BLA and CeA. Intra-BLA CB1 receptor agonist administered immediately after stress exposure reduces GABA release in BLA interneurons, thereby reducing their inhibition of the GABAergic neurons of the intercalated nuclei, which, in turn, increases their inhibition of the pyramidal neurons of the CeA. Hence, CB1 receptor agonists can reduce hypothalamic-pituitary-adrenal axis activation (and corticosterone release) and modulate the effects of stress on emotional memory. Hence, cannabinoid receptor activation after stress exposure prevents the stress-induced increase in corticosterone levels. The BLA is reciprocally connected with the hippocampal formation. Hence, the amygdala may modulate hippocampal-dependent memory processes directly or indirectly via its effects on the hypothalamic-pituitary-adrenal axis (e.g. as corticosterone readily enters the brain and binds to glucocorticoid receptors in the hippocampus to affect memory). (Data was published by [95] in Neurosci Biobehav Rev)

a reduction in the stress-induced increase in corticosterone levels. Corticosterone easily re-enters the brain to affect glucocorticoid receptors in brain areas that are highly involved in memory processes (e.g. the hippocampal formation). Hence, the reduction in hypothalamic-pituitary-adrenal axis activity may prevent the enhancing or the impairing effects of stress on emotional memory. In support, it has been shown that CB1 receptor agonists decrease the excitability of projection neurons in the rat basolateral amygdala [108]. Several studies have shown that activating CB1 receptors or increasing AEA signaling, prevents some of the effects of stress in the amygdala and hippocampus and can reduce stress-induced hypothalamic-pituitary-adrenal axis activation [25, 30–31, 73, 87]. Also, a study using imaging in humans has shown that a cannabinoid agonist significantly reduced amygdala reactivity to social signals of threats [109]. We have shown that WIN55,212–2 administered into the basolateral amygdala prevented the stress-induced elevation in corticosterone levels [71]. Finally, it has been suggested that the inhibition of GABA release from axon terminals of local-circuit GABAergic interneurons in the basolateral amygdala by presynaptic CB1 receptors may constitute an important aspect of the neurobiological substrates of cannabinoid-induced emotional responses [40].

Taken together, the data suggest that one possible mechanism underlying cannabinoid modulation of emotional learning after stress is the decrease in the activity of the hypothalamic-pituitary-adrenal axis, through the basolateral amygdala GABAergic system.

Conclusions

Mounting evidence supports a role for the ECS in emotionality. From a neuroanatomical perspective, CB1 receptors are abundantly located throughout corticolimbic regions implicated in proper emotional responding. Consequently, facilitating endocannabinoid signaling within this corticolimbic network may someday prove to be an effective strategy in combating the pathophysiological development of emotional disorders that are precipitated by stress.

The diversity of the effects of cannabinoids on memory suggest that by enhancing endocannabinoids signaling we may change the impact of environmental influences on emotional and cognitive behavior. The data provide a rationale for exploring novel therapeutic strategies that target the cannabinoid system for disorders of anxiety and stress-related diseases. Specifically, this may have potential implications to developing pharmacological approaches to correct the prolonged retention of memories of negative events in depression and other stress-related states.

References

1. Lichtman AH, Varvel SA, Martin BR. Endocannabinoids in cognition and dependence. Prostaglandins Leukot Essent Fatty Acids. 2002;66(2–3):269–85.
2. Puighermanal E, Marsicano G, Busquets-Garcia A, Lutz B, Maldonado R, Ozaita A. Cannabinoid modulation of hippocampal long-term memory is mediated by mTOR signaling. Nat Neurosci. 2009;12(9):1152–8.
3. Riedel G, Davies SN. Cannabinoid function in learning, memory and plasticity. Handb Exp Pharmacol. 2005;168:445–77.
4. Varvel SA, Lichtman AH. Evaluation of CB1 receptor knockout mice in the Morris water maze. J Pharmacol Exp Ther. 2002;301(3):915–24.
5. Zanettini C, Panlilio LV, Alicki M, Goldberg SR, Haller J, Yasar S. Effects of endocannabinoid system modulation on cognitive and emotional behavior. Front Behav Neurosci. 2011;5:57.

6. Bolla KI, Brown K, Eldreth D, Tate K, Cadet JL. Dose-related neurocognitive effects of marijuana use. Neurology. 2002;59(9):1337–43.
7. Castellano C, Rossi-Arnaud C, Cestari V, Costanzi M. Cannabinoids and memory: animal studies. Curr Drug Targets CNS Neurol Disord. 2003;2(6):389–402.
8. Solowij N, Battisti R. The chronic effects of cannabis on memory in humans: a review. Curr Drug Abuse Rev. 2008;1(1):81–98.
9. Crean RD, Crane NA, Mason BJ. An evidence based review of acute and long-term effects of cannabis use on executive cognitive functions. J Addict Med. 2011;5(1):1–8.
10. Solowij N, Stephens RS, Roffman RA, Babor T, Kadden R, Miller M, et al. Cognitive functioning of long-term heavy cannabis users seeking treatment. JAMA. 2002;287(9):1123–31.
11. Puighermanal E, Busquets-Garcia A, Maldonado R, Ozaita A. Cellular and intracellular mechanisms involved in the cognitive impairment of cannabinoids. Philos Trans R Soc Lond B Biol Sci. 2012;367(1607):3254–63.
12. de Oliveira Alvares L Genro BP Diehl F Quillfeldt JA. Differential role of the hippocampal endocannabinoid system in the memory consolidation and retrieval mechanisms. Neurobiol Learn Mem. 2008;90(1):1–9.
13. Pamplona FA, Prediger RD, Pandolfo P, Takahashi RN. The cannabinoid receptor agonist WIN 55,212–2 facilitates the extinction of contextual fear memory and spatial memory in rats. Psychopharmacology. 2006;188(4):641–9.
14. Abush H, Akirav I. Cannabinoids modulate hippocampal memory and plasticity. Hippocampus. 2010;20(10):1126–38.
15. Marsicano G, Wotjak CT, Azad SC, Bisogno T, Rammes G, Cascio MG, et al. The endogenous cannabinoid system controls extinction of aversive memories. Nature. 2002;418(6897):530–4.
16. Marsicano G, Lafenêtre P. Roles of the endocannabinoid system in learning and memory. Curr Top Behav Neurosci. 2009;1:201–30.
17. Bisaz R, Conboy L, Sandi C. Learning under stress: a role for the neural cell adhesion molecule NCAM. Neurobiol Learn Mem. 2009;91(4):333–42.
18. Sandi C, Woodson JC, Haynes VF, Park CR, Touyarot K, Lopez-Fernandez MA, et al. Acute stress-induced impairment of spatial memory is associated with decreased expression of neural cell adhesion molecule in the hippocampus and prefrontal cortex. Biol Psychiatry. 2005;57(8):856–64.
19. Shors TJ. Acute stress rapidly and persistently enhances memory formation in the male rat. Neurobiol Learn Mem. 2001;75(1):10–29.
20. Howland JG, Wang YT. Synaptic plasticity in learning and memory: stress effects in the hippocampus. Prog Brain Res. 2008;169:145–58.
21. Matsuda LA, Lolait SJ, Brownstein MJ, Young AC, Bonner TI. Structure of a cannabinoid receptor and functional expression of the cloned Cdna. Nature. 1990;346(6284):561–4.
22. Devane WA, Hanus L, Breuer A, Pertwee RG, Stevenson LA, Griffin G, et al. Isolation and structure of a brain constituent that binds to the cannabinoid receptor. Science. 1992;258(5090):1946–9.
23. Sugiura S, Kondo A, Sukagawa S, Nakane A, Shinoda K, Itoh K, et al. 2-Arachidonoylglycerol: a possible endogenous cannabinoid receptor ligand in brain. Biochem Biophys Res Commun. 1995;215(1):89–97A.
24. Fraser GA. The use of a synthetic cannabinoid in the management of Treatment-Resistant nightmares in posttraumatic stress disorder (PTSD). CNS Neurosci Ther. 2009;15(1):84–8.
25. Gorzalka BB, Hill MN, Hillard CJ. Regulation of endocannabinoid signaling by stress: implications for stress-related affective disorders. Neurosci Biobehav Rev. 2008;32(6):1152–60.
26. Moreira FA, Wotjak CT. Cannabinoids and anxiety. Curr Top Behav Neurosci. 2010;2:429–50.
27. Riebe CJ, Wotjak CT. Endocannabinoids and stress. Stress. 2011;14(4):384–97.
28. Abush H, Akirav I. Cannabinoids ameliorate impairments induced by chronic stress to synaptic plasticity and short-term memory. Neuropsychopharmacology. 2013;38(8):1521–34 (Feb 20).

29. Bortolato M, Mangieri RA, Fu J, Kim JH, Arguello O, Duranti A, et al. Antidepressant-like activity of the fatty acid amide hydrolase inhibitor URB597 in a rat model of chronic mild stress. Biol Psychiatry. 2007;62(10):1103–10.
30. Ganon-Elazar E, Akirav I. Cannabinoids prevent the development of behavioral and endocrine alterations in a rat model of intense stress. Neuropsychopharmacology. 2012;37(2):456–66.
31. Ganon-Elazar E, Akirav I. Cannabinoids and traumatic stress modulation of contextual fear extinction and GR expression in the amygdala-hippocampal-prefrontal circuit. Psychoneuroendocrinology. 2013;38(9):1675–87.
32. Segev A, Rubin AS, Abush H, Richter-Levin G, Akirav I. Cannabinoid receptor activation prevents the effects of chronic mild stress on emotional learning and ltp in a rat model of depression. Neuropsychopharmacology. 2013;39(4):919–33 (Oct 21).
33. Viveros MP, Marco EM, File SE. Endocannabinoid system and stress and anxiety responses. Pharmacol Biochem Behav. 2005;81(2):331–42.
34. Drevets WC, Price JL, Furey ML. Brain structural and functional abnormalities in mood disorders: implications for neurocircuitry models of depression. Brain Struct Funct. 2008;213 (1–2):93–118.
35. McEwen BS. Physiology and neurobiology of stress and adaptation: central role of the brain. Physiol Rev. 2007;87(3):873–904.
36. McEwen BS. The brain is the central organ of stress and adaptation. Neuroimage. 2009;47(3):911–3.
37. Ahima RS, Harlan RE. Charting of type II glucocorticoid receptor-like immunoreactivity in the rat central nervous system. Neuroscience. 1990;39(3):579–604.
38. Fuxe K, Agnati LF. Receptor-receptor interactions in the central nervous system. A new integrative mechanism in synapses. Med Res Rev. 1985;5(4):441–82.
39. Herkenham M, Lynn AB, Little MD, Johnson MR, Melvin LS, de Costa BR, Rice KC. Cannabinoid receptor localization in brain. Proc Natl Acad Sci U S A. 1990;87(5):1932–1936.
40. Katona I, Rancz EA, Acsady L, Ledent C, Mackie K, Hajos N, Freund TF. Distribution of CB1 cannabinoid receptors in the amygdala and their role in the control of GABAergic transmission. J Neurosci. 2001;21(23):9506–18.
41. Sah P, Lopez de Armentia M. Excitatory synaptic transmission in the lateral and central amygdala. Ann N Y Acad Sci. 2003;985:67–77.
42. Shin LM, Liberzon I. The neurocircuitry of fear, stress, and anxiety disorders. Neuropsychopharmacology. 2010;35(1):169–91.
43. Herman JP, Figueiredo H, Mueller NK, Ulrich-Lai Y, Ostrander MM, Choi DC, Cullinan WE. Central mechanisms of stress integration: hierarchical circuitry controlling hypothalamo-pituitary-adrenocortical responsiveness. Front Neuroendocrinol. 2003;24(3):151–80.
44. Amat J, Baratta MV, Paul E, Bland ST, Watkins LR, Maier SF. Medial prefrontal cortex determines how stressor controllability affects behavior and dorsal raphe nucleus. Nat Neurosci. 2005;8(3):365–71.
45. Maier SF, Amat J, Baratta MV, Paul E, Watkins LR. Behavioral control, the medial prefrontal cortex, and resilience. Dialogues Clin Neurosci. 2006;8(4):397–406.
46. Jankord R, Herman JP. Limbic regulation of hypothalamo-pituitary-adrenocortical function during acute and chronic stress. Ann N Y Acad Sci. 2008;1148:64–73.
47. Akirav I. The role of cannabinoids in modulating emotional and non-emotional memory processes in the hippocampus. Front Behav Neurosci. 2011;5:34.
48. Black MD, Stevens RJ, Rogacki N, Featherstone RE, Senyah Y, Giardino O, et al. AVE1625, a cannabinoid CB1 receptor antagonist, as a co-treatment with antipsychotics for schizophrenia: improvement in cognitive function and reduction of antipsychotic-side effects in rodents. Psychopharmacology (Berl). 2011;215(1):149–63.
49. Block RI, Ghoneim MM. Effects of chronic marijuana use on human cognition. Psychopharmacology (Berl). 1993;110(1–2):219–28.
50. Curran HV, Brignell C, Fletcher S, Middleton P, Henry J. Cognitive and subjective dose–response effects of acute oral delta9-tetrahydrocannabinol (THC) in infrequent cannabis users. Psychopharmacology (Berl). 2002;164(1):61–70.

51. Ferrari F, Ottani A, Vivoli R, Giuliani D. Learning impairment produced in rats by the cannabinoid agonist HU 210 in a water-maze task. Pharmacol Biochem Behav. 1999;64(3):555–6.
52. Molina-Holgado F, Gonzalez MI, Leret ML. Effect of delta9-tetrahydrocannabinol on short-term memory in the rat. Physiol Behav. 1995;57(1):177–9.
53. Pamplona FA, Bitencourt RM, Takahashi RN. Short- and long-term effects of cannabinoids on the extinction of contextual fear memory in rats. Neurobiol Learn Mem. 2008;90(1):290–3.
54. Pamplona FA, Takahashi RN. WIN 55212–2 impairs contextual fear conditioning through the activation of CB1 cannabinoid receptors. Neurosci Lett. 2006;397(1–2):88–92.
55. Ranganathan M, D'Souza DC. The acute effects of cannabinoids on memory in humans: a review. Psychopharmacology (Berl). 2006;188(4):425–44.
56. Campolongo P, Roozendaal B, Trezza V, Hauer D, Schelling G, McGaugh JL, Cuomo V. Endocannabinoids in the rat basolateral amygdala enhance memory consolidation and enable glucocorticoid modulation of memory. Proc Natl Acad Sci U S A. 2009;106(12):4888–93.
57. Hauer D, Ratano P, Morena M, Scaccianoce S, Briegel I, Palmery M, et al. Propofol enhances memory formation via an interaction with the endocannabinoid system. Anesthesiology. 2011;114:1380–8.
58. Ennaceur A, Delacour J. A new one-trial test for neurobiological studies of memory in rats. 1: behavioral data. Behav Brain Res. 1988;31(1):47–59.
59. Quinn HR, Matsumoto I, Callaghan PD, Long LE, Arnold JC, Gunasekaran N, et al. Adolescent rats find repeated delta-THC less aversive than adult rats but display greater residual cognitive deficits and changes in hippocampal protein expression following exposure. Neuropsychopharmacology. 2008;33(5):1113–26.
60. Schneider M, Schömig E, Leweke FM. Preclinical study: acute and chronic cannabinoid treatment differentially affects recognition memory and social behavior in pubertal and adult rats. Addict Biol. 2008;13(3–4):345–57.
61. Suenaga T, Ichitani Y. Effects of hippocampal administration of a cannabinoid receptor agonist WIN55,212–2 on spontaneous object and place recognition in rats. Behav Brain Res. 2008;190(2):248–52.
62. Campolongo P, Morena M, Scaccianoce S, Trezza V, Chiarotti F, Schelling G, et al. Novelty-induced emotional arousal modulates cannabinoid effects on recognition memory and adrenocortical activity. Neuropsychopharmacology. 2013;38(7):1276–86.
63. Schneider M, Koch M. The cannabinoid agonist WIN 55,212–2 reduces sensorimotor gating and recognition memory in rats. Behav Pharmacol. 2002;13(1):29–37.
64. Segev A, Akirav I. Differential effects of cannabinoid receptor agonist on social discrimination and contextual fear in amygdala and hippocampus. Learn Mem. 2011;18(4):254–9.
65. Da Silva GE Takahashi RN. SR 141716A prevents [delta]9-tetrahydrocannabinol-induced spatial learning deficit in a Morris-type water maze in mice. Prog Neuropsychopharmacol Biol Psychiatry. 2002;26(2):321–5.
66. Robinson L, Goonawardena AV, Pertwee R, Hampson RE, Platt B, Riedel G. WIN55, 212–2 induced deficits in spatial learning are mediated by cholinergic hypofunction. Behav Brain Res. 2010;208:584–92.
67. Mishima K, Egashira N, Hirosawa N, Fujii M, Matsumoto Y, Iwasaki K, Fujiwara M. Characteristics of learning and memory impairment induced by delta9-tetrahydrocannabinol in rats. Jpn J Pharmacol. 2001;87(4):297–308.
68. Varvel SA, Hamm RJ, Martin BR, Lichtman AH. Differential effects of delta9-THC on spatial reference and working memory in mice. Psychopharmacology. 2001;157(2):142–50.
69. Moore NLT, Greenleaf ALR, Acheson SK, Wilson WA, Swartzwelder HS, Kuhn CM. Role of cannabinoid receptor type 1 desensitization in greater tetrahydrocannabinol impairment of memory in adolescent rats. J Pharmacol Exp Ther. 2010;335(2):294–301.
70. Diana G, Malloni M, Pieri M. Effects of the synthetic cannabinoid nabilone on spatial learning and hippocampal neurotransmission. Pharmacol Biochem Behav. 2003;75(3):585–91.
71. Atsak P, Roozendaal B, Campolongo P. Role of the endocannabinoid system in regulating glucocorticoid effects on memory for emotional experiences. J Neurosci. 2012;1:104–16.

72. Nasehi M, Piri M, Jamali-Raeufy N, Zarrindast MR. Influence of intracerebral administration of NO agents in dorsal hippocampus (CA1) on cannabinoid state-dependent memory in the step-down passive avoidance test. Physiol Behav. 2010;100(4):297–304.
73. Ganon-Elazar E, Akirav I. Cannabinoid receptor activation in the basolateral amygdala blocks the effects of stress on the conditioning and extinction of inhibitory avoidance. J Neurosci. 2009;29(36):11078–88.
74. Lin HC, Mao SC, Su CL, Gean PW. The role of prefrontal cortex CB1 receptors in the modulation of fear memory. Cereb Cortex. 2009;19(1):165–75.
75. Bitencourt RM, Pamplona FA, Takahashi RN. Facilitation of contextual fear memory extinction and anti-anxiogenic effects of AM404 and cannabidiol in conditioned rats. Eur Neuropsychopharmacol. 2008;18(12):849–59.
76. Chhatwal JP, Davis M, Maguschak KA, Ressler KJ. Enhancing cannabinoid neurotransmission augments the extinction of conditioned fear. Neuropsychopharmacology. 2005;30(3):516–24.
77. Lemos JI, Resstel LB, Guimarães FS. Involvement of the prelimbic prefrontal cortex on cannabidiol-induced attenuation of contextual conditioned fear in rats. Behav Brain Res. 2010;207(1):105–11.
78. Varvel SA, Wise LE, Niyuhire F, Cravatt BF, Lichtman AH. Inhibition of fatty-acid amide hydrolase accelerates acquisition and extinction rates in a spatial memory task. Neuropsychopharmacology. 2007;32(5):1032–41.
79. Harloe JP, Thorpe AJ, Lichtman AH. Differential endocannabinoid regulation of extinction in appetitive and aversive Barnes maze tasks. Learn Mem. 2008;15(11):806–9.
80. Holter SM, Kallnik M, Wurst W, Marsicano G, Lutz B, Wotjak CT. Cannabinoid CB1 receptor is dispensable for memory extinction in an appetitively-motivated learning task. Eur J Pharmacol. 2005;510(1–2):69–74.
81. Niyuhire F, Varvel SA, Thorpe AJ, Stokes RJ, Wiley JL, Lichtman AH. The disruptive effects of the CB 1 receptor antagonist rimonabant on extinction learning in mice are task-specific. Psychopharmacology. 2007;191(2):223–31.
82. Ramot A, Akirav I. Cannabinoid receptors activation and glucocorticoid receptors deactivation in the amygdala prevent the stress-induced enhancement of a negative learning experience. Neurobiol Learn Mem. 2012;97(4):393–401.
83. Segev A, Ramot A, Akirav I. Stress hormones receptors in the amygdala mediate the effects of stress on the consolidation, but not the retrieval, of a non aversive spatial task. PLoS One. 2012;7(1):e29988.
84. Abush H, Akirav I. Short- and long-term cognitive effects of chronic cannabinoids administration in late-adolescence rats. PLoS One. 2012;7(2):e31731.
85. Hill MN, McEwen BS. Involvement of the endocannabinoid system in the neurobehavioural effects of stress and glucocorticoids. Prog Neuropsychopharmacol Biol Psychiatry. 2010;34(5):791–7.
86. Hill MN, McLaughlin RJ, Bingham B, Shrestha L, Lee TT, Gray JM, et al. Endogenous cannabinoid signaling is essential for stress adaptation. PNAS. 2010;107(20):9406–11.
87. Patel S, Roelke CT, Rademacher DJ, Cullinan WE, Hillard CJ. Endocannabinoid signaling negatively modulates stress-induced activation of the hypothalamic-pituitary-adrenal axis. Endocrinology. 2004;145(12):5431–8.
88. Patel S, Hillard CJ. Adaptations in endocannabinoid signaling in response to repeated homotypic stress: a novel mechanism for stress habituation. Eur J Neurosci. 2008;27(11):2821–9.
89. Hill MN, McEwen BS. Endocannabinoids: the silent partner of glucocorticoids in the synapse. Proc Natl Acad Sci U S A. 2009;106(12):4579–80.
90. de Bitencourt R, Pamplona FA, Takahashi RN. A current overview of cannabinoids and glucocorticoids in facilitating extinction of aversive memories: potential extinction enhancers. Neuropharmacology. 2013;64:389–95.
91. Clay R, Hebert M, Gill G, Stapleton LA, Pridham A, Coady M, et al. Glucocorticoids are required for extinction of predator stress-induced hyperarousal. Neurobiol Learn Mem. 2011;96(2):367–77.

92. Yang YL, Chao PK, Lu KT. Systemic and intra-amygdala administration of glucocorticoid agonist and antagonist modulate extinction of conditioned fear. Neuropsychopharmacology. 2006;31(5):912–24.
93. Di S, Malcher-Lopes R, Halmos KC, Tasker JG. Nongenomic glucocorticoid inhibition via endocannabinoid release in the hypothalamus: a fast feedback mechanism. J Neurosci. 2003;23(12):4850–7.
94. Hill MN, Ho WS, Meier SE, Gorzalka BB, Hillard CJ. Chronic corticosterone treatment increases the endocannabinoid 2-arachidonylglycerol in the rat amygdala. Eur J Pharmacol. 2005;528(1–3):99–102.
95. Patel S, Roelke CT, Rademacher DJ, Hillard CJ. Inhibition of restraint stress-induced neural and behavioural activation by endogenous cannabinoid signalling. Eur J Neurosci. 2005;21(4):1057–69.
96. Akirav I. Cannabinoids and glucocorticoids modulate emotional memory after stress. Neurosci Biobehav Rev. 2013;37:2554–63 (Aug 14).
97. Hill MN, McLaughlin RJ, Morrish AC, Viau V, Floresco SB, Hillard CJ, Gorzalka BB. Suppression of amygdalar endocannabinoid signaling by stress contributes to activation of the hypothalamic–pituitary–adrenal axis. Neuropsychopharmacology. 2009;34(13):2733–45.
98. Akirav I, Sandi C, Richter-Levin G. Differential activation of hippocampus and amygdala following spatial learning under stress. Eur J Neurosci. 2001;14(4):719–25.
99. Akirav I, Kozenicky M, Tal D, Sandi C, Venero C, Richter-Levin G. A facilitative role for corticosterone in the acquisition of a spatial task under moderate stress. Learn Mem. 2004;11(2):188–95.
100. Diamond DM, Campbell AM, Park CR, Woodson JC, Conrad CD, Bachstetter AD, Mervis RF. Influence of predator stress on the consolidation versus retrieval of long-term spatial memory and hippocampal spinogenesis. Hippocampus. 2006;16(7):571–6.
101. De Kloet ER Oitzl MS Joels M. Stress and cognition: are corticosteroids good or bad guys? Trends Neurosci. 1999;22(10):422–6.
102. Sandi C, Loscertales M, Guaza C. Experience-dependent facilitating effect of corticosterone on spatial memory formation in the water maze. Eur J Neurosci. 1997;9(4):637–42.
103. Roozendaal B. Glucocorticoids and the regulation of memory consolidation. Psychoneuroendocrinology. 2000;25(3):213–38.
104. Rodríguez de Fonseca F, Carrera MR, Navarro M, Koob GF, Weiss F. Activation of corticotropin-releasing factor in the limbic system during cannabinoid withdrawal. Science. 1997;276(5321):2050–4.
105. Ehrlich I, Humeau Y, Grenier F, Ciocchi S, Herry C, Lüthi A. Amygdala inhibitory circuits and the control of fear memory. Neuron. 2009;62(6):757–71.
106. LeDoux JE. Emotion circuits in the brain. Annu Rev Neurosci. 2000;23:155–84.
107. Paré D, Smith Y. The intercalated cell masses project to the central and medial nuclei of the amygdala in cats. Neuroscience. 1993;57(4):1077–90.
108. Pistis M, Perra S, Pillolla G, Melis M, Gessa GL, Muntoni AL. Cannabinoids modulate neuronal firing in the rat basolateral amygdala: evidence for CB1- and non-CB1-mediated actions. Neuropharmacology. 2004;46(1):115–25.
109. Phan KL, Angstadt M, Golden J, Onyewuenyi I, Popovska A, de Wit H. Cannabinoid modulation of amygdala reactivity to social signals of threat in humans. J Neurosci. 2008;28(10):2313–9.

Chapter 3
The Hippocampal Endocannabinoid System in Different Memory Phases: Unveiling the CA1 Circuitry

Jorge Alberto Quillfeldt and Lucas de Oliveira Alvares

Abstract CB1 Cannabinoid receptors are widely expressed throughout the brain, particularly in areas involved in learning and memory, such as the hippocampus. In the CA1 area, they are mainly present at the presynaptic terminals of both GABAergic and glutamatergic neurons. The antagonist/inverse agonist AM251 is a useful pharmacological tool due to its selectivity and ability to tap into the endocannabinoid system (ECS). When infused into the brain, it interferes with the natural functioning of the local pool of endocannabinoids present in each memory phase, and by suppressing the natural course of events, exposes its function in each situation. Anandamide (AEA) was also studied, but results were less consistent. In a large set of experiments spanning several years we have shown that different memory phases are modulated in opposite, complementary ways: AM251 was amnestic (and AEA, facilitatory) when infused into CA1 both after training (consolidation) or after a long reactivation session (extinction), suggesting that ECS modulates positively these phases. On the other hand, AM251 facilitated (and AEA frequently disrupted) memory before test (retrieval) or after a short reactivation session (reconsolidation), suggesting a negative modulatory role. Thus, simmetrically opposed actions are the rule for the ECS in the CA1 area, suggesting both plastic events and complex selective roles taking place under its control, e.g. "switching" between extinction and reconsolidation. Results were interpreted according to known CA1 circuitry and the most probable position of cannabinoid CB1 receptors, pointing to a complex, multifunctional modulatory system that is perfectly consistent with its ubiquity in mammal brains.

Keywords Endocannabinoid system (ECS) · Hippocampus · Memory phases · Memory consolidation · Memory retrieval · Memory extinction · Memory reconsolidation · AM251 · CA1 neural circuit · Switching mechanism

J. A. Quillfeldt (✉) · L. de Oliveira Alvares
Psychobiology and Neurocomputing Lab (LPBNC), Dept. of Biophysics, IB & Neurosciences Graduate Program, ICBS, Universidade Federal do Rio Grande do Sul (UFRGS), Porto Alegre, Brazil
e-mail: quillfe@ufrgs.br

P. Campolongo, L. Fattore (eds.), *Cannabinoid Modulation of Emotion, Memory, and Motivation,* DOI 10.1007/978-1-4939-2294-9_3

The plant Cannabis has been used by humans for thousands of years for religious, medicinal and recreational proposes [1]. Extensive evidence from animals to humans indicate that cannabinoids affect cognitive performance mainly by modifying the brain endocannabinoid system (ECS) [2]. This endogenous modulatory system is involved in a plethora of physiological functions, including pain, appetite control, motricity, cognition—and one particularly important cognitive function is memory.

ECS is composed by a machinery that involves endocannabinoid ligands, such as anandamide and 2-arachidonoyl glycerol (2-AG), the whole set of enzymes responsible for their synthesis, degradation and reuptake, and the receptors, both canonical (CB1 and CB2), and putative (GPR55, GPR119 and GPR18)—notwithstanding the fact that some endocannabinoids, such as anandamide, can also bind to the transient receptor potential vanilloid 1 cation channel (TRPV1) and other targets [3–7].

Cannabinoid receptors are widely expressed throughout the brain, and noticeably higher levels of CB1 are expressed in brain areas involved in learning and memory, such as the hippocampus and amygdala [8–10]. At the cellular level, CB1 receptors are mainly present at the presynaptic axon terminals of both GABAergic and glutamatergic neurons [11, 12]. Endocannabinoids such as anandamide or 2-AG are synthesized on demand, and released as retrograde messengers from post-synaptic neurons into the synaptic cleft, inhibiting the neurotransmitter release [11, 13].

Ample evidence indicates that either exogenously injected or endogenously released cannabinoids—or their interaction—may have pronounced, yet contradictory effects on learning and memory. In this chapter we will focus on the distinct modulatory roles played by the hippocampal ECS on memory consolidation, retrieval, reconsolidation and extinction.

Temporal Phases of Memory

Memory is a non-instantaneous, complex physiological process that evolves in time—from seconds to minutes, hours, days and even months or years—recruiting different molecular agents in order to produce more or less durable plastic changes in the connections among neural cells of different brain areas, i.e. "recording" the experience in an *engram*. Different phases can be identified according to the temporal position relative to the behavioral experience, and the molecular markers and brain areas engaged [14].

Acquisition

Acquisition, also known as *learning* or *encoding*: takes place during the behavioral experience itself, and integrates information from a whole set of cognitive functions—sensory data, emotional response, attention, motivation, motricity and exploratory strategy [14]. Since each one of those are different functions with different mechanisms and brain location, treatment before acquisition—"training"—session usually produces results difficult to interpret because it affects several confounding variables, requiring additional control groups.

Consolidation

Right after the acquisition session, memory is labile and can be easily modified by any competitor experience, in which it may be either strengthened or disrupted. The experience begins to amalgamate into a memory trace that stabilizes in a process called *consolidation* [15, 16]. At first the so called *synaptic consolidation* [17, 18] takes place at cellular and subcellular levels, especially in temporal areas such as the hippocampus [19] and the amygdala [20], but also in several cortical areas. After a few weeks, the engram establishes itself in the neocortex in a more persistent way (would this be the real "memory storage"?), becoming less and less dependent on hippocampus—a phenomenon known as *systems consolidation* [18, 21, 22].

Retrieval or Recall

The only actual way to assure that a memory trace was retained is the animal exhibiting a consistent behavioral change in a posterior "test" session [23, 24]. This change must be somewhat measurable, e.g., in an apparatus designed to evince the memory class, be it spatial, emotional/aversive, recognition or appetitive. Retrieval may be tried at any time after training, but if done just seconds or minutes later, it will show only *working memory;* if tests are performed from 2 to 6 h after training, engrams are not considered fully consolidated, and only a *short-term memory* can be expressed; after that, and up to days, weeks or months, a consolidated, *long-term memory* should be accessible. However, even well-consolidated memories will fade away with time, the phenomenon known as *forgetting.*

Post-retrieval Memory Phases

In the recent years it became largely recognized that consolidated traces are not in any sense "unchangeable", but may undergo modifications if some "boundary conditions" are achieved (the right re-exposure time, etc) during its retrieval. Then, a previously consolidated memory may undergo *extinction* (be replaced by a second, different trace that blocks the first) [25] or *reconsolidation* (change its original qualities in different ways, including erasure). Particularly in the case of reconsolidation, the re-exposure to the training context (in the absence of the Unconditioned Stimulus (US—e.g., a shock) seems to destabilize/relabilize the memory trace, that later re-consolidates in a process that is dependent on protein synthesis

[26]. Extinction involves a "consolidation of extinction" phase that also depends on protein synthesis, with a dynamics quite similar to that of a first-time consolidation.

Memory *Consolidation* is Positively Modulated by the ECS in the CA1 Area: the Cannabinoid Antagonist/Inverse Agonist AM251 is Disruptive

Memory consolidation can be selectively targeted in a simple experimental design requiring only one learning trial (e.g., the inhibitory avoidance paradigm in rodents): by injecting/infusing the drug *immediately after training* it is possible to selectively affect the memory consolidation phase, since all the intervening cognitive aspects of learning—attention, motivation, motricity, emotions and information acquisition itself—have been already completed at the time of drug administration. This approach results in "cleaner" interpretation of the obtained data [14], since any observed effect can only be attributed to an interference with whatever takes place in the first hours after training—i.e., the consolidation process [15, 17], avoiding all the other possible confounding variables.

Consolidation plays an important and adaptive role in defining the fate of a memory according to the relevance of its content. For instance, stress hormones released in a threatening situation, such as glucocorticoids or noradrenaline, enhance memory consolidation [27, 28]. The mechanisms behind the memory consolidation processes rely on several cellular and molecular events that occur in brain structures such as the hippocampus. Important steps are the activation of NMDA and AMPA glutamatergic and other metabotropic receptors which leads to an increase of calcium intracellular levels, the consecutive mobilization of a collection of second messengers and the activation of several important enzymes such as PKA, PKC, CAMKII, and MAPK. This will affect both receptors' sensibility and signaling cascades, with some enzymes entering to the nucleus and inducing changes such as the phosphorylation of CREB, promoting gene expression and a change in protein synthesis [17]. In this view, it is important to employ a clear terminology: the endogenous cannabinoid system operates as a *neuromodulatory system*, whose molecular agents perform the fine-tuning of the *effector* excitatory (glutamatergic) and inhibitory (GABAergic) synapses present at the local circuits, resembling, for instance, the operation of the cholinergic, dopaminergic or glucocorticoid systems.

The involvement of the hippocampal endocannabinoid signaling in memory consolidation has been extensively demonstrated, but results may be quite contradictory. Several studies have demonstrated that CB1 receptor agonists impair hippocampus-dependent memory consolidation of inhibitory avoidance, contextual fear conditioning or Morris water maze tasks when injected systemically [29–31] or infused directly into the hippocampus [30, 32]. Moreover, electrophysiological studies show that CB1 receptor activation inhibits long-term potentiation (LTP) induction [33–35].

Our behavioral-pharmacological screening began in 2004–2005, when we first showed that the intra-CA1 infusion of the selective antagonist/inverse agonist[1] AM251, *disrupts memory consolidation* of the Step-Down Inhibitory Avoidance (IA), but not of the Open Field Habituation (HAB), a less aversive task (see Figs 2 and 3 of ref. [36]). This behavioral effect was later confirmed and extended to another aversive task, Contextual Fear Conditioning (CFC) [37].

Learning situations are known to be highly sensitive to a great deal of non-mnemonic factors [34], including animal strain, basal stress levels and/or differences in protocol details, for instance, employed doses, time elapsing between infusion and the task, etc. This is why we carefully controlled for possible locomotor/exploratory unexpected effects in all the above experiments in order to avoid false positives (or negatives) due to behavioural measures misinterpreted for reasons other than mnemonic.

We employed the antagonist/inverse agonist AM251 to interfere with the endocannabinoid system in order to understand its physiological function. Once *selectivity* is the true "Holy Grail" of neuropsychopharmacology, we opted to use AM251, known to be more selective for CB1 receptors than, for instance SR141716A [38–40]. This aspect was further emphasized by the beautiful "U" dose-effect curve we found (see Fig. 2 in ref. [36]), where the maximum effect took place with an intermediate low concentration of AM251, probably the one more selective for CB1 receptors [36].

Notice that AM251, in the exact same concentration found effective on the behavioral experiments, was able to suppress LTP induction in the CA1 region (see Fig. 1 of ref. [41]) in a specially designed ex vivo slice preparation [41].

Consistently with the AM251 findings, we were able to show a weak, yet reproducible *facilitatory* effect of a small concentration of anandamide (AEA) infused immediately after IA training [37]. The effect was absent at higher concentrations, which is consistent with a CB1 receptor-mediated effect. Indeed, anandamide is a poorly selective ligand that can bind to different molecular targets, from the vanilloid TRPV1 receptors [42, 43] to other important targets such as NMDA [44] and muscarinic receptors [45], and possibly also acting as a neurotrophic blocker [46] and/or a modulator of glycine transport [47]. Again, the absence of effect of any drug at higher doses may be explained by its binding to other, "non-specific" sites, where the effect could be different, even opposite to the first one, neutralizing or compensating for it. Being so unspecific, AEA effects may be hard to interpret, except when low doses are used, because binding to lower affinity targets is less probable in this case, favoring activation of the CB1 receptors.

Since neither AM251 nor anandamide were effective when infused before *training* in this task (see Figs. 2 and 3 in ref. [37]), we are confident that, at least in our experimental setup, the CB1-mediated plasticity events underlying memory formation were confined to the post-training, consolidation phase, certainly ignited

[1] AM251 is an inverse agonist [126–129], however, since it also acts as a competitive antagonist displacing endocannabinoids at CB1 receptors and because it is not possible to discriminate between both effects *in vivo* [130], we prefer to classify AM251 as some authors already do, calling it an "antagonist/ inverse agonist".

by the new experience. The absence of pre-training effects contradicted some previous reports [31, 48–50], but considering how hard it may be to separate mnemonic from other cognitive/emotional/motor aspects in this experimental design, we decided not to pursue the *acquisition* avenue much longer.

Taken together, the results with post-training-infused AM251 or AEA in two different aversive (and one non-aversive) tasks suggest that, in the CA1 region of the dorsal hippocampus, memory consolidation mechanisms depend on the integrity of the local endocannabinoid system. The infusion of an antagonist/inverse agonist such as AM251 effectively disrupted the aversive memory, suggesting an action *in the opposite direction of any endogenous agonism*, possibly by blocking the binding of endogenously released cannabinoids to their CB1 targets. The AEA low dose result nicely fits to this suggestion, showing that some exogenously supplemented molecules may be "reinforcing" (within certain concentration limits), possibly by pooling themselves to the endogenously released molecules—the "physiological" pool. There is still no direct proof of endocannabinoids being released in CA1 as consequence of inhibitory avoidance or context fear training, but concentrations of both AEA and 2-AG peaked after a similarly aversive auditory fear conditioning task [72, suppl.]. Also, endocannabinoids were shown to be released in response to stressful/alerting factors at least in the periaqueductal grey substance [50] and the amygdala [9].

A Classical Source of Conflicting Data: Systemic vs. Local Infusions

Some years ago (2004–2005), our findings were *apparently* at odds with the literature, since most authors were used to find just the opposite: amnestic effects with cannabinoid agonists, and facilitatory with antagonists. However, ours was one of the first few works to investigate AM251 effects employing intra-cerebral infusions, i.e. we administrated the drug bilaterally directly into the CA1 area of the dorsal hippocampus, while most of the previous studies employed *systemic* treatments (i.p.). As discussed elsewhere [36, 41] the two approaches should show converging results only if the systemic effect would be relayed by the exact brain structure being targeted, which clearly seems not to be the case here. Indeed, our results strongly suggest that the hippocampus *is not* the mediator of the observed systemic effects. In spite of some evidence supporting the contrary [34], due to the high concentration of CB1 receptors and their dispersion among different brain areas, the systemic effects may probably be the result of a "multitarget" action.

This is not to say that systemic studies are too consistent with the "amnestic-with-agonist/facilitating-with-antagonist" cannabinoid rule: although most agonists exhibit an amnestic effect [34, 51–53], the antagonist/inverse agonist SR141716A (Rimonabant) was shown to produce either facilitatory [53–56] or no effect upon memory [34, 57].

Concerned with this contradictory scenario in those years, we performed some intracerebroventricular infusions of both AM251 and AEA. Our results were very

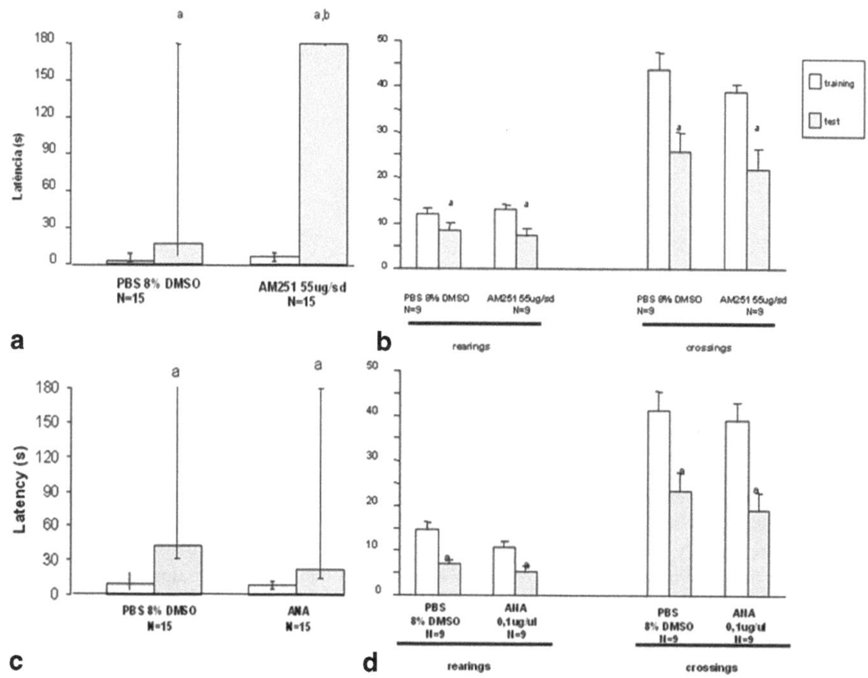

Fig. 3.1 Effect of post-training intracerebroventricular infusions of AM251 (55 ng/uL) or AEA (0.1 ug/uL) upon the performance on IA (**a**, **c**) or HAB (**b**, **d**) tasks. Since IA data (**a**, **c**) were not normally distributed, they were expressed as median (interquartile range) and analyzed with nonparametric statistics (Mann-Whitney for independent groups and Wilcoxon for repeated measures). HAB data were expressed as mean + S.E.M. and analyzed with One-Way ANOVA (independent groups) and Student's t test (dependent groups). **a** Significantly different from training of the same group, $P<0.05$ (Wilcoxon in **a** and **c**; t test in **b** and **d**). **b** Significantly different from the control group test (Mann-Whitney, $P=0.015$). Vehicle x Drug test values did not differ significantly between groups in the HAB task for any of the tested drugs (**b** and **d**) and also in the IA task for AEA (**c**)

different from the intra-hippocampal infusions: a significant facilitatory effect of AM251, but not of AEA, upon IA consolidation (none of the drugs caused any effect upon the less aversive task HAB). Notice that, since these IA results did not conform to a normal distribution, they are expressed as median (interquartile range), but regardless of the dispersion in the test groups, the effect was statistically significant (Fig. 3.1).

We brought the above data in support of a principled stand: we should refrain from straightforward comparison of systemic to local infusion results. No matter how important any pharmacological study is, these two approaches carry very different information most of the time. Besides, the more we delve into more anatomically restricted studies in order to understand the exact role of each brain structure and/or neuromodulator system, the less generalizable information becomes. Systemic studies continue to be fundamental, especially when looking for possible clinical applications, but we should be careful to analyze them in the light of intra-

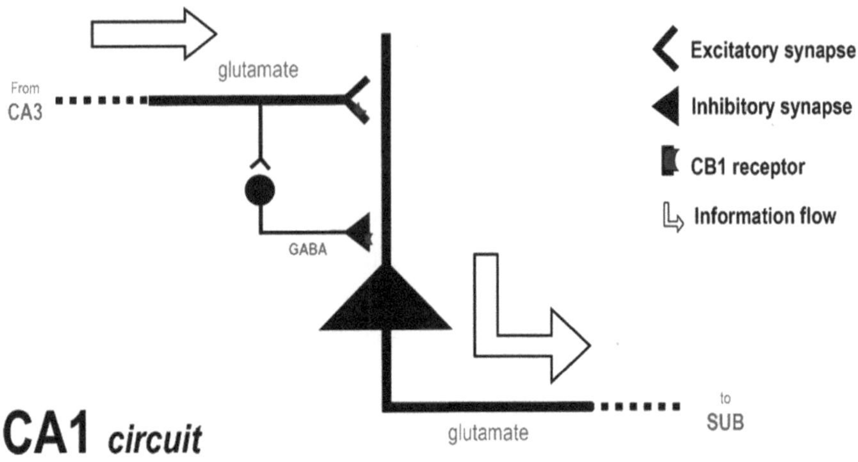

Fig. 3.2 Intrinsic circuitry of CA1 area with CB1 receptor positioned in glutamatergic cells and CCK-expressing GABAergic interneurons. The general hypothesis here is that CA1 glutamatergic efferences are directly involved in the building of the engram. Since there are 20 times more CB1 receptors in the presynaptic terminals of the inhibitory cells than in the excitatory neurons, these CB1 receptors are considered the "target" that explain our results—AM251 amnestic and AEA facilitatory effects

cerebral findings. Hopefully, complex synergic interactions between multiple areas and/or systems seem not to be the rule, otherwise things may be even more complicate experimentally.

Interpreting the Basic Findings According to CA1 Inner Circuitry I: Algebra Doesn't Lie

Since CB1 receptors are present mainly in presynaptic terminals of a subclass (CCK-expressing cells) of the local GABAergic/inhibitory interneurons and, with lower density, in glutamatergic/excitatory "main" neurons [12, 58], we believe it is much simpler to interpret our results in terms of circuitry operation. The trisynaptic circuitry of the hippocampus is well-described since the 60s [59, 60] and brings both frameworks together—CB1 positions upon the trisynaptic circuitry constituting cells, in particular, the CA1 local circuitry including one main excitatory cell and different local interneurons, allowing us to integrate the first findings nicely (Fig. 3.2). The inspiration for this approach comes from the outstanding work of James McGaugh in deciphering the functional (micro)neuroanatomy of the basolateral amygdala, a much more intricate structure, where a putative "circuit" was patiently built over the years through a puzzle-solving strategy that integrated psychopharmacological findings with local infusions of selective drugs [61].

The hippocampal role of these supposedly learning-recruited endocannabinoids is purely speculative at this point: we hypothesize that their endogenous role

Table 3.1 Algebraic model to interpret the present findings in the light of the hypothetical anatomo-functional circuitry depicted in Fig. 3.2

	A	B	C	A X B X C *the "algebra"*	
Cell type in CA1	Natural role on engram recording	ECS tonus acting upon this cell[a]	AM251 on CB1 receptors	Memory	LTP
GABAergic interneuron	−	−	−	−	−
Glutamatergic neuron	+	−	−	+	+
Mechanism?	Excitatory/ Inhibitory Action	ECs activate inhibitory CB1 receptors	Antagonists displace ECs or inv.agon. = inv.response		

OBS: + enhancement/facilitation, − inhibition/blocking
[a] same (or absence of) effect obtained by exogenously infused agonist AEA

could modulate the strength of the memory trace (or engram) under consolidation. Endocannabinoids are retrograde messengers released *on demand* that act to decrease the activity of inhibitory networks, which leads to the *disinhibition* of pyramidal neurons [62–65]. These glutamatergic cells project important hippocampal efference putatively involved in feeding the engram formation network, and represent one of the many intra-hippocampal sites where long-term potentiation (LTP) takes place. LTP, or any similar/derived glutamatergic-driven, calcium-dependent process, is a neural plasticity phenomenon long considered a putative mechanism of memory formation [66–69].

The CA1 circuit-based hypothesis is tailored to explain the post-training results, especially the AM251 data, as this antagonist/inverse agonist directly taps into the local ECS, suppressing any natural endogenous functions going on, and thus telling us what might be their nature. An amnestic effect of the antagonist means that the ECS was acting as a natural disinhibitor of local plasticity, since CB1 is an inhibitory G-coupled metabotropic receptor. This is also fully consistent with the observation that endocannabinoids might facilitate hippocampal long-term potentiation through the DSI phenomenon of retrograde inhibition of presynaptic GABA release [70], and in support of that involvement, we have demonstrated that AM251 also suppresses LTP induction in the CA1 region in an *ex vivo* hippocampal slice preparation, with the exact same concentration being effective in the behavioral tasks [41].

Considering that these receptors are located mainly in pre-synaptic terminals of both excitatory and inhibitory cells, in order to interpret the observed drug effects, we will have to resort to some simple algebra (Table 3.1). Considering activation/agonism/facilitation as equivalent to a positive (+) sign, and inhibition/antagonism/amnesia, to a negative (−) one, Table 3.1 synthetizes most of the results covered in this chapter.

Consistency, however, is not enough for a model to be considered proven. Hopefully, the functional circuitry hypothesis depicted in Fig. 3.2 and Table 3.1 seems fully *testable*, something that assures its scientificity. The challenge includes, of course, accommodating also the different findings from other authors.

Notwithstanding the fact that a CB1 receptor-sensitive hippocampal LTP appears to be necessary for the consolidation of the aversive IA memory [41], since AM251 did not affect HAB, LTP seems NOT necessary for the memory consolidation of this less aversive task (see Fig. 3 of ref. [36]), at least in this particular brain region.

Aversive Memory Consolidation Requires a Crosstalk Between Stress Hormones and Endocannabinoids

From the first experiments, it was clear that some degree of aversiveness or alertness was necessary for the cannabinoid drugs to produce any effect, since the Open Field Habituation task (HAB) remained refractory to both AM251 and AEA [36, 41], and the same effects observed in the IA task [36, 37, 41] were also confirmed for CFC [71]. Regardless some reports with knockout (KO) animals on ECS involvement in less aversive situations [72, 73] the aversive/ECS-sensitive-nonaversive/ECS-insensitive pattern was already observed by others [31, 74, 75]. That aversive/emotional aspects play a critical role in memory formation is not a novelty [61, 76–78]. For instance, either corticosterone or stress promote the enhancement of memory consolidation [76, 78]. Another suggestive clue comes from the fact that cannabinoids can influence synaptic events taking place in areas such as the hippocampus, particularly after an aversive stimulation [65, 70]. Finally, several studies demonstrate both the endocannabinoid and the glucocorticoid systems independently improving memory consolidation and suppressing memory retrieval [77, 79], reinforcing the connection that can be either of causality (one promotes the other) or complementarity (one compensates for or substitutes the other).

We then decided to investigate deeper this putative connection between stress and the ECS. The fact that IA and CFC are sensitive to post-training AM251 raises the possibility that the endogenous cannabinoids competing with the antagonist/inverse agonist AM251 were somewhat being recruited by the concomitant stress-related hormones. Indeed, Kamprath et al. [80] have demonstrated that endocannabinoid action depends on the intensity of the footshock in a fear conditioning task that associates tone-response with previous shock treatment. Moreover, several studies prove that both stress and glucocorticoids increase endocannabinoid levels in areas, such as the hypothalamus [81, 82] and the periaqueductal grey matter [50], involved in the fear response. Later, an increase in endocannabinoids released in response to behavioral factors was demonstrated in memory-related areas such as the amygdala [9] and the hippocampus [28, 72], and 2-AG was the main endocannabinoid mobilized in these cases.

In our experiments we did not measure the release of endocannabinoids directly, but managed to approach this possibility in an indirect way. First we showed that CFC memory can be retained after a 0.7 mA, but not a 0.3 mA footshock alone, and that the first result was—exactly like the previous IA consolidation results—reversed by post-training intra-hippocampal AM251 administration. Then we found that the effect of the weak shock was somewhat enhanced by a pre-training procedure, either a single stress session or an i.p. infusion of the corticosteroid dexamethasone, reaching a freezing level compatible with the one observed with the strong shock. In this case, the response was again blocked by AM251 [27], suggesting that the hippocampal ECS interacts with the glucocorticoid system. Intensity of the stimulus was not the only factor, since both reinforcing procedures were effective when presented immediately, but not 30 min before the training session, a time interval that also fits to the kinetics of corticosterone release [83].

These hippocampal findings are consistent with two similar findings in the amygdala [84, 85]. In the first paper, the memory-enhancing effect of the glucocorticoid system (GCs) was attributed to an ECS-mediated disinhibitory influence on noradrenaline release, a mechanism that facilitated the formation of the aversive memory trace [84, 86]. In the second work, intra-BLA AM251 disrupted the extinction of an avoidance memory, and the agonist WIN55,212–2 (locally or systemically infused) was able to modulate the behavioral enhancement effect of stress, prompting a small increase in plasma corticosterone levels [85]. Thus, not only the dorsal hippocampus and the basolateral amygdala may share an analogous mechanism, but they might even be functionally connected in order to modulate a cognitive process such as the formation of a new contextual fear memory trace [87–89], an important avenue of research still open to contributions.

What is sure to this point is that the activation of the hippocampal ECS requires a certain level of aversiveness in order to exert its modulatory role on fear memory consolidation. The negative emotional state may be provided by [1] the task stimulus itself (e.g., a strong shock), [2] a previous stress session, or, alternatively, [3] the availability of stress hormones such as glucocorticoids. Glucocorticoids may be the putative endogenous mediators of this aversive-dependent hippocampal ECS recruitment, even though it's not clear if this functional link would be taking place in the same or nearby brain structures.

More on the interaction between stress and endocannabinoid modulation of memory can be found in chap. 1 and 5.

Memory Retrieval is Negatively Modulated by the ECS in the CA1 Area: AM251 is Facilitatory

When AM251 was infused before the test session, the effect was exactly the opposite of the post-training one, i.e., we observed a clear-cut *facilitatory* effect on memory retrieval (see Fig. 6 of ref. [37]). AEA, on the other hand, was ineffective. Other authors had also failed to find an effect with AEA [32], possibly due to the low selectivity of AEA as a cannabinoid agonist or to its rapid degradation because of its endogenous nature. Synthetic agonists produce different results; for instance, WIN55,212–2 impaired retrieval when infused into the hippocampus [28]. We should also consider the fact that, for technical reasons, our infusion was made 15 min before test and it is possible that with the longer delay, the already low dose of exogenous AEA had diffused away and was not effective anymore. Again, additional behavioral tests exclude the possibility of locomotor, non-cognitive actions of the employed pretest concentrations, both for AM251 and AEA. The possibility of a state-dependent phenomenon for AM251 was not fully investigated here.

Although the synthetic cannabinoid agonist WIN55,212–2 did not cause state-dependency [31], this phenomenon has been detected with the infusion of endogenous agonists [90, 91]: the confirmation of an endogenous state-dependency can be understood because AM251 is an antagonist/inverse agonist somehow interfering with the natural functioning of the endogenous pool of endocannabinoids. Independent of any AEA result, we believe the findings of major relevance here are the AM251 effects due to its ability to tap into the ECS.

Memory retrieval is known to share some molecular mechanisms with consolidation—after all, the same cells and brain structures are demanded—but there are also important differences. In particular, retrieval requires neither protein synthesis nor the activation of NMDA receptors and CaMKII, even though there is also a role for CREB phosphorylation [23]. The opposite action of AM251 in memory retrieval vs. consolidation asks for an explanation, and the first thing that comes to mind is that, wherever the drug is acting upon, *something changed.* Thus, following the lead of the circuitry hypothesis outlined above (Fig. 3.2 and Table 3.1), we suppose that some durable changes took place in the CA1 region between the consolidation process and the test moment, that modified the logical attributes of the pharmacological response, changing from disruption to facilitation. Of course, this remains to be proven.

Same Drug, Opposite Effects Demands Explanation: May They Predict Plastic Changes?

At this point, it is interesting to note that while several pharmacological studies show the same effect with the same drug in the two different memory phases described above (see Table 3.2 [92–102]), there is cumulative evidence on the contrary for at least two other important neuromodulatory systems—the cholinergic and the glucocorticoid systems.

Thus, a phenomenon similar to what was found with AM251 was previously described by our lab for a different endogenous modulatory system, the cholinergic system acting though the M4 muscarinic receptors, in the very same structure and behavioral task [92]. The biological nature of such "change" is under investigation in both cases, but we have good reasons to believe they both are of neural circuitry nature, and can be interpreted in the same framework depicted in Fig. 3.2 and Table 3.1. It may well involve either an increased expression of CB1 receptors over the excitatory/pyramidal output pathway as consequence of the learning process—as similarly suggested by ref. [92], or a down-regulation of the CB1 receptors modulating the GABAergic interneurons, or even a combination of such plastic

Table 3.2 Similar vs. opposite effects of drug treatments upon memory consolidation and retrieval phases involving different neurotransmitter and neuromodulator systems[a]

Same treatment, *same* effects			
Treatment		Memory phase	
Drug	Target system	Consolidation	Retrieval
Muscimol (GABA-A agonist)	*GABAergic (GABA$_A$)*	Impairement [93]	Impairement [94]
Nicotine (NAChR agonist)	*Cholinergic (nicotinic)*	Enhancement [95]	Enhancement 95
Scopolamine (MAChR antagonist)	*Cholinergic (muscarinic)*	Impairement [96][b]	Impairement [97]
Noradrenaline (Adrenergic agonist)	*Cathecolaminergic*	Enhancement [98]	Enhancement [99]
Same treatment, *opposite* effects			
MT3 (ACh M4 antagonist)	*Cholinergic (muscarinic)*	Impairement [100]	Enhancement [92]
AM251 (CB1 antagonist/inverse agonist)	*ECS*	Impairement [27 36, 41]	Enhancement [37]
Win55,212-2 (CB1 agonist)	*ECS*	Enhancement [84]	Impairement [102]
RU28362 (GR agonist)	Glicocorticoid	Enhancement [76]	Impairement [79]

[a] A non-exhaustive compilation, only for illustrative reasons. Reference numbers are informed above

[b] The same effect was observed with pirenzepine, a selective muscarinic M1 receptor antagonist, the most common muscarinic receptor and the probable target of less selective muscarinic ligands [101]

events at the same time. Any of these possibilities could equally well explain our results, and fortunately, notwithstanding the technical difficulties, this is a fully testable hypothesis.

In the same sense, both stress and glucocorticoids were shown to promote an analogous response pattern, enhancing memory consolidation and impairing retrieval [79]. Based on this, it has been suggested that the ECS and the glucocorticoid system interact in order to modulate memory retrieval. Indeed, a recent paper has shown that corticosterone increases the endocannabinoids levels in the hippocampus [28], and, complementarily, the local infusion of AM251 into the hippocampus blocks the impairing effect of corticosterone in contextual fear conditioning [102]. Hence, under a stressful situation such as the retrieval of a fear memory, glucocorticoids are released from the adrenal cortex, acting in the hippocampus through the activation of the glucocorticoid receptors that, somehow, induce the synthesis of endocannabinoids that, for their turn, bind to the CB1 receptors, inhibiting the release of neurotransmitters.

Taken together, the *same-drug-opposite-effect* class of pharmacological findings may provide support for a promising circuit-based theoretical framework useful to evince mechanisms behind relevant plastic phenomena related to learning and memory. Given the fact that Table 3.2 is far from exhaustive, we cannot assure the generality of this phenomenon, but the suggestion is fertile enough to deserve further investigation.

Memory Extinction is Positively Modulated by the ECS in the CA1 Area: AM251 is Disruptive

The extinction phenomenon was first described by Ivan Pavlov, using dogs as subjects [103]. He paired a sound (conditioning stimulus, CS), with the presentation of a piece of meat (unconditioning stimulus, US). The CS alone did not promote any particular response in the animals, however, after being paired with the US, the sound elicits a salivatory response. The prolonged exposure to the conditioned environment without the US led to a gradual reduction of the conditioned response, a process called extinction [25, 104, 105]. In the last decades, memory extinction has been studied by several laboratories, with a dominant use of rodents in fear conditioning to the CS (the context or a tone), with US consisting mostly represented by foot-shocks. In our case, a single, prolonged re-exposure to the context without reinforcement was enough to promote extinction [71].

Thus, when AM251 was locally infused in CA1 after a 25 min re-exposure to the CFC context, fear response in a posterior test was higher than that exhibited by the controls, suggesting that this drug has disrupted the consolidation of extinction [106–108]. Complementarily, intra-hippocampal administration of exogenous AEA has facilitated extinction, an effect reversed by a concomitant sub-effective concentration of AM251 (see Figs. 3 and 6 of ref. [71]). Chhatwal et al. [109] also reported, for Conditioned Fear, a disruption of extinction with the antagonist/inverse agonist SR141716A, and facilitation with the AEA uptake inhibitor and a TRPV1 receptor agonist AM404, that increased the endogenous cannabinoids tonus, an effect then reversed by a concomitant sub-effective concentration of SR141716A. There are other converging studies reporting from extinction attenuation in CB1 KO mice to CB1 receptor blockers disrupting the extinction of a variety of tasks [9, 31, 49, 110, 111]. Although most studies indicate an essential role played by the ECS in memory extinction, its effects seem not to be ubiquitous. There are some types of memory that seem to be independent of CB1 activation, such as those memories with lower levels of aversiveness [36, 41, 74, 75].

The process of fear extinction demands great attention of psychologists and neuroscientists due to its implications in treating PTSD and phobias, as extinction is a core component of most major behavioral-cognitive therapies that approach those devastating, resilient pathologies, despite being plagued by drawbacks like spontaneous recovery and rapid reacquisition. In this scenario, reconsolidation represents new hope, due to its intrinsic qualities whose comprehension is necessary to consolidate it as a promising therapeutically alternative. For more details about the possible therapeutic potential of the endocannabinoid system in PTSD see Chap. 1.

Memory Reconsolidation is Negatively Modulated by the ECS in the CA1 Area: AM251 is Facilitatory

Memory consolidation refers to the gradual stabilization process that occurs after acquisition. Then, when memory is consolidated, it becomes somewhat resistant to interference. However, following retrieval, an already established memory might undergo a labile state again, requiring reconsolidation in order to persist [26]. Reconsolidation is a widespread phenomenon and has been reported in a variety of species and types of memory, indicating that it is a fundamental memory process. The functional role of reconsolidation goes beyond a simple restabilization of the retrieved memory: it allows for memory updating, maintaining its predictive and adaptive relevance. In a recent paper we showed that when memory is reactivated in a situation that does not match the original information, content is qualitatively modified, or "updated"; however, when the contextual condition matches to some extent the original one, memory reactivation operates to either strengthen the trace or maintain the precision of its content over time [112]. Hence, reconsolidation allows for memory modification, such as the integration of new background information into the originally established memory trace.

To this point, data on the effects of endocannabinoid modulation of reconsolidation in the hippocampus region is relatively scarce and controversial. In our experiments AM251 infused intra-hippocampally right after a 3 min re-exposure—a "reactivation" session with the omission of the US—increased the freezing time in the test, implying that memory was labilized and re-consolidated in a *facilitated* form (see Fig. 2 of ref. [71]), an effect that persisted for almost a week. We knew the drug was interfering with reconsolidation because, consistent with previous works of others [26, 110], the suppression of protein synthesis by the transcriptional blocker 5,6-dichloro-1-bold beta-D-ribofuranosylbenzimidazole (DRB) effectively blocked memory expression in the test [113]. This means that a labile state, with its new protein synthesis window of opportunity, was somewhat induced by the reactivation session. Complementarily, AEA *disrupted* memory reconsolidation when infused into the CA1 area, an effect reversible by the concomitant administration of a subthreshold concentration of AM251, in agreement with the idea that the effect was mediated by CB1 receptors (see Fig. 5 in ref. [71]).

The reactivation session was short-lasting, and this time was decisive to define the fate of the memory. None of the treatments affected fear memory when the reactivation session was omitted, meaning not only that *reactivation* was a necessary condition, but that the treatment did not influence fear *per se*.

The disruption of reconsolidation observed with the local infusion of AEA could be alternatively interpreted as a facilitatory effect of AEA upon memory extinction after re-exposure, an effect that could mimic a "reconsolidation disruption". However, this possibility seems unlikely because the response of the control (vehicle-infused) group did not change between reactivation and test sessions (in a real extinction, latencies should decrease). A re-exposure of 3 min only is probably too brief and not "intense" enough to be able to initiate a new memory trace necessary to extinguish the previous one, dissociating CS from the US (see Fig. 2 of ref. [71]).

This argument reminds the classic argument against the existence of reconsolidation as a phenomenon in itself, a flawed criticism in view not only of the two aspects above, but also of the absence of any of other four well known characteristics of an extinction memory, namely spontaneous recovery, rapid reacquisition, renewal, and reinstatement [25].

In apparent contrast to these findings, SR141716A, another CB1 antagonist/inverse agonist, was shown to cause no effect by itself when infused into the hippocampus on memory reconsolidation of two different tasks, but was able to reverse the amnestic effect of the protein synthesis inhibitor anisomycin, suggesting that the CB1 receptor activation has an important role on the destabilization of the memory trace [110, 114]. Importantly, SR141716A is less selective for the CB1 receptor than AM251 [40]. Also, Lin and colleagues [107] has shown that two different CB1 receptor agonists infused in the amygdala impaired reconsolidation, an effect prevented by concomitant administration of AM251. Kobilo et al. [111] also reported that the CB1 receptor agonist WIN55212–2 infused into the insular cortex disrupted memory reconsolidation in the conditioning taste aversion. Recently, Stern and colleagues [115] has shown that the systemic injection of cannabidiol (the main non-psychotomimetic compound of marijuana) was able to disrupt memory reconsolidation in CFC, a hippocampus-dependent task, an effect prevented by co-injection of AM251. Accordingly, De Carvalho and colleagues [116] reported that systemic SR141716A impaired reconsolidation in the morphine-conditioning place preference.

Our results support a role of the dorsal hippocampus ECS in the memory updating process, one of the many demonstrated uses for the process of reconsolidation [112] that allows for the incorporation of new information during a re-exposure/reactivation session.

Opposite Actions of the ECS in Extinction and Reconsolidation: Would ECS be the Switching Mechanism?

Consistently with previous studies, our findings with reconsolidation and extinction showed that the duration of the *reactivation* session is a crucial variable that determines subsequent memory [110, 117, 118]: thus, a brief re-exposure leads to reconsolidation whereas a prolonged reactivation session induces extinction. As commented above, there are relevant functional differences between both processes, reconsolidation being an updating mechanism dedicated to incorporate new information to an already consolidated memory [110, 119], and extinction, the establishment of a new memory that competes with and temporarily suppresses the one formed during the original association [25, 104].

The symmetrically opposed effects of AM251 and AEA in these two post- retrieval phases contrasts with the previous report of Lee et al. 2006 [120] on the role of NMDA in CFC in the very same memory phases. In this study, MK-801 blocked both extinction and reconsolidation, and DCS enhanced both phases in the same way. As NMDA receptors—besides protein synthesis itself—are key agents sustaining the plastic events behind any new memory formation process, such as "first-time" consolidation, the important results of Lee et al. [120] may imply that they are a necessary component behind the plasticity events necessary either for trace updating (in reconsolidation) or new trace consolidation (in extinction).

ECS modulation, on the other hand, may have a more subtle, differentiated role: by "directly" modulating the effector synapses (by the glutamatergic main cells), or "indirectly" controlling them (through CCK-expressing GABAergic interneurons), this system might act as a selector of the "final" response. Considering the symmetrically opposed effects here described for AM251 and AEA we propose that hippocampal ECS may be acting as a putative *switching mechanism* that decides the fate of a memory trace after its reactivation. This would be the case at least for aversive memories, since there is growing consensus that the ECS needs some stress/glucocorticoids level in order to be recruited [81, 82]. Context matching and time of re-exposure are two important elements to be computed in whatever mechanism exists that determine one or other response, be it reconsolidation or extinction. At this point, however, this is all very conjectural and demands more investigation.

ECS Modulation of CA1 Circuitry in Each Memory Phase: Symmetrically Opposed Actions are the Rule

Table 3.3 summarizes the main findings here reviewed about the observed effects for AM251 and AEA bilaterally infused into the CA1 area of the rat dorsal hippocampus. Except for the pre-training treatment and the infusion without the reactivation session, there is a noticeable pattern of symmetrically opposed actions for each memory phase observed. Remember also that two different aversive tasks were studied here, IA and CFC, and that no effect was ever found in HAB, the only non-aversive task studied.

Our original choice of agonist, AEA, may not have been the best due to the discussed lack of selectivity. However, a low, more CB1 receptor-selective concentration allowed us to observe the asymmetric effects in three of the four memory phases—consolidation, reconsolidation and extinction. In memory retrieval, the *addition* of a low concentration of exogenously infused AEA to the already present endogenous pool of recently-released AEA molecules may have led to two possible situations: either there is not enough AEA in order to reach any effectiveness threshold or, on the contrary, there was too much of it. In this latter case, the extra molecules provided by our exogenous infusion added to the existing pool of AEA may reach a concentration high enough to promote non-specific actions, not mediated by CB1 receptors, which may offset each other and result in no measurable behavioral effects. To decide which of these hypotheses is true, we need to conduct further experiments, either by measuring the amount of AEA locally present, or employing other agonists with a "cleaner" effect. Both possibilities are under scrutiny in our lab.

Table 3.3 A synthesis of our basic findings with AM251 and AEA bilaterally infused into the CA1 area of the dorsal hippocampus in two aversive tasks, IA and CFC, at different stages of memory

Memory phase	CB1 antagonist (AM251– blocks ECS tonus)	CB1 agonist (adds exogenous AEA)	Behavioral task	Ref.
Pre-training (acquisition phase)	Ø	Ø	IA	[37]
Post-training (consolidation phase)	–	+	IA & CFC (AEA in IA only)	[36, 37, 41] (IA) [27] (CFC)
Pretest (retrieval phase only)	+	Ø	IA	[37]
Post-reactivation (of 3 min) (reconsolidation phase)	+	–	CFC	[71]
Post-reactivation (of 25 min) (extinction phase)	–	+	CFC	[71]
No reactivation (neither 3, nor 25 min)	Ø	Ø	CFC	[71]

OBS: – amnestic effect, + facilitatory effect, Ø no observed effect

The problem of adding an exogenous amount of a molecule that exists endogenously and the way they accumulate and/or interact, reminds us of a classic set of findings using beta-endorphin back in the 1980's: the first studies infused a huge amount of substance that was adding to the internal pool to produce one type of effect. However, when the endogenous levels of those opioids were finally measured, much smaller quantities of exogenous substance were infused, and new, unexpected effects, discovered, which later lead to the first demonstration of an endogenous state-dependency [91]. We may be facing the same kind of problem here, which favors the use of different, more selective synthetic agonists.

In an attempt to understand the complexity of AEA effects in the literature, we also performed a set of experiments on the role of TRPV1 receptors in the memory consolidation of two aversive tasks in order to understand something about this "second identity" of AEA as an endovanilloid—as it should be the case under higher concentrations of the ligand. Thus, the TRPV1 antagonist capsazepine was able to impair memory consolidation of both IA and CFC aversive tasks, but the agonist capsaicin, on the other side, had no effect in any of the two tasks [121]. However, the amnestic effect took place only with a strong shock, showing that endovanilloid system also depends on the emotional level in order to be recruited, somewhat resembling the stress-induced recruitment of the ECS. This means that, when studying aversive memories, AEA has "multiple personalities" and may act either as an endocannabinoid (on CB1 receptors) or as an endovanilloid (on TRPV1 receptors), or (more probably) on both targets. So, no easy "separation" is possible, for two reasons: both endogenous systems seem to depend in the same way on the stress level of the task, and, in both cases, AEA physiological action points to a *facilitation* of the memory consolidation—AEA was facilitatory and AM251 or capsazepine were amnestic post-training. This also suggests that, in terms of memory mechanisms, the endovanilloid system is a modulatory system that somewhat complements the ECS.

Interpreting the Findings According to CA1 Inner Circuitry II: Fitting all the Gang

The cannabinoid antagonist/inverse agonist AM251, when infused into the CA1 hippocampal region, interferes in some way with the natural functioning of the endogenous pool of endocannabinoids present at each memory phase studied here: thus, notwithstanding AEA results, we believe the findings of major relevance here are the AM251 effects due to its ability to *tap into the ECS* and show us exactly what it is doing in every particular moment assayed. To this point, the main empirical conclusions are the following:

- Memory *consolidation* is **positively** modulated by the ECS in the CA1 area, since AM251 was **amnestic**;
- Memory *retrieval* is **negatively** modulated by the ECS in the CA1 area, since AM251 was **facilitatory**;
- Memory *extinction* is **positively** modulated by the ECS in the CA1 area, since AM251 was **amnestic**;
- Memory *reconsolidation* is **negatively** modulated by the ECS in the CA1 area, since AM251 was **facilitatory.**

To these four conclusions, we could add that the ECS in the CA1 area *do not interfere* with memory *acquisition*, since AM251 had not caused any observable pre-training effect in the studied concentration.

From now on, we will be highly speculative. Let's suppose—for the purpose of this model—that only the CA1 area is relevant to decide the fate of a new memory in the whole brain, and that the circuit depicted in Fig. 3.2, and algebraically "explained" in Table 3.1, is a good model to explain those results. Remember that in the "normal" situation, hippocampal CB1 receptors are much more concentrated in GABAergic interneurons than in glutamatergic terminals [11, 12, 122, 123]. We may suggest the following:

- The first conclusion above implies that the targeted CB1 receptors during *consolidation* consist mainly of those present upon the GABAergic interneurons, since, according to our CA1 circuit hypothesis, only in this case an amnestic effect is conceivable;
- If this is correct, the second conclusion could be explained by some plasticity taking place in the CA1 circuitry during *retrieval*, after the memory has been consolidated, with the ECS modulation changed to be more effective on the glutamatergic, excitatory cells (see Table 3.4).

Table 3.4 Fitting the present findings with different memory phases into the algebraic/anatomo-functional circuitry model depicted in Fig. 3.2

	A	B	C	A X B X C *the "algebra"*			
Cell type in CA1	Natural role onengram recording	ECS tonus acting upon this cell[a]	AM251 on CB1 receptors	Memory	LTP	Observed in phase	Switching role?
GABAergic interneuron	–	–	–	–	–	*Consolidation*	*Extinction*
Glutamatergic neuron	+	–	–	+	+	*Retrieval*	*Reconsolidation*
Mechanism?	Excitatory/ Inhibitory Action	ECs activate inhibitory CB1 receptors	Antagonists displace ECs or inv. agon. = inv. response			*Plastic change after learning?*	*ECS assists decision on trace fate?*

OBS: + enhancement/facilitation, – inhibition/blocking

[a] Same (or absence of) effect obtained by exogenously infused agonist AEA

Despite being both testable—however hard this would be—these are two complex and mutually interdependent hypotheses. In this framework, how can we fit conclusions 3 and 4?

- The third conclusion, being qualitatively identical to the first, suggests a similar explanation: it could just be derived from the fact that extinction is the onset of a new memory—despite weaker. Thus, "consolidation of extinction" should naturally share the very same mechanisms of any "first-time" consolidation, including the positive endogenous cannabinoid modulation here verified. It remains to be understood why this new learning ends up being "second class", i.e. subject to such drawbacks as spontaneous recovery, but a strong association with the original memory trace could somewhat help us to solve this riddle;
- Finally, what about the negative ECS modulation of *reconsolidation* inferred in the fourth conclusion? The simplest way to approach this may be to consider it as just a continuation of the retrieval ECS state, since in this case, different plastic mechanisms are being summoned, more or less the same mechanisms specifically involved in the retrieval session.

In an insightful paper of Ivan Izquierdo's Lab concerning memory retrieval [23] it was stated that "the molecular mechanisms that generate extinction are initiated at the time of retrieval in the CAl region of the hippocampus, and include some of those that are involved in retrieval. Therefore, it may be said that retrieval 'plants' the seeds of its own extinction in CA1. Some of the 'seeds' (the MAPK and PKA signaling pathways) are required for retrieval itself; the others (NMDA receptors and CaMKII) are not". This may be true, but we have reasons to suppose that the *seeded* processes should include, above all, those pointing to *reconsolidation*, that may depend both on the mentioned enzymes *and* on mechanisms more recently proposed as being specifically pertinent to the reconsolidation/updating process underlying an *already-consolidated-just-being-retuned* memory trace, namely those involving L-type VGCCs [114, 124] and the early gene Zif268 [125].

Concerning the dynamical, real life situations, the capacity to update memory seems to be essential for individual survival. The process of reconsolidation might be the most important property of memory, aside from the obvious function of remembering, and understand it is a great challenge to the area of behavioral psychopharmacology in the following years. Table 3.4 summarizes this interpretative scenario.

Concluding Remarks

A substantial amount of research has been undertaken over the past decade reporting the role of the endocannabinoid system on cognitive functions, and it is not possible to adhere to simplifications such as the proposition of ECS as a mere positive or negative modulatory system. There are many factors that can influence the effect of the endocannabinoid on memory, such as the nature of the cognitive task, memory phase moment, drug specificity, context complexity, concentration/dose, level of arousal/aversiveness, and many others. It seems that the endogenous levels of endocannabinoids are fine-tuned to select memory processes in order to warrant regular optimal conditions: the infusion of an extra amount of exogenous cannabinoid, or an antagonist, may break this balance and displace the memory process.

All things taken together, the findings point to a complex, multifunctional modulatory system that performs several functions aimed at shaping the destiny of a memory trace. This would be a good reason for the CB1 receptors to be the most common metabotropic receptors in the brain: they are badly needed to perform such complex switching/directioning functions.

Acknowledgements JAQ and LOA would like to acknowledge the great contribution made, in particular, by Carla Dalmaz and Sidia C. Jacques, and co-authors Jaderson C da Costa (PUCRS), and Victor A. Molina (UNC, Argentina). Other collaborators over the years include Lucas Fürstenau de Oliveira, Clarissa Camboim, Felipe Diehl, Bruna P. Genro, Vanusa B. Lanziotti, Ricardo V. Breda (PUCRS), Michele F. Pedroso (PUCRS) Douglas S. Engelke, Robson Scheffer-Teixeira, Josué Haubrich, and Lindsey F. Cassini. We have also had the valuable assistance of Mariane Castro da Silva, Marco Alexandre S. da Silva, Ney T. Ferreira Jr, Fabrício H. M. do Monte (UFSC), Andréa A. Tavares, Thiago P. Henriques, and Fabrício Pamplona (UFSC), and the professional and kind technical assistance of Ms. Zelma Regina de Almeida. Over the years, this research received the support of grants and fellowships from several national and international, specially: CAPES (MEC), CNPq & FINEP (MCTI), FAPERGS, PROPESQ/UFRGS, PRPID/PUCRS, SECyT & FONCyT (Argentina) and IFS (Sweden).

References

1. Russo EB. History of cannabis and its preparations in saga, science, and sobriquet. Chem Biodivers. 2007;4(8):1614–48.
2. Riedel G, Davies SN. Cannabinoid function in learning, memory and plasticity. Handb Exp Pharmacol. 2005;(168):445–77.
3. Herkenham M, Lynn AB, Johnson MR, Melvin LS, de Costa BR, Rice KC. Characterization and localization of cannabinoid receptors in rat brain: a quantitative in vitro autoradiographic study. J Neurosci. 1991;11(2):563–83.
4. Devane WA, Hanus L, Breuer A, Pertwee RG, Stevenson LA, Griffin G, Gibson D, Mandelbaum A, Etinger A, Mechoulam R. Isolation and structure of a brain constituent that binds to the cannabinoid receptor. Science. 1992;258(5090):1946–9.
5. Sugiura T, Kondo S, Sukagawa A, Nakane S, Shinoda A, Itoh K, Yamashita A, Waku K. 2-Arachidonoylglycerol: a possible endogenous cannabinoid receptor ligand in brain. Biochem Biophys Res Commun. 1995;215(1):89–97.
6. Hillard CJ. Biochemistry and pharmacology of the endocannabinoids arachidonylethanolamide and 2-arachidonylglycerol. Prostaglandins Other Lipid Mediat. 2000;61(1–2):3–18.
7. Han J-H, Kushner SA, Yiu AP, Cole CJ, Matynia A, Brown RA, Neve RL, Guzowski JF, Silva AJ, Josselyn SA. Neuronal competition and selection during memory formation. Science. 2007;316(5823):457–60.
8. Herkenham M, Lynn AB, Little MD, Johnson MR, Melvin LS, de Costa BR, Rice KC. Cannabinoid receptor localization in brain. Proc Natl Acad Sci U S A. 1990;87(5):1932–6.
9. Marsicano G, Wotjak CT, Azad SC, Bisogno T, Rammes G, Cascio MG, Hermann H, Tang J, Hofmann C, Zieglgänsberger W, Di Marzo V, Lutz B. The endogenous cannabinoid system controls extinction of aversive memories. Nature. 2002;418(6897):530–4.
10. Marsicano G, Kuner R. Anatomical distribution of receptors, ligands and enzymes in the brain and in the spinal cord. Circuitries and neurochemistry. Cannabinoids and the Brain. New York: Springer; 2008.
11. Katona I, Sperlágh B, Sík A, Käfalvi A, Vizi ES, Mackie K, Freund TF. Presynaptically located CB1 cannabinoid receptors regulate GABA release from axon terminals of specific hippocampal interneurons. J Neurosci. 1999;19(11):4544–58.
12. Kawamura Y, Fukaya M, Maejima T, Yoshida T, Miura E, Watanabe M, Ohno-Shosaku T, Kano M. The CB1 cannabinoid receptor is the major cannabinoid receptor at excitatory presynaptic sites in the hippocampus and cerebellum. J Neurosci. 2006;26(11):2991–3001.
13. Hájos N, Katona I, Naiem SS, MacKie K, Ledent C, Mody I, Freund TF. Cannabinoids inhibit hippocampal GABAergic transmission and network oscillations. Eur J Neurosci. 2000;12(9):3239–49.
14. McGaugh JL. Time-dependent processes in memory storage. Science. 1966;153(3742):1351–8.
15. McGaugh JL. Memory-a century of consolidation. Science. 2000;287(5451):248–51.
16. Müller G, Pilzecker A. Experimentelle Beitraege zur Lehre vom Gedaechtnis. Z Psychol. 1900.
17. Izquierdo I, Bevilaqua LRM, Rossato JI, Bonini JS, Medina JH, Cammarota M. Different molecular cascades in different sites of the brain control memory consolidation. Trends Neurosci. 2006;29(9):496–505.
18. Dudai Y, Morris R. To consolidate or not to consolidate: what are the questions? Brain, perception, memory advances in cognitive sciences. Oxford: Oxford University Press; 2000.
19. Squire LR, Stark CEL, Clark RE. The medial temporal lobe. Annu Rev Neurosci 2004;27:279–306.
20. McIntyre CK, McGaugh JL, Williams CL. Interacting brain systems modulate memory consolidation. Neurosci Biobehav Rev. 2012;36(7):1750–62.
21. Kim JJ, Fanselow MS. Modality-specific retrograde amnesia of fear. Science. 1992;256(5057):675–7.

22. Quillfeldt JA, Zanatta MS, Schmitz PK, Quevedo J, Schaeffer E, Lima JB, Medina JH, Izquierdo I. Different brain areas are involved in memory expression at different times from training. Neurobiol Learn Mem. 1996;66(2):97–101.
23. Szapiro G, Galante JM, Barros DM, Levi de Stein M, Vianna MRM, Izquierdo LA, Izquierdo I, Medina JH. Molecular mechanisms of memory retrieval. Neurochem Res. 2002; 27(11):1491–8.
24. Quillfeldt JA. Behavioral methods to study learning and memory in rats. Animal models as tools in ethical biomedical research. São Paulo: Associação Fundo de Incentivo à Psicofarmacologia; 2010. p. 227–69.
25. Bouton ME. Context and behavioral processes in extinction. Learn Mem. 2004;11(5):485–94.
26. Nader K, Schafe GE, Le Doux JE. Fear memories require protein synthesis in the amygdala for reconsolidation after retrieval. Nature. 2000;406(6797):722–6.
27. De Oliveira Alvares L, Engelke DS, Diehl F, Scheffer-Teixeira R, Haubrich J, de Freitas Cassini L, Molina VA, Quillfeldt JA. Stress response recruits the hippocampal endocannabinoid system for the modulation of fear memory. Learn Mem. 2010;17(4):202–9.
28. Atsak P, Hauer D, Campolongo P, Schelling G, McGaugh JL, Roozendaal B. Glucocorticoids interact with the hippocampal endocannabinoid system in impairing retrieval of contextual fear memory. Proc Natl Acad Sci U S A. 2012;109(9):3504–9.
29. Castellano C, Cabib S, Palmisano A, Di Marzo V, Puglisi-Allegra S. The effects of anandamide on memory consolidation in mice involve both D1 and D2 dopamine receptors. Behav Pharmacol. 1997;8(8):707–12.
30. Abush H, Akirav I. Cannabinoids modulate hippocampal memory and plasticity. Hippocampus. 2010;20(10):1126–38.
31. Pamplona FA, Takahashi RN. WIN 55212-2 impairs contextual fear conditioning through the activation of CB1 cannabinoid receptors. Neurosci Lett. 2006;397(1–2):88–92.
32. Lichtman AH, Dimen KR, Martin BR. Systemic or intrahippocampal cannabinoid administration impairs spatial memory in rats. Psychopharmacology (Berl). 1995;119(3):282–90.
33. Collins DR, Pertwee RG, Davies SN. Prevention by the cannabinoid antagonist, SR141716A, of cannabinoid-mediated blockade of long-term potentiation in the rat hippocampal slice. Br J Pharmacol. 1995;115(6):869–70.
34. Davies SN, Pertwee RG, Riedel G. Functions of cannabinoid receptors in the hippocampus. Neuropharmacology. 2002;42(8):993–1007.
35. Terranova JP, Michaud JC, Le Fur G, Soubrié P. Inhibition of long-term potentiation in rat hippocampal slices by anandamide and WIN55212-2: reversal by SR141716 A, a selective antagonist of CB1 cannabinoid receptors. Naunyn Schmiedebergs Arch Pharmacol. 1995;352(5):576–9.
36. De Oliveira Alvares L, de Oliveira LF, Camboim C, Diehl F, Genro BP, Lanziotti VB, Quillfeldt JA. Amnestic effect of intrahippocampal AM251, a CB1-selective blocker, in the inhibitory avoidance, but not in the open field habituation task, in rats. Neurobiol Learn Mem. 2005;83(2):119–24.
37. De Oliveira Alvares L, Genro BP, Diehl F, Quillfeldt JA. Differential role of the hippocampal endocannabinoid system in the memory consolidation and retrieval mechanisms. Neurobiol Learn Mem. 2008;90(1):1–9.
38. Breivogel CS, Griffin G, Di Marzo V, Martin BR. Evidence for a new G protein-coupled cannabinoid receptor in mouse brain. Mol Pharmacol. 2001;60(1):155–63.
39. Hájos N, Ledent C, Freund TF. Novel cannabinoid-sensitive receptor mediates inhibition of glutamatergic synaptic transmission in the hippocampus. Neuroscience. 2001;106(1):1–4.
40. Hájos N, Freund TF. Pharmacological separation of cannabinoid sensitive receptors on hippocampal excitatory and inhibitory fibers. Neuropharmacology. 2002;43(4):503–10.
41. De Oliveira Alvares L, Genro BP, Vaz Breda R, Pedroso MF, Da Costa JC, Quillfeldt JA. AM251, a selective antagonist of the CB1 receptor, inhibits the induction of long-term potentiation and induces retrograde amnesia in rats. Brain Res. 2006;1075(1):60–7.
42. Ross RA, Gibson TM, Brockie HC, Leslie M, Pashmi G, Craib SJ, Di Marzo V, Pertwee RG. Structure-activity relationship for the endogenous cannabinoid, anandamide, and certain of its analogues at vanilloid receptors in transfected cells and vas deferens. Br J Pharmacol. 2001

43. Di Marzo V Bisogno T De Petrocellis L. Anandamide: some like it hot. Trends Pharmacol Sci. 2001;22(7):346–9.
44. Hampson AJ, Bornheim LM, Scanziani M, Yost CS, Gray AT, Hansen BM, Leonoudakis DJ, Bickler PE. Dual effects of anandamide on NMDA receptor-mediated responses and neurotransmission. J Neurochem. 1998;70(2):671–6.
45. Christopoulos A, Wilson K. Interaction of anandamide with the M(1) and M(4) muscarinic acetylcholine receptors. Brain Res. 2001;915(1):70–8.
46. Sancho R, Calzado MA, Di Marzo V, Appendino G, Muñoz E. Anandamide inhibits nuclear factor-kappaB activation through a cannabinoid receptor-independent pathway. Mol Pharmacol. 2003;63(2):429–38.
47. Pearlman RJ, Aubrey KR, Vandenberg RJ. Arachidonic acid and anandamide have opposite modulatory actions at the glycine transporter, GLYT1a. J Neurochem. 2003;84(3):592–601.
48. Barros DM, Carlis V, Maidana M, Silva ES, Baisch ALM, Ramirez MR, Izquierdo I. Interactions between anandamide-induced anterograde amnesia and post-training memory modulatory systems. Brain Res. 2004;1016(1):66–71.
49. Varvel SA, Cravatt BF, Engram AE, Lichtman AH. Fatty acid amide hydrolase (-/-) mice exhibit an increased sensitivity to the disruptive effects of anandamide or oleamide in a working memory water maze task. J Pharmacol Exp Ther. 2006;317(1):251–7.
50. Hohmann AG, Suplita RL, Bolton NM, Neely MH, Fegley D, Mangieri R, Krey JF, Walker JM, Holmes PV, Crystal JD, Duranti A, Tontini A, Mor M, Tarzia G, Piomelli D. An endocannabinoid mechanism for stress-induced analgesia. Nature. 2005;435(7045):1108–12.
51. Ameri A. The effects of cannabinoids on the brain. Prog Neurobiol. 1999;58(4):315–48.
52. Hernández-Tristán R, Arévalo C, Canals S, Leret ML. The effects of acute treatment with delta9-THC on exploratory behaviour and memory in the rat. J Physiol Biochem. 2000;56(1):17–24.
53. Lichtman AH. SR 141716A enhances spatial memory as assessed in a radial-arm maze task in rats. Eur J Pharmacol. 2000;404(1–2):175–9.
54. Takahashi RN, Pamplona FA, Fernandes MS. The cannabinoid antagonist SR141716A facilitates memory acquisition and consolidation in the mouse elevated T-maze. Neurosci Lett. 2005;380(3):270–5.
55. Terranova JP, Storme JJ, Lafon N, Péŕio A, Rinaldi-Carmona M, Le Fur G, Soubrié P. Improvement of memory in rodents by the selective CB1 cannabinoid receptor antagonist, SR 141716. Psychopharmacology (Berl). 1996;126(2):165–72.
56. Wolff MC, Leander JD. SR141716A, a cannabinoid CB1 receptor antagonist, improves memory in a delayed radial maze task. Eur J Pharmacol. 2003;477(3):213–7.
57. Da Silva GE, Takahashi RN. SR 141716A prevents delta9-tetrahydrocannabinol-induced spatial learning deficit in a Morris-type water maze in mice. Prog Neuropsychopharmacol Biol Psychiatry. 2002;26(2):321–5.
58. Xu J-Y, Chen C. Endocannabinoids in synaptic plasticity and neuroprotection. Neuroscientist. 2014.
59. Andersen P, Blackstad TW, Lömo T. Location and identification of excitatory synapses on hippocampal pyramidal cells. Exp Brain Res. 1966;1(3):236–48.
60. Witter MP, Canto CB, Couey JJ, Koganezawa N, O'Reilly KC. Architecture of spatial circuits in the hippocampal region. Philos Trans R Soc Lond B Biol Sci. 2014;369(1635):20120515.
61. Okuda S, Roozendaal B, McGaugh JL. Glucocorticoid effects on object recognition memory require training-associated emotional arousal. Proc Natl Acad Sci U S A. 2004.
62. Kreitzer AC, Regehr WG. Retrograde inhibition of presynaptic calcium influx by endogenous cannabinoids at excitatory synapses onto Purkinje cells. Neuron. 2001;29(3):717–27.
63. Ohno-Shosaku T, Maejima T, Kano M. Endogenous cannabinoids mediate retrograde signals from depolarized postsynaptic neurons to presynaptic terminals. Neuron. 2001;29(3):729–38.
64. Wilson RI, Nicoll RA. Endogenous cannabinoids mediate retrograde signalling at hippocampal synapses. Nature. 2001;410(6828):588–92.
65. Wilson RI, Nicoll RA. Endocannabinoid signaling in the brain. Science. 2002;296(5568):678–82.
66. Bliss TV, Collingridge GL. A synaptic model of memory: long-term potentiation in the hippocampus. Nature. 1993;361(6407):31–9.
67. Frankland PW, Bontempi B. The organization of recent and remote memories. Nat Rev Neurosci. 2005;6(2):119–30.

68. Izquierdo I, Medina JH. Correlation between the pharmacology of long-term potentiation and the pharmacology of memory. Neurobiol Learn Mem. 1995;63(1):19–32.
69. Lamprecht R, LeDoux J. Structural plasticity and memory. Nat Rev Neurosci. 2004;5(1):45–54.
70. Carlson G, Wang Y, Alger BE. Endocannabinoids facilitate the induction of LTP in the hippocampus. Nat Neurosci. 2002;5(8):723–4.
71. De Oliveira Alvares L, Pasqualini Genro B, Diehl F, Molina VA, Quillfeldt JA. Opposite action of hippocampal CB1 receptors in memory reconsolidation and extinction. Neuroscience. 2008;154(4):1648–55.
72. Kamprath K. Cannabinoid CB1 receptor mediates fear extinction via habituation-like processes. J Neurosci. 2006;26(25):6677–86.
73. Jacob W, Yassouridis A, Marsicano G, Monory K, Lutz B, Wotjak CT. Endocannabinoids render exploratory behaviour largely independent of the test aversiveness: role of glutamatergic transmission. Genes Brain Behav. 2009;8(7):685–98.
74. Hölter SM, Kallnik M, Wurst W, Marsicano G, Lutz B, Wotjak CT. Cannabinoid CB1 receptor is dispensable for memory extinction in an appetitively-motivated learning task. Eur J Pharmacol. 2005;510(1–2):69–74.
75. Niyuhire F, Varvel SA, Thorpe AJ, Stokes RJ, Wiley JL, Lichtman AH. The disruptive effects of the CB1 receptor antagonist rimonabant on extinction learning in mice are task-specific. Psychopharmacology (Berl). 2007;191(2):223–31.
76. Roozendaal B, McGaugh JL. Amygdaloid nuclei lesions differentially affect glucocorticoid-induced memory enhancement in an inhibitory avoidance task. Neurobiol Learn Mem. 1996;65(1):1–8.
77. De Quervain DJ, Roozendaal B, McGaugh JL. Stress and glucocorticoids impair retrieval of long-term spatial memory. Nature. 1998;394(6695):787–90.
78. Roozendaal B, Nguyen BT, Power AE, McGaugh JL. Basolateral amygdala noradrenergic influence enables enhancement of memory consolidation induced by hippocampal glucocorticoid receptor activation. Proc Natl Acad Sci U S A. 1999;96(20):11642–7.
79. Roozendaal B. Systems mediating acute glucocorticoid effects on memory consolidation and retrieval. Prog Neuropsychopharmacol Biol Psychiatry. 2003;27(8):1213–23.
80. Kamprath K, Plendl W, Marsicano G, Deussing JM, Wurst W, Lutz B, Wotjak CT. Endocannabinoids mediate acute fear adaptation via glutamatergic neurons independently of corticotropin-releasing hormone signaling. Genes Brain Behav. 2009;8(2):203–11.
81. Di S, Malcher-Lopes R, Halmos KC, Tasker JG. Nongenomic glucocorticoid inhibition via endocannabinoid release in the hypothalamus: a fast feedback mechanism. J Neurosci. 2003;23(12):4850–7.
82. Di S, Malcher-Lopes R, Marcheselli VL, Bazan NG, Tasker JG. Rapid glucocorticoid-mediated endocannabinoid release and opposing regulation of glutamate and gamma-aminobutyric acid inputs to hypothalamic magnocellular neurons. Endocrinology. 2005;146(10):4292–301.
83. Stahn C, Löwenberg M, Hommes DW, Buttgereit F. Molecular mechanisms of glucocorticoid action and selective glucocorticoid receptor agonists. Mol Cell Endocrinol. 2007;275(1–2):71–8.
84. Campolongo P, Roozendaal B, Trezza V, Hauer D, Schelling G, McGaugh JL, Cuomo V. Endocannabinoids in the rat basolateral amygdala enhance memory consolidation and enable glucocorticoid modulation of memory. Proc Natl Acad Sci U S A. 2009;106(12):4888–93.
85. Ganon-Elazar E, Akirav I. Cannabinoid receptor activation in the basolateral amygdala blocks the effects of stress on the conditioning and extinction of inhibitory avoidance. J Neurosci. 2009;29(36):11078–88.
86. Hill MN, McEwen BS. Endocannabinoids: the silent partner of glucocorticoids in the synapse. Proc Natl Acad Sci U S A. 2009;106(12):4579–80.
87. Akirav I, Sandi C, Richter-Levin G. Differential activation of hippocampus and amygdala following spatial learning under stress. Eur J Neurosci. 2001;14(4):719–25.
88. Kim JJ, Diamond DM. The stressed hippocampus, synaptic plasticity and lost memories. Nat Rev Neurosci. 2002;3(6):453–62.
89. Akirav I, Richter-Levin G. Factors that determine the non-linear amygdala influence on hippocampus-dependent memory. Dose Response. 2006;4(1):22–37.

90. Izquierdo I, Dias RD. Endogenous state-dependency-memory regulation by post-training and pre-testing administration of ACTH, beta-endorphin, adrenaline and tyramine. Braz J Med Biol Res. 1983;16(1):55–64.
91. De Almeida MA Izquierdo I. Effect of the intraperitoneal and intracerebroventricular administration of ACTH, epinephrine, or beta-endorphin on retrieval of an inhibitory avoidance task in rats. Behav Neural Biol. 1984;40(1):119–22.
92. Diehl F, Fürstenau de Oliveira L, Sánchez G, Camboim C, de Oliveira Alvares L, Lanziotti VB, Cerveñansky C, Kornisiuk E, Jerusalinky D, Quillfeldt JA. Facilitatory effect of the intra-hippocampal pre-test administration of MT3 in the inhibitory avoidance task. Behav Brain Res. 2007;177(2):227–31.
93. Rossato JI, Bonini JS, Coitinho AS, Vianna MRM, Medina JH, Cammarota M, Izquierdo I. Retrograde amnesia induced by drugs acting on different molecular systems. Behav Neurosci. 2004;118(3):563–8.
94. De Oliveira Alvares L Einarsson EÖ Santana F Crestani AP Haubrich J Cassini LF Nader K Quillfeldt JA. Periodically reactivated context memory retains its precision and dependence on the hippocampus. Hippocampus. 2012;22(5):1092–5.
95. Martí Barros D Ramirez MR Dos Reis EA Izquierdo I. Participation of hippocampal nicotinic receptors in acquisition, consolidation and retrieval of memory for one trial inhibitory avoidance in rats. Neuroscience. 2004;126(3):651–6.
96. Izquierdo I, da Cunha C, Rosat R, Jerusalinsky D, Ferreira MB, Medina JH. Neurotransmitter receptors involved in post-training memory processing by the amygdala, medial septum, and hippocampus of the rat. Behav Neural Biol. 1992;58(1):16–26.
97. Barros DM, Mello e Souza T, De David T, Choi H, Aguzzoli A, Madche C, Ardenghi P, Medina JH, Izquierdo I. Simultaneous modulation of retrieval by dopaminergic D(1), beta-noradrenergic, serotonergic-1A and cholinergic muscarinic receptors in cortical structures of the rat. Behav Brain Res. 2001;124(1):1–7.
98. Bevilaqua L, Ardenghi P, Schröder N, Bromberg E, Schmitz PK, Schaeffer E, Quevedo J, Bianchin M, Walz R, Medina JH, Izquierdo I. Drugs acting upon the cyclic adenosine monophosphate/protein kinase A signalling pathway modulate memory consolidation when given late after training into rat hippocampus but not amygdala. Behav Pharmacol. 1997;8(4):331–8.
99. Cammarota M, Bevilaqua LRM, Medina JH, Izquierdo I. Retrieval does not induce reconsolidation of inhibitory avoidance memory. Learn Mem. 2004;11(5):572–8.
100. Ferreira AR, Fürstenau L, Blanco C, Kornisiuk E, Sánchez G, Daroit D, Castro e Silva M, Cerveñansky C, Jerusalinsky D, Quillfeldt JA. Role of hippocampal M1 and M4 muscarinic receptor subtypes in memory consolidation in the rat. Pharmacol Biochem Behav. 2003;74(2):411–5.
101. Jerusalinsky D, Kornisiuk E, Alfaro P, Quillfeldt J, Alonso M, Verde ER, Cerveñansky C, Harvey A. Muscarinic toxin selective for m4 receptors impairs memory in the rat. Neuroreport. 1998;9(7):1407–11.
102. Atsak P, Roozendaal B, Campolongo P. Role of the endocannabinoid system in regulating glucocorticoid effects on memory for emotional experiences. Neuroscience. 2012;204:104–16.
103. Pavlov I. Conditioned Reflexes. London: Oxford University. Press; 1927.
104. Myers KM, Davis M. Systems-level reconsolidation: reengagement of the hippocampus with memory reactivation. Neuron. 2002;36(3):340–3.
105. Myers KM, Davis M. Mechanisms of fear extinction. Mol Psychiatry. 2007;12(2):120–50.
106. Duvarci S, Mamou C Ben, Nader K. Extinction is not a sufficient condition to prevent fear memories from undergoing reconsolidation in the basolateral amygdala. Eur J Neurosci. 2006;24(1):249–60.
107. Lin H-C, Mao S-C, Gean P-W. Effects of intra-amygdala infusion of CB1 receptor agonists on the reconsolidation of fear-potentiated startle. Learn Mem. 2006;13(3):316–21.
108. Santini E, Muller RU, Quirk GJ. Consolidation of extinction learning involves transfer from NMDA-independent to NMDA-dependent memory. J Neurosci. 2001;21(22):9009–17.
109. Chhatwal JP, Davis M, Maguschak KA, Ressler KJ. Enhancing cannabinoid neurotransmission augments the extinction of conditioned fear. Neuropsychopharmacology. 2005;30(3):516–24.

110. Suzuki A, Josselyn SA, Frankland PW, Masushige S, Silva AJ, Kida S. Memory reconsolidation and extinction have distinct temporal and biochemical signatures. J Neurosci. 2004;24(20):4787–95.
111. Kobilo T, Hazvi S, Dudai Y. Role of cortical cannabinoid CB1 receptor in conditioned taste aversion memory. Eur J Neurosci. 2007;25(11):3417–21.
112. De Oliveira Alvares L, Crestani AP, Cassini LF, Haubrich J, Santana F, Quillfeldt JA. Reactivation enables memory updating, precision-keeping and strengthening: exploring the possible biological roles of reconsolidation. Neuroscience. 2013;244:42–8.
113. Igaz LM, Vianna MRM, Medina JH, Izquierdo I. Two time periods of hippocampal mRNA synthesis are required for memory consolidation of fear-motivated learning. J Neurosci. 2002;22(15):6781–9.
114. Suzuki A, Mukawa T, Tsukagoshi A, Frankland PW, Kida S. Activation of LVGCCs and CB1 receptors required for destabilization of reactivated contextual fear memories. Learn Mem. 2008;15(6):426–33.
115. Stern CAJ, Gazarini L, Takahashi RN, Guimarães FS, Bertoglio LJ. On disruption of fear memory by reconsolidation blockade: evidence from cannabidiol treatment. Neuropsychopharmacology. 2012;37(9):2132–42.
116. De Carvalho CR, Pamplona FA, Cruz JS, Takahashi RN. Erratum to: Endocannabinoids underlie reconsolidation of hedonic memories in Wistar rats. Psychopharmacology (Berl). 2014;231(7):1427.
117. Boccia MM, Acosta GB, Blake MG, Baratti CM. Memory consolidation and reconsolidation of an inhibitory avoidance response in mice: effects of i.c.v. injections of hemicholinium-3. Neuroscience. 2004;124(4):735–41.
118. Pedreira ME, Maldonado H. Protein synthesis subserves reconsolidation or extinction depending on reminder duration. Neuron. 2003;38(6):863–9.
119. Tronson NC, Taylor JR. Molecular mechanisms of memory reconsolidation. Nat Rev Neurosci. 2007;8(4):262–75.
120. Lee JLC, Milton AL, Everitt BJ. Reconsolidation and extinction of conditioned fear: inhibition and potentiation. J Neurosci. 2006;26(39):10051–6.
121. Genro BP, de Oliveira Alvares L, Quillfeldt JA. Role of TRPV1 in consolidation of fear memories depends on the averseness of the conditioning procedure. Neurobiol Learn Mem. 2012;97(4):355–60.
122. Takahashi KA, Castillo PE. The CB1 cannabinoid receptor mediates glutamatergic synaptic suppression in the hippocampus. Neuroscience. 2006;139(3):795–802.
123. Domenici MR, Azad SC, Marsicano G, Schierloh A, Wotjak CT, Dodt HU, Zieglgänsberger W, Lutz B, Rammes G. Cannabinoid receptor type 1 located on presynaptic terminals of principal neurons in the forebrain controls glutamatergic synaptic transmission. J Neurosci. 2006;26(21):5794–9.
124. Flavell CR, Barber DJ, Lee JLC. Behavioural memory reconsolidation of food and fear memories. Nat Commun. 2011;2:504.
125. Lee JLC. Memory reconsolidation mediates the updating of hippocampal memory content. Front Behav Neurosci. 2010;4:168.
126. Gatley SJ, Gifford AN, Volkow ND, Lan R, Makriyannis A. 123I-labeled AM251: a radioiodinated ligand which binds in vivo to mouse brain cannabinoid CB1 receptors. Eur J Pharmacol. 1996;307(3):331–8.
127. Landsman RS, Burkey TH, Consroe P, Roeske WR, Yamamura HI. SR141716A is an inverse agonist at the human cannabinoid CB1 receptor. Eur J Pharmacol. 1997;334(1):R1–2.
128. Savinainen JR, Saario SM, Niemi R, Järvinen T, Laitinen JT. An optimized approach to study endocannabinoid signaling: evidence against constitutive activity of rat brain adenosine A1 and cannabinoid CB1 receptors. Br J Pharmacol. 2003;140(8):1451–9.
129. Pertwee RG. Inverse agonism and neutral antagonism at cannabinoid CB1 receptors. Life Sci. 2005;76(12):1307–24.
130. Storr MA, Bashashati M, Hirota C, Vemuri VK, Keenan CM, Duncan M, Lutz B, Mackie K, Makriyannis A, Macnaughton WK, Sharkey KA. Differential effects of CB(1) neutral antagonists and inverse agonists on gastrointestinal motility in mice. Neurogastroenterol Motil. 2010;22(7):787–96, e223.

Chapter 4
Interactions Between Cannabinoid Signaling and Anxiety: A Comparative Analysis of Intervention Tools and Behavioral Effects

Mano Aliczki and Jozsef Haller

Abstract Cannabinoid signaling is believed to decrease anxiety, albeit the conflicting nature of evidence is generally acknowledged. Here we provide a comprehensive overview of available findings by grouping them according to the tools that have been used to modulate cannabinoid signaling. The systemic administration of cannabinoid receptor agonists and antagonists led to the most conflicting findings; such treatments may increase, decrease, or leave anxiety unaffected. In addition, antagonists and agonists had similar effects in many instances including their biphasic effects. The effects of genetic manipulations, cannabinoid synthesis or reuptake inhibition as well as the effects of local brain treatments with cannabinoid ligands appear more consistent. We suggest that systemically administered receptor ligands affect cannabinoid signaling globally and as such lack the spatial and temporal specificity of endocannabinoid signaling. By contrast, gene disruption and the indirect modulation of endocannabinoid availability affect ongoing (natural) processes and lead to more specific and consistent effects. Local brain treatments whit receptor ligands are spatially restricted which increases the consistency of findings, but also reveals that cannabinoids affect anxiety in a brain area-specific manner, which further explains the inconsistency of findings with systemically injected ligands. Environmental conditions have a large impact on effects with all techniques, suggesting that endocannabinoid signaling affects coping with environmental challenges rather than unconditionally decreasing anxiety. The relationship between cannabinoid signaling, anxiety and coping styles is largely understudied, but holds great promise for understanding the roles of cannabinoids in behavioral control and may broaden their therapeutic implications.

Keywords 2-arachidonoylglycerol · Anandamide · Anxiety · Coping styles · Endocannabinoids · Rodents

J. Haller (✉) ·
Institute of Experimental Medicine, Hungarian Academy of Sciences, Budapest, Hungary
e-mail: haller.jozsef@koki.mta.hu

M. Aliczki · J. Haller
Department of Behavioural Neurobiology, Institute of Experimental Medicine, Budapest, Hungary

P. Campolongo, L. Fattore (eds.), *Cannabinoid Modulation of Emotion, Memory, and Motivation,* DOI 10.1007/978-1-4939-2294-9_4

Introduction

The totality of scientific evidence obtained so far suggests that cannabinoids do play a role in the inhibitory control of anxiety, but findings are highly contradictory both within and between the techniques employed to manipulate cannabinoid signaling. Inhibition by various means (gene disruption, receptor antagonism) can increase anxiety, decrease anxiety and may be without effect, and the same applies to the enhancement of cannabinoid signaling by cannabinoid receptor agonists, reuptake blockers or by the inhibition of enzymes involved in their degradation. While there seem to be more studies attributing an anxiolytic role to cannabinoids, conflicting evidence is too many to be attributable to experimental error. Contradictions were explained in various ways, and led to several hypotheses. A thorough review of these makes it clear that theoretical approaches are based on *partial* evidence and none of them is comprehensive enough to create a consistent picture. The goal of the present chapter is to provide a full review of the evidence contained by the PubMed database and to evaluate the reasons of contradictions with the ultimate aim of disentangling the roles played by endocannabinoid signaling in anxiety. We are aware of the fact that neither goal is realistic in absolute terms, because:

- The particularities of the search engine of PubMed do not rule out that some studies remained hidden to this review. The search was performed with the search term "(cannabinoid OR endocannabinoid OR THC OR arachidonoyle-thanolamide OR anandamide OR AEA OR 2-arachydonoylglycerol OR 2-AG OR WIN 55, 212–2 OR HU 210 OR JWH 133 OR CP 55,940 OR URB 597 OR PF 622 OR PF 3845 OR PF 750 OR JZL 184 OR FAAH OR MAGL OR AM 404 OR AM 1172 OR VDM-11 OR rimonabant OR SR 141716 OR AM 251 OR NESS 0327 OR CB1 KO OR CB2 KO) AND (anxiety OR anxiolytic OR anxiogenic OR anxiolysis OR anxiogenesis or anxious)". This term resulted in 1017 hits out of which 186 original research studies were identified as relevant for the present study. While the overwhelming majority of studies were likely identified, the database created by this search is probably incomplete. This figure does not include studies on the phytocannabinoid cannabidiol[1].
- A full understanding of the role played by endocannabinoids in anxiety may not be achievable at present stage. Reasons are multiple and range from the variability of research techniques and conditions, through species, strain, and even individual differences in the particularities of the endocannabinoid system, to yet unraveled or poorly known epiphenomena of research tools used to manipulate endocannabinoid signaling. In addition to these primarily technical reasons, one cannot rule out that the anxiety-related effects of endocannabinoids

[1] This compound has anxiolytic properties [1–14], and as such it is highly relevant to anxiety research in general. However, cannabidiol binds to cannabinoid receptors with very low affinity [15], and its mechanisms are either indirectly related to endocannabinoid signaling [16] or involve direct effects on other neurotransmitter systems [17]. Therefore, data on cannabidiol were not reviewed here.

are reflections of more general effects on emotions, emotional responsiveness or coping styles. If this was true –and many recent findings point to this possibility–, then the anxiety-related effects of endocannabinoid signaling are inherently complex and condition-dependent, and rule out the possibility of answering simple questions of the type "does endocannabinoid signaling increase or decrease anxiety"?

The next section briefly reviews the main findings of the search described above. This section is free of interpretations or explanations which constitute the subject of the third chapter. The last, concluding section is an attempt to integrate data and views.

Findings

Systemic Effects

Decreased Endocannabinoid Activity

Decreasing endocannabinoid activity *via* the genetic disruption of the type 1 cannabinoid receptor (CB_1R) resulted in an anxious phenotype in most studies employing well-validated tests of anxiety (e.g. the elevated plus-maze, light/dark, social interaction tests [18–25]. The anxiety enhancing effect of CB_1R disruption seemed to be specific to young mice in one [26] and to aversive conditions in another study [27]. Two studies did not detect anxiety-like behavior in CB_1R knockout (KO) mice tested in the elevated plus-maze [28, 29], while others may suggest that CB_1R gene disruption decreases anxiety. For instance, CB_1 KO mice showed decreased burying in the shock-prod burying test which was interpreted as an anxiolytic effect [30]. In the cue-induced conditioned fear test, CB_1R KO mice did show increased anxiety, but this *decreased* when mice were socially stressed [29], suggesting that stress exposure paradoxically ameliorates the anxious phenotype of CB_1R KOs.

The role of other cannabinoid receptors was poorly studied by transgenic techniques. Two studies suggest that the disruption of the type 2 cannabinoid receptor (CB_2R) increases anxiety in the elevated plus-maze [31], light/dark [31] and open-field tests [32]. The disruption of the G protein-coupled receptor 55, a novel cannabinoid receptor [33–35], had no effects on anxiety in the only study available so far [36].

The down-regulation of endocannabinoid signaling by the CB_1R antagonist rimonabant (SR141716A) results in biphasic effects. Low doses (0.3–3 mg/kg) reduced anxiety in several models, e.g. the elevated plus-maze [37–39], light/dark [40] and Vogel tests [38], while higher doses (3–10 mg/kg) exerted anxiogenic effects in the elevated plus-maze [28, 41–47], light/dark [48], open-field [28], novelty induced hypophagia [49], elevated T-maze [28], defensive withdrawal [45], social interaction [50] and footshock-induced ultra sound vocalization [51] tests. Such large doses also increased cue-induced conditioned fear after both acute [52] and chronic treatment [29]; in addition, rimonabant inhibited the extinction of this response [53–55]. Rodent findings are supported by human studies, where both acute and chronic treatment with rimonabant exerted anxiogenic effects [56–60].

This ostensibly clear picture is obscured by a large body of conflicting evidence. Firstly, rimonabant did not always produce the effects presented above. Low doses –that decreased anxiety in the aforementioned studies– were sometimes without effect [28, 38, 61, 62]. High doses –anxiogenic in the studies presented above– were anxiolytic in the shock prod-burying paradigm [30]. Effects in humans were not replicated either [63]. Secondly, the effects of other antagonists were not always in line with those obtained with rimonabant. For instance, the CB_1R blocker AM251 did not show the biphasic effect seen with rimonabant. This antagonist proved to be anxiogenic over a wide range of doses (0.3–8 mg/kg) [21, 25, 50, 64–68]. In addition, AM251 reduced urocortin1 microinjection- and nicotine abstinence-induced anxieties[69, 70]. Other antagonists (AM281, AM4113, and AVE1625) did not affect anxiety [66, 71–73].

Data on CB_2R antagonists are sparse. Acute treatment with AM630, a CB_2R antagonist, led to anxiogenic effects, while chronic treatment attenuated anxiety in the same paradigm [74].

Taken together, the findings briefly reviewed above are in line with expectations and show that the effects of inhibited endocannabinoid signaling are highly variable (for a summary see Table 4.1).

Increased Endocannabinoid Activity

Similar to the antagonist rimonabant, CB_1R agonists have biphasic effects on anxiety. Surprisingly, however, the effects are not only biphasic but entirely similar to those seen with rimonabant (but not other antagonists): low doses decrease, while high doses increase anxiety. Anxiolytic effects were shown for low doses of the phytocannabinoid Δ^9-tetrahydrocannabinol (THC; 0.075–2 mg/kg), the endocannabinoid anandamide (AEA; 0.1–1.25 mg/kg) and synthetic cannabinoids (WIN55,212–2: 0.5–3 mg/kg; CP55,940: <0.1 mg/kg; HU210: 0.01 mg/kg) [21, 43, 53, 65, 67, 68, 75–90]. Higher doses of the same agonists (THC: 2.5–10 mg/kg; AEA: 10 mg/kg; WIN55,212–2: 3–5 mg/kg; CP55,940: >0.1 mg/kg; HU210: 0.05–0.1 mg/kg) were anxiogenic [40, 44, 71, 78, 83, 89–104]. High doses of THC increased anxiety in humans as well [105–111].

This apparently consistent picture is blurred by a large body of conflicting evidence. Low doses of agonists –anxiolytic in the above studies– increased anxiety under specific conditions, such as repeated treatments in adults, perinatal administration and in rats that were chronically treated with vehicle before drug administration [112–115]. High doses of CB_1R agonists –anxiogenic in the above studies—decreased anxiety in cocaine-self-administering subjects, in the 3,4-methylenedioxy-N-methylamphetamine-induced anxiety model, after chronic vehicle pretreatment and in adolescent subjects [116]. The biphasic effect was also overturned by species,

Table 4.1 Effects of decreased endocannabinoid activity on anxiety

Assessment tools		Effects on anxiety (references)	Number of studies
CB_1 KO		Anxiogenesis	9
		Condition-dependent effects	2
		No effects	2
		Anxiolysis	2
CB_2 KO		Anxiogenesis	2
GPR55 KO		No effects	1
CB_1R antagonist	findings compatible with biphasic effects[a] (total No. of studies: 25)	Rimonabant	25
		AM251	0
		Other antagonists	0
	findings incompatible with biphasic effects[b] (total No. of studies: 20)	Rimonabant	6
		AM251	10
		Other antagonists	4
CB_2R antagonists		*acute*: anxiogenesis; *chronic*: anxiolysis	1

[a] the general consensus is that low doses of CB_1R antagonists decrease, while large doses increase anxiety. The dose ranges for these effects were indicated in the text

[b] the hypothesis on the biphasic nature of effects was considered challenged by studies where either the low or the large those did not produce the expected effect

strain, gender, and experimental conditions (e.g. enriched environment, treatments received in adolescence) [117]. Additionally, there is a large set of studies, in which doses that effectively altered anxiety in the above studies were without effects [54, 98, 101, 114, 118–123]. Inefficacy was sometimes seen under specific conditions, like stress-induced anxiety [124] or alcohol-withdrawal [125].

The enhancement of endocannabinoid signaling *via* the selective blockade of their degrading enzymes is a novel approach for the up-regulation of endocannabinoid activity [126–129]. Endocannabinoids are synthesized "on-demand"; therefore the blockade of their breakdown promotes ongoing signaling processes, i.e. their effects are more specific than those of agonists, which activate cannabinoid receptors throughout the brain. Both genetic and pharmacological blockade of the anandamide metabolizing enzyme, fatty acid amide hydrolase (FAAH), led to anxiolytic effects in a number of reports [43, 46, 67, 71, 86, 100, 126, 128, 130–138]. In other cases, however, no effects were seen either after genetic [139] or pharmacological blockade of FAAH activity [138–140]. FAAH inhibition was anxiogenic in one study [50]. Strong dependence on environmental conditions was reported in two studies [138, 139].

Studies on the specific role of 2-AG signaling were only recently made possible by the synthesis of the first selective, specific monoacylglycerol lipase (MAGL) blocker. This compound decreased anxiety in a number of studies [132, 134, 141,

Table 4.2 Effects of increased endocannabinoid activity on anxiety

Assessment tools	Remarks	Effect on anxiety (references)	Number of studies
CB_1R agonists	Findings compatible with biphasic effects[a]	Low doses anxiolytic, large doses anxiogenic	35
	Findings incompatible with biphasic effects[b]	Effect altered/reverted by specific experimental conditions	13
		No effects on anxiety	10
Blockade of AEA degradation		Anxiolysis	17
		Condition-dependent effects	2
		No effects	1
		Anxiogenesis	1
Blockade of 2-AG degradation		Anxiolysis	4
		Condition-dependent effects	2
Endocannabinoid reuptake inhibition		Anxiolysis	6
		no effect	2

[a] the general consensus is that low doses of CB_1R antagonists decrease, while large doses increase anxiety. The dose ranges for these effects were indicated in the text

[b] the hypothesis on the biphasic nature of effects was considered challenged by studies where either the low or the large those did not produce the expected effect

142]. A few studies suggest that these effects depend on environmental aversiveness, and HPA-axis activity [143, 144].

Endocannabinoid signaling can also be stimulated by the inhibition of endocannabinoid transport. This treatment led to anxiolytic effects in a number of reports [43, 54, 87, 145–148]; no effects were seen in two studies [71, 149].

Taken together, the effects of pharmacological enhancement of endocannabinoid activity have variable effects on anxiety-like behavior. Findings are summarized in Table 4.2.

Local Brain Treatments

Neuron Type-Specific Effects

In this type of studies, transgenic animals were used; the selective disruption of CB_1Rs in glutamatergic, dopaminergic and serotonergic neurons all increased anxiety [20, 29, 150]. The same manipulation in GABA-ergic neurons did not cause such changes [29]. One study suggests that cannabinoid signaling in serotonergic neurons ameliorates conditioned fear, despite the fact that the same transgenic animals showed anxiety in the elevated plus-maze [29]. By contrast, dopamine neuron-specific gene disruptions had congruent effects in the social interaction test of anxiety and conditioned fear [150].

Brain Area-Specific Effects

General effects. Blockade of CB_1Rs in the brain by the intracerebroventricular injection of the CB_1R antagonist AM251 increased anxiety [151], while the enhancement of endocannabinoid activity by FAAH administered *via* the same route was anxiolytic [133]. The effects of AM251 were reversed in animals treated with corticotrophin releasing hormone and in those submitted to cocaine-withdrawal [151]. Mice expressing CB_1Rs only in the dorsal telencephalon showed reduced anxiety compared to CB_1R KO mice [152].

Prefrontal cortex. The enhancement of cannabinoid signaling by cannabinoid agonists and FAAH inhibition had biphasic effects in this brain area; small doses decreased, while large doses increased anxiety [153, 154]. The genetic over-expression of the CB_1Rs in the same area mimicked the effects of large doses, i.e. it increased anxiety [155]. Thus, the studies performed so far provide a congruent picture. Interestingly, the biphasic effects seen after systemic treatments were replicated by local agonist infusions into the prefrontal cortex. Similar biphasic effects were seldom reported in other brain regions.

Amygdala. We found only one study where local treatments were suggested to cover the whole amygdala; in this case, the cannabinoid agonist arachidonylcyclopropylamide (ACPA) reduced anxiety [156]. This effect was replicated by the infusion of agonists into the basolateral amygdala but not by local treatments targeting specifically the central amygdala. In the former region, agonists (Δ9-THC, WIN55,212–2) and N-arachidonoyl-serotonin (a combined FAAH inhibitor/TRPV1 antagonist) reduced anxiety; the effect was valid to certain doses and particular conditions only, but no anxiogenic effects were observed at any dose [153, 157]. It is worth to note that no similar effects were observed with anandamide and pure FAAH inhibitors [158, 159], while Δ9-THC administration into the central amygdala increased anxiety [160].

Cannabinoid antagonists were administered into the basolateral, central, and medial amygdala. In the basolateral amygdala, where agonists decreased anxiety, antagonists increased it [69, 159, 161]; thus, the two types of treatments led to congruent effects in this brain region. In the central amygdala, antagonists (rimonabant, AM251) increased anxiety [161], similar to the agonist Δ9-THC. Thus, in this amygdala region, findings are incongruent. One study suggests that the local disruption of CB_1R expression in the medial amygdala decreases anxiety [162].

Taken together, the studies reviewed above suggest that cannabinoid signaling in the basolateral amygdala decreases anxiety. Reports on other amygdalar subregions are disparate, but suggest that the effects of cannabinoid signaling are amygdala subarea-specific.

Hippocampus. Agonists or FAAH inhibitors were infused into the CA1 region in three studies: effects were contrasting as anxiogenic effects [163], no effects [164] or anxiolytic effects [165] were observed. CB_1R blockade in the very same brain region either decreased or increased anxiety [163, 166]. In the ventral hippocampus, the enhancement of endocannabinoid signaling by agonists (Δ9-THC, high doses), as well as by FAAH or reuptake blockade resulted in anxiogenesis [153, 167, 168], while the blockade of CB_1Rs did not affect anxiety [167]. Noteworthy, the effects of Δ^9-THC were biphasic, while the effects of reuptake blockade were reversed by stress exposure [153, 167, 168].

Periaqueductal gray. Cannabinoid receptor agonists (2-AG, AEA, ACEA), the blockade of MAGL, as well as the inhibition of cannabinoid reuptake in the dorsal and dorsolateral periaqueductal gray decreased anxiety [141, 169–172]. The CB_1R antagonist AM251 was without effect [172]. Except for this latter finding, the anxiolytic roles of cannabinoid signaling in the dorsal/dorsolateral periaqueductal gray appear well supported.

Other brain regions. The local deletion of CB_1Rs in the posterior hypothalamus, the paraventricular and supraoptic nuclei increased anxiety [162]. The microinjection of AM251 into the enteropeduncular nucleus also increased anxiety [160].

Conclusions

The number of studies on neuron type-specific and brain area-specific roles of cannabinoid signaling in anxiety are clearly insufficient to draw definite conditions. Nevertheless, the findings obtained so far suggest that cannabinoids have anxiolytic effects in most brain regions. As exception, they appear to have biphasic effects in the prefrontal cortex, and anxiogenic effects in the ventral hippocampus. Data in the dorsal hippocampus and medial amygdala are sparse. Findings appear to be rather congruent in many brain regions, and neuron types. The brain area-specific effects of cannabinoids on anxiety are summarized in Table 4.3.

Interpretation

Clearly, data on the anxiety-related effects of cannabinoids are conflicting, but the thorough overview of the available findings leads to a series of interesting conclusions:

- The less reliable findings were obtained with cannabinoid agonists and antagonists. The most blatant dissimilarities relate to the biphasic effect of such treatments. Biphasic effects are not particularly unusual in pharmacology, but in the case of cannabinoid ligands, antagonists and agonists have highly similar effect profiles: small doses of both decrease anxiety, while large doses of both increase anxiety. In addition, the largest number of conflicting findings was obtained with these experimental tools.

Table 4.3 Brain area-specific effects of cannabinoid signaling on anxiety

Brain area		Most frequently reported effect	Number of studies	
Prefrontal cortex		Biphasic (low doses are anxiolytic; high doses are anxiogenic)	Supporting	4
			Not supporting	0
			Opposite effect	0
Amygdala	Whole	Anxiolysis	Supporting	1
			Not supporting	0
			Opposite effect	0
	Basolateral nucleus	Anxiolysis	Supporting	6
			Not supporting	2
			Opposite effect	0
	Central nucleus	Anxiolysis	Supporting	2
			Not supporting	0
			Opposite effect	1
	Medial nucleus	Anxiogenesis	Supporting	1
			Not supporting	0
			Opposite effect	0
Hippocampus, dorsal		Anxiogenesis		1
		No effects		1
		Anxiolysis		1
Hippocampus, ventral		Anxiogenesis	Supporting	3
			Not supporting	1
			Opposite effect	0
Periaqueductal gray, dorsal/dorsolateral		Anxiolysis	Supporting	6
			Not supporting	1
			Opposite effect	0
Hypothalamus		Anxiolysis	Supporting	1
			Not supporting	0
			Opposite effect	0
Enteropeduncular nucleus		Anxiolysis	Supporting	1
			Not supporting	0
			Opposite effect	0

- The selective genetic disruption of cannabinoid receptors provided more congruent findings: this procedure increased anxiety in the overwhelming majority reports. One study reported no effects, while another reported context-dependent effects which included anxiogenesis under particular conditions and no effects under other conditions. In addition, one of the reports where anxiolytic effects were observed employed the shock-prod burying paradigm, a mixed anxiety and coping test [173]. Effects on coping will be discussed below. In conclusion, the anxiogenic effects of CB_1R disruption is contradicted by one single study, and no effects were obtained in another.

- Findings obtained with agents that indirectly modulate endocannabinoid signaling (FAAH, MAGL, and reuptake blockers) are not devoid of contradictions, but again the overwhelming majority of findings suggest that such agents decrease anxiety. This statement is supported by 27 studies. Condition-dependent effects were obtained in 4 studies (usually implicating anxiolysis under particular conditions) and no effects were obtained in 3 studies. Anxiogenic effects were obtained in one study only.
- Local brain treatments with cannabinoid agents provided the most consistent sets of data. There are virtually no contradictions in the case of certain brain areas, while opposing effects are missing in other cases (e.g. discrepancies are between effects and no effects).

The perspective summarized above raise a series of questions; the following sections are attempts to answer them.

Why are the Effects of Receptor Ligands Less Reliable than Those of Indirect Modulators?

The characteristics of endocannabinoid signaling and those of receptor ligands decrease the reliability of the latter as experimental tools. Endocannabinoids are secreted from the post-synaptic membrane and retrogradely inhibit the synaptic neurotransmission that triggered their release [174]. Although a probably low level of tonic activation cannot be excluded, the endocannabinoid signal occurs phasically i.e. when the intensity of anterograde synaptic communication reaches certain levels [175–178]. As such, the main role of endocannabinoid signaling appears to be the blockade of excessive neuronal activation [179].

Agonists overrule this finely tuned mechanism by inhibiting neurotransmission in synapses where this is not justified by its intensity, i.e. where retrograde signaling is not activated under normal conditions. As such, the effects of agonists are broader than those of endocannabinoids, and instead of mimicking natural activity they extend effects to synapses, neurons and brain areas where such activity normally does not take place.

Antagonists on their turn (especially those extensively used in anxiety research), have inverse agonist properties, by which they also overrule the above-described mechanism. Instead of inhibiting endocannabinoid signaling, their inverse agonist effects inhibit neuronal discharges in areas where endocannabinoids are normally not released. Thus, their effects are also extended to synapses, neurons and brain areas where endocannabinoids are not active.

In addition, many of the tools regularly used to affect receptor function affect both CB_1Rs and CB_2Rs. Originally, this was not perceived as a problem, but relatively recent findings demonstrate that CB_2Rs are expressed in the brain and have roles in behavior control [180]. In addition, receptor ligands also bind to other receptors, for instance to the still poorly known "third" cannabinoid receptor as well as to the GPR55 and TRPV1 receptors [181]. Naturally, endocannabinoids also bind to these receptors non-selectively; however, they affect the function of these mechanisms in spatially and temporally selective ways, while exogenous receptor ligands act indiscriminately.

A third problem with exogenous ligands is that their brain distribution is not uniform; moreover, different receptor ligands have specific patterns of brain distribution. For instance, two times more WIN55,212–2 was found in the hypothalamus than in the amygdala after the systemic administration of the compound; by contrast, the amounts of the antagonist rimonabant (administered by the same route) were similar in these two brain regions [182]. While the issue remains understudied, the available findings strongly suggest that compound-specific brain distribution patterns constitute an additional confounding factor in the elucidation of the roles of endocannabinoids in behavioral control. Furthermore, cannabinoid receptor ligands may show species- and neuron type-specific ligand sensitivity. Electrophysiological studies showed for instance that WIN-55,212–2 preferentially affected GABA-ergic neurotransmission in mice, while the same compound appeared to affect glutamatergic neurotransmission in rats, which together with species- and neuron type-specific effects of AM251 led to large species differences in the behavioral effects of these ligands and marked differences in their interaction [23].

The use of indirect modulators circumvents most these problems. Metabolic enzyme inhibitors and reuptake blockers enhance and prolong naturally occurring endocannabinoid release. Consequently, the up-regulation of endocannabinoid signaling is restricted to synapses, neurons and brain regions where the system is activated by the behavioral paradigm investigated. The enhanced activation of natural endocannabinoid signaling also eliminates problems related to receptor specificity, brain distribution and ligand sensitivity.

Why are Gene Disruption and Local Treatments More Reliable than Receptor Ligands?

The problems related to the use of receptor ligands are also circumvented by the genetic disruption of the endocannabinoid receptor and by the local brain administration of compounds. The gene disruption technique has its own flaws, among which the development of compensatory mechanisms are believed to have the largest impact on experimental findings. At the same time, however, most of the problems raised by the use of receptor ligands are avoided by this technique. The reason is the spatio-temporal overlap of networks activated by a behavioral context and the lack of receptors in these networks. While receptors are eliminated throughout the brain, the consequences of this are manifested only at those synapses which are activated under the conditions of a particular study. The effects of gene disruption on networks that are unrelated to the context (i.e. are not "working" when a particular behavior is expressed) remain "silent" because they do not contribute to the execution of the behavioral act. Therefore, gene disruption eliminates naturally occurring cannabinoid signaling without having effects on other mechanisms. The same holds true for selectivity: while receptor ligands act on more than one receptor, gene disruptions are selective in this respect. Finally, problems associated by ligand-specific brain distribution patterns and ligand specificity are not present in receptor knockouts, where the ligands are the natural ones, i.e. endocannabinoids.

The local application of receptor ligands involves all the problems associated with direct receptor modulation, but these are spatially restricted, and by this their consequences are minimized. In other words, nonspecific effects at the targeted brain area are not amplified by nonspecific effects at other brain sites. Moreover, local applications eliminate the problem of differential effects exerted in certain brain regions. As shown above, the local administration of cannabinoids results in anxiolysis in some but not all brain regions. Systemically administered cannabinoids activate in parallel biphasic effects in the prefrontal cortex, anxiolytic effects in the amygdala, and anxiogenic effects in the hippocampus, while local administration activate only one of these mechanisms, which leads to clearer findings.

Why are Effects Condition-Dependent?

It is a common observation that the condition of subjects and experimental conditions have a large impact on how cannabinoids affect anxiety; examples were outlined above and will not be reiterated here. One possible interpretation of such condition-dependent effects is that cannabinoids do not affect particular behaviors but affect the way in which the organism responds to challenges, i.e. they affect coping styles. We identified four papers addressing the effects of cannabinoids from this perspective [25, 183–185]. Taken together, these studies suggest that cannabinoids promote active coping, which is associated with anxiolytic-like and antidepressant-like effects in particular tests.

Active and passive coping styles are two distinct behavioral phenotypes which differ in the way challenges are dealt with, and which show a bimodal distribution [186, 187]. Behavior is internally driven and problem oriented in active copers. In contrast, passive copers are governed by environmental stimuli and tend to respond challenges by avoidant behavior. These temporally stable behavioral phenotypes have adaptive significance in animals, while in humans, active (type "A") and passive (particularly type "C") coping styles influence disease susceptibility and resilience under adverse conditions [187–189]. Moreover, coping styles are believed to reliably predict disease-induced decreases in quality of life [190, 191]. Consequently, interventions promoting active coping styles -which are associated more favorably with resilience- have been proposed as therapeutic goals for a variety of physical diseases and mental disorders [190, 192, 193]. Thus, the putative effects of endocannabinoid signaling on coping styles are highly relevant from a therapeutic point of view.

The relationships between cannabinoids and coping on one side, and cannabinoids and anxiety on the other side have not been elucidated so far. There are several scenarios that may be considered: (1) cannabinoids affect anxiety in the first place, and promote active coping by decreasing anxiety; (2) cannabinoids affect coping in the first place, and their anxiolytic effects are context-dependent consequences of the shift in coping styles; (3) effects on coping and anxiety are mediated by different cannabinoid-dependent mechanisms that interact under specific conditions.

Conclusions

Overall, the findings suggest that cannabinoid signaling decreases anxiety. The number of conflicting findings is large. A comparison of different technologies demonstrates that the reliability of findings is rather low with receptor ligands (agonists and antagonists). Considerably more consistent findings were obtained with gene knockouts, the indirect enhancement of endocannabinoid signaling (e.g. enzyme inhibitors), and local brain treatments. The anxiolytic effects of cannabinoid signaling are more robustly shown by the latter three as compared with the former approach, but notably, the effects of cannabinoids is not uniform across brain areas. In the prefrontal cortex, biphasic effects were noticed (anxiolysis at low and anxiogenesis at large doses), while in the amygdala and hippocampus cannabinoids seem to decrease and increase, respectively, anxiety-like behavior. The condition of subjects and experimental conditions have a strong impact on the effects of cannabinoids, and this seems to be independent from the technique employed to manipulate endocannabinoid signaling. Recent findings demonstrate that cannabinoids promote a shift from passive to active coping with challenges, which may explain the context-dependence of their anxiety-related effects, and may broaden their therapeutic implications. The relationship and directionality of the triple association between cannabinoid signaling, anxiety and coping styles is largely understudied, but holds great promise for the understanding of the roles of cannabinoids in behavioral control, and the therapeutic potentials of cannabinoid modulators.

References

1. Uribe-Marino A, Francisco A, Castiblanco-Urbina MA, Twardowschy A, Salgado-Rohner CJ, Crippa JA, Hallak JE, Zuardi AW, Coimbra NC. Anti-aversive effects of cannabidiol on innate fear-induced behaviors evoked by an ethological model of panic attacks based on a prey vs the wild snake Epicrates cenchria crassus confrontation paradigm. Neuropsychopharmacology. 2012;37(2):412–21.
2. Casarotto PC, Gomes FV, Resstel LB, Guimaraes FS. Cannabidiol inhibitory effect on marble-burying behaviour: involvement of CB1 receptors. Behav Pharmacol. 2010;21(4):353–58.

3. Deiana S, Watanabe A, Yamasaki Y, Amada N, Arthur M, Fleming S, Woodcock H, Dorward P, Pigliacampo B, Close S, Platt B, Riedel G. Plasma and brain pharmacokinetic profile of cannabidiol (CBD), cannabidivarine (CBDV), Delta(9)-tetrahydrocannabivarin (THCV) and cannabigerol (CBG) in rats and mice following oral and intraperitoneal administration and CBD action on obsessive-compulsive behaviour. Psychopharmacology (Berl). 2012;219(3):859–73.
4. Campos AC, Ferreira FR, Guimaraes FS. Cannabidiol blocks long-lasting behavioral consequences of predator threat stress: possible involvement of 5HT1A receptors. J Psychiatr Res. 2012;46(11):1501–10.
5. Guimaraes FS, Chiaretti TM, Graeff FG, Zuardi AW. Antianxiety effect of cannabidiol in the elevated plus-maze. Psychopharmacology (Berl). 1990;100(4):558–9.
6. Resstel LB, Tavares RF, Lisboa SF, Joca SR, Correa FM, Guimaraes FS. 5-HT1A receptors are involved in the cannabidiol-induced attenuation of behavioural and cardiovascular responses to acute restraint stress in rats. Br J Pharmacol. 2009;156(1):181–88.
7. Campos AC, Guimaraes FS. Evidence for a potential role for TRPV1 receptors in the dorsolateral periaqueductal gray in the attenuation of the anxiolytic effects of cannabinoids. Prog Neuropsychopharmacol Biol Psychiatry. 2009;33(8):1517–21.
8. Campos AC, de Paula Soares V, Carvalho MC, Ferreira FR, Vicente MA, Brandao ML, Zuardi AW, Zangrossi H, Jr., Guimaraes FS. Involvement of serotonin-mediated neurotransmission in the dorsal periaqueductal gray matter on cannabidiol chronic effects in panic-like responses in rats. Psychopharmacology (Berl). 2013;226(1):13–24.
9. Guimaraes FS, de Aguiar JC, Mechoulam R, Breuer A. Anxiolytic effect of cannabidiol derivatives in the elevated plus-maze. Gen Pharmacol. 1994;25(1):161–64.
10. Gomes FV, Resstel LB, Guimaraes FS. The anxiolytic-like effects of cannabidiol injected into the bed nucleus of the stria terminalis are mediated by 5-HT1A receptors. Psychopharmacology (Berl). 2011;213(2–3):465–73.
11. Granjeiro EM, Gomes FV, Guimaraes FS, Correa FM, Resstel LB. Effects of intracisternal administration of cannabidiol on the cardiovascular and behavioral responses to acute restraint stress. Pharmacol Biochem Behav. 2011;99(4):743–48.
12. Long LE, Chesworth R, Huang XF, McGregor IS, Arnold JC, Karl T. A behavioural comparison of acute and chronic Delta9-tetrahydrocannabinol and cannabidiol in C57BL/6JArc mice. Int J Neuropsychopharmacol. 2010;13(7):861–76.
13. Moreira FA, Aguiar DC, Guimaraes FS. Anxiolytic-like effect of cannabidiol in the rat Vogel conflict test. Prog Neuropsychopharmacol Biol Psychiatry. 2006;30(8):1466–71.
14. Soares Vde P Campos AC Bortoli VC Zangrossi H Jr. Guimaraes FS Zuardi AW. Intra-dorsal periaqueductal gray administration of cannabidiol blocks panic-like response by activating 5-HT1A receptors. Behav Brain Res. 2010;213(2):225–29.
15. Pertwee RG. The diverse CB1 and CB2 receptor pharmacology of three plant cannabinoids: delta9-tetrahydrocannabinol, cannabidiol and delta9-tetrahydrocannabivarin. Br J Pharmacol. 2008;153(2):199–215.
16. Hayakawa K, Mishima K, Hazekawa M, Sano K, Irie K, Orito K, Egawa T, Kitamura Y, Uchida N, Nishimura R, Egashira N, Iwasaki K, Fujiwara M. Cannabidiol potentiates pharmacological effects of Delta(9)-tetrahydrocannabinol via CB(1) receptor-dependent mechanism. Brain Res. 2008;1188:157–64.
17. Russo EB, Burnett A, Hall B, Parker KK. Agonistic properties of cannabidiol at 5-HT1a receptors. Neurochem Res. 2005;30(8):1037–43.
18. Derks NM, Pinter O, Zuure W, Ledent C, Watanabe M, Molnar CS, Wei Y, Roubos EW, Wu S, Hrabovszky E, Zelena D, Kozicz T. Cannabinoid modulation of midbrain urocortin 1 neurones during acute and chronic stress. J Neuroendocrinol. 2012;24(11):1447–61.
19. Hill MN, Hillard CJ, McEwen BS. Alterations in corticolimbic dendritic morphology and emotional behavior in cannabinoid CB1 receptor-deficient mice parallel the effects of chronic stress. Cereb Cortex. 2011;21(9):2056–64.

20. Jacob W, Yassouridis A, Marsicano G, Monory K, Lutz B, Wotjak CT. Endocannabinoids render exploratory behaviour largely independent of the test aversiveness: role of glutamatergic transmission. Genes Brain Behav. 2009;8(7):685–98.
21. Haller J, Varga B, Ledent C, Freund TF. CB1 cannabinoid receptors mediate anxiolytic effects: convergent genetic and pharmacological evidence with CB1-specific agents. Behav Pharmacol. 2004;15(4):299–304.
22. Uriguen L, Perez-Rial S, Ledent C, Palomo T, Manzanares J. Impaired action of anxiolytic drugs in mice deficient in cannabinoid CB1 receptors. Neuropharmacology. 2004;46(7):966–73.
23. Haller J, Bakos N, Szirmay M, Ledent C, Freund TF. The effects of genetic and pharmacological blockade of the CB1 cannabinoid receptor on anxiety. Eur J Neurosci. 2002;16(7):1395–98.
24. Martin M, Ledent C, Parmentier M, Maldonado R, Valverde O. Involvement of CB1 cannabinoid receptors in emotional behaviour. Psychopharmacology. 2002;159(4):379–87.
25. Litvin Y, Phan A, Hill MN, Pfaff DW, McEwen BS. CB1 receptor signaling regulates social anxiety and memory. Genes Brain Behav. 2013;12(5):479–89.
26. Maccarrone M, Valverde O, Barbaccia ML, Castane A, Maldonado R, Ledent C, Parmentier M, Finazzi-Agro A. Age-related changes of anandamide metabolism in CB1 cannabinoid receptor knockout mice: correlation with behaviour. Eur J Neurosci. 2002;15(7):1178–86.
27. Haller J, Varga B, Ledent C, Barna I, Freund TF. Context-dependent effects of CB1 cannabinoid gene disruption on anxiety-like and social behaviour in mice. Eur J Neurosci. 2004;19(7):1906–12.
28. Thiemann G, Watt CA, Ledent C, Molleman A, Hasenohrl RU. Modulation of anxiety by acute blockade and genetic deletion of the CB(1) cannabinoid receptor in mice together with biogenic amine changes in the forebrain. Behav Brain Res. 2009;200(1):60–7.
29. Dubreucq S, Matias I, Cardinal P, Haring M, Lutz B, Marsicano G, Chaouloff F. Genetic dissection of the role of cannabinoid type-1 receptors in the emotional consequences of repeated social stress in mice. Neuropsychopharmacology. 2012;37(8):1885–1900.
30. Degroot A, Nomikos GG. Genetic deletion and pharmacological blockade of CB1 receptors modulates anxiety in the shock-probe burying test. Eur J Neurosci. 2004;20(4):1059–64.
31. Ortega-Alvaro A, Aracil-Fernandez A, Garcia-Gutierrez MS, Navarrete F, Manzanares J. Deletion of CB2 cannabinoid receptor induces schizophrenia-related behaviors in mice. Neuropsychopharmacology. 2011;36(7):1489–1504.
32. Flake NM, Zweifel LS. Behavioral effects of pulp exposure in mice lacking cannabinoid receptor 2. J Endod. 2012;38(1):86–90.
33. Ryberg E, Larsson N, Sjogren S, Hjorth S, Hermansson NO, Leonova J, Elebring T, Nilsson K, Drmota T, Greasley PJ. The orphan receptor GPR55 is a novel cannabinoid receptor. Br J Pharmacol. 2007;152(7):1092–101.
34. Lauckner JE, Jensen JB, Chen HY, Lu HC, Hille B, Mackie K. GPR55 is a cannabinoid receptor that increases intracellular calcium and inhibits M current. Proc Natl Acad Sci USA. 2008;105(7):2699–704.
35. Sharir H, Abood ME. Pharmacological characterization of GPR55, a putative cannabinoid receptor. Pharmacol Ther. 2010;126(3):301–13.
36. Wu CS, Chen H, Sun H, Zhu J, Jew CP, Wager-Miller J, Straiker A, Spencer C, Bradshaw H, Mackie K, Lu HC. GPR55, a G-protein coupled receptor for lysophosphatidylinositol, plays a role in motor coordination. PloS ONE. 2013;8(4):e60314.
37. Rodgers RJ, Haller J, Halasz J, Mikics E. 'One-trial sensitization' to the anxiolytic-like effects of cannabinoid receptor antagonist SR141716A in the mouse elevated plus-maze. Eur J Neurosci. 2003;17(6):1279–86.
38. Griebel G, Stemmelin J, Scatton B. Effects of the cannabinoid CB1 receptor antagonist rimonabant in models of emotional reactivity in rodents. Biol Psychiatry. 2005;57(3):261–67.
39. Rubio M, Fernandez-Ruiz J, de Miguel R, Maestro B, Michael Walker J, Ramos JA. CB1 receptor blockade reduces the anxiogenic-like response and ameliorates the neurochemical imbalances associated with alcohol withdrawal in rats. Neuropharmacology. 2008;54(6):976–88.

40. Akinshola BE, Chakrabarti A, Onaivi ES. In-vitro and in-vivo action of cannabinoids. Neurochem Res. 1999;24(10):1233–40.
41. Takahashi RN, Pamplona FA, Fernandes MS. The cannabinoid antagonist SR141716A facilitates memory acquisition and consolidation in the mouse elevated T-maze. Neurosci Lett. 2005;380(3):270–75.
42. Huang P, Liu-Chen LY, Kirby LG. Anxiety-like effects of SR141716-precipitated delta9-tetrahydrocannabinol withdrawal in mice in the elevated plus-maze. Neurosci Lett. 2010;475(3):165–68.
43. Patel S, Hillard CJ. Pharmacological evaluation of cannabinoid receptor ligands in a mouse model of anxiety: further evidence for an anxiolytic role for endogenous cannabinoid signaling. J Pharmacol Exp Ther. 2006;318(1):304–11.
44. Arevalo C, de Miguel R, Hernandez-Tristan R. Cannabinoid effects on anxiety-related behaviours and hypothalamic neurotransmitters. Pharmacol Biochem Behav. 2001;70(1):123–31.
45. Navarro M, Hernandez E, Munoz RM, del Arco I, Villanua MA, Carrera MR, Rodriguez de Fonseca F. Acute administration of the CB1 cannabinoid receptor antagonist SR 141716A induces anxiety-like responses in the rat. Neuroreport. 1997;8(2):491–96.
46. Moise AM, Eisenstein SA, Astarita G, Piomelli D, Hohmann AG. An endocannabinoid signaling system modulates anxiety-like behavior in male Syrian hamsters. Psychopharmacology. 2008;200(3):333–46.
47. Blasio A, Iemolo A, Sabino V, Petrosino S, Steardo L, Rice KC, Orlando P, Iannotti FA, Di Marzo V, Zorrilla EP, Cottone P. Rimonabant precipitates anxiety in rats withdrawn from palatable food: role of the central amygdala. Neuropsychopharmacology. 2013;38(12):2498–507.
48. O'Brien LD, Wills KL, Segsworth B, Dashney B, Rock EM, Limebeer CL, Parker LA. Effect of chronic exposure to rimonabant and phytocannabinoids on anxiety-like behavior and saccharin palatability. Pharmacol Biochem Behav. 2013;103(3):597–602.
49. Gamble-George JC, Conger JR, Hartley ND, Gupta P, Sumislawski JJ, Patel S. Dissociable effects of CB1 receptor blockade on anxiety-like and consummatory behaviors in the novelty-induced hypophagia test in mice. Psychopharmacology (Berl). 2013;228(3):401–9.
50. Seillier A, Martinez AA, Giuffrida A. Phencyclidine-induced social withdrawal results from deficient stimulation of cannabinoid CB(1) receptors: implications for schizophrenia. Neuropsychopharmacology. 2013;38(9):1816–24.
51. Varga B, Kassai F, Gyertyan I. Interactions of CB1 and mGlu5 receptor antagonists in food intake, anxiety and memory models in rats. Pharmacol Biochem Behav. 2012;103(2):425–30.
52. Hofelmann D, di Benedetto B, Azad SC, Micale V, Wotjak CT, Rammes G. Lack of interaction of endocannabinoids and 5-HT(3) neurotransmission in associative fear circuits of the amygdala: evidence from electrophysiological and behavioural experiments. Brain Res. 2013;1527:47–56.
53. Pamplona FA, Prediger RD, Pandolfo P, Takahashi RN. The cannabinoid receptor agonist WIN 55,212–2 facilitates the extinction of contextual fear memory and spatial memory in rats. Psychopharmacology (Berl). 2006;188(4):641–49.
54. Chhatwal JP, Davis M, Maguschak KA, Ressler KJ. Enhancing cannabinoid neurotransmission augments the extinction of conditioned fear. Neuropsychopharmacology. 2005;30(3):516–24.
55. Chhatwal JP, Gutman AR, Maguschak KA, Bowser ME, Yang Y, Davis M, Ressler KJ. Functional interactions between endocannabinoid and CCK neurotransmitter systems may be critical for extinction learning. Neuropsychopharmacology. 2009;34(2):509–21.
56. Van Gaal L Pi-Sunyer X Despres JP McCarthy C Scheen A. Efficacy and safety of rimonabant for improvement of multiple cardiometabolic risk factors in overweight/obese patients: pooled 1-year data from the Rimonabant in Obesity (RIO) program. Diabetes Care. 2008;31(2):229–40.

57. Dibble CT, Gelfand EV, Cannon CP. Rimonabant: the role of endocannabinoid type 1 receptor antagonism in modulating the weight and lipid profile of obese patients. Curr Atheroscler Rep. 2007;9(5):359–66.
58. Hollander PA, Amod A, Litwak LE, Chaudhari U. Effect of rimonabant on glycemic control in insulin-treated type 2 diabetes: the ARPEGGIO trial. Diabetes Care. 2010;33(3):605–7.
59. Doggrell SA. Is rimonabant efficacious and safe in the treatment of obesity? Expert Opin Pharmacother. 2008;9(15):2727–31.
60. Christensen R, Kristensen PK, Bartels EM, Bliddal H, Astrup A. Efficacy and safety of the weight-loss drug rimonabant: a meta-analysis of randomised trials. Lancet. 2007;370(9600):1706–13.
61. Lopez-Moreno JA, Gonzalez-Cuevas G, Navarro M. The CB1 cannabinoid receptor antagonist rimonabant chronically prevents the nicotine-induced relapse to alcohol. Neurobiol Dis. 2007;25(2):274–83.
62. Assareh N, ElBatsh MM, Marsden CA, Kendall DA. The effects of chronic administration of tranylcypromine and rimonabant on behaviour and protein expression in brain regions of the rat. Pharmacol Biochem Behav. 2012;100(3):506–12.
63. Horder J, Browning M, Di Simplicio M, Cowen PJ, Harmer CJ. Effects of 7 days of treatment with the cannabinoid type 1 receptor antagonist, rimonabant, on emotional processing. J Psychopharmacol. 2012;26(1):125–32.
64. Rodgers RJ, Evans PM, Murphy A. Anxiogenic profile of AM-251, a selective cannabinoid CB1 receptor antagonist, in plus-maze-naive and plus-maze-experienced mice. Behav Pharmacol. 2005;16(5–6):405–13.
65. Haller J, Matyas F, Soproni K, Varga B, Barsy B, Nemeth B, Mikics E, Freund TF, Hajos N. Correlated species differences in the effects of cannabinoid ligands on anxiety and on GABAergic and glutamatergic synaptic transmission. Eur J Neurosci. 2007;25(8):2445–6.
66. Sink KS, Segovia KN, Sink J, Randall PA, Collins LE, Correa M, Markus EJ, Vemuri VK, Makriyannis A, Salamone JD. Potential anxiogenic effects of cannabinoid CB1 receptor antagonists/inverse agonists in rats: comparisons between AM4113, AM251, and the benzodiazepine inverse agonist FG-7142. Eur Neuropsychopharmacol. 2010;20(2):112–22.
67. Gobira PH, Aguiar DC, Moreira FA. Effects of compounds that interfere with the endocannabinoid system on behaviors predictive of anxiolytic and panicolytic activities in the elevated T-maze. Pharmacol Biochem Behav. 2013;110:33–9.
68. Cagni P, Barros M. Cannabinoid type 1 receptor ligands WIN 55,212–2 and AM 251 alter anxiety-like behaviors of marmoset monkeys in an open-field test. Behav Brain Res. 2013;240:91–4.
69. Dono LM, Currie PJ. The cannabinoid receptor CB(1) inverse agonist AM251 potentiates the anxiogenic activity of urocortin I in the basolateral amygdala. Neuropharmacology. 2012;62(1):192–99.
70. Aydin C, Oztan O, Isgor C. Nicotine-induced anxiety-like behavior in a rat model of the novelty-seeking phenotype is associated with long-lasting neuropeptidergic and neuroplastic adaptations in the amygdala: effects of the cannabinoid receptor 1 antagonist AM251. Neuropharmacology. 2012;63(8):1335–45.
71. Rutkowska M, Jamontt J, Gliniak H. Effects of cannabinoids on the anxiety-like response in mice. Pharmacol Rep. 2006;58(2):200–06.
72. Jiang W, Zhang Y, Xiao L, Van Cleemput J, Ji SP, Bai G, Zhang X. Cannabinoids promote embryonic and adult hippocampus neurogenesis and produce anxiolytic- and antidepressant-like effects. J Clin Invest. 2005;115(11):3104–16.
73. Black MD, Stevens RJ, Rogacki N, Featherstone RE, Senyah Y, Giardino O, Borowsky B, Stemmelin J, Cohen C, Pichat P, Arad M, Barak S, De Levie A, Weiner I, Griebel G, Varty GB. AVE1625, a cannabinoid CB1 receptor antagonist, as a co-treatment with antipsychotics for schizophrenia: improvement in cognitive function and reduction of antipsychotic-side effects in rodents. Psychopharmacology (Berl). 2011;215(1):149–63.

74. Garcia-Gutierrez MS, Garcia-Bueno B, Zoppi S, Leza JC, Manzanares J. Chronic blockade of cannabinoid CB2 receptors induces anxiolytic-like actions associated with alterations in GABA(A) receptors. Br J Pharmacol. 2012;165(4):951–64.
75. Fokos S, Panagis G. Effects of delta9-tetrahydrocannabinol on reward and anxiety in rats exposed to chronic unpredictable stress. J Psychopharmacol. 2010;24(5):767–77.
76. Braida D, Limonta V, Malabarba L, Zani A, Sala M. 5-HT1A receptors are involved in the anxiolytic effect of Delta9-tetrahydrocannabinol and AM 404, the anandamide transport inhibitor, in Sprague-Dawley rats. Eur J Pharmacol. 2007;555(2–3):156–63.
77. Rubino T, Sala M, Vigano D, Braida D, Castiglioni C, Limonta V, Guidali C, Realini N, Parolaro D. Cellular mechanisms underlying the anxiolytic effect of low doses of peripheral Delta9-tetrahydrocannabinol in rats. Neuropsychopharmacology. 2007;32(9):2036–45.
78. Harte-Hargrove LC, Dow-Edwards DL. Withdrawal from THC during adolescence: sex differences in locomotor activity and anxiety. Behav Brain Res. 2012;231(1):48–59.
79. Berrendero F, Maldonado R. Involvement of the opioid system in the anxiolytic-like effects induced by Delta(9)-tetrahydrocannabinol. Psychopharmacology (Berl). 2002;163(1):111–17.
80. Valjent E, Mitchell JM, Besson MJ, Caboche J, Maldonado R. Behavioural and biochemical evidence for interactions between Delta9-tetrahydrocannabinol and nicotine. Br J Pharmacol. 2002;135(2):564–78.
81. Spano MS, Fattore L, Cadeddu F, Fratta W, Fadda P. Chronic cannabinoid exposure reduces phencyclidine-induced schizophrenia-like positive symptoms in adult rats. Psychopharmacology (Berl). 2013;225(3):531–42.
82. Barba-Escobedo PA, Gould GG. Visual social preferences of lone zebrafish in a novel environment: strain and anxiolytic effects. Genes Brain Behav. 2012;11(3):366–73.
83. Hill MN, Gorzalka BB. Enhancement of anxiety-like responsiveness to the cannabinoid CB(1) receptor agonist HU-210 following chronic stress. Eur J Pharmacol. 2004;499(3):291–95.
84. Ribeiro A, Ferraz-de-Paula V, Pinheiro ML, Palermo-Neto J. Dose-response effects of systemic anandamide administration in mice sequentially submitted to the open field and elevated plus-maze tests. Braz J Med Biol Res. 2009;42(6):556–60.
85. Umathe SN, Manna SS, Utturwar KS, Jain NS. Endocannabinoids mediate anxiolytic-like effect of acetaminophen via CB1 receptors. Prog Neuropsychopharmacol Biol Psychiatry. 2009;33(7):1191–99.
86. Scherma M, Medalie J, Fratta W, Vadivel SK, Makriyannis A, Piomelli D, Mikics E, Haller J, Yasar S, Tanda G, Goldberg SR. The endogenous cannabinoid anandamide has effects on motivation and anxiety that are revealed by fatty acid amide hydrolase (FAAH) inhibition. Neuropharmacology. 2008;54(1):129–40.
87. Naderi N, Haghparast A, Saber-Tehrani A, Rezaii N, Alizadeh AM, Khani A, Motamedi F. Interaction between cannabinoid compounds and diazepam on anxiety-like behaviour of mice. Pharmacol Biochem Behav. 2008;89(1):64–75.
88. Panlilio LV, Solinas M, Matthews SA, Goldberg SR. Previous exposure to THC alters the reinforcing efficacy and anxiety-related effects of cocaine in rats. Neuropsychopharmacology. 2007;32(3):646–57.
89. Rey AA, Purrio M, Viveros MP, Lutz B. Biphasic effects of cannabinoids in anxiety responses: CB1 and GABA(B) receptors in the balance of GABA-ergic and glutamatergic neurotransmission. Neuropsychopharmacology. 2012;37(12):2624–34.
90. Marco EM, Perez-Alvarez L, Borcel E, Rubio M, Guaza C, Ambrosio E, File SE, Viveros MP. Involvement of 5-HT1A receptors in behavioural effects of the cannabinoid receptor agonist CP 55,940 in male rats. Behav Pharmacol. 2004;15(1):21–7.
91. Onaivi ES, Green MR, Martin BR. Pharmacological characterization of cannabinoids in the elevated plus maze. J Pharmacol Exp Ther. 1990;253(3):1002–9.
92. Trezza V, Campolongo P, Cassano T, Macheda T, Dipasquale P, Carratu MR, Gaetani S, Cuomo V. Effects of perinatal exposure to delta-9-tetrahydrocannabinol on the emotional reactivity of the offspring: a longitudinal behavioral study in Wistar rats. Psychopharmacology (Berl). 2008;198(4):529–37.

93. Schramm-Sapyta NL, Cha YM, Chaudhry S, Wilson WA, Swartzwelder HS, Kuhn CM. Differential anxiogenic, aversive, and locomotor effects of THC in adolescent and adult rats. Psychopharmacology (Berl). 2007;191(4):867–77.
94. Page ME, Oropeza VC, Sparks SE, Qian Y, Menko AS, Van Bockstaele EJ. Repeated cannabinoid administration increases indices of noradrenergic activity in rats. Pharmacol Biochem Behav. 2007;86(1):162–68.
95. Lin HC, Mao SC, Chen PS, Gean PW. Chronic cannabinoid administration in vivo compromises extinction of fear memory. Learn Mem. 2008;15(12):876–84.
96. Giuliani D, Ferrari F, Ottani A. The cannabinoid agonist HU 210 modifies rat behavioural responses to novelty and stress. Pharmacol Res. 2000;41(1):47–53.
97. McLaughlin PJ, Lu D, Winston KM, Thakur G, Swezey LA, Makriyannis A, Salamone JD. Behavioral effects of the novel cannabinoid full agonist AM 411. Pharmacol Biochem Behav. 2005;81(1):78–88.
98. Panlilio LV, Mazzola C, Medalie J, Hahn B, Justinova Z, Drago F, Cadet JL, Yasar S, Goldberg SR. Anandamide-induced behavioral disruption through a vanilloid-dependent mechanism in rats. Psychopharmacology (Berl). 2009;203(3):529–38.
99. Macri S, Ceci C, Altabella L, Canese R, Laviola G. The Directive 2010/63/EU on animal experimentation may skew the conclusions of pharmacological and behavioural studies. Sci Rep. 2013;3:2380.
100. Bambico FR, Cassano T, Dominguez-Lopez S, Katz N, Walker CD, Piomelli D, Gobbi G. Genetic deletion of fatty acid amide hydrolase alters emotional behavior and serotonergic transmission in the dorsal raphe, prefrontal cortex, and hippocampus. Neuropsychopharmacology. 2010;35(10):2083–100.
101. Hill MN, Gorzalka BB. Increased sensitivity to restraint stress and novelty-induced emotionality following long-term, high dose cannabinoid exposure. Psychoneuroendocrinology. 2006;31(4):526–36.
102. Marin S, Marco E, Biscaia M, Fernandez B, Rubio M, Guaza C, Schmidhammer H, Viveros MP. Involvement of the kappa-opioid receptor in the anxiogenic-like effect of CP 55,940 in male rats. Pharmacol Biochem Behav. 2003;74(3):649–56.
103. Arnold JC, Dielenberg RA, McGregor IS. Cannabinoids increase conditioned ultrasonic vocalisations and cat odour avoidance in rats: strain differences in drug-induced anxiety. Life Sci. 2010;87(17–18):572–78.
104. Bambico FR, Nguyen NT, Katz N, Gobbi G. Chronic exposure to cannabinoids during adolescence but not during adulthood impairs emotional behaviour and monoaminergic neurotransmission. Neurobiol Dis. 2010;37(3):641–55.
105. Morrison PD, Zois V, McKeown DA, Lee TD, Holt DW, Powell JF, Kapur S, Murray RM. The acute effects of synthetic intravenous Delta9-tetrahydrocannabinol on psychosis, mood and cognitive functioning. Psychol Med. 2009;39(10):1607–16.
106. Martin-Santos R, Crippa JA, Batalla A, Bhattacharyya S, Atakan Z, Borgwardt S, Allen P, Seal M, Langohr K, Farre M, Zuardi AW, McGuire PK. Acute effects of a single, oral dose of d9-tetrahydrocannabinol (THC) and cannabidiol (CBD) administration in healthy volunteers. Curr Pharm Des. 2012;18(32):4966–79.
107. Karschner EL, Darwin WD, McMahon RP, Liu F, Wright S, Goodwin RS, Huestis MA. Subjective and physiological effects after controlled Sativex and oral THC administration. Clin Pharmacol Ther. 2011;89(3):400–7.
108. D'Souza DC, Perry E, MacDougall L, Ammerman Y, Cooper T, Wu YT, Braley G, Gueorguieva R, Krystal JH. The psychotomimetic effects of intravenous delta-9-tetrahydrocannabinol in healthy individuals: implications for psychosis. Neuropsychopharmacology. 2004;29(8):1558–72.
109. Zuardi AW, Shirakawa I, Finkelfarb E, Karniol IG. Action of cannabidiol on the anxiety and other effects produced by delta9-THC in normal subjects. Psychopharmacology (Berl). 1982;76(3):245–50.
110. Fusar-Poli P, Crippa JA, Bhattacharyya S, Borgwardt SJ, Allen P, Martin-Santos R, Seal M, Surguladze SA, O'Carrol C, Atakan Z, Zuardi AW, McGuire PK. Distinct effects of

{delta}9-tetrahydrocannabinol and cannabidiol on neural activation during emotional processing. Arch Gen Psychiatry. 2009;66(1):95–105.

111. Bhattacharyya S, Fusar-Poli P, Borgwardt S, Martin-Santos R, Nosarti C, O'Carroll C, Allen P, Seal ML, Fletcher PC, Crippa JA, Giampietro V, Mechelli A, Atakan Z, McGuire P. Modulation of mediotemporal and ventrostriatal function in humans by Delta9-tetrahydrocannabinol: a neural basis for the effects of Cannabis sativa on learning and psychosis. Arch Gen Psychiatry. 2009;66(4):442–51.
112. Newsom RJ, Kelly SJ. Perinatal delta-9-tetrahydrocannabinol exposure disrupts social and open field behavior in adult male rats. Neurotoxicol Teratol. 2008;30(3):213–219.
113. Schneider M, Drews E, Koch M. Behavioral effects in adult rats of chronic prepubertal treatment with the cannabinoid receptor agonist WIN 55,212–2. Behav Pharmacol. 2005;16(5–6):447–54.
114. Malone DT, Jongejan D, Taylor DA. Cannabidiol reverses the reduction in social interaction produced by low dose Delta(9)-tetrahydrocannabinol in rats. Pharmacol Biochem Behav. 2009;93(2):91–6.
115. Franklin JM, Mathew M, Carrasco GA. Cannabinoid-induced upregulation of serotonin 2A receptors in the hypothalamic paraventricular nucleus and anxiety-like behaviors in rats. Neurosci Lett. 2013;548:165–69.
116. Biscaia M, Marin S, Fernandez B, Marco EM, Rubio M, Guaza C, Ambrosio E, Viveros MP. Chronic treatment with CP 55,940 during the peri-adolescent period differentially affects the behavioural responses of male and female rats in adulthood. Psychopharmacology (Berl). 2003;170(3):301–08.
117. Macri S, Lanuzza L, Merola G, Ceci C, Gentili S, Valli A, Macchia T, Laviola G. Behavioral responses to acute and sub-chronic administration of the synthetic cannabinoid JWH-018 in adult mice prenatally exposed to corticosterone. Neurotox Res. 2013;24(1):15–28.
118. Delatte MS, Paronis CA. Evaluation of cannabinoid agonists using punished responding and midazolam discrimination procedures in squirrel monkeys. Psychopharmacology (Berl). 2008;198(4):521–28.
119. Klumpers F, Denys D, Kenemans JL, Grillon C, van der Aart J, Baas JM. Testing the effects of Delta9-THC and D-cycloserine on extinction of conditioned fear in humans. J Psychopharmacol. 2012;26(4):471–78.
120. Pandolfo P, Pamplona FA, Prediger RD, Takahashi RN. Increased sensitivity of adolescent spontaneously hypertensive rats, an animal model of attention deficit hyperactivity disorder, to the locomotor stimulation induced by the cannabinoid receptor agonist WIN 55,212–2. Eur J Pharmacol. 2007;563(1–3):141–48.
121. Llorente-Berzal A, Fuentes S, Gagliano H, Lopez-Gallardo M, Armario A, Viveros MP, Nadal R. Sex-dependent effects of maternal deprivation and adolescent cannabinoid treatment on adult rat behaviour. Addiction biology. 2011;16(4):624–37.
122. Higuera-Matas A, Botreau F, Miguens M, Del Olmo N, Borcel E, Perez-Alvarez L, Garcia-Lecumberri C, Ambrosio E. Chronic periadolescent cannabinoid treatment enhances adult hippocampal PSA-NCAM expression in male Wistar rats but only has marginal effects on anxiety, learning and memory. Pharmacol Biochem Behav. 2009;93(4):482–90.
123. Gonzalez-Cuevas G, Aujla H, Martin-Fardon R, Lopez-Moreno JA, Navarro M, Weiss F. Subchronic cannabinoid agonist (WIN 55,212–2) treatment during cocaine abstinence alters subsequent cocaine seeking behavior. Neuropsychopharmacology. 2007;32(11):2260–66.
124. Abush H, Akirav I. Cannabinoids ameliorate impairments induced by chronic stress to synaptic plasticity and short-term memory. Neuropsychopharmacology. 2013;38(8):1521–34.
125. Alen F, Santos A, Moreno-Sanz G, Gonzalez-Cuevas G, Gine E, Franco-Ruiz L, Navarro M, Lopez-Moreno JA. Cannabinoid-induced increase in relapse-like drinking is prevented by the blockade of the glycine-binding site of N-methyl-D-aspartate receptors. Neuroscience. 2009;158(2):465–73.

126. Kathuria S, Gaetani S, Fegley D, Valino F, Duranti A, Tontini A, Mor M, Tarzia G, La Rana G, Calignano A, Giustino A, Tattoli M, Palmery M, Cuomo V, Piomelli D. Modulation of anxiety through blockade of anandamide hydrolysis. Nat Med. 2003;9(1):76–81.
127. Gobbi G, Bambico FR, Mangieri R, Bortolato M, Campolongo P, Solinas M, Cassano T, Morgese MG, Debonnel G, Duranti A, Tontini A, Tarzia G, Mor M, Trezza V, Goldberg SR, Cuomo V, Piomelli D. Antidepressant-like activity and modulation of brain monoaminergic transmission by blockade of anandamide hydrolysis. Proc Natl Acad Sci U S A. 2005;102(51):18620–5.
128. Piomelli D, Tarzia G, Duranti A, Tontini A, Mor M, Compton TR, Dasse O, Monaghan EP, Parrott JA, Putman D. Pharmacological profile of the selective FAAH inhibitor KDS-4103 (URB597). CNS Drug Rev. 2006;12(1):21–38.
129. Long JZ, Li W, Booker L, Burston JJ, Kinsey SG, Schlosburg JE, Pavon FJ, Serrano AM, Selley DE, Parsons LH, Lichtman AH, Cravatt BF. Selective blockade of 2-arachidonoylglycerol hydrolysis produces cannabinoid behavioral effects. Nat Chem Biol. 2009;5(1):37–44.
130. Hill MN, Kumar SA, Filipski SB, Iverson M, Stuhr KL, Keith JM, Cravatt BF, Hillard CJ, Chattarji S, McEwen BS. Disruption of fatty acid amide hydrolase activity prevents the effects of chronic stress on anxiety and amygdalar microstructure. Mol Psychiatr. 2013;18(10):1125–35.
131. Hermanson DJ, Hartley ND, Gamble-George J, Brown N, Shonesy BC, Kingsley PJ, Colbran RJ, Reese J, Marnett LJ, Patel S. Substrate-selective COX-2 inhibition decreases anxiety via endocannabinoid activation. Nat Neurosci. 2013;16(9):1291–98.
132. Kinsey SG, O'Neal ST, Long JZ, Cravatt BF, Lichtman AH. Inhibition of endocannabinoid catabolic enzymes elicits anxiolytic-like effects in the marble burying assay. Pharmacol Biochem Behav. 2011;98(1):21–7.
133. Rossi S, De Chiara V, Musella A, Sacchetti L, Cantarella C, Castelli M, Cavasinni F, Motta C, Studer V, Bernardi G, Cravatt BF, Maccarrone M, Usiello A, Centonze D. Preservation of striatal cannabinoid CB1 receptor function correlates with the antianxiety effects of fatty acid amide hydrolase inhibition. Mol Pharmacol. 2010;78(2):260–68.
134. Busquets-Garcia A, Puighermanal E, Pastor A, de la Torre R, Maldonado R, Ozaita A. Differential role of anandamide and 2-arachidonoylglycerol in memory and anxiety-like responses. Biol Psychiatry. 2011;70(5):479–86.
135. Moreira FA, Kaiser N, Monory K, Lutz B. Reduced anxiety-like behaviour induced by genetic and pharmacological inhibition of the endocannabinoid-degrading enzyme fatty acid amide hydrolase (FAAH) is mediated by CB1 receptors. Neuropharmacology. 2008;54(1):141–50.
136. Cippitelli A, Astarita G, Duranti A, Caprioli G, Ubaldi M, Stopponi S, Kallupi M, Sagratini G, Rodriguez de Fonseca F, Piomelli D, Ciccocioppo R. Endocannabinoid regulation of acute and protracted nicotine withdrawal: effect of FAAH inhibition. PloS ONE. 2011;6(11):e28142.
137. Cippitelli A, Cannella N, Braconi S, Duranti A, Tontini A, Bilbao A, Defonseca FR, Piomelli D, Ciccocioppo R. Increase of brain endocannabinoid anandamide levels by FAAH inhibition and alcohol abuse behaviours in the rat. Psychopharmacology (Berl). 2008;198(4):449–60.
138. Haller J, Barna I, Barsvari B, Gyimesi Pelczer K, Yasar S, Panlilio LV, Goldberg S. Interactions between environmental aversiveness and the anxiolytic effects of enhanced cannabinoid signaling by FAAH inhibition in rats. Psychopharmacology. 2009;204(4):607–16.
139. Naidu PS, Varvel SA, Ahn K, Cravatt BF, Martin BR, Lichtman AH. Evaluation of fatty acid amide hydrolase inhibition in murine models of emotionality. Psychopharmacology. 2007;192(1):61–70.
140. Zaitone SA, El-Wakeil AF, Abou-El-Ela SH. Inhibition of fatty acid amide hydrolase by URB597 attenuates the anxiolytic-like effect of acetaminophen in the mouse elevated plus-maze test. Behav Pharmacol. 2012;23(4):417–25.

141. Almeida-Santos AF, Gobira PH, Rosa LC, Guimaraes FS, Moreira FA, Aguiar DC. Modulation of anxiety-like behavior by the endocannabinoid 2-arachidonoylglycerol (2-AG) in the dorsolateral periaqueductal gray. Behav Brain Res. 2013;252:10–7.
142. Sumislawski JJ, Ramikie TS, Patel S. Reversible gating of endocannabinoid plasticity in the amygdala by chronic stress: a potential role for monoacylglycerol lipase inhibition in the prevention of stress-induced behavioral adaptation. Neuropsychopharmacology. 2011;36(13):2750–61.
143. Sciolino NR, Zhou W, Hohmann AG. Enhancement of endocannabinoid signaling with JZL184, an inhibitor of the 2-arachidonoylglycerol hydrolyzing enzyme monoacylglycerol lipase, produces anxiolytic effects under conditions of high environmental aversiveness in rats. Pharmacol Res. 2011;64(3):226–34.
144. Aliczki M, Balogh Z, Tulogdi A, Haller J. The temporal dynamics of the effects of monoacylglycerol lipase blockade on locomotion, anxiety, and body temperature. Behav Pharmacol. 2012;23(4):348–57.
145. Bortolato M, Campolongo P, Mangieri RA, Scattoni ML, Frau R, Trezza V, La Rana G, Russo R, Calignano A, Gessa GL, Cuomo V, Piomelli D. Anxiolytic-like properties of the anandamide transport inhibitor AM404. Neuropsychopharmacology. 2006;31(12):2652–9.
146. Bitencourt RM, Pamplona FA, Takahashi RN. Facilitation of contextual fear memory extinction and anti-anxiogenic effects of AM404 and cannabidiol in conditioned rats. Eur Neuropsychopharmacol. 2008;18(12):849–59.
147. Pamplona FA, Bitencourt RM, Takahashi RN. Short- and long-term effects of cannabinoids on the extinction of contextual fear memory in rats. Neurobiol Learn Mem. 2008;90(1):290–93.
148. Powers MS, Barrenha GD, Mlinac NS, Barker EL, Chester JA. Effects of the novel endocannabinoid uptake inhibitor, LY2183240, on fear-potentiated startle and alcohol-seeking behaviors in mice selectively bred for high alcohol preference. Psychopharmacology (Berl). 2010;212(4):571–83.
149. Scherma M, Justinova Z, Zanettini C, Panlilio LV, Mascia P, Fadda P, Fratta W, Makriyannis A, Vadivel SK, Gamaleddin I, Le Foll B, Goldberg SR. The anandamide transport inhibitor AM404 reduces the rewarding effects of nicotine and nicotine-induced dopamine elevations in the nucleus accumbens shell in rats. Br J Pharmacol. 2012;165(8):2539–48.
150. Terzian AL, Drago F, Wotjak CT, Micale V. The dopamine and cannabinoid interaction in the modulation of emotions and cognition: assessing the role of cannabinoid cb1 receptor in neurons expressing dopamine d1 receptors. Front Behav Neurosci. 2011;5:49.
151. Kupferschmidt DA, Newman AE, Boonstra R, Erb S. Antagonism of cannabinoid 1 receptors reverses the anxiety-like behavior induced by central injections of corticotropin-releasing factor and cocaine withdrawal. Neuroscience. 2012;204:125–33.
152. Ruehle S, Remmers F, Romo-Parra H, Massa F, Wickert M, Wortge S, Haring M, Kaiser N, Marsicano G, Pape HC, Lutz B. Cannabinoid CB1 receptor in dorsal telencephalic glutamatergic neurons: distinctive sufficiency for hippocampus-dependent and amygdala-dependent synaptic and behavioral functions. J Neurosci. 2013;33(25):10264–277.
153. Rubino T, Guidali C, Vigano D, Realini N, Valenti M, Massi P, Parolaro D. CB1 receptor stimulation in specific brain areas differently modulate anxiety-related behaviour. Neuropharmacology. 2008;54(1):151–60.
154. Rubino T, Realini N, Castiglioni C, Guidali C, Vigano D, Marras E, Petrosino S, Perletti G, Maccarrone M, Di Marzo V, Parolaro D. Role in anxiety behavior of the endocannabinoid system in the prefrontal cortex. Cereb Cortex. 2008;18(6):1292–301.
155. Klugmann M, Goepfrich A, Friemel CM, Schneider M. AAV-Mediated Overexpression of the CB1 Receptor in the mPFC of Adult Rats Alters Cognitive Flexibility, Social Behavior, and Emotional Reactivity. Front Behav Neurosci. 2011;5:37.
156. Zarrindast MR, Mahboobi S, Sadat-Shirazi MS, Ahmadi S. Anxiolytic-like effect induced by the cannabinoid CB1 receptor agonist, arachydonilcyclopropylamide (ACPA), in the rat amygdala is mediated through the D1 and D2 dopaminergic systems. J Psychopharmacol. 2011;25(1):131–40.

157. Ganon-Elazar E, Akirav I. Cannabinoid receptor activation in the basolateral amygdala blocks the effects of stress on the conditioning and extinction of inhibitory avoidance. J Neurosci. 2009;29(36):11078–88.
158. John CS, Currie PJ. N-arachidonoyl-serotonin in the basolateral amygdala increases anxiolytic behavior in the elevated plus maze. Behav Brain Res. 2012;233(2):382–88.
159. Munguba H, Cabral A, Leao AH, Barbosa FF, Izidio GS, Ribeiro AM, Silva RH. Pre-training anandamide infusion within the basolateral amygdala impairs plus-maze discriminative avoidance task in rats. Neurobiol Learn Mem. 2011;95(4):527–33.
160. Mendez-Diaz M, Caynas-Rojas S, Arteaga Santacruz V, Ruiz-Contreras AE, Aguilar-Roblero R, Prospero-Garcia O. Entopeduncular nucleus endocannabinoid system modulates sleep-waking cycle and mood in rats. Pharmacol Biochem Behav. 2013;107:29–35.
161. Kamprath K, Romo-Parra H, Haring M, Gaburro S, Doengi M, Lutz B, Pape HC. Short-term adaptation of conditioned fear responses through endocannabinoid signaling in the central amygdala. Neuropsychopharmacology. 2011;36(3):652–63.
162. Dubreucq S, Kambire S, Conforzi M, Metna-Laurent M, Cannich A, Soria-Gomez E, Richard E, Marsicano G, Chaouloff F. Cannabinoid type 1 receptors located on single-minded 1-expressing neurons control emotional behaviors. Neuroscience. 2012;204:230–44.
163. Roohbakhsh A, Moghaddam AH, Massoudi R, Zarrindast MR. Role of dorsal hippocampal cannabinoid receptors and nitric oxide in anxiety like behaviours in rats using the elevated plus-maze test. Clin Exp Pharmacol Physiol. 2007;34(3):223–29.
164. Clarke JR, Rossato JI, Monteiro S, Bevilaqua LR, Izquierdo I, Cammarota M. Posttraining activation of CB1 cannabinoid receptors in the CA1 region of the dorsal hippocampus impairs object recognition long-term memory. Neurobiol Learn Mem. 2008;90(2):374–81.
165. Hakimizadeh E, Oryan S, Hajizadeh Moghaddam A, Shamsizadeh A, Roohbakhsh A. Endocannabinoid System and TRPV1 Receptors in the Dorsal Hippocampus of the Rats Modulate Anxiety-like Behaviors. Iran J Basic Med Sci. 2012;15(3):795–802.
166. Zarrindast MR, Nasehi M, Piri M, Bina P. Anxiety-like behavior induced by histaminergic agents can be prevented by cannabinoidergic WIN55,212–2 injected into the dorsal hippocampus in mice. Pharmacol Biochem Behav. 2010;94(3):387–96.
167. Roohbakhsh A, Keshavarz S, Hasanein P, Rezvani ME, Moghaddam AH. Role of endocannabinoid system in the ventral hippocampus of rats in the modulation of anxiety-like behaviours. Basic Clin Pharmacol Toxicol. 2009;105(5):333–38.
168. Campos AC, Ferreira FR, Guimaraes FS, Lemos JI. Facilitation of endocannabinoid effects in the ventral hippocampus modulates anxiety-like behaviors depending on previous stress experience. Neuroscience. 2010;167(2):238–46.
169. Resstel LB, Lisboa SF, Aguiar DC, Correa FM, Guimaraes FS. Activation of CB1 cannabinoid receptors in the dorsolateral periaqueductal gray reduces the expression of contextual fear conditioning in rats. Psychopharmacology. 2008;198(3):405–11.
170. Casarotto PC, Terzian AL, Aguiar DC, Zangrossi H, Guimaraes FS, Wotjak CT, Moreira FA. Opposing roles for cannabinoid receptor type-1 (CB(1)) and transient receptor potential vanilloid type-1 channel (TRPV1) on the modulation of panic-like responses in rats. Neuropsychopharmacology. 2012;37(2):478–86.
171. Lisboa SF, Resstel LB, Aguiar DC, Guimaraes FS. Activation of cannabinoid CB1 receptors in the dorsolateral periaqueductal gray induces anxiolytic effects in rats submitted to the Vogel conflict test. Eur J Pharmacol. 2008;593(1–3):73–8.
172. Moreira FA, Aguiar DC, Guimaraes FS. Anxiolytic-like effect of cannabinoids injected into the rat dorsolateral periaqueductal gray. Neuropharmacology. 2007;52(3):958–65.
173. De Boer SF Koolhaas JM. Defensive burying in rodents: ethology, neurobiology and psychopharmacology. Eur J Pharmacol. 2003;463(1–3):145–61.
174. Wilson RI, Nicoll RA. Endogenous cannabinoids mediate retrograde signalling at hippocampal synapses. Nature. 2001;410(6828):588–92.
175. Di Marzo V De Petrocellis L Bisogno T Melck D. Metabolism of anandamide and 2-arachidonoylglycerol: an historical overview and some recent developments. Lipids. 1999;34 Suppl:S319–25.

176. Marsicano G, Goodenough S, Monory K, Hermann H, Eder M, Cannich A, Azad SC, Cascio MG, Gutierrez SO, van der Stelt M, Lopez-Rodriguez ML, Casanova E, Schutz G, Zieglgansberger W, Di Marzo V, Behl C, Lutz B. CB1 cannabinoid receptors and on-demand defense against excitotoxicity. Science. 2003;302(5642):84–8.
177. Adermark L, Lovinger DM. Retrograde endocannabinoid signaling at striatal synapses requires a regulated postsynaptic release step. Proc Natl Acad Sci U S A. 2007;104(51):20564–69.
178. Lutz B. On-demand activation of the endocannabinoid system in the control of neuronal excitability and epileptiform seizures. Biochem Pharmacol. 2004;68(9):1691–98.
179. Zanettini C, Panlilio LV, Alicki M, Goldberg SR, Haller J, Yasar S. Effects of endocannabinoid system modulation on cognitive and emotional behavior. Front Behav Neurosci. 2011;5:57.
180. Onaivi ES, Ishiguro H, Gong JP, Patel S, Perchuk A, Meozzi PA, Myers L, Mora Z, Tagliaferro P, Gardner E, Brusco A, Akinshola BE, Liu QR, Hope B, Iwasaki S, Arinami T, Teasenfitz L, Uhl GR. Discovery of the presence and functional expression of cannabinoid CB2 receptors in brain. Ann NY Acad Sci. 2006;1074:514–36.
181. Begg M, Pacher P, Batkai S, Osei-Hyiaman D, Offertaler L, Mo FM, Liu J, Kunos G. Evidence for novel cannabinoid receptors. Pharmacol Ther. 2005;106(2):133–45.
182. Barna I, Till I, Haller J. Blood, adipose tissue and brain levels of the cannabinoid ligands WIN-55,212 and SR-141716A after their intraperitoneal injection in mice: compound-specific and area-specific distribution within the brain. Eur Neuropsychopharmacol. 2009;19(8):533–41.
183. McLaughlin RJ, Hill MN, Bambico FR, Stuhr KL, Gobbi G, Hillard CJ, Gorzalka BB. Prefrontal cortical anandamide signaling coordinates coping responses to stress through a serotonergic pathway. Eur Neuropsychopharmacol. 2012;22(9):664–71.
184. Haller J, Aliczki M, Pelczer KG, Spitzer K, Balogh Z, Kantor S. Effects of the fatty acid amide hydrolase inhibitor URB597 on coping behavior under challenging conditions in mice. Psychopharmacology (Berl). 2014;231(3):593–601.
185. Haller J, Goldberg SR, Pelczer KG, Aliczki M, Panlilio LV. The effects of anandamide signaling enhanced by the FAAH inhibitor URB597 on coping styles in rats. Psychopharmacology (Berl). 2013;230(3):353–62.
186. Koolhaas JM, Korte SM, De Boer SF, Van Der Vegt BJ, Van Reenen CG, Hopster H, De Jong IC, Ruis MA, Blokhuis HJ. Coping styles in animals: current status in behavior and stress-physiology. Neurosci Biobehav Rev. 1999;23(7):925–35.
187. Koolhaas JM. Coping style and immunity in animals: making sense of individual variation. Brain Behav Immun. 2008;22(5):662–67.
188. Kessler RC, Price RH, Wortman CB. Social factors in psychopathology: stress, social support, and coping processes. Annu Rev Psychol. 1985;36:531–72.
189. Temoshok LR. Complex coping patterns and their role in adaptation and neuroimmunomodulation. Theory, methodology, and research. Ann NY Acad Sci. 2000;917:446–55.
190. Westerhuis W, Zijlmans M, Fischer K, van Andel J, Leijten FS. Coping style and quality of life in patients with epilepsy: a cross-sectional study. J Neurol. 2011;258(1):37–43.
191. Pucheu S, Consoli SM, D'Auzac C, Francais P, Issad B. Do health causal attributions and coping strategies act as moderators of quality of life in peritoneal dialysis patients? J Psychosom Res. 2004;56(3):317–22.
192. Cooke M, Peters E, Fannon D, Anilkumar AP, Aasen I, Kuipers E, Kumari V. Insight, distress and coping styles in schizophrenia. Schizophr Res. 2007;94(1–3):12–22.
193. Tiemensma J, Kaptein AA, Pereira AM, Smit JW, Romijn JA, Biermasz NR. Coping strategies in patients after treatment for functioning or nonfunctioning pituitary adenomas. J Clin Endocrinol Metab. 2011;96(4):964–71.

Chapter 5
Role of the Endocannabinoid System in Depression: from Preclinical to Clinical Evidence

Vincenzo Micale, Katarina Tabiova, Jana Kucerova and Filippo Drago

Abstract The endogenous cannabinoid system (ECS) works as pro-homeostatic and pleiotropic signaling system activated in a time- and tissue-specific way during physiological conditions, which include cognitive, emotional and motivational processes. It is composed of two G protein-coupled receptors (the cannabinoid receptors types 1 and 2 [CB1 and CB2] for marijuana's psychoactive ingredient Δ9-tetrahydrocannabinol [Δ9-THC]), their endogenous small lipid ligands (anandamide [AEA] and 2-arachidonoylglycerol [2-AG], also known as endocannabinoids), and the proteins for endocannabinoid biosynthesis and deactivation. Data from preclinical and clinical studies have reported that a hypofunction of the endocannabinoid signaling could induce a depressive-like phenotype; consequently, enhancement of endocannabinoid signaling could be a novel therapeutic avenue for the treatment of depression. To this aim there have been proposed cannabinoid receptor agonists or synthetic molecules that inhibit endocannabinoid degradation. The latter ones do not induce the psychotropic side effects by direct CB1 receptor activation, but rather elicit antidepressant-like effects by enhancing the monoaminergic neurotransmission, promoting hippocampal neurogenesis and normalizing the hyperactivity of hypothalamic-pituitary-adrenal axis, similarly as the standard antidepressants. The dysfunction of elements belonging to the ECS and the possible therapeutic use of endocannabinoid deactivation inhibitors and phytocannabinoids in depression is discussed in this chapter.

Keywords Endocannabinoid system · CB1 and CB2 receptors · TRPV1 channels · Animal models · Depression · Antidepressants · Δ9-THC · Cannabidiol

---

V. Micale (✉) · K. Tabiova · J. Kucerova
Department of Pharmacology, CEITEC (Central European Institute of Technology)
Masaryk University, Brno, Czech Republic
e-mail: vincenzomicale@inwind.it

F. Drago
Department of Clinical and Molecular Biomedicine, Section of Pharmacology and Biochemistry, Medical School, University of Catania, Catania, Italy

P. Campolongo, L. Fattore (eds.), *Cannabinoid Modulation of Emotion, Memory, and Motivation,* DOI 10.1007/978-1-4939-2294-9_5

Introduction

Current Pharmacological Approach for the Treatment of Depression

Depression is one of the most common mental illness with a lifetime prevalence of about 15–20 %, resulting in enormous personal suffering, as well as social and economic burden [1]. The major depressive disorder is characterized by episodes of depressed mood lasting for more than 2 weeks often associated with feelings of guilt, decreased interest in pleasurable activities and inability to experience pleasure (named anhedonia), low self-esteem and worthlessness, high anxiety, disturbed sleep patterns and appetite, impairment in memory and suicidal ideation [2].

The treatment of depression was revolutionized in the 1950s with the introduction of two classes of pharmacological agents to the clinical practice: the monoamine oxidase inhibitors "MAOIs" and the tricyclic antidepressants "TCAs". The discovery was based on the serendipitous finding that enhancement of the synaptic levels of monoamines improves the symptoms of depression, leading to the *monoamine hypothesis of depression* [3]. Thus, the introduction of antidepressant drugs had a profound impact on the way depression was viewed: if chemicals can reverse most depressive symptomatologies, then depression itself may be caused by chemical abnormalities in the brain. However first generation antidepressants, due to their toxic and poorly tolerated profile, were largely replaced by the selective serotonin reuptake inhibitors (SSRIs), norepinephrine reuptake inhibitors and serotonin norepinephrine reuptake inhibitors and by atypical antidepressants (i.e. nefazodone and mirtazapine), which are not more effective than MAOIs or TCAs but show an improved safety profile [4].

Recently, some atypical antipsychotics such as olanzapine, quetiapine or aripiprazole, used either as monotherapy or in combination with venlafaxine or sertraline, have also shown efficacy at ameliorating symptoms of bipolar disorder and treatment-resistant major depression and received approval from the FDA (US Food and Drug Administration) for these indications [5]. Since disruptions of circadian and sleep-wake cycles have been recognized as major contributor to mood disturbance, and agomelatine (a melatonergic agonist and a serotonin 5-HT_{2C} receptor antagonist) was found to be very effective in ameliorating depressive symptoms with a good tolerability and safety profile, a new concept for the treatment of mood disorders has recently emerged [6].

However, the past decade has witnessed a driven focus on the rational discovery of highly selective drugs, acting at novel non monoamine based targets such as GABAergic and glutamatergic neurotransmission, neuroendocrine system or neuropeptide signaling, which in turn could affect intracellular signal transduction pathways. Yet, except for the N-methyl-D-aspartate (NMDA) receptor antagonist ketamine [7], none of these drugs has reached the market [8–11]. Thus, the dominant hypothesis of depression is still based on the monoamine model, which comprises the primary target for current antidepressants. Although today's treatments are generally safe and effective, 30 % of depressed patients treated with the conventional antidepressants are pharmacoresistant. In addition, the medication has to be administered for weeks or months to see appreciable clinical benefit [12]. Therefore, there is still a great need to update the current level of knowledge with regard to the pathophysiological mechanisms underlying depressive disorders in order to develop safer, more effective, and faster acting pharmacotherapies. The partial efficacy of current drugs raises the central question to be addressed in this chapter: Does the alteration of the endocannabinoid system (ECS) have a crucial role in the pathophysiology of depressive disorders and is the ECS consequently able to provide a promising therapeutic approach for their treatment?

The Endocannabinoid System (ECS)

The ECS is a neuromodulatory system, which plays a role in a variety of physiological processes both in the central nervous system (CNS) and in the periphery, mediating the effects of the psychoactive constituent of Cannabis Δ9-tetrahydrocannabinol (Δ9-THC) [13]. Multiple lines of evidence have shown that its dysregulation is associated with several pathological conditions such as pain and inflammation [14, 15], obesity, metabolic [16, 17], gastrointestinal [18], hepatic [19], neurodegenerative [20–22] and psychiatric disorders [23–25]. However, the exact pathophysiological mechanisms through which the ECS controls these functions are not fully elucidated yet. The ECS is comprised of: (1) the cannabinoid receptors type CB1 and CB2 [26–28], (2) their endogenous ligands anandamide (N-arachidonoyl-ethanolamine, AEA) and 2-arachidonylglycerol (2-AG) [29, 30], (3) a specific and not yet identified cellular uptake mechanism [31, 32], and (4) the enzymes for endocannabinoid biosynthesis, N-acyl-phosphatidylethanolamine-selective phosphodiesterase or glycerophosphodiesterase E1 and diacylglycerol lipase α or β [33, 34], or their inactivation, fatty acid amide hydrolase (FAAH) and monoacylglycerol lipase (MAGL) [35, 36], respectively for AEA and 2-AG. However, additional "players" which are described as potential members of the ECS include the TRPV1 channels, the putative CB1 receptor antagonist peptides like hemopressins, peroxisome proliferator-activated receptor-α (PPAR-α) and γ (PPAR-γ) ligands, such as oleoylethanolamide (OEA) or palmitoylethanolamide (PEA), and N-arachidonoyl-dopamine (NADA), which activates both TRPV1 and CB1 receptors. Although the existence of a third cannabinoid receptor subtype has also been suggested [37], to date only CB1 and CB2 receptors are recognized as G protein-coupled receptors for endocannabinoids [38].

The cannabinoid CB1 and CB2 receptors are established as mediators of the biological effects induced by cannabinoids, either plant derived, synthetic, or endogenously produced. These receptors are encoded by two different genes on human chromosomes: 6q14-q15 (CNR1) and 1p36.11 (CNR2). They are 7 transmembrane Gi/o coupled receptors that share 44 % protein identity and display different pharmacological profiles and patterns of expression, a dichotomy that provides a unique opportunity to develop pharmaceutical approaches.

The CB1 receptors are ubiquitously expressed in the CNS where they are predominantly found at high densities in the basal ganglia, frontal cortex, hippocampus and cerebellum. They are present at a moderate/low densities in the periaqueductal gray, amygdala, nucleus accumbens, thalamus and medulla. However, the CB1 receptors are also found in non-neuronal cells of the brain such as microglia, oligodendrocytes and astrocytes [39]. Within these cortical areas there are two major neuronal subpopulations expressing the CB1 receptors: the GABAergic interneurons (with high CB1 receptor levels) and glutamatergic neurons (with relatively low CB1 receptor levels) [40], which represent the two major opposing players regulating the excitation state of the brain, GABAergic interneurons being inhibitory and glutamatergic neurons being excitatory. CB1 receptors are also located in neurons of the dorsal raphe nucleus (DRN) and in the *locus coeruleus* (LC) which are the major sources of serotonin (5-HT) and noradrenalin (NE) in the brain [41, 42]. Thus, the direct or indirect modulation of monoamine activity or of GABA and glutamate neurons, respectively, could underlie the psychotropic and non-psychotropic effects of CB1 receptor activation.

The cannabinoid CB2 receptors, which are also activated by AEA and 2-AG, are mainly distributed in immune tissues and inflammatory cells, although they are also detected in glial cells, and to a much lesser extent, in neurons of several brain regions such as cerebral cortex, hippocampus, amygdala, hypothalamus and cerebellum [43, 44]. While their role in pain and inflammation has been extensively reported, recently their involvement in emotional processes has been suggested [45]. The observation that the elements belonging to the ECS are prevalent throughout the neuroanatomical structures and circuits implicated in emotionality, including prefrontal cortex (PFC), hippocampus, amygdala, hypothalamus and forebrain monoaminergic circuits, provides a rationale for the preclinical development of agents targeting this system to treat affective diseases.

Cannabis, Endocannabinoid System and Depression: Clinical and Preclinical Evidence

Cannabis sativa is the most commonly used illicit "recreational" drug worldwide, its popularity being due to its capacity to increase sociability, to induce euphoria and to alter sensory perception. Although the association between *Cannabis sativa* and psychopathologic conditions has been known for thousands of years, only in the last 50 years the identification of the chemical structure of marijuana components, the cloning of specific cannabinoid receptors and the discovery of the ECS in the brain have triggered an exponential growth of studies to explore its real effects on mental health [46].

The Cannabis plant contains over 100 terpenophenolic pharmacologically active compounds, known as cannabinoids. Of these, Δ9-THC, characterized in 1964 by Mechoulam's team [47], was identified as the primary psychoactive component of Cannabis, and later shown to act as a direct agonist of CB1 and CB2 receptors. Other cannabinoids include cannabichromene, cannabigerol and cannabidiol (CBD), which do not seem to induce the psychotropic side effects of Δ9-THC. They act on several levels in the CNS, including modulation of endocannabinoid tone [48–50], interaction with transient receptor potential vanilloid 1 (TRPV1) channels [48] and serotonin 5-HT_{1A} receptors [51], and enhancement of adenosine signaling [52, 53]. The above mentioned mechanisms could underlie the positive effects induced by CBD treatment in preclinical studies of several psychiatric as well as other disorders [54, 55].

Although elevation of mood is one of the commonly cited motivations for the use of Cannabis, in addition to its recreational actions, data from clinical trials in the 1970's failed to show any antidepressant effects of Δ9-THC [56, 57]. Additionally, the hypothesis that depressed individuals use Cannabis as a mean of self-medication proposed by preclinical studies [58] has not been fully supported by clinical data yet [59, 60]. By contrast, some data support the hypothesis that Cannabis use precipitates depression [61–65], where genetic and environmental factors could play a pivotal role [66–68]. However, a recent study has shown that depressive symptoms are indirectly related to Cannabis use through positive, but not negative, expectancies [69]. It is not to be excluded that other factors such as the dose, route of administration, baseline emotional states, personality, environment and the setting, during which the drug is used, could be involved in Δ9-THC effects on mood.

Despite preclinical data supporting an altered endocannabinoid signaling as a molecular underpinning of several psychiatric disorders [70], to date only few direct investigations have assessed endocannabinoid activity in depressed patients, as reviewed in Table 5.1. A significant increase of CB1 receptor density has been found in the dorsolateral prefrontal cortex (dlPFC) of depressed suicide victims, possibly suggesting a hyperfunctionality of the ECS in this population [71]. By contrast, a down-regulation of the ECS activity was suggested by Koethe et al. [72] and Hill et al. [73, 74], showing a decreased CB1 receptor density in grey matter glial cells and lower serum concentration of 2-AG in patients with major depression. However, an increase of endocannabinoid tissue content in the dlPFC of alcoholic depressed patients as well as a significantly enhanced serum level of AEA in patients suffering of minor depression were also reported [73, 75]. Furthermore, in two recent clinical studies, a positive correlation was found among high blood pressure and serum contents of endocannabinoids in depressed females [76] and among intense physical exercise, AEA and brain-derived neurotrophic factor (BDNF) levels [77], suggesting that an interrelationship among endocannabinoids, depression and cardiovascular risk factors in women and an increase in peripheral BDNF levels could be a mechanism by which AEA intervenes in the neuroplastic and antidepressant effects of exercise.

Thus, considering the recent preclinical evidence relating the effects of enhanced endocannabinoid signaling to the promotion of neurogenesis, it is not to exclude that its activation exerts antidepressant properties through mechanisms that resemble the ones triggered by conventional antidepressants on synaptic plasticity [78, 79]. However, the increasing interest concerning ECS dysfunction in depressive disorders was engendered after the clinical use of the CB1 receptor antagonist

Table 5.1 Schematic representation of the changes of the endocannabinoid system (ECS) elements in clinical studies of depression

ECS elements	Sex (number of cases)	Diagnosis	Tissue sample[a]	Molecular readout	References
CB1	♂♀ (n=10)	Major depression	dlPFC	↑ density	[71]
	♂♀ (n=11)	Alcohol dependence	dlPFC/occipital cortex	↑ density (dlPFC)	[75]
	♂♀ (n=15)	Major depression	Anterior-cingulate cortex	↓ density	[72]
AEA	♂♀ (n=11)	Alcohol dependence	dlPFC	↑ level	[75]
	♀ (n=16)	Major depression	Serum	No effect	[73]
	♀ (n=12)	Minor depression	Serum	↑ level	[73]
	♀ (n=15)	Major depression	Serum	↓ level	[74]
	♀ (n=28)	Major/Minor depression	Serum	↑ level	[76]
2-AG	♂♀ (n=11)	Alcohol dependence	dlPFC	↑ level	[75]
	♀ (n=16)	Major depression	Serum	↓ level	[73]
	♀ (n=12)	Minor depression	Serum	No effect	[73]
	♀ (n=15)	Major depression	Serum	↓ level	[74]
	♀ (n=28)	Major/Minor depression	Serum	↑ level	[76]
Palmitoylethanolamide (PEA)	♀ (n=15)	Major depression	Serum	No effect	[74]
Oleoylethanolamide (OEA)	♀ (n=15)	Major depression	Serum	No effect	[74]

[a] *dlPFC* dorsolateral prefrontal cortex

rimonabant for the treatment of obesity was interrupted. In line with the theory that a deficiency in CB1 receptor signaling could be involved in depression, rimonabant was withdrawn from the market because of undesirable psychiatric side effects such as anxiety, depression and suicidal ideations [80]. Although no controlled clinical trials concerning endocannabinoid signaling in depression are available, opposite changes in endocannabinoid activity could underlie the different forms of depressive illness.

As recently suggested, genetic variations in CB1 receptor function could also facilitate the development of mood disorders in humans [81]. The human CB1 receptor gene (CNR1), which is located on the chromosome 6q14–15, seems to play a role in a broad spectrum of psychiatric disorders such as substance abuse disorders, schizophrenia and autism spectrum conditions [82–84]. With regard to depression, while Barrero et al. [85] showed a significant association between polymorphisms in CNR1 and depression only in Parkinson's disease patients, recent studies support that genetic variations in CB1 receptor function and in FAAH could influence both

Table 5.2 Schematic representation of the antidepressants effects on the endocannabinoid system (ECS) elements

Drug class	Effective medication	Brain region[a]	Molecular readout	References
Tricyclic antidepressants	Desipramine	Hippocampus, Hypothalamus	↑ CB1 receptor binding	[95]
	Imipramine	Hypothalamus, Hippocampus, Midbrain, vStriatum, Amygdala	↓ CB1 receptor binding (Hypothalamus, Midbrain, vStriatum) ↑ CB1 receptor binding (Amygdala)	[96]
MAO (A-B) inhibitors	Tranylcypromine	PFC, Hippocampus, Hypothalamus	↑ CB1 receptor binding ↑ 2-AG content (PFC) ↓ AEA content	[92]
Selective serotonin reuptake inhibitors (SSRI)	Fluoxetine	PFC	↑ CB1 receptor binding	[92, 93]
	Citalopram	Hippocampus, Hypothalamic paraventricular nucleus	↓ CB1 receptor binding	[94]

[a] *PFC* prefrontal cortex, *vStriatum* ventral striatum

the development of depressive symptoms and the antidepressant treatment response [86–88]. However, a significant genetic interaction among the polymorphism in the serotonin transporter gene 5-HTTLPR, variants in the CNR1 gene, anxiety or stress adaptation have also been found [89, 90]. Thus, the identification of individuals with a high-risk of psychiatric disorders through genetic testing could be a promising strategy for the development of safer drugs [91].

The putative role of the ECS in depression is supported by evidence showing that the majority of available antidepressants also modify CB1 receptor expression and endocannabinoid content in brain regions related to mood disorders (Table 5.2). While fluoxetine increased CB1 receptor binding and/or signaling in the limbic region [92, 93], citalopram reduced CB1 receptor signaling in the hippocampus and hypothalamic paraventricular nucleus [94], suggesting a region-specific effect of SSRI on CB1 receptor-mediated signaling. Similarly, TCAs elicited different effects based on various brain regions: desipramine increased hippocampal and hypothalamic CB1 receptor binding [95], while imipramine reduced it within the hypothalamus, midbrain and ventral striatum and increased it within the amygdala [96]. However, no difference has been found in the AEA content. The MAOI tranylcypromine enhanced CB1 receptor binding and 2-AG level in PFC and hippocampus, while reducing AEA content within the PFC, hippocampus and hypothalamus [92]. Despite the conflicting panorama, these findings suggest that the antidepressants modify the endocannabinoid tone in different ways, depending both on the class of drugs and on the different brain regions considered.

Changes in ECS elements have also been reported in several stress related animal models (Table 5.3), in accordance with the clinical data described above. In

Table 5.3 Schematic representation of the changes of the endocannabinoid system (ECS) elements in preclinical studies of depression

ECS elements	Experimental model	Animals	Behavioural response[a]	Brain region[a]	Molecular readout	Positive control	References
CB1	CMS	Wistar rats	↓ sucrose preference ↓ body weight	PFC Midbrain	↑ expression ↓ expression	Imipramine	[97]
		Sprague-Dawley rats	↓ body weight ♂♀ ↓ sucrose preference ♂	Hippocampus	↓ expression ♂ ↑ expression ♀	ND	[102]
	Chronic unpredictable stress	Long-Evans rats	Cognitive deficit in the MWM	Hippocampus Limbic forebrain	↓ expression No effects	ND	[101]
			↓ sexual motivation	PFC Hippocampus Hypothalamus vStriatum	↑ binding ↓ binding ↓ binding ↓ binding	Imipramine	[96]
		Sprague-Dawley rats	↑ immobility time in the FST	vmPFC dmPFC	↑ binding (vmPFC)	ND	[100]
			↑ immobility time in the FST ↓ sucrose preference ↓ locomotor activity in the OFT	Hippocampus	↓ expression	Transcranial magnetic stimulation	[103]
	OBX	Sprague-Dawley rats	↑ locomotor activity in the OFT	PFC	↑ binding	Fluoxetine	[98]
	Restraint stress	Sprague-Dawley rats	ND	Amygdala Hippocampus PFC	↑ binding (adolescent) ↓ binding (adult) ↑ binding (adolescent/adult)	ND	[99]
CB2	Chronic unpredictable mild stress	Wild type mice of CB2 overexpressing mice	↑ immobility time in the FST ↓ sucrose preference	Hippocampus	↓ expression	ND	[106]

Table 5.3 (continued)

ECS elements	Experimental model	Animals	Behavioural response[a]	Brain region[a]	Molecular readout	Positive control	References
TRPV1	Restraint stress	Wistar rats	↑ immobility time in the FST	Hippocampus	↑ expression	Clomipramine	[158]
FAAH	Restraint stress	Wistar rats	↑ immobility time in the FST	Hippocampus	↑ expression	Clomipramine	[158]
AEA	CMS	Wistar rats	↓ sucrose preference ↓ body weight	PFC, Midbrain, Hippocampus, Striatum, Thalamus	No effect	Imipramine	[97]
	Restraint stress	ICR mice	ND	Amygdala	↓ content	ND	[108]
				Amygdala vStriatum, mPFC	↓ content (Amygdala and mPFC) ↑ content (vStriatum)	ND	[111]
		Sprague-Dawley rats	ND	PFC, Hippocampus, Hypothalamus, Amygdala	↓ content	ND	[109]
		Bl6 mice	ND	Amygdala	↓ content	ND	[110]
		Wistar rats	↑ immobility time in the FST	PFC, Hippocampus	No effect	Clomipramine	[158]
	Chronic unpredictable stress	Long-Evans rats	↓ sexual motivation	PFC, Hippocampus, Hypothalamus, vStriatum, Amygdala, Midbrain	↓ content	Imipramine	[96]

Table 5.3 (continued)

ECS elements	Experimental model	Animals	Behavioural response[a]	Brain region[a]	Molecular readout	Positive control	References
2-AG	Chronic unpredictable stress	Long-Evans rats	Cognitive deficit in the MWM	Hippocampus Limbic forebrain	↓ content No effect	ND	[101]
			↓ sexual motivation	PFC, Hippocampus, Hypothalamus, vStriatum, Amygdala, Midbrain	↑ content (Hypothalamus, Midbrain)	Imipramine	[96]
	Restraint stress	ICR mice	ND	Amygdala, Forebrain	↑ content	ND	[108]
	CMS	Wistar rats	↓sucrose preference ↓ body weight	PFC, Hippocampus, Striatum, Midbrain, Thalamus	↑ content (Thalamus)	Imipramine	[97]
	Restraint stress	ICR mice	ND	Amygdala, vStriatum mPFC	↑ content (Amygdala, mPFC) ↓ content (vStriatum)	ND	[111]
				Amygdala	↑ content	ND	[112]
		Sprague-Dawley rats	ND	Amygdala	↑ content	ND	[109]
		Bl6 mice	ND	Amygdala	No effect	ND	[110]
		Wistar rats	↑ immobility time in the FST	PFC, Hippocampus	No effect	Clomipramine	[158]

[a] *FST* forced swim test, *CMS* chronic mild stress, *MWM* Morris water maze, *ND* not determined, *OBX* bilateral olfactory bulbectomy, *OFT* open field test, *PFC* prefrontal cortex, *mPFC* medial prefrontal cortex, *dmPFC* dorsomedial prefrontal cortex, *vmPFC* ventromedial prefrontal cortex, *vStriatum* ventral striatum

well validated animal models of depression such as the chronic mild stress (CMS) paradigm or the bilateral olfactory bulbectomy (OBX) model, which produce behavioural and neurochemical changes similar to those in human depression, a significant increase of CB1 receptor density and binding has been found in the PFC [96–100], together with a significant decrease in the ventral striatum, hypothalamus [96], midbrain [97] and hippocampus [99, 101–103]. This latter seems to be associated with a significant alteration of the hippocampal endocannabinoid-mediated neurotransmission and synaptic plasticity [104]. Collectively, the effects of experimental stress procedures on brain CB1 receptor expression seem to be region dependent.

Although the presence of CB2 receptors in stress responsive brain regions suggests their involvement in the regulation of mood, to date there is no evidence concerning their modification in the brain of depressed patients. More data come from preclinical studies, which reported a reduction of CB2 receptors in the hippocampus, striatum and midbrain in animal models of depression. Similarly, an increase of CB2 receptor expression counteracts behavioural and neurochemical features related to a depressive-like state [105–107]. Other controversial data about the endocannabinoid brain content in depression have also been recorded. While Bortolato et al. [97] did not find a change in AEA levels in different brain regions of rats subjected to CMS, others reported a significant reduction of AEA content following different chronic stress paradigms [96, 108–111]. The effects of stress procedure on 2-AG levels are confusing as well, since a reduction in the hippocampus and an increase in thalamus, hypothalamus and amygdala has been shown [96, 97, 101, 109, 112], or no such effects [97, 110]. Although the discrepancy may be due to numerous factors, such as the nature and duration of the stress, the species (rats vs. mice) or strain (Wistar vs. Sprague-Dawley rats), differences in response to stress procedure, or the time and tissue of extraction, the data described above supports the general hypothesis that a deficiency in the functioning of the endocannabinoid signaling, both in depressed patients and in animal models of depression, may directly lead to a vulnerability in development of the illness. Thus, it seems reasonable to hypothesize that its pharmacological facilitation would produce certain antidepressant effects.

Current Status of Animal Models of Depression and Antidepressant Responsive Tests

Due to the limited efficacy of antidepressant treatments, a better understanding of the pathophysiology of mental health disorders and the development of novel, improved therapeutic treatments would fill a considerable unmet medical need [113]. Due to the enormous cost of clinical trials, pharmaceutical companies make all efforts at testing new chemicals designed to alter the function of a specific target of disease in a predictable and safe manner [114]. Thus, of central importance to this approach is the availability of valid preclinical animal models for the evaluation of the potential efficacy of novel compounds and the further understanding of the neuropathology that underlies the idiopathic disease state of depression [115].

Ideally, an experimental animal model should reflect the human psychiatric disease in terms of face validity (i.e. reproduce the symptoms of depression observed in humans), construct validity (the same neurochemical mechanisms in humans as in the animal model) and predictive validity (chronic antidepressant treatment must reverse the phenotype of the animal model) [116]. In the case of depression, an animal model which perfectly includes the etiology, the pathophysiology and the symptoms of depression whilst allowing evaluating the responses to treatments remains impossible to fully envisage. However, different models, each with specific limitations, are able to reproduce most of the etiological factors and many symptoms of depression or possess a satisfactory predictive value for identifying new compounds. For this purpose, the forced swim test (FST) or the tail suspension test (TST) and the CMS or the OBX seem to be good experimental approaches for screening potential new antidepressants and shape the underlying disease etiology [117].

The most widely used paradigm to assess antidepressant-like behaviour is the FST also known as Porsolt's test [118]. In the FST rodents are forced to swim in an inescapable cylinder filled with water and eventually adopt a characteristic immobile posture which is interpreted as a passive stress-coping strategy or depressive-like behaviour (behavioural despair). The FST has shown its ability to detect a broad spectrum of substances with antidepressant efficacy, as these drugs shift from passive stress-coping towards active coping, which is detected as reduced immobility. Furthermore, the quantity of different movements such as climbing and swimming behaviour has predictive value to differentiate between NEergic and 5-HTergic activity. Some of the most representative potential antidepressants with different mechanisms of action have been submitted to this test [23, 119].

Similar assumptions and interpretations as the FST is the TST [120]. In this test, mice are suspended by their tails for a defined period of time and their immobility is decreased by several antidepressants. A major drawback of the TST is that its application is restricted to mice and limited to strains which do not tend to climb their tail, a behaviour that would otherwise confuse the interpretations of the results [121]. The test however is sensitive to acute treatment only and its validity for non-monoamine antidepressants is uncertain [119, 122].

A different model is the CMS paradigm, which is based on reduced sweet fluid intake as an index of anhedonia, induced by repeated (at least 2 weeks) exposure to unpredictable stressors (i.e. wet bedding, disruption of dark-light cycle and food or water deprivation) [123]. This model induces various long-term behavioural and neurochemical alterations resembling some of the dysfunctions observed in depressed patients, which are reversed only by chronic treatment with a broad spectrum of antidepressants. As compared to other experimental models of depression, it has been evaluated as a high perspective research approach, despite its procedural complexity and poor inter-laboratory reliability.

The OBX, a lesion model of depression is based on surgical removal of olfactory bulbs by aspiration [124] and results in a disruption of the limbic hypothalamic axis followed by neurochemical (i.e. changes in all major neurotransmitter systems) and behavioural (e.g. hyperactive response in the open field paradigm and anhedonia) alterations, which resemble changes seen in depressed patients and are reversed only by chronic administration of antidepressants [125, 126]. In most of the models described above, locomotor activity in the open field test must be also monitored to ensure that motor depression rather than emotional behaviour is not influencing animal responses [126].

Although none of the available experimental paradigms are able to model all aspects of depression disorders in terms of etiological factors and symptoms, and most likely never will, the paradigms described above have proven extremely useful both in the identification of potential new antidepressants and in the validation of neurobiological concepts. More specifically, they have been extensively used for assessing the potential antidepressant-like activity of compounds modulating the endocannabinoid signaling in rodents.

Effects of Pharmacological Manipulation of the Endocannabinoid Signaling in Preclinical Studies of Depression

After discovering the ECS members (CB1 and CB2 receptors, endocannabinoids AEA and 2-AG and enzymes for their degradation, FAAH and MAGL) several pharmacological tools, which vary from direct agonists or antagonists (Fig. 5.1) to endocannabinoid enhancers have been evaluated in several *in vitro* and *in vivo* studies to assess their therapeutic potential in stress-related neuropsychiatric disorders [23] (Table 5.4). Based on the hypothesis that a reduction of endocannabinoid signaling could underlie depressive disorders, it has been seen that acute or repeated treatment with different compounds which activate directly cannabinoid receptors, such as the main pharmacologically active principle of *Cannabis sativa* Δ9-THC [98, 127–130], the endogenous cannabinoid AEA [131, 132], the synthetic nonspecific CB1/CB2 receptor agonists CP55,940 [133], WIN55,212–2 [134, 135] and HU-210 [136–139] or the selective CB1 receptor agonist arachidonoyl 2'-chloroethylamide (ACEA) [140, 141] elicited antidepressant-like effects through CB1 and 5-HTergic or NEergic receptor-mediated mechanisms.

However, chronic exposure to Δ9-THC or WIN55,212–2 in adolescence led to a depressive-like phenotype in adulthood, further supporting the fact that adolescence is a critical period in which protracted direct CB1 receptor activation may influence mood control [142–146] (see also Chap. 12). Although the CB1 receptor antagonist rimonabant, which was introduced into clinical practice as antiobesity agent, was withdrawn from the market due to the higher incidence of psychiatric side effects [147], preclinical studies have reported an antidepressant-like activity of rimonabant in rodents [129, 130, 148–151]. Using a genetic approach controversial results regarding the effects of CB1 receptor signaling inhibition on stress coping

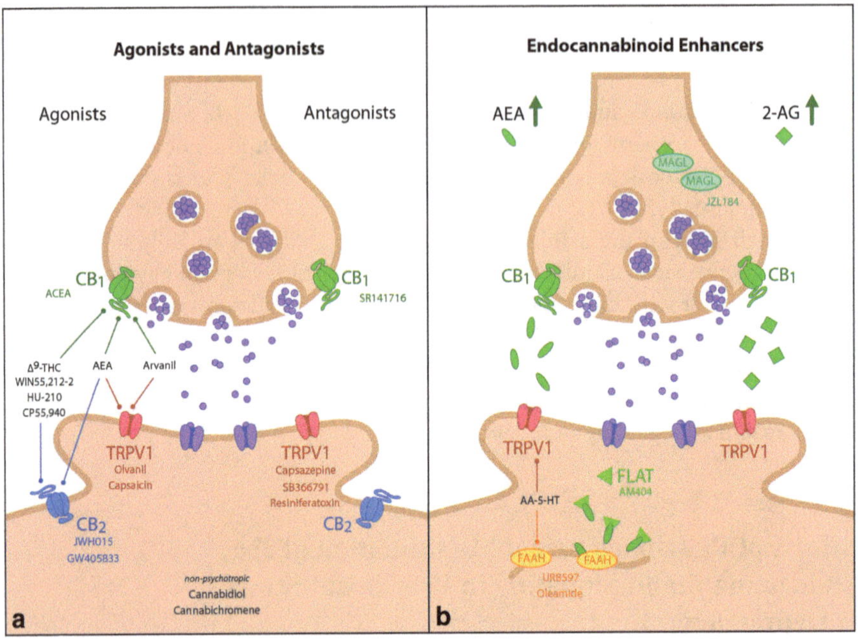

Fig. 5.1 Schematic illustration of the pharmacological modulation (i.e. agonists, antagonists and endocannabinoid enhancers) of the endocannabinoid system in preclinical studies of depression. For details about the different drugs see the main text and Table 5.4

behaviour have been obtained indicating that they could depend on specific deletion of CB1 receptors in some neuronal subpopulations [129, 152, 153]. However, compensatory mechanisms which develop in mutant mice could underlie the discrepancies between pharmacological and genetic inhibition of CB1 receptor signaling.

Although CB2 receptor ligands might be potentially safer due to the lack of psychoactive effects, controversial evidence concerning the effects of CB2 receptor signaling modulation on depressive-like behaviour has been recently described [23]. Thus, further clinical and preclinical investigations are required to define the role of CB2 receptors in the pathophysiology and treatment of depression. Despite the fact that vanilloid TRPV1 channels, due to their co-localization with CB1 receptors in several brain regions [154], seem to represent "the other side of the coin" in the regulation of anxiety, a similar function in depression is still ambiguous, since both TRPV1 agonists [155, 156] and pharmacological [155–158] or genetic TRPV1 blockade [159] elicited antidepressant-like effects. Thus, further studies are necessary to assess the role of TRPV1 channels as additional ECS "players" in mood regulation. Based on the assumption that direct activation of CB1 receptors elicited psychotropic side effects, several compounds have been developed that reinforce the effects of AEA and 2-AG by inhibiting their degradative enzymes FAAH and MAGL, or by blocking their cellular reuptake. Since CB1 receptors, FAAH and MAGL are not equally distributed in the brain; the indirect stimulation of CB1 receptors by endocannabinoid breakdown blockers could modulate the endocannabinoid signaling in selected brain areas which control mood [160].

Table 5.4 Schematic representation of the effects of the pharmacological modulation of the endocannabinoid system (ECS) in preclinical studies of depression

Drugs	Mechanism of action	Experimental model[a]	Animals	Behavioural response[a]	Positive control	References
Δ9-THC	Non selective CB1/CB2 receptor agonist	OBX	Sprague-Dawley rats	↓ locomotor activity	Fluoxetine	[98]
			Lister hooded rats	↓ locomotor activity	ND	[130]
		FST/TST	Swiss-DBA/2 mice	↓ immobility time	Fluoxetine, Desipramine	[127]
			Sprague-Dawley rats	↓ immobility time	Citalopram	[128]
			Bl6N mice	↓ immobility time	ND	[129]
AEA	Non selective CB1/CB2 receptor agonist	FST/TST/CMS	ICR mice	No effect on immobility time/↑ sucrose consumption	Clomipramine	[131]
		FST	Swiss mice	↓ immobility time	Fluoxetine	[132]
CP,55940	Non selective CB1/CB2 receptor agonist	FST	Wistar rats	↓ immobility time	ND	[133]
WIN55,212–2	Non selective CB1/CB2 receptor agonist	FST	Sprague-Dawley rats	↓ immobility time	Citalopram, Desipramine	[134]
		CMS	Sprague-Dawley rats	↓ immobility time/↑ extinction of avoidance behaviour/ No effect on sucrose consumption	ND	[135]
HU-210	Non selective CB1/CB2 receptor agonist	FST	Long-Evans rats	↓ immobility time	Desipramine	[136]
					ND	[137]
			Sprague-Dawley rats	↓ immobility time	ND	[138]
					Desipramine	[139]
Arachidonoyl 2'-chloro-ethylamide (ACEA)	Selective CB1 receptor agonist	FST	BALB/c mice	↓ immobility time	Fluoxetine	[140]
		CMS	Sprague-Dawley rats	↑ extinction of aversive memories	ND	[141]

Table 5.4 (continued)

Drugs	Mechanism of action	Experimental model[a]	Animals	Behavioural response[a]	Positive control	References
JWH015	Selective CB2 receptor agonist	CMS	BALB/c mice	↑ sucrose consumption	ND	[105]
GW405833	Selective CB2 receptor agonist	FST	Wistar rats	↓ immobility time	Desipramine	[224]
Olvanil	Selective TRPV1 agonist	FST/TST	ICR mice	↓ immobility time	ND	[156]
Capsaicin	Selective TRPV1 agonist	FST/TST	ICR mice	↓ immobility time	ND	[156]
			Swiss mice	↓ immobility time	Fluoxetine	[155]
Arvanil	Nonselective TRPV1/ CB1 receptor agonist	FST/TST	ICR mice	↓ immobility time	ND	[156]
Rimonabant (SR141716)	Selective CB1 receptor antagonist/inverse agonist	FST	Swiss mice	↓ immobility time	ND	[148]
		CMS/FST	Wistar rats/ BALB/c mice	↓ immobility time	Fluoxetine	[149]
		FST	Bl6 N mice	↓ immobility time	ND	[129]
					Desipramine	[150]
			ICR mice	↓ immobility time	Imipramine	[151]
		OBX	Lister hooded rats	↓ locomotor activity	ND	[130]
Capsazepine	selective TRPV1 antagonist	FST/TST	Swiss mice	↓ immobility time	Fluoxetine	[155]
Resiniferatoxin	selective TRPV1 antagonist	FST	Swiss mice	↓ immobility time (26 °C) ↑ immobility time (41 °C)	Amitriptyline, Ketamine	[157]
SB366791	selective TRPV1 antagonist	FST	Wistar rats	↓ immobility time in STR rats	Clomipramine	[158]

Table 5.4 (continued)

Drugs	Mechanism of action	Experimental model[a]	Animals	Behavioural response[a]	Positive control	References
URB597	FAAH inhibitor	FST	Long-Evans rats	↓ immobility time	ND	[161]
			Wistar rats	↓ immobility time	ND	[133]
			Swiss mice	↓ immobility time	Fluoxetine	[132]
			Sprague-Dawley rats	↓ immobility time	ND	[162]
		TST	Bl6J mice	↓ immobility time	Desipramine	[163]
		CMS	Wistar rats	↑ sucrose consumption	Imipramine	[97]
			ICR mice	↑ sucrose consumption	ND	[164]
Oleamide	FAAH inhibitor	FST	Long-Evans rats	↓ immobility time	Desipramine	[136]
			Albino mice	↓ immobility time	ND	[166]
AA-5-HT	FAAH inhibitor/TRPV1 antagonist	FST	Wistar rats	↓ immobility time in STR rats	Clomipramine	[158]
AM404	AEA uptake inhibitor	FST	Long-Evans rats	↓ immobility time	Desipramine	[136]
			Wistar rats	↓ immobility time	Imipramine	[133]
			Swiss mice	↓ immobility time	ND	[172]
			Swiss mice	↓ immobility time	Fluoxetine	[132]
JZL184	MAGL inhibitor	Chronic unpredictable mild stress	Bl6J mice	↑ sucrose consumption ↓ immobility time	ND	[176]
Cannabidiol	CB1-CB2 receptor antagonist/inverse agonist, 5-HT1 A receptor agonist, TRPV1 agonist, AEA uptake inhibitor, FAAH inhibitor	FST	Swiss mice	↓ immobility time	Fluoxetine, Desipramine	[127]
			Swiss mice	↓ immobility time	Imipramine	[180]
Cannabichromene	TRPV1 agonist, AEA uptake inhibitor	FST/TST	Swiss-DBA/2 mice	↓ immobility time	Fluoxetine, Desipramine	[127]

[a] *CMS* chronic mild stress, *FST* forced swim test, *ND* not determined, *OBX* bilateral olfactory bulbectomy, *TST* tail suspension test, *STR* stressed group

The FAAH inhibitor URB597 has shown CB1 receptor-mediated antidepressant-like effects by enhancing AEA signaling in several experimental models such as FST [132, 133, 161, 162], TST [163], CMS paradigm [97, 164], adolescent Δ9-THC exposure [146] and tail-pinch test [165]. Another FAAH inhibitor, oleamide, elicited antidepressant-like effects through a CB1 receptor-mediated mechanism [136, 166]. In agreement with the pharmacological approach, transgenic mice lacking FAAH, which exhibit more than 10-fold higher levels of AEA as compared to wild-type mice, have shown a less depressive-like phenotype [145].

A particularly innovative approach in the treatment of mood disorders could be the use of compounds with the capability to combine inhibition of AEA hydrolysis with antagonism of TRPV1 channels. One such dual FAAH/TRPV1 blocker is N-arachidonoyl-serotonin (AA-5-HT) [167, 168], which elicited anxiolytic- [169–171] and antidepressant-like activity [158], suggesting the potential therapeutic use of dual FAAH/TRPV1 inhibitors in stress-related disorders. A different strategy to enhance AEA signaling at the receptor is to block its uptake into pre- and/or post-synaptic terminals, thereby promoting the indirect activation of CB1 receptors. The prototypical endocannabinoid transport inhibitor AM404 has improved the behavioural performance of rodents in the FST, through a CB1 receptor-mediated mechanism [132, 133, 136, 172]. However, the exact mechanism of action of endocannabinoid uptake inhibitors as well as the molecular identity of the transporter itself still remains to be characterized. Therefore, further biomolecular studies will have to be performed in this direction.

Collectively, this evidence supports the clinical potential of endocannabinoid level modulators as new therapeutic tools for the treatment of mood disorders. Recent data have suggested that 2-AG could act in the brain modulating behavioural responses in stress-related conditions [173–175]. In this context the prototypical MAGL inhibitor JZL184, by inducing an 8-fold increase in 2-AG, but not AEA, brain content reversed the depressive-like behaviour via activation of both CB1 receptor and mTor signaling [176]. However, contrary to FAAH blockade, a potential drawback in the use of MAGL inhibitors could be the development of tetrad effects which are typical of CB1 receptor agonists [177] as well as of tolerance with chronic use [178, 179].

In conclusion, while endocannabinoids are rapidly metabolized in vivo, limiting the potential efficacy of their exogenous administration, the data described above supports more FAAH than MAGL as a potential therapeutic target for the identification of new pharmacotherapies for affective disorders [160]. In addition to the pharmacological modulation of the endocannabinoid signaling, a different approach to reduce the psychotropic side effects of Cannabis is the use of plant-derived cannabinoids with very weak or no psychotropic effects such as CBD, cannabichromene, cannabigerol, cannabidivarin and Δ9-Tetrahydrocannabinol, some of which show potential as therapeutic agents in preclinical models of CNS disorders [55]. Special emphasis is given to CBD, which exerts several positive pharmacological effects in preclinical and clinical studies to the point of making it a highly attractive therapeutic entity in several diseases. We still do not know the exact mechanism(s) of action underlying the mood-elevating effect of CBD, as it may act not only through the ECS, but also by directly or indirectly activating the metabotropic receptors for 5-HT or adenosine or by targeting nuclear receptors of the PPAR family as well as modulating ion channels including TRPV1 [18]. Contrary to the extensive research done regarding the potential therapeutic effects of CBD in anxiety [23] or schizophrenia [24], only few studies have examined its antidepressant-like effects. In the FST, which represents a standard preclinical test to assess the effects of potential antidepressants, cannabichromene and CBD decreased the immobility time, the latter acting through a 5-HT1 A receptor–mediated mechanism [127, 180]. However, further studies are necessary to establish the efficacy and safety profile of phytocannabinoids for the treatment of stress-related disorders.

Endocannabinoid Signaling and Antidepressant-Like Effects: Potential Molecular Underpinning

As described above, based on the monoaminergic hypothesis of depression, the actual antidepressants act by enhancing the central 5-HTergic and/or NEergic neurotransmission through the inhibition of the synaptic re-uptake or enzymatic degradation, and the desensitization or sensitization of specific receptors [4]. Several lines of evidence suggest that modulation of endocannabinoid signaling could facilitate 5-HTergic neurotransmission through an enhancement of 5-HT neuronal activity, an increased 5-HT efflux or modulation of 5-HT receptors (i.e. 5-HT_{1A} and 5-$HT_{2A/C}$). Both direct and indirect activation of CB1 receptors (the latter acting through pharmacological or genetic inhibition of FAAH activity) increased firing activity of 5-HTergic neurons in the DRN [128, 134, 162, 181], and enhanced basal 5-HT efflux in several brain regions such as nucleus accumbens, striatum, hippocampus and PFC [181–183]. However, chronic exposure to the CB1 receptor agonist WIN55,212–2 during adolescence attenuated 5-HTergic activity and elicited a depressive-like phenotype in adulthood, further supporting the importance of adolescence as a highly sensitive developmental window within which the disruptive effects of cannabinoid exposure increase the risk for developing psychiatric disorders [145]. Interestingly, inhibition of CB1 receptor signaling induced a depressive-like phenotype in mice, which was mediated by an impairment of 5-HTergic neural activity [152, 153, 184–186], strenghening the role of the endocannabinoid tone in emotional behaviour through the modulation of the 5-HTergic neurotransmission. As described for conventional antidepressants, which induce a desensitization of the 5-$HT_{2A/C}$ autoreceptors and/or an enhancement of the tonic activity of 5-HT_{1A} receptors [187], the antidepressant-like effects elicited by cannabinoids could be due to changes in the expression and function of these receptors [128, 188]. However, further 5-HT receptor subtypes (i.e. 5-HT_3 or 5-HT_4) could also be involved in the emotional responses induced by the endocannabinoid tone modulation [189–192].

A dysregulation of NEergic system seems to be implicated in the pathophysiology of depression, as supported by the primary action of antidepressants to enhance central NEergic transmission. In this context, a strong interaction between the endocannabinoid and NEergic systems could participate in the antidepressant effects of endocannabinoid signaling enhancement, based on the expression of CB1 receptors in the LC (the major NEergic nucleus). More specifically, CB1 receptor activation could directly or indirectly, by modulating inhibitory and/or excitatory inputs to LC, increase the firing activity of NEergic neurons and consequently the release of NE in the forebrain. This indicates the existence of a functional interaction between these two systems in the action of antidepressants [181, 193, 194]. However, *in vitro* studies have shown the capacity of cannabinoids to inhibit monoamine reuptake and metabolism, sharing some pharmacological properties with antidepressants [195–198].

Increasing evidence links stress to depression and antidepressant action, and suggests that stressors act by inducing a disruption in cellular mechanisms governing neuronal plasticity and disturbances in the hypothalamic-pituitary adrenal (HPA) axis [199, 200]. Hence, current and potential antidepressants exert neurotrophic activity, by increasing the hippocampal expression of factors such as cyclic adenosine monophosphate-response element binding protein (CREB) and BDNF, and also affect HPA axis hyperactivity [201–205]. The endogenous cannabinoids AEA and 2-AG [206] and the synthetic nonspecific cannabinoid CB1/CB2 receptor agonists HU-210 [137] or WIN55,212–2 [207, 208] stimulate neurogenesis, which is inhibited by pharmacological [151, 206] or genetic [209–212] CB1 receptor blockade. The enhanced AEA signaling also stimulates hippocampal cell proliferation, through a CB1 receptor-mediated mechanism [158, 213, 214]

Based on the recent detection of CB2 receptors in the brain [43], their potential mechanisms underlying emotional responses are under investigation. So far, it has been seen that pharmacological activation or genetic inactivation of CB2 receptors enhanced or reduced hippocampal neuronal plasticity, respectively [215, 216]. Similarly, the CMS procedure did not alter BDNF expression in mice overexpressing CB2 receptors [106], suggesting their potential protective role. On one hand the controversial *in vivo* data does not give us a coherent picture concerning the role of CB2 receptors in depression, on the other hand, however, the molecular data further strengthens the rationale for the development of selective CB2 receptor agonists as promising candidates to target neurogenesis, thus bypassing the undesired psychoactive effects of central CB1 receptor activation.

Taken together the data presented herein suggests that facilitation of the endocannabinoid signaling through CB1 and/or CB2 receptors activation seems to mimic the effects of current antidepressants on hippocampal neuroplasticity. The HPA axis acts as a neuroendocrine bridge, regulating the stress response by controlling the secretion of corticotrophin-releasing hormone, adrenocorticotropic and glucocorticoidhormones. Additionally, it is controlled by a negative feedback inhibition loop which involves mineralocorticoid and glucocorticoid receptors [217]. Depressive disorders are also characterized by an inability of glucocorticoids to bind their receptors, which in turn can lead to HPA axis hyperactivity and increased levels of circulating glucocorticoids. Treatment with the current antidepressants results in reduction of glucocorticoid release, suggesting that the attenuation of HPA axis hyper-responsivity could be one of the long-term adaptations in response to antidepressants that contributes to their therapeutic efficacy [218]. Several evidence highlights the role of the endocannabinoid signaling to regulate the HPA axis both during basal conditions and after stress exposure [133, 219] (see also Chap. 1). While CB1 receptor activation inhibits HPA axis activity, as a part of the HPA axis negative feedback inhibition loop, impairment in the CB1 receptor signaling increases HPA axis activity under both basal conditions and following stress exposure [152, 220–222]. Collectively the data described above suggests that the antidepressant-like effects of different classes of cannabinoids may in part be due to molecular mechanisms which resemble the ones triggered by antidepressants.

Future Perspective and Conclusive Remarks

In conclusion, the current evidence suggests a strong link between ECS and depressive disorders. A deficiency in the endocannabinoid tone leads to a depressive-like phenotype in experimental animal models of depression (Table 5.3), which is in line with clinical findings where depressed patients have reduced levels of endogenous cannabinoids (Table 5.1). Hence, facilitation of the endocannabinoid signaling could be the target for developing potential new antidepressants. Supporting this hypothesis is preclinical data which has shown that elevated endocannabinoid signaling is able to produce behavioural and biochemical effects as the conventional antidepressant treatment (Table 5.4), and that many antidepressants alter endogenous cannabinoid tone (Table 5.2). However, whilst the direct activation of CB1 receptors is hampered by unwanted psychotropic effects, and the possibly safer direct modulation of CB2 receptors still lacks sufficient experimental evidence to justify its use, the indirect activation of cannabinoid receptors with agents that inhibit endocannabinoids deactivation has produced very promising results in experimental animal models of depression. Yet, this approach is not devoid of intrinsic problems, mostly due to the fact that endocannabinoid-deactivating proteins also recognize other non-endocannabinoid mediators as substrates which then activate different receptors—a property also shared to some extent by endocannabinoids like AEA and NADA. Thus, inhibition of enzymes like FAAH or of the putative endocannabinoid transporter might lead to the activation of these alternative receptors. This complication and the possible compensatory action of co-occurring deactivation routes and enzymes for endocannabinoids [223] may render this approach not sufficiently efficacious or safe. In view of these potential problems and of the fact that genetic studies have revealed a relationship between depression and polymorphisms of cannabinoid receptors and/or degradative enzymes, only time will tell if targeting the ECS may result in effective pharmacotherapies for major depression and other affective-related disorders.

Acknowledgments The research of the authors is supported by the project "CEITEC—Central European Institute of Technology" (CZ.1.05/1.1.00/02.0068) from European Regional Development Fund. We thank Caitlin Riebe (independent scientific illustrator, Vancouver, Canada) for the artwork and Vanessa Raileanu (Toronto, Canada) for the proof-reading.

References

1. Wittchen HU, Jacobi F, Rehm J, Gustavsson A, Svensson M, Jönsson B, Olesen J, Allgulander C, Alonso J, Faravelli C, Fratiglioni L, Jennum P, Lieb R, Maercker A, van Os J, Preisig M, Salvador-Carulla L, Simon R, Steinhausen H-C. The size and burden of mental disorders and other disorders of the brain in Europe 2010. Eur Neuropsychopharmacol. 2011;21(9):655–79.
2. American Psychiatric Association. Diagnostic and statistical manual of mental disorders. 5th ed. A.P. Publishing: Arlington;2013.
3. Schildkraut JJ. The catecholamine hypothesis of affective disorders: a review of supporting evidence. Am J Psychiatry. 1965;122(5):509–22.
4. Li X, Frye MA, Shelton RC. Review of pharmacological treatment in mood disorders and future directions for drug development. Neuropsychopharmacology. 2012;37(1):77–101.
5. Han C, Wang S-M, Kato M, Lee S-J, Patkar AA, Masand PS, Pae C-U. Second-generation antipsychotics in the treatment of major depressive disorder: current evidence. Expert Rev Neurother. 2013;13(7):851–70.
6. Srinivasan V, De Berardis D, Shillcutt SD, Brzezinski A. Role of melatonin in mood disorders and the antidepressant effects of agomelatine. Expert Opin Investig Drugs. 2012;21(10):1503–22.
7. Krystal JH, Sanacora G, Duman RS. Rapid-acting glutamatergic antidepressants: the path to ketamine and beyond. Biol Psychiatry. 2013;73(12):1133–41.
8. Machado-Vieira R, Salvadore G, Diazgranados N, Zarate CA Jr. Ketamine and the next generation of antidepressants with a rapid onset of action. Pharmacol Ther. 2009;123(2):143–50.
9. Kehne JH, Cain CK. Therapeutic utility of non-peptidic CRF1 receptor antagonists in anxiety, depression, and stress-related disorders: evidence from animal models. Pharmacol Ther. 2010;128(3):460–87.
10. Wong EHF, Tarazi FI, Shahid M. The effectiveness of multi-target agents in schizophrenia and mood disorders: Relevance of receptor signature to clinical action. Pharmacol Ther. 2010;126(2):173–85.
11. Engin E, Liu J, Rudolph U. α2-containing GABA(A) receptors: a target for the development of novel treatment strategies for CNS disorders. Pharmacol Ther. 2012;136(2):142–52.
12. Connolly KR, Thase ME. Emerging drugs for major depressive disorder. Expert Opin Emerg Drugs. 2012;17(1):105–26.
13. Isbell H, Gorodetzsky CW, Jasinski D, Clausseln U, von Spulak F, Korte F. Effects of (—) delta-9-trans-tetrahydrocannabinol in man. Psychopharmacologia. 1967;11(2):184–8.
14. Luongo L, Maione S, Di Marzo V. Endocannabinoids and neuropathic pain: focus on neuron-glia and endocannabinoid-neurotrophin interactions. Eur J Neurosci. 2014;39(3):401–8.
15. Starowicz K, Makuch W, Korostynski M, Malek N, Slezak M, Zychowska M, Petrosino S, De Petrocellis L, Cristino L, Przewlocka B, Di Marzo V. Full inhibition of spinal FAAH leads to TRPV1-mediated analgesic effects in neuropathic rats and possible lipoxygenase-mediated remodeling of anandamide metabolism. PLoS ONE. 2013;8(4):e60040.
16. Di Marzo V. "De-liver-ance" from CB(1): a way to counteract insulin resistance? Gastroenterology. 2012;142(5):1063–6.
17. Silvestri C, Di Marzo V. The endocannabinoid system in energy homeostasis and the etiopathology of metabolic disorders. Cell Metab. 2013;17(4):475–90.
18. Izzo AA, Sharkey KA. Cannabinoids and the gut: new developments and emerging concepts. Pharmacol Ther. 2010;126(1):21–38.
19. Silvestri C, Ligresti A, Di Marzo V. Peripheral effects of the endocannabinoid system in energy homeostasis: adipose tissue, liver and skeletal muscle. Rev Endocr Metab Disord. 2011;12(3):153–62.
20. Mazzola C, Micale V, Drago F. Amnesia induced by beta-amyloid fragments is counteracted by cannabinoid CB1 receptor blockade. Eur J Pharmacol. 2003;477(3):219–25.
21. Micale V, Mazzola C, Drago F. Endocannabinoids and neurodegenerative diseases. Pharmacol Res. 2007;56(5):382–92.

22. Micale V, Cristino L, Tamburella A, Petrosino S, Leggio GM, Di Marzo V, Drago F. Enhanced cognitive performance of dopamine D3 receptor "knock-out" mice in the step-through passive-avoidance test: assessing the role of the endocannabinoid/endovanilloid systems. Pharmacol Res. 2010;61(6):531–6.
23. Micale V, Di Marzo V, Sulcova A, Wotjak CT, Drago F. Endocannabinoid system and mood disorders: priming a target for new therapies. Pharmacol Ther. 2013;138(1):18–37.
24. Kucerova J, Tabiova K, Drago F, Micale V. Therapeutic potential of cannabinoids in schizophrenia. Recent Pat CNS Drug Discov. 2014;9(1):13–25.
25. Terzian ALB, Micale V, Wotjak CT. Cannabinoid receptor type 1 receptors on GABAergic vs. glutamatergic neurons differentially gate sex-dependent social interest in mice. Eur J Neurosci. 2014;40(1):2293–8.
26. Howlett AC, Bidaut-Russell M, Devane WA, Melvin LS, Johnson MR, Herkenham M. The cannabinoid receptor: biochemical, anatomical and behavioral characterization. Trends Neurosci. 1990;13(10):420–3.
27. Matsuda LA, Lolait SJ, Brownstein MJ, Young AC, Bonner TI. Structure of a cannabinoid receptor and functional expression of the cloned cDNA. Nature. 1990;346(6284):561–4.
28. Munro S, Thomas KL, Abu-Shaar M. Molecular characterization of a peripheral receptor for cannabinoids. Nature. 1993;365(6441):61–5.
29. Devane WA, Hanus L, Breuer A, Pertwee RG, Stevenson LA, Griffin G, Gibson D, Mandelbaum A, Etinger A, Mechoulam R. Isolation and structure of a brain constituent that binds to the cannabinoid receptor. Science. 1992;258(5090):1946–9.
30. Mechoulam R, Ben-Shabat S, Hanus L, Ligumsky M, Kaminski NE, Schatz AR, Gopher A, Almog S, Martin BR, Compton DR. Identification of an endogenous 2-monoglyceride, present in canine gut, that binds to cannabinoid receptors. Biochem Pharmacol. 1995;50(1):83–90.
31. Lovinger DM. Endocannabinoid liberation from neurons in transsynaptic signaling. J Mol Neurosci. 2007;33(1):87–93.
32. Marnett LJ. Decoding endocannabinoid signaling. Nat Chem Biol. 2009;5(1):8–9.
33. Di Marzo V Petrosino S. Endocannabinoids and the regulation of their levels in health and disease. Curr Opin Lipidol. 2007;18(2):129–40.
34. Liu J, Wang L, Harvey-White J, Huang BX, Kim H-Y, Luquet S, Palmiter RD, Krystal G, Rai R, Mahadevan A, Razdan RK, Kunos G. Multiple pathways involved in the biosynthesis of anandamide. Neuropharmacology. 2008;54(1):1–7.
35. Cravatt BF, Giang DK, Mayfield SP, Boger DL, Lerner RA, Gilula NB. Molecular characterization of an enzyme that degrades neuromodulatory fatty-acid amides. Nature. 1996;384(6604):83–7.
36. Dinh TP, Carpenter D, Leslie FM, Freund TF, Katona I, Sensi SL, Kathuria S, Piomelli D. Brain monoglyceride lipase participating in endocannabinoid inactivation. Proc Natl Acad Sci USA. 2002;99(16):10819–24.
37. Begg M, Pacher P, Bátkai S, Osei-Hyiaman D, Offertáler L, Mo FM, Liu J, Kunos G. Evidence for novel cannabinoid receptors. Pharmacol Ther. 2005;106(2):133–45.
38. Pertwee RG. Receptors and channels targeted by synthetic cannabinoid receptor agonists and antagonists. Curr Med Chem. 2010;17(14):1360–81.
39. Mackie K. Distribution of cannabinoid receptors in the central and peripheral nervous system. Handb Exp Pharmacol. 2005;(168):299–325.
40. Marsicano G, Lutz B. Expression of the cannabinoid receptor CB1 in distinct neuronal subpopulations in the adult mouse forebrain. Eur J Neurosci. 1999;11(12):4213–25.
41. Häring M, Marsicano G, Lutz B, Monory K. Identification of the cannabinoid receptor type 1 in serotonergic cells of raphe nuclei in mice. Neuroscience. 2007;146(3):1212–9.
42. Oropeza VC, Mackie K, Van Bockstaele EJ. Cannabinoid receptors are localized to noradrenergic axon terminals in the rat frontal cortex. Brain Res. 2007;1127(1):36–44.
43. Van Sickle MD Duncan M Kingsley PJ Mouihate A Urbani P Mackie K Stella N Makriyannis A Piomelli D Davison JS Marnett LJ Di Marzo V Pittman QJ Patel KD Sharkey KA. Identification and functional characterization of brainstem cannabinoid CB2 receptors. Science. 2005;310(5746):329–32.

44. Gong JP, Onaivi ES, Ishiguro H, Liu QR, Tagliaferro PA, Brusco A, Uhl GR. Cannabinoid CB2 receptors: immunohistochemical localization in rat brain. Brain Res. 2006;1071(1):10–23.
45. Marco EM, García-Gutiérrez MS, Bermúdez-Silva F-J, Moreira FA, Guimarães F, Manzanares J, Viveros M-P. Endocannabinoid system and psychiatry: in search of a neurobiological basis for detrimental and potential therapeutic effects. Front Behav Neurosci. 2011;5:63.
46. Pacher P, Bátkai S, Kunos G. The endocannabinoid system as an emerging target of pharmacotherapy. Pharmacol Rev. 2006;58(3):389–462.
47. Gaoni Y, Mechoulam R. Isolation, structure, and partial synthesis of an active constituent of hashish. J Am Chem Soc. 1964;86(8):1646–7.
48. Bisogno T, Hanus L, De Petrocellis L, Tchilibon S, Ponde DE, Brandi I, Moriello AS, Davis JB, Mechoulam R, Di Marzo V. Molecular targets for cannabidiol and its synthetic analogues: effect on vanilloid VR1 receptors and on the cellular uptake and enzymatic hydrolysis of anandamide. Br J Pharmacol. 2001;134(4):845–52.
49. Carrier EJ, Auchampach JA, Hillard CJ. Inhibition of an equilibrative nucleoside transporter by cannabidiol: a mechanism of cannabinoid immunosuppression. Proc Natl Acad Sci USA. 2006;103(20):7895–900.
50. De Petrocellis L Ligresti A Moriello AS AllarÃ M Bisogno T Petrosino S Stott CG Di Marzo V. Effects of cannabinoids and cannabinoid-enriched Cannabis extracts on TRP channels and endocannabinoid metabolic enzymes. Br J Pharmacol. 2011;163(7):1479–94.
51. Russo EB, Burnett A, Hall B, Parker KK. Agonistic properties of cannabidiol at 5-HT1a receptors. Neurochem Res. 2005;30(8):1037–43.
52. Magen I, Avraham Y, Ackerman Z, Vorobiev L, Mechoulam R, Berry EM. Cannabidiol ameliorates cognitive and motor impairments in mice with bile duct ligation. J Hepatol. 2009;51(3):528–34.
53. Cascio MG, Gauson LA, Stevenson LA, Ross RA, Pertwee RG. Evidence that the plant cannabinoid cannabigerol is a highly potent alpha2-adrenoceptor agonist and moderately potent 5HT1 A receptor antagonist. Br J Pharmacol. 2010;159(1):129–41.
54. Izzo AA, Borrelli F, Capasso R, Di Marzo V, Mechoulam R. Non-psychotropic plant cannabinoids: new therapeutic opportunities from an ancient herb. Trends Pharmacol Sci. 2009;30(10):515–27.
55. Hill AJ, Williams CM, Whalley BJ, Stephens GJ. Phytocannabinoids as novel therapeutic agents in CNS disorders. Pharmacol Ther. 2012;133(1):79–97.
56. Ablon SL, Goodwin FK. High frequency of dysphoric reactions to tetrahydrocannabinol among depressed patients. Am J Psychiatry. 1974;131(4):448–53.
57. Kotin J, Post RM, Goodwin FK. 9-Tetrahydrocannabinol in depressed patients. Arch Gen Psychiatry. 1973;28(3):345–8.
58. Amchova P, Kucerova J, Giugliano V, Babinska Z, Zanda MT, Scherma M, Dusek L, Fadda P, Micale V, Sulcova A, Fratta W, Fattore L. Enhanced self-administration of the CB1 receptor agonist WIN55, 212–2 in olfactory bulbectomized rats: evaluation of possible serotonergic and dopaminergic underlying mechanisms. Front Pharmacol. 2014;5:44.
59. Patton GC, Coffey C, Carlin JB, Degenhardt L, Lynskey M, Hall W. Cannabis use and mental health in young people: cohort study. BMJ. 2002;325(7374):1195–8.
60. Arendt M, Rosenberg R, Fjordback L, Brandholdt J, Foldager L, Sher L, Munk-Jørgensen P. Testing the self-medication hypothesis of depression and aggression in cannabis-dependent subjects. Psychol Med. 2007;37(7):935–45.
61. Bovasso GB. Cannabis abuse as a risk factor for depressive symptoms. Am J Psychiatry. 2001;158(12):2033–7.
62. Rey JM, Sawyer MG, Raphael B, Patton GC, Lynskey M. Mental health of teenagers who use cannabis. Results of an Australian survey. Br J Psychiatry. 2002;180:216–21.
63. Degenhardt L, Hall W, Lynskey M. Exploring the association between cannabis use and depression. Addiction. 2003;98(11):1493–504.
64. Harder VS, Morral AR, Arkes J. Marijuana use and depression among adults: testing for causal associations. Addiction. 2006;101(10):1463–72.

65. Tziraki S. [Mental disorders and neuropsychological impairment related to chronic use of cannabis]. Rev Neurol. 2012;54(12):750–60.
66. Fu Q, Heath AC, Bucholz KK, Nelson E, Goldberg J, Lyons MJ, True WR, Jacob T, Tsuang MT, Eisen SA. Shared genetic risk of major depression, alcohol dependence, and marijuana dependence: contribution of antisocial personality disorder in men. Arch Gen Psychiatry. 2002;59(12):1125–32.
67. Lynskey MT, Heath AC, Nelson EC, Bucholz KK, Madden PAF, Slutske WS, Statham DJ, Martin NG. Genetic and environmental contributions to cannabis dependence in a national young adult twin sample. Psychol Med. 2002;32(2):195–207.
68. Lynskey MT, Glowinski AL, Todorov AA, Bucholz KK, Madden PAF, Nelson EC, Statham DJ, Martin NG, Heath AC. Major depressive disorder, suicidal ideation, and suicide attempt in twins discordant for cannabis dependence and early-onset cannabis use. Arch Gen Psychiatry. 2004;61(10):1026–32.
69. Farris SG, Zvolensky MJ, Boden MT, Bonn-Miller MO. Cannabis use expectancies mediate the relation between depressive symptoms and cannabis use among cannabis-dependent veterans. J Addict Med. 2014;8(2):130–6.
70. Parolaro D, Realini N, Vigano D, Guidali C, Rubino T. The endocannabinoid system and psychiatric disorders. Exp Neurol. 2010;224(1):3–14.
71. Hungund BL, Vinod KY, Kassir SA, Basavarajappa BS, Yalamanchili R, Cooper TB, Mann JJ, Arango V. Upregulation of CB1 receptors and agonist-stimulated [35S]GTPgammaS binding in the prefrontal cortex of depressed suicide victims. Mol Psychiatry. 2004;9(2):184–90.
72. Koethe D, Llenos IC, Dulay JR, Hoyer C, Torrey EF, Leweke FM, Weis S. Expression of CB1 cannabinoid receptor in the anterior cingulate cortex in schizophrenia, bipolar disorder, and major depression. J Neural Transm. 2007;114(8):1055–63.
73. Hill MN, Miller GE, Ho W-SV, Gorzalka BB, Hillard CJ. Serum endocannabinoid content is altered in females with depressive disorders: a preliminary report. Pharmacopsychiatry. 2008;41(2):48–53.
74. Hill MN, Miller GE, Carrier EJ, Gorzalka BB, Hillard CJ. Circulating endocannabinoids and N-acyl ethanolamines are differentially regulated in major depression and following exposure to social stress. Psychoneuroendocrinology. 2009;34(8):1257–62.
75. Vinod KY, Arango V, Xie S, Kassir SA, Mann JJ, Cooper TB, Hungund BL. Elevated levels of endocannabinoids and CB1 receptor-mediated G-protein signaling in the prefrontal cortex of alcoholic suicide victims. Biol Psychiatry. 2005;57(5):480–6.
76. Ho WS, Hill MN, Miller GE, Gorzalka BB, Hillard CJ. Serum contents of endocannabinoids are correlated with blood pressure in depressed women. Lipids Health Dis. 2012;11:32.
77. Heyman E, Gamelin F-X, Goekint M, Piscitelli F, Roelands B, Leclair E, Di Marzo V, Meeusen R. Intense exercise increases circulating endocannabinoid and BDNF levels in humans—possible implications for reward and depression. Psychoneuroendocrinology. 2012;37(6):844–51.
78. Duman RS, Monteggia LM. A neurotrophic model for stress-related mood disorders. Biol Psychiatry. 2006;59(12):1116–27.
79. Hill MN, Hillard CJ, Bambico FR, Patel S, Gorzalka BB, Gobbi G. The therapeutic potential of the endocannabinoid system for the development of a novel class of antidepressants. Trends Pharmacol Sci. 2009;30(9):484–93.
80. Moreira FA, Crippa JAS. The psychiatric side-effects of rimonabant. Rev Bras Psiquiatr. 2009;31(2):145–53.
81. Lazary J, Juhasz G, Hunyady L, Bagdy G. Personalized medicine can pave the way for the safe use of CBâ receptor antagonists. Trends Pharmacol Sci. 2011;32(5):270–80.
82. Levinson DF, Holmans P, Straub RE, Owen MJ, Wildenauer DB, Gejman PV, Pulver AE, Laurent C, Kendler KS, Walsh D, Norton N, Williams NM, Schwab SG, Lerer B, Mowry BJ, Sanders AR, Antonarakis SE, Blouin JL, DeLeuze JF, Mallet J. Multicenter linkage study of schizophrenia candidate regions on chromosomes 5q, 6q, 10p, and 13q: schizophrenia linkage collaborative group III. Am J Hum Genet. 2000;67(3):652–63.

83. Chen X, Williamson VS, An S-S, Hettema JM, Aggen SH, Neale MC, Kendler KS. Cannabinoid receptor 1 gene association with nicotine dependence. Arch Gen Psychiatry. 2008;65(7):816–24.
84. Chakrabarti B, Baron-Cohen S. Variation in the human cannabinoid receptor CNR1 gene modulates gaze duration for happy faces. Mol Autism. 2011;2(1):10.
85. Barrero FJ, Ampuero I, Morales B, Vives F, de Dios Luna Del Castillo J, Hoenicka J, García Yébenes J. Depression in Parkinson's disease is related to a genetic polymorphism of the cannabinoid receptor gene (CNR1). Pharmacogenomics J. 2005;5(2):135–41.
86. Domschke K, Dannlowski U, Ohrmann P, Lawford B, Bauer J, Kugel H, Heindel W, Young R, Morris P, Arolt V, Deckert J, Suslow T, Baune BT. Cannabinoid receptor 1 (CNR1) gene: impact on antidepressant treatment response and emotion processing in major depression. Eur Neuropsychopharmacol. 2008;18(10):751–9.
87. Juhasz G, Chase D, Pegg E, Downey D, Toth ZG, Stones K, Platt H, Mekli K, Payton A, Elliott R, Anderson IM, Deakin JFW. CNR1 gene is associated with high neuroticism and low agreeableness and interacts with recent negative life events to predict current depressive symptoms. Neuropsychopharmacology. 2009;34(8):2019–27.
88. Monteleone P, Bifulco M, Maina G, Tortorella A, Gazzerro P, Proto MC, Di Filippo C, Monteleone F, Canestrelli B, Buonerba G, Bogetto F, Maj M. Investigation of CNR1 and FAAH endocannabinoid gene polymorphisms in bipolar disorder and major depression. Pharmacol Res. 2010;61(5):400–4.
89. Lazary J, Lazary A, Gonda X, Benko A, Molnar E, Hunyady L, Juhasz G, Bagdy G. Promoter variants of the cannabinoid receptor 1 gene (CNR1) in interaction with 5-HTTLPR affect the anxious phenotype. Am J Med Genet B Neuropsychiatr Genet. 2009;150B(8):1118–27.
90. Agrawal A, Nelson EC, Littlefield AK, Bucholz KK, Degenhardt L, Henders AK, Madden PAF, Martin NG, Montgomery GW, Pergadia ML, Sher KJ, Heath AC, Lynskey MT. Cannabinoid receptor genotype moderation of the effects of childhood physical abuse on anhedonia and depression. Arch Gen Psychiatry. 2012;69(7):732–40.
91. Mitjans M, Gastó C, Catalán R, Fañanás L, Arias B. Genetic variability in the endocannabinoid system and 12-week clinical response to citalopram treatment: the role of the CNR1, CNR2 and FAAH genes. J Psychopharmacol. 2012;26(10):1391–8.
92. Hill MN, Ho W-SV, Hillard CJ, Gorzalka BB. Differential effects of the antidepressants tranylcypromine and fluoxetine on limbic cannabinoid receptor binding and endocannabinoid contents. J Neural Transm. 2008;115(12):1673–9.
93. Mato S, Vidal R, Castro E, Díaz A, Pazos A, Valdizán EM. Long-term fluoxetine treatment modulates cannabinoid type 1 receptor-mediated inhibition of adenylyl cyclase in the rat prefrontal cortex through 5-hydroxytryptamine 1 A receptor-dependent mechanisms. Mol Pharmacol. 2010;77(3):424–34.
94. Hesketh SA, Brennan AK, Jessop DS, Finn DP. Effects of chronic treatment with citalopram on cannabinoid and opioid receptor-mediated G-protein coupling in discrete rat brain regions. Psychopharmacology (Berl). 2008;198(1):29–36.
95. Hill MN, Ho W-SV, Sinopoli KJ, Viau V, Hillard CJ, Gorzalka BB. Involvement of the endocannabinoid system in the ability of long-term tricyclic antidepressant treatment to suppress stress-induced activation of the hypothalamic-pituitary-adrenal axis. Neuropsychopharmacology. 2006;31(12):2591–9.
96. Hill MN, Carrier EJ, McLaughlin RJ, Morrish AC, Meier SE, Hillard CJ, Gorzalka BB. Regional alterations in the endocannabinoid system in an animal model of depression: effects of concurrent antidepressant treatment. J Neurochem. 2008;106(6):2322–36.
97. Bortolato M, Mangieri RA, Fu J, Kim JH, Arguello O, Duranti A, Tontini A, Mor M, Tarzia G, Piomelli D. Antidepressant-like activity of the fatty acid amide hydrolase inhibitor URB597 in a rat model of chronic mild stress. Biol Psychiatry. 2007;62(10):1103–10.
98. Rodríguez-Gaztelumendi A, Rojo ML, Pazos A, Díaz A. Altered CB receptor-signaling in prefrontal cortex from an animal model of depression is reversed by chronic fluoxetine. J Neurochem. 2009;108(6):1423–33.
99. Lee TTY, Hill MN. Age of stress exposure modulates the immediate and sustained effects of repeated stress on corticolimbic cannabinoid CBâ receptor binding in male rats. Neuroscience. 2013;249:106–14.

100. McLaughlin RJ, Hill MN, Dang SS, Wainwright SR, Galea LAM, Hillard CJ, Gorzalka BB. Upregulation of CBâ receptor binding in the ventromedial prefrontal cortex promotes proactive stress-coping strategies following chronic stress exposure. Behav Brain Res. 2013;237:333–7.
101. Hill MN, Patel S, Carrier EJ, Rademacher DJ, Ormerod BK, Hillard CJ, Gorzalka BB. Downregulation of endocannabinoid signaling in the hippocampus following chronic unpredictable stress. Neuropsychopharmacology. 2005;30(3):508–15.
102. Reich CG, Taylor ME, McCarthy MM. Differential effects of chronic unpredictable stress on hippocampal CB1 receptors in male and female rats. Behav Brain Res. 2009;203(2):264–9.
103. Wang H, Wang L, Zhang R, Chen Y, Liu L, Gao F, Nie H, Hou W, Peng Z, Tan Q. Antidepressive mechanism of repetitive transcranial magnetic stimulation in rat: the role of the endocannabinoid system. J Psychiatr Res. 2014;51:79–87.
104. Reich CG, Mihalik GR, Iskander AN, Seckler JC, Weiss MS. Adolescent chronic mild stress alters hippocampal CB1 receptor-mediated excitatory neurotransmission and plasticity. Neuroscience. 2013;253:444–54.
105. Onaivi ES, Ishiguro H, Gong JP, Patel S, Meozzi PA, Myers L, Perchuk A, Mora Z, Tagliaferro PA, Gardner E, Brusco A, Akinshola BE, Liu Q-R, Chirwa SS, Hope B, Lujilde J, Inada T, Iwasaki S, Macharia D, Teasenfitz L, Arinami T, Uhl GR. Functional expression of brain neuronal CB2 cannabinoid receptors are involved in the effects of drugs of abuse and in depression. Ann N Y Acad Sci. 2008;1139:434–49.
106. García-Gutiérrez MS, Pérez-Ortiz JM, Gutiérrez-Adán A, Manzanares J. Depression-resistant endophenotype in mice overexpressing cannabinoid CB(2) receptors. Br J Pharmacol. 2010;160(7):1773–84.
107. Zoppi S, Madrigal JL, Caso JR, García-Gutiérrez MS, Manzanares J, Leza JC, García-Bueno B. Regulatory role of the cannabinoid CB2 receptor in stress-induced neuroinflammation in mice. Br J Pharmacol. 2014;171(11):2814–26.
108. Patel S, Roelke CT, Rademacher DJ, Hillard CJ. Inhibition of restraint stress-induced neural and behavioural activation by endogenous cannabinoid signalling. Eur J Neurosci. 2005;21(4):1057–69.
109. Hill MN, McLaughlin RJ, Bingham B, Shrestha L, Lee TTY, Gray JM, Hillard CJ, Gorzalka BB, Viau V. Endogenous cannabinoid signaling is essential for stress adaptation. Proc Natl Acad Sci USA. 2010;107(20):9406–11.
110. Hill MN, Kumar SA, Filipski SB, Iverson M, Stuhr KL, Keith JM, Cravatt BF, Hillard CJ, Chattarji S, McEwen BS. Disruption of fatty acid amide hydrolase activity prevents the effects of chronic stress on anxiety and amygdalar microstructure. Mol Psychiatry. 2013;18(10):1125–35
111. Rademacher DJ, Meier SE, Shi L, Ho WV, Jarrahian A, Hillard CJ. Effects of acute and repeated restraint stress on endocannabinoid content in the amygdala, ventral striatum, and medial prefrontal cortex in mice. Neuropharmacology. 2008;54(1):108–16.
112. Patel S, Kingsley PJ, Mackie K, Marnett LJ, Winder DG. Repeated homotypic stress elevates 2-arachidonoylglycerol levels and enhances short-term endocannabinoid signaling at inhibitory synapses in basolateral amygdala. Neuropsychopharmacology. 2009;34(13):2699–709.
113. Millan MJ. Dual- and triple-acting agents for treating core and co-morbid symptoms of major depression: novel concepts, new drugs. Neurotherapeutics. 2009;6(1):53–77.
114. DiMasi JA, Hansen RW, Grabowski HG. The price of innovation: new estimates of drug development costs. J Health Econ. 2003;22(2):151–85.
115. Micale V, Kucerova J, Sulcova A. Leading compounds for the validation of animal models of psychopathology. Cell Tissue Res. 2013;354(1):309–30.
116. McKinney WT Jr, Bunney WE Jr. Animal model of depression. I. Review of evidence: implications for research. Arch Gen Psychiatry. 1969;21(2):240–8.
117. Nestler EJ, Gould E, Manji H, Buncan M, Duman RS, Greshenfeld HK, Hen R, Koester S, Lederhendler I, Meaney M, Robbins T, Winsky L, Zalcman S. Preclinical models: status of basic research in depression. Biol Psychiatry. 2002;52(6):503–28.

118. Porsolt RD, Le Pichon M, Jalfre M. Depression: a new animal model sensitive to antidepressant treatments. Nature. 1977;266(5604):730–2.
119. Cryan JF, Valentino RJ, Lucki I. Assessing substrates underlying the behavioral effects of antidepressants using the modified rat forced swimming test. Neurosci Biobehav Rev. 2005;29(4–5):547–69.
120. Steru L, Chermat R, Thierry B, Simon P. The tail suspension test: a new method for screening antidepressants in mice. Psychopharmacology (Berl). 1985;85(3):367–70.
121. Cryan JF, Mombereau C, Vassout A. The tail suspension test as a model for assessing antidepressant activity: review of pharmacological and genetic studies in mice. Neurosci Biobehav Rev. 2005;29(4–5):571–625.
122. Berrocoso E, Ikeda K, Sora I, Uhl GR, Sánchez-Blázquez P, Mico JA. Active behaviours produced by antidepressants and opioids in the mouse tail suspension test. Int J Neuropsychopharmacol. 2013;16(1):151–62.
123. Willner P, Towell A, Sampson D, Sophokleous S, Muscat R. Reduction of sucrose preference by chronic unpredictable mild stress, and its restoration by a tricyclic antidepressant. Psychopharmacology (Berl). 1987;93(3):358–64.
124. Kelly JP, Wrynn AS, Leonard BE. The olfactory bulbectomized rat as a model of depression: an update. Pharmacol Ther. 1997;74(3):299–316.
125. Song C, Leonard BE. The olfactory bulbectomised rat as a model of depression. Neurosci Biobehav Rev. 2005;29(4–5):627–47.
126. Cryan JF, Markou A, Lucki I. Assessing antidepressant activity in rodents: recent developments and future needs. Trends Pharmacol Sci. 2002;23(5):238–45.
127. El-Alfy AT, Ivey K, Robinson K, Ahmed S, Radwan M, Slade D, Khan I, ElSohly M, Ross S. Antidepressant-like effect of delta9-tetrahydrocannabinol and other cannabinoids isolated from Cannabis sativa L. Pharmacol Biochem Behav. 2010;95(4):434–42.
128. Bambico FR, Hattan PR, Garant JP, Gobbi G. Effect of delta-9-tetrahydrocannabinol on behavioral despair and on pre- and postsynaptic serotonergic transmission. Prog Neuropsychopharmacol Biol Psychiatry. 2012;38(1):88–96.
129. Häring M, Grieb M, Monory K, Lutz B, Moreira FA. Cannabinoid CBâ receptor in the modulation of stress coping behavior in mice: the role of serotonin and different forebrain neuronal subpopulations. Neuropharmacology. 2013;65:83–9.
130. Elbatsh MM, Moklas MAA, Marsden CA, Kendall DA. Antidepressant-like effects of Δ9-tetrahydrocannabinol and rimonabant in the olfactory bulbectomised rat model of depression. Pharmacol Biochem Behav. 2012;102(2):357–65.
131. Hayase T. Depression-related anhedonic behaviors caused by immobilization stress: a comparison with nicotine-induced depression-like behavioral alterations and effects of nicotine and/or “antidepressant” drugs. J Toxicol Sci. 2011;36(1):31–41.
132. Umathe SN, Manna SS, Jain NS. Involvement of endocannabinoids in antidepressant and anti-compulsive effect of fluoxetine in mice. Behav Brain Res. 2011;223(1):125–34.
133. Adamczyk P, Gołda A, McCreary AC, Filip M, Przegaliński E. Activation of endocannabinoid transmission induces antidepressant-like effects in rats. J Physiol Pharmacol. 2008;59(2):217–28.
134. Bambico FR, Katz N, Debonnel G, Gobbi G. Cannabinoids elicit antidepressant-like behavior and activate serotonergic neurons through the medial prefrontal cortex. J Neurosci. 2007;27(43):11700–11.
135. Segev A, Rubin AS, Abush H, Richter-Levin G, Akirav I. Cannabinoid receptor activation prevents the effects of chronic mild stress on emotional learning and LTP in a rat model of depression. Neuropsychopharmacology. 2014;39(4):919–33.
136. Hill MN, Gorzalka BB. Pharmacological enhancement of cannabinoid CB1 receptor activity elicits an antidepressant-like response in the rat forced swim test. Eur Neuropsychopharmacol. 2005;15(6):593–9.
137. Jiang W, Zhang Y, Xiao L, Van Cleemput J, Ji SP, Bai G, Zhang X. Cannabinoids promote embryonic and adult hippocampus neurogenesis and produce anxiolytic- and antidepressant-like effects. J Clin Invest. 2005;115(11):3104–16.

138. McLaughlin RJ, Hill MN, Morrish AC, Gorzalka BB. Local enhancement of cannabinoid CB1 receptor signalling in the dorsal hippocampus elicits an antidepressant-like effect. Behav Pharmacol. 2007;18(5–6):431–8.
139. Morrish AC, Hill MN, Riebe CJN, Gorzalka BB. Protracted cannabinoid administration elicits antidepressant behavioral responses in rats: role of gender and noradrenergic transmission. Physiol Behav. 2009 Aug 4;98(1–2):118–24.
140. Rutkowska M, Jachimczuk O. Antidepressant—like properties of ACEA (arachidonyl-2-chloroethylamide), the selective agonist of CB1 receptors. Acta Pol Pharm. 2004;61(2):165–7.
141. Reich CG, Iskander AN, Weiss MS. Cannabinoid modulation of chronic mild stress-induced selective enhancement of trace fear conditioning in adolescent rats. J Psychopharmacol. 2013;27(10):947–55.
142. Rubino T, Vigano' D, Realini N, Guidali C, Braida D, Capurro V, Castiglioni C, Cherubino F, Romualdi P, Candeletti S, Sala M, Parolaro D. Chronic delta9-tetrahydrocannabinol during adolescence provokes sex-dependent changes in the emotional profile in adult rats: behavioral and biochemical correlates. Neuropsychopharmacology. 2008;33(11):2760–71.
143. Rubino T, Realini N, Braida D, Alberio T, Capurro V, Viganò D, Guidali C, Sala M, Fasano M, Parolaro D. The depressive phenotype induced in adult female rats by adolescent exposure to THC is associated with cognitive impairment and altered neuroplasticity in the prefrontal cortex. Neurotox Res. 2009;15(4):291–302.
144. Rubino T, Zamberletti E, Parolaro D. Adolescent exposure to cannabis as a risk factor for psychiatric disorders. J Psychopharmacol. 2012;26(1):177–88.
145. Bambico FR, Nguyen N-T, Katz N, Gobbi G. Chronic exposure to cannabinoids during adolescence but not during adulthood impairs emotional behaviour and monoaminergic neurotransmission. Neurobiol Dis. 2010;37(3):641–55.
146. Realini N, Vigano' D, Guidali C, Zamberletti E, Rubino T, Parolaro D. Chronic URB597 treatment at adulthood reverted most depressive-like symptoms induced by adolescent exposure to THC in female rats. Neuropharmacology. 2011;60(2–3):235–43.
147. Van Gaal LF, Rissanen AM, Scheen AJ, Ziegler O, Rössner S, RIO-Europe Study Group. Effects of the cannabinoid-1 receptor blocker rimonabant on weight reduction and cardiovascular risk factors in overweight patients: 1-year experience from the RIO-Europe study. Lancet. 2005;365(9468):1389–97.
148. Tzavara ET, Davis RJ, Perry KW, Li X, Salhoff C, Bymaster FP, Witkin JM, Nomikos GG. The CB1 receptor antagonist SR141716 A selectively increases monoaminergic neurotransmission in the medial prefrontal cortex: implications for therapeutic actions. Br J Pharmacol. 2003;138(4):544–53.
149. Griebel G, Stemmelin J, Scatton B. Effects of the cannabinoid CB1 receptor antagonist rimonabant in models of emotional reactivity in rodents. Biol Psychiatry. 2005;57(3):261–7.
150. Steiner MA, Marsicano G, Nestler EJ, Holsboer F, Lutz B, Wotjak CT. Antidepressant-like behavioral effects of impaired cannabinoid receptor type 1 signaling coincide with exaggerated corticosterone secretion in mice. Psychoneuroendocrinology. 2008;33(1):54–67.
151. Lee S, Kim DH, Yoon SH, Ryu JH. Sub-chronic administration of rimonabant causes loss of antidepressive activity and decreases doublecortin immunoreactivity in the mouse hippocampus. Neurosci Lett. 2009;467(2):111–6.
152. Steiner MA, Marsicano G, Wotjak CT, Lutz B. Conditional cannabinoid receptor type 1 mutants reveal neuron subpopulation-specific effects on behavioral and neuroendocrine stress responses. Psychoneuroendocrinology. 2008;33(8):1165–70.
153. Steiner MA, Wanisch K, Monory K, Marsicano G, Borroni E, Bächli H, Holsboer F, Lutz B, Wotjak CT. Impaired cannabinoid receptor type 1 signaling interferes with stress-coping behavior in mice. Pharmacogenomics J. 2008;8(3):196–208.
154. Cristino L, de Petrocellis L, Pryce G, Baker D, Guglielmotti V, Di Marzo V. Immunohistochemical localization of cannabinoid type 1 and vanilloid transient receptor potential vanilloid type 1 receptors in the mouse brain. Neuroscience. 2006;139(4):1405–15.

155. Manna SS, Umathe SN. A possible participation of transient receptor potential vanilloid type 1 channels in the antidepressant effect of fluoxetine. Eur J Pharmacol. 2012;685 (1–3):81–90.
15. Hayase T. Differential effects of TRPV1 receptor ligands against nicotine-induced depression-like behaviors. BMC Pharmacol. 2011;11:6.
157. Abdelhamid RE, Kovács KJ, Nunez MG, Larson AA. Depressive behavior in the forced swim test can be induced by TRPV1 receptor activity and is dependent on NMDA receptors. Pharmacol Res. 2014;79:21–7.
158. Navarria A, Tamburella A, Iannotti FA, Micale V, Camillieri G, Gozzo L, Verde R, Imperatore R, Leggio GM, Drago F, Di Marzo V. The dual blocker of FAAH/TRPV1 N-arachidonoylserotonin reverses the behavioral despair induced by stress in rats and modulates the HPA-axis. Pharmacol Res. 2014;87:151–9.
159. You IJ, Jung YH, Kim MJ, Kwon SH, Hong SI, Lee SY, Jang CG. Alterations in the emotional and memory behavioral phenotypes of transient receptor potential vanilloid type 1-deficient mice are mediated by changes in expression of 5-HTâA, GABA(A), and NMDA receptors. Neuropharmacology. 2012;62(2):1034–43.
160. Petrosino S, Di Marzo V. FAAH and MAGL inhibitors: therapeutic opportunities from regulating endocannabinoid levels. Curr Opin Investig Drugs. 2010;11(1):51–62.
161. Hill MN, Karacabeyli ES, Gorzalka BB. Estrogen recruits the endocannabinoid system to modulate emotionality. Psychoneuroendocrinology. 2007;32(4):350–7.
162. McLaughlin RJ, Hill MN, Bambico FR, Stuhr KL, Gobbi G, Hillard CJ, Gorzalka BB. Prefrontal cortical anandamide signaling coordinates coping responses to stress through a serotonergic pathway. Eur Neuropsychopharmacol. 2012;22(9):664–71.
163. Naidu PS, Varvel SA, Ahn K, Cravatt BF, Martin BR, Lichtman AH. Evaluation of fatty acid amide hydrolase inhibition in murine models of emotionality. Psychopharmacology (Berl). 2007;192(1):61–70.
164. Rademacher DJ, Hillard CJ. Interactions between endocannabinoids and stress-induced decreased sensitivity to natural reward. Prog Neuropsychopharmacol Biol Psychiatry. 2007;31(3):633–41.
165. Haller J, Goldberg SR, Pelczer KG, Aliczki M, Panlilio LV. The effects of anandamide signaling enhanced by the FAAH inhibitor URB597 on coping styles in rats. Psychopharmacology (Berl). 2013;230(3):353–62.
166. Akanmu MA, Adeosun SO, Ilesanmi OR. Neuropharmacological effects of oleamide in male and female mice. Behav Brain Res. 2007;182(1):88–94.
167. Bisogno T, Melck D, De Petrocellis L, Bobrov MYu, Gretskaya NM, Bezuglov VV, Sitachitta N, Gerwick WH, Di Marzo V. Arachidonoylserotonin and other novel inhibitors of fatty acid amide hydrolase. Biochem Biophys Res Commun. 1998;248(3):515–22.
168. Maione S, De Petrocellis L, de Novellis V, Moriello AS, Petrosino S, Palazzo E, Rossi FS, Woodward DF, Di Marzo V. Analgesic actions of N-arachidonoyl-serotonin, a fatty acid amide hydrolase inhibitor with antagonistic activity at vanilloid TRPV1 receptors. Br J Pharmacol. 2007;150(6):766–81.
169. Micale V, Cristino L, Tamburella A, Petrosino S, Leggio GM, Drago F, Di Marzo V. Anxiolytic effects in mice of a dual blocker of fatty acid amide hydrolase and transient receptor potential vanilloid type-1 channels. Neuropsychopharmacology. 2009;34(3):593–606.
170. Micale V, Cristino L, Tamburella A, Petrosino S, Leggio GM, Drago F, Di Marzo V. Altered responses of dopamine D3 receptor null mice to excitotoxic or anxiogenic stimuli: Possible involvement of the endocannabinoid and endovanilloid systems. Neurobiol Dis. 2009;36(1):70–80.
171. John CS, Currie PJ. N-arachidonoyl-serotonin in the basolateral amygdala increases anxiolytic behavior in the elevated plus maze. Behav Brain Res. 2012;233(2):382–8.
172. Mannucci C, Navarra M, Pieratti A, Russo GA, Caputi AP, Calapai G. Interactions between endocannabinoid and serotonergic systems in mood disorders caused by nicotine withdrawal. Nicotine Tob Res. 2011;13(4):239–47.

173. Sütt S, Raud S, Areda T, Reimets A, Kõks S, Vasar E. Cat odour-induced anxiety—a study of the involvement of the endocannabinoid system. Psychopharmacology. 2008;198(4):509–20.
174. Suárez J, Rivera P, Llorente R, Romero-Zerbo SY, Bermúdez-Silva FJ, de Fonseca FR, Viveros M-P. Early maternal deprivation induces changes on the expression of 2-AG biosynthesis and degradation enzymes in neonatal rat hippocampus. Brain Res. 2010;1349:162–73.
175. Eisenstein SA, Clapper JR, Holmes PV, Piomelli D, Hohmann AG. A role for 2-arachidonoylglycerol and endocannabinoid signaling in the locomotor response to novelty induced by olfactory bulbectomy. Pharmacol Res. 2010;61(5):419–29.
176. Zhong P, Wang W, Pan B, Liu X, Zhang Z, Long JZ, Zhang H, Cravatt BF, Liu Q. Monoacylglycerol lipase inhibition blocks chronic stress-induced depressive-like behaviors via activation of mTOR signaling. Neuropsychopharmacology. 2014;39(7):1763–76.
177. Long JZ, Li W, Booker L, Burston JJ, Kinsey SG, Schlosburg JE, Pavón FJ, Serrano AM, Selley DE, Parsons LH, Lichtman AH, Cravatt BF. Selective blockade of 2-arachidonoylglycerol hydrolysis produces cannabinoid behavioral effects. Nat Chem Biol. 2009;5(1):37–44.
178. Schlosburg JE, Blankman JL, Long JZ, Nomura DK, Pan B, Kinsey SG, Nguyen PT, Ramesh D, Booker L, Burston JJ, Thomas EA, Selley DE, Sim-Selley LJ, Liu Q, Lichtman AH, Cravatt BF. Chronic monoacylglycerol lipase blockade causes functional antagonism of the endocannabinoid system. Nat Neurosci. 2010;13(9):1113–9.
179. Busquets-Garcia A, Puighermanal E, Pastor A, de la Torre R, Maldonado R, Ozaita A. Differential role of anandamide and 2-arachidonoylglycerol in memory and anxiety-like responses. Biol Psychiatry. 2011;70(5):479–86.
180. Zanelati TV, Biojone C, Moreira FA, Guimarães FS, Joca SRL. Antidepressant-like effects of cannabidiol in mice: possible involvement of 5-HT1 A receptors. Br J Pharmacol. 2010;159(1):122–8.
181. Gobbi G, Bambico FR, Mangieri R, Bortolato M, Campolongo P, Solinas M, Cassano T, Morgese MG, Debonnel G, Duranti A, Tontini A, Tarzia G, Mor M, Trezza V, Goldberg SR, Cuomo V, Piomelli D. Antidepressant-like activity and modulation of brain monoaminergic transmission by blockade of anandamide hydrolysis. Proc Natl Acad Sci USA. 2005;102(51):18620–5.
182. Bambico FR, Duranti A, Tontini A, Tarzia G, Gobbi G. Endocannabinoids in the treatment of mood disorders: evidence from animal models. Curr Pharm Des. 2009;15(14):1623–46.
183. Cassano T, Gaetani S, Macheda T, Laconca L, Romano A, Morgese MG, Cimmino CS, Chiarotti F, Bambico FR, Gobbi G, Cuomo V, Piomelli D. Evaluation of the emotional phenotype and serotonergic neurotransmission of fatty acid amide hydrolase-deficient mice. Psychopharmacology (Berl). 2011;214(2):465–76.
184. Martin M, Ledent C, Parmentier M, Maldonado R, Valverde O. Involvement of CB1 cannabinoid receptors in emotional behaviour. Psychopharmacology (Berl). 2002;159(4):379–87.
185. Aso E, Renoir T, Mengod G, Ledent C, Hamon M, Maldonado R, Lanfumey L, Valverde O. Lack of CB1 receptor activity impairs serotonergic negative feedback. J Neurochem. 2009;109(3):935–44.
186. Burokas A, Martín-García E, Gutiérrez-Cuesta J, Rojas S, Herance JR, Gispert JD, Serra MÁ, Maldonado R. Relationships between serotonergic and cannabinoid system in depressive-like behavior: a PET study with [(11) C]-DASB. J Neurochem. 2014;130(1):126–35.
187. Mahar I, Bambico FR, Mechawar N, Nobrega JN. Stress, serotonin, and hippocampal neurogenesis in relation to depression and antidepressant effects. Neurosci Biobehav Rev. 2014;38:173–92.
188. Bambico FR, Nguyen NT, Gobbi G. Decline in serotonergic firing activity and desensitization of 5-HT1 A autoreceptors after chronic unpredictable stress. Eur Neuropsychopharmacol. 2009;19(3):215–28.

189. Höfelmann D, di Benedetto B, Azad SC, Micale V, Wotjak CT, Rammes G. Lack of interaction of endocannabinoids and 5-HT(3) neurotransmission in associative fear circuits of the amygdala: evidence from electrophysiological and behavioural experiments. Brain Res. 2013;1527:47–56.
190. Walstab J, Rappold G, Niesler B. 5-HT(3) receptors: role in disease and target of drugs. Pharmacol Ther. 2010;128(1):146–69.
191. Chegini HR, Nasehi M, Zarrindast MR. Differential role of the basolateral amygdala 5-HT3 and 5-HT4 serotonin receptors upon ACPA-induced anxiolytic-like behaviors and emotional memory deficit in mice. Behav Brain Res. 2014;261:114–26.
192. Xiong W, Koo B, Morton R, Zhang L. Psychotropic and nonpsychotropic cannabis derivatives inhibit human 5-HT(3A) receptors through a receptor desensitization-dependent mechanism. Neuroscience. 2011;184:28–37.
193. Muntoni AL, Pillolla G, Melis M, Perra S, Gessa GL, Pistis M. Cannabinoids modulate spontaneous neuronal activity and evoked inhibition of locus coeruleus noradrenergic neurons. Eur J Neurosci. 2006;23(9):2385–94.
194. Esteban S, García-Sevilla JA. Effects induced by cannabinoids on monoaminergic systems in the brain and their implications for psychiatric disorders. Prog Neuropsychopharmacol Biol Psychiatry. 2012;38(1):78–87.
195. Steffens M, Feuerstein TJ. Receptor-independent depression of DA and 5-HT uptake by cannabinoids in rat neocortex—involvement of Na(+)/K(+)-ATPase. Neurochem Int. 2004;44(7):529–38.
196. Velenovská M, Fisar Z. Effect of cannabinoids on platelet serotonin uptake. Addict Biol. 2007;12(2):158–66.
197. Fisar Z, Hroudová J, Raboch J. Inhibition of monoamine oxidase activity by antidepressants and mood stabilizers. Neuro Endocrinol Lett. 2010;31(5):645–56.
198. Fišar Z. Cannabinoids and monoamine neurotransmission with focus on monoamine oxidase. Prog Neuropsychopharmacol Biol Psychiatry. 2012;38(1):68–77.
199. Kunugi H, Hori H, Adachi N, Numakawa T. Interface between hypothalamic-pituitary-adrenal axis and brain-derived neurotrophic factor in depression. Psychiatry Clin Neurosci. 2010;64(5):447–59.
200. Schmidt HD, Duman RS. The role of neurotrophic factors in adult hippocampal neurogenesis, antidepressant treatments and animal models of depressive-like behavior. Behav Pharmacol. 2007;18(5–6):391–418.
201. Pittenger C, Duman RS. Stress, depression, and neuroplasticity: a convergence of mechanisms. Neuropsychopharmacology. 2008;33(1):88–109.
202. Grady MM, Stahl SM. Novel agents in development for the treatment of depression. CNS Spectr. 2013;18(Suppl 1):37–40.
203. Tamburella A, Micale V, Navarria A, Drago F. Antidepressant properties of the 5-HT4 receptor partial agonist, SL65.0155: behavioral and neurochemical studies in rats. Prog Neuropsychopharmacol Biol Psychiatry. 2009;33(7):1205–10.
204. Tamburella A, Micale V, Leggio GM, Drago F. The beta3 adrenoceptor agonist, amibegron (SR58611 A) counteracts stress-induced behavioral and neurochemical changes. Eur Neuropsychopharmacol. 2010;20(10):704–13.
205. Tamburella A, Leggio GM, Micale V, Navarria A, Bucolo C, Cicirata V, Drago F, Salomone S. Behavioural and neurochemical changes induced by stress-related conditions are counteracted by the neurokinin-2 receptor antagonist saredutant. Int J Neuropsychopharmacol. 2013;16(4):813–23.
206. Aguado T, Carracedo A, Julien B, Velasco G, Milman G, Mechoulam R, Alvarez L, Guzmán M, Galve-Roperh I. Cannabinoids induce glioma stem-like cell differentiation and inhibit gliomagenesis. J Biol Chem. 2007;282(9):6854–62.
207. Aguado T, Monory K, Palazuelos J, Stella N, Cravatt B, Lutz B, Marsicano G, Kokaia Z, Guzmán M, Galve-Roperh I. The endocannabinoid system drives neural progenitor proliferation. FASEB J. 2005;19(12):1704–6.

208. Marchalant Y, Brothers HM, Wenk GL. Cannabinoid agonist WIN-55,212–2 partially restores neurogenesis in the aged rat brain. Mol Psychiatry. 2009;14(12):1068–9.
209. Jin K, Xie L, Kim SH, Parmentier-Batteur S, Sun Y, Mao XO, Childs J, Greenberg DA. Defective adult neurogenesis in CB1 cannabinoid receptor knockout mice. Mol Pharmacol. 2004;66(2):204–8.
210. Kim SH, Won SJ, Mao XO, Jin K, Greenberg DA. Molecular mechanisms of cannabinoid protection from neuronal excitotoxicity. Mol Pharmacol. 2006;69(3):691–6.
211. Aso E, Ozaita A, Valdizán EM, Ledent C, Pazos A, Maldonado R, Valverde O. BDNF impairment in the hippocampus is related to enhanced despair behavior in CB1 knockout mice. J Neurochem. 2008;105(2):565–72.
212. Beyer CE, Dwyer JM, Piesla MJ, Platt BJ, Shen R, Rahman Z, Chan K, Manners MT, Samad TA, Kennedy JD, Bingham B, Whiteside GT. Depression-like phenotype following chronic CB1 receptor antagonism. Neurobiol Dis. 2010;39(2):148–55.
213. Derkinderen P, Valjent E, Toutant M, Corvol JC, Enslen H, Ledent C, Trzaskos J. Regulation of extracellular signal-regulated kinase by cannabinoids in hippocampus. J Neurosci. 2003;23(6):2371–82.
214. Ferreira-Vieira TH, Bastos CP, Pereira GS, Moreira FA, Massensini AR. A role for the endocannabinoid system in exercise-induced spatial memory enhancement in mice. Hippocampus. 2014;24(1):79–88.
215. García-Gutiérrez MS, Ortega-Álvaro A, Busquets-García A, Pérez-Ortiz JM, Caltana L, Ricatti MJ, Brusco A, Maldonado R, Manzanares J. Synaptic plasticity alterations associated with memory impairment induced by deletion of CB2 cannabinoid receptors. Neuropharmacology. 2013;73:388–96.
216. Avraham HK, Jiang S, Fu Y, Rockenstein E, Makriyannis A, Zvonok A, Masliah E, Avraham S. The cannabinoid CBâ receptor agonist AM1241 enhances neurogenesis in GFAP/Gp120 transgenic mice displaying deficits in neurogenesis. Br J Pharmacol. 2014;171(2):468–79.
217. De Kloet ER Sibug RM Helmerhorst FM Schmidt MV Schmidt M. Stress, genes and the mechanism of programming the brain for later life. Neurosci Biobehav Rev. 2005;29(2):271–81.
218. Pariante CM, Lightman SL. The HPA axis in major depression: classical theories and new developments. Trends Neurosci. 2008;31(9):464–8.
219. Riebe CJ, Wotjak CT. Endocannabinoids and stress. Stress. 2011;14(4):384–97.
220. Patel S, Roelke CT, Rademacher DJ, Cullinan WE, Hillard CJ. Endocannabinoid signaling negatively modulates stress-induced activation of the hypothalamic-pituitary-adrenal axis. Endocrinology. 2004;145(12):5431–8.
221. Hill MN, Tasker JG. Endocannabinoid signaling, glucocorticoid-mediated negative feedback, and regulation of the hypothalamic-pituitary-adrenal axis. Neuroscience. 2012;204:5–16.
222. Roberts CJ, Stuhr KL, Hutz MJ, Raff H, Hillard CJ. Endocannabinoid signaling in hypothalamic-pituitary-adrenocortical axis recovery following stress: effects of indirect agonists and comparison of male and female mice. Pharmacol Biochem Behav. 2014;117:17–24.
223. Piscitelli F, Di Marzo V. "Redundancy" of endocannabinoid inactivation: new challenges and opportunities for pain control. ACS Chem Neurosci. 2012;3(5):356–63.
224. Hu B, Doods H, Treede RD, Ceci A. Depression-like behaviour in rats with mononeuropathy is reduced by the CB2-selective agonist GW405833. Pain. 2009;143(3):206–12.

Chapter 6
Cannabinoid Control of Fear Responses

Mathilde Metna-Laurent, Giovanni Marsicano and Edgar Soria-Gómez

Abstract In 2002, a landmark study showed that the endogenous activity of the cannabinoid type-1 (CB_1) receptors is necessary for extinction of conditioned freezing in mice (Marsicano G, Wotjak CT, Azad SC, Bisogno T, Rammes G, Cascio MG, Hermann H, Tang J, Hofmann C, Zieglgänsberger W, Di Marzo V, Lutz B. The endogenous cannabinoid system controls extinction of aversive memories. Nature. 2002 Aug 1;418(6897):530–4.). Since extinction of conditioned freezing is an important indicator of fear adaptation in animals and because our ability to control emotional responses is important to ensure adapted behaviors, the potential function of the endocannabinoid system (ECS) in such processes has generated a large interest. In this chapter, we will provide pieces of information linking the activity of the ECS and fear modulation.

Keywords CB_1 receptor · Learned fear responses · Coping strategies · Conditional CB_1 knock-out mice

Fear and Its Regulation: Theoretical Framework

Definitions

Fear can be defined as a subjective unpleasant emotional state elicited by the presence of a threat [1]. Fear is a primary survival mechanism because it prepares the organism to effectively avoid potential dangers. Fear is accompanied by a number of physiological responses leading to the adoption of behavioral responses aiming at removing the threat situation.

Despite these words are often confused and used almost as synonymous, it is important to differentiate the concept of fear from other types of aversive responses

E. Soria-Gómez (✉) · M. Metna-Laurent · G. Marsicano
University of Bordeaux, INSERM U862 NeuroCentre Magendie, Bordeaux, France
e-mail: edgar.soria@inserm.fr

P. Campolongo, L. Fattore (eds.), *Cannabinoid Modulation of Emotion, Memory, and Motivation,* DOI 10.1007/978-1-4939-2294-9_6

such as anxiety. According to the last version of the Diagnostic and Statistical Manual of Mental Disorders [2], anxiety denotes an "*apprehensive anticipation of future danger or misfortune accompanied by a feeling of dysphoria or somatic symptoms*". Indeed, both fear and anxiety can be regarded as emotional states. However, fear is proposed to differ from anxiety in having an identifiable eliciting stimulus [1]. Moreover, anxiety mostly denotes a pathological state, whereas fear is a natural reaction to threatening stimuli [1].

Fear is a physiological emotional response that helps avoiding present or future threats. However, if fear responses do not adapt to the environmental changes, the individuals' ability to exert normal life activities may be compromised. Indeed, whereas the ability to produce fear responses is fundamental for individuals' survival, the control of fear responses is also necessary to maintain normal mental and physical activities [3]. Many authors used the terms of coping behaviors to define the behavioral repertoire engaged by individuals in aversive situations [4, 5]. Even though the precise definitions of coping behaviors differ among authors, Wechsler [6] proposed that "*coping is a behavioral response aiming at reducing the effect of aversive stimuli on fitness or physiological measures related to fitness*". Indeed, effective coping implies that the engaged responses successfully remove the aversive stimuli or decrease the physiological consequences of aversive stimuli that cannot be removed. As we will see below, coping behaviors are strongly subjected to individual differences [6–9].

Studying Fear in Animals

Emotions in general and fear in particular are highly preserved across species evolution [10, 11]. Certain emotions are expressed similarly in people around the world, mostly independently of possible cultural transmissions. Moreover, certain emotions are expressed similarly across closely related species further suggesting that they are phylogenetically conserved. Thus, it is likely that emotion circuits in the body are conserved across mammalian species [12]. Hence, exploring emotional mechanisms in the nonhuman mammalian brain can be considered an acceptable approximation to understand human emotions. Following the definition proposed above, fear would be characterized by three main components: (i) physiological changes, (ii) behavioral responses, and (iii) affective experiences (feelings). For obvious reasons, it is not possible to objectively measure feelings in animals other than humans. However, it is possible to assess animals' emotional behaviors and their related physiological responses that allow them to deal with challenges in their environments. For instance, rodents faced with predator odors or stimuli that predict potential injury will adopt defensive responses [13]. In animal models, these threats can be innately recognized or learned [8, 13–15]. In particular, the fear conditioning paradigms, especially in rodents, have received large interest for studying the biological basis of fear.

Aversive Learning Paradigms in Animals

Classical Fear Conditioning

Behavioral Procedure

In classical fear conditioning, an initially neutral stimulus, that does not elicit observable emotional response by itself (for instance an auditory cue of mild intensity), is temporally associated with an aversive stimulus or unconditioned stimulus (US; such as the delivery of an electric footshock). Following this pairing, the presentation of the tone (conditioned stimulus, CS) in the absence of US becomes able to induce the fear responses (conditioned fear responses, CR). In this task, subjects learn an association that allows a novel stimulus to become a "warning of danger", eliciting the defensive responses in anticipation of the inescapable danger [14, 16–18]. Classical fear conditioning can be induced in many species from invertebrate to mammals, including humans [14, 19, 20]. In rodents, several types of conditioned fear responses were described. These include autonomous arousal (e.g. an increase of heart rate and blood pressure), endocrine responses (hormone release), reflex potentiation and hypoalgesia. Conditioned fear is also expressed by several defensive behavioral responses [13, 21, 22]. In particular, animals readily respond to CS with the absence of any movements except those dedicated to breathing. This particular behavior is common to several species, but it is particularly strong in rodents and has been defined as "freezing" [14, 23], and likely reflects the need of the individual to hide from potential dangers, such as predators. Fear conditioning procedures produce rapid and robust learning as a single footshock can produce high levels of freezing that can be retained for months. Indeed, research from many laboratories used the classical fear conditioning procedure to study the neuronal mechanisms of fear and memory [18, 24–26]. Importantly, in the early 1990s, critical observations were made about different roles played by several brain regions in fear conditioning. From these studies, two regions emerged, among others, as key players in the brain processes mediating conditioned fear responses, the amygdala and the hippocampus, which were also shown to subserve different fear-related components of freezing behavior. Whereas the amygdala was found to be necessary for learning about both contextual (i.e. learning about where shocks were delivered) and discrete (i.e. cues) stimuli [27], the hippocampus was found to have a selective role in fear to contextual stimuli [27, 28], although the putative necessity of the hippocampus in discrete fear conditioning is still highly debated [29, 30].

Once the conditioned freezing response is acquired, it is possible to inhibit it by a procedure called extinction. Extinction of conditioned freezing is induced by a prolonged or repeated CS presentation in the absence of the US, which lead to a progressive decrease of the fear response(s) observed [16, 31, 32]. Extinction of conditioned fear is generally not considered as forgetting of the previously acquired CS-US association. Indeed, a previously extinguished fear response can recover with the passage of time (spontaneous recovery). Moreover, extinction is context-specific; if extinction is induced in a different context than acquisition, the freezing response is again high when the animals are re-exposed to the acquisition context (renewal). In the same way, if the US is again delivered after extinction completion, the conditioned fear response is again expressed at successive tests (reinstatement) [33]. Therefore, extinction would rely on the identification of the decreased contingency between the CS and the US. However, an important question in fear extinction mechanisms is its associative nature; although it is argued that extinction involves the formation of a new competing inhibitory CS-US association (new learning), non-associative mechanisms have been also proposed to account for the freezing inhibition observed following the extinction procedure [33]. For instance, mice can exhibit strong freezing response to a neutral, unconditioned tone if they previously experienced a strong footshock in a distinct context (sensitized fear). The freezing response then declines during tone re-exposure in a way similar to that observed during extinction of conditioned freezing. In this sense, extinction of freezing would also involve habituation processes [34]. This is an important point as CB_1 receptors have been particularly involved in this component of fear adaptation [35].

The Conditioned Freezing Response

When exposed to threatening stimuli, rodents primarily tend to freeze in the absence of an escape route [23]. The measurement of freezing behavior is non-invasive and easy to perform. Indeed, visual observation and scoring of freezing behavior is a reliable and commonly used index of conditioned fear. However, intense US or many conditioning trials have been shown to reduce the expression of freezing [36]. It has been thus proposed that, in wild conditions, the level of fear determines the nature of the defensive behavior that is engaged in response to a threat following the "*predatory imminence continuum*" [37]. Accordingly, moderate levels of fear associated with a distal predator might induce freezing behavior to allow threat detection. However, particularly high levels of fear associated with contact with the predator might induce active defensive behaviors (fighting/ escape attempt), thereby reducing freezing behaviors. Shock probability in fear conditioning settings might essentially model the predator distance in the "*predatory imminence continuum*" [37]. Previous work has implicated projections from the acoustic thalamus to the amygdala in the classical conditioning of emotional responses to auditory stimuli. The purpose of those studies was to determine whether the lateral amygdaloid nucleus (LA), which is a major subcortical target of projections from the acoustic thalamus, might be the sensory interface of the amygdala in emotional conditioning. Lesions were performed in the LA of rats and the effects on emotional conditioning were examined [38]. Lesions of the LA, but not lesions of the adjacent striatum or cortex, interfered with emotional conditioning. Lesions that only partially destroyed LA or lesions placed too ventrally that completely missed LA had no effect. LA lesions did not affect the responses elicited following non-associative (random) training. LA is thus an essential link in the circuitry through which auditory stimuli are endowed with affective properties and may function as the sensory interface of the amygdala during emotional learning. Recently, selective inhibition of a subset of neurons within the CeA was shown to switch the response of conditioned mice form freezing to intense digging and rearing behaviors. This switch was interpreted as an increase of active coping behaviors [39]. Hence, an absence or reduction of freezing associated with a neural manipulation may not necessarily imply a loss of fear or modulation of the original fear memory, but rather a shift of the nature of the fear response engaged by the animal.

Neuronal Mechanisms

The neurobiological literature on fear conditioning and fear extinction is impressively rich and comprises regional, cellular and molecular description of the brain mechanisms supporting these phenomena [18, 25, 33, 40–44]. In sake of brevity, in this section, we will specifically focus on the mechanisms that will be useful for understanding the role of the ECS in the regulation of fear responses.

The amygdala is a critical center for both the integration of relevant fearful stimuli and for the organization of the neuronal outputs leading to the expression of the fear responses (Fig. 6.1). The amygdalar activity is also important for proper extinction of conditioned freezing. In the case of auditory fear conditioning, both CS and US sensory inputs mainly converge to the lateral nucleus of the amygdala (LA) either directly from the somatosensory cortex or indirectly from the thalamus [38, 45]. Indeed, the LA is a site of convergence of CS and US information, and lesions confined to the LA abolish auditory fear conditioning [38]. Accordingly, fear conditioning is accompanied by an enhancement of synaptic transmission at excitatory auditory input synapses in the LA. Auditory stimuli elicit field potentials in the LA of behaving rats, and induction of long-term potentiation (LTP) in the thalamic input to the LA leads to an enhancement of these auditory responses [46]. Additionally, fear conditioning in freely behaving rats results in a potentiation of excitatory field potentials recorded from LA [47]. These studies suggest that associative auditory fear learning occurs, at least in part, through changes in synaptic activity in the LA. The LA projects to the CeA, both directly and through the basal nucleus of the amygdala (BA). The intrinsic connection between the LA and CeA involve complex local excitatory and inhibitory circuits [25, 48]. Damage to the CeA interferes with the expression of conditioned freezing [48, 49]. The CeA is mainly composed of GABAergic neurons, which are spatially and functionally organized to encode acquisition and expression of conditioned freezing. Indeed, the lateral subdivision of the CeA is necessary for acquisition of conditioned freezing, whereas its expression is mediated by GABAergic neurons of the medial subdivision of the CeA [48]. BA-restricted lesions do not prevent the acquisition or expression of conditioned freezing [50, 51]. In turn, the CeA projects to brainstem areas controlling the expression of conditioned fear responses, including autonomous, endocrine and behavioral (e.g. freezing) responses.

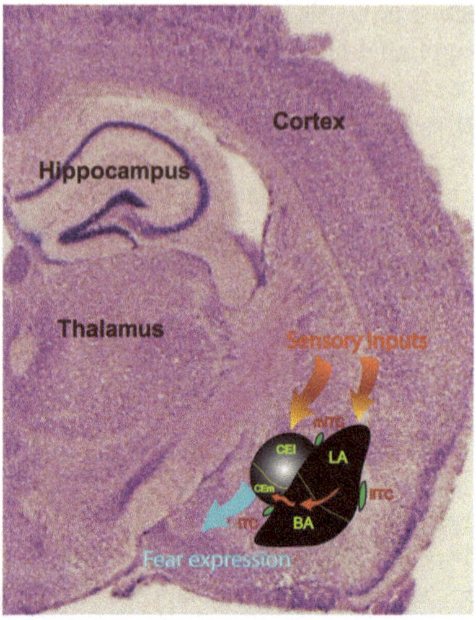

Fig. 6.1 Simplified schematic representation of amygdalar circuits participating in fear conditioning and expression. The figure shows the localization of the amygdalar complex in a coronal section of a mouse brain stained with cresyl violet. The flow of information starts with the sensory inputs arriving to both amygdalar subregions, central (*CEl*, lateral subdivision; *CEm*, medial subdivision) and basolateral (*LA*, lateral; *BA*, basal). The basolateral region sends information to the central amygdala, which is the main amygdalar output targeting several brain structures resulting in fear expression, either passive or active. ITC, intercalated cells; lITC, lateral intercalated cells; mITC, medial intercalated cells

Amygdala nuclei are also under the control of other brain structures, including the mPFC and the hippocampus that are thought to ensure contextual, temporal and mnemonic modulations of conditioned fear expression [18]. For instance, the activity of the prelimbic portion of the mPFC is necessary for conditioned freezing expression and has been proposed to participate in the encoding and integration of emotionally salient information [41, 52].

Extinction of conditioned freezing also requires amygdalar activity. Its implication in extinction has been mainly shown by acute pharmacological inactivation of targeted nuclei since amygdalar lesions primarily prevent conditioned fear acquisition and expression [33]. Inhibition of mitogen-activated protein kinases (MAPK) within the BLA prevented within-session freezing extinction [53] and both group I metabotropic glutamate receptors (mGluR) and N-Methyl-D-Aspartate receptors (NMDAR) signaling in the BLA is also necessary for acquisition of extinction [54, 55], suggesting that synaptic plasticity events occur in the BLA during extinction. Morgan et al. [56] observed in rats that mPFC lesions delayed extinction of conditioned freezing. Accordingly, the infralimbic (IL) region of the mPFC is a potential site of extinction consolidation, as IL lesions left within-session acquisition of extinction intact, but impaired extinction retrieval on the following day [57].

Electrophysiological evidence suggests that the IL inhibits the expression of conditioned freezing during extinction through reciprocal connections with the amygdala [58, 59]. Therefore, inhibition of conditioned freezing during extinction may occur through IL activation of the amygdala. Evidence also suggests that hippocampal projections to the mPFC and the amygdala mediate the context-dependent expression of extinction [60]. Interestingly, Herry et al. could discriminate, in behaving mice, specific circuitries in the BA, whose activities correlate with acquisition of extinction of auditory fear conditioning, or with high fear maintenance, respectively. These circuits depicted specific electrophysiological signatures, revealing distinct connections with both the hippocampus and the mPFC [60]. In particular, only "extinction neurons" are characterized by reciprocal connections with the mPFC, emphasizing the importance of its bidirectional connections with the amygdala to ensure extinction [58, 59, 61]. These results also suggest that close local microcircuits can switch between distinct behavioral states.

Avoidance Learning

Behavioral Procedures

One of the most common behaviors correlated with the occurrence of aversive event is the tendency of the individuals to escape from stimuli associated with those events. Avoidance learning paradigms were used well before classical fear conditioning to study fear in laboratory animals [12, 14, 62]. Originally, Estes and Skinner [62] developed the conditioned suppression of feeding behavior task in which food restricted rats trained to press a lever for food reinforcement decreased their rate of responding when a warning cue, i.e. CS, was presented and predicted the delivery of an electric shock, i.e. US. Later on, Sidman and colleagues observed that animals rapidly learn to engage behaviors to escape from a place where they were presented to a conditioned fear stimulus (active avoidance) [63]. The symmetric procedure, passive avoidance, in which animals learn to avoid approaching places that were previously associated with an aversive stimulus, was then established [64]. Indeed, avoidance learning is considered as the behavioral consequence of an instrumental (operant) conditioning in which a predictable aversive event (e.g. electric shock) does not occur contingent upon the occurrence or non-occurrence of a specified response [13]. In the active form, the avoidance contingency depends on the occurrence of a specific, ongoing response; in the passive form, the avoidance contingency depends on the suppression of a specified response. In both forms the conditioned avoidance behavior results in the prevention of the punishment [64]. In each case, the conditioned avoidance response is considered as a measurement of fear learning [15, 62–64].

Because conditioned avoidance tasks can be carried out in various ways (active, passive, signaled, unsignaled), each primarily involving the learning of a Pavlovian (CS-US) and/or an instrumental association [50, 65–67], the behavioral complexity of avoidance conditioning is not fully understood yet. Thus, classical conditioning is the initial phase of avoidance conditioning. After the subjects rapidly undergo Pavlovian conditioning, they then learn avoidance responding using the CS as a warning signal. Failures to separate these components probably impeded the understanding of the brain mechanisms of conditioned avoidance. From the 1960s, Pavlovian conditioning has been considered as a more direct way to study fear processing, leading to its progressive success to assess the brain mechanisms of fear memories, at the expense of conditioned avoidance paradigms [12].

Neuronal Mechanisms

Another reason accounting for the relative neglect of avoidance learning models is that no consensus about the role of the amygdala was found [68], likely due to the little appreciation of the anatomical complexity of the amygdalar subnuclei at that time [45]. In 2000, Amorapanth et al. performed selective electrolytic lesions on different amygdalar subnuclei in rats and evaluated the consequences of such lesions on both the freezing response induced by classical fear conditioning and active avoidance induced by the same CS-US pairing [50]. Whereas lesions of the LA prevented acquisition of both conditioned freezing and active avoidance, CeA lesion specifically impaired acquisition of conditioned freezing, leaving active avoidance learning intact. Conversely, BA lesion had no effect on conditioned freezing whereas it blocked acquisition of active avoidance. Both LA and BA, but not CeA, are also necessary for the expression of already learned active avoidance responding [66]. These findings indicate that the neural pathways within the amygdala mediating the ability of a CS to elicit conditioned freezing responses and those enabling active avoidance learning can be dissociated. The LA would be necessary for the acquisition and expression of the CS-US association in fear conditioning settings; the LA-CeA projections would be part of the output system that responds to stimuli predicting danger by eliciting freezing response; and the LA-BA pathway would be part of an output system, through which active fear responses are acquired and maintained to minimize exposure to a threatening stimulus [50, 66].

Individual variability in the rate of active avoidance learning has been often reported in rodents [7, 9, 14, 67, 69–71]. Animals that show poor active avoidance performances tend to express persistent freezing responses even though freezing fails to avoid the aversive US, suggesting that conditioned freezing could be a competing defensive behavior that can interfere with active avoidance responses. Moreover, rat lines have been bred and behaviorally characterized on the basis of higher and lower performances in active avoidance tasks, suggesting that the ability to acquire and perform active avoidance is a trait-like characteristic [69]. Interestingly, the CeA is not only dispensable for acquisition of active avoidance, but its focal lesion also rescued active avoidance learning in poor performer rats that express high levels of freezing during training, suggesting that CeA activity can constrain active avoidance learning in these animals [66, 67]. The amygdala sends neuronal projections to both dorsal and ventral portions of the striatum [72]. Recent fMRI studies in humans have correlated active avoidance learning with strong activation of the striatum, and it has been thus suggested that BA-striatal output would play a role in active avoidance learning [26, 73–78].

How CB_1 Receptors Modulate Aversive Memories

Role of the Endogenous CB_1 Receptor Signaling in Learned Fear Responses

In classical cued fear conditioning, most of the published literature indicate that the systemic pharmacological blockade of CB_1 receptors with SR141716A, a CB_1 receptor antagonist (3 mg/kg), or their genetic deletion in constitutive *CB_1*-KO mice deriving from C57BL/6N background induce little or no effect on acquisition or retrieval of the conditioned freezing response when tested 24 h after the conditioning phase [35, 76, 79, 80]. However, another CB_1 receptor antagonist (AM251, 3 mg/kg) administered prior to conditioning enhances acquisition of the freezing response in rat [81]. In contextual fear conditioning, administration of AM251 prior to conditioning is able to decrease conditioned freezing in rats [81, 82]. Similarly, constitutive *CB_1*-KO mice deriving from CD1 background [83] do not display any freezing response when re-exposed to the conditioning context while *CB_1*-KO mice deriving from C57BL/6N background show a sustained freezing response to conditioned context [84]. However, SR141716A (1–10 mg/kg) does not alter acquisition of the conditioned freezing response to context [85, 86]. Indeed, the endogenous role of CB_1 receptors in acquisition of the conditioned freezing response is not clear and the results seem to vary according to the experimental conditions. Noteworthy, these results indicate different consequences following CB_1 receptor blockade using either SR141617A or AM251. Interestingly, these compounds can differentially affect glutamatergic and GABAergic transmissions in a species-dependent manner (i.e. rats *versus* mice) [87], suggesting that these discrepancies could be explained, at least partially, by the different pharmacological properties of the drugs or different sites of action. This is suggested by a recent study showing that the effects induced by SR141716A on fear expression depend on the activation of peripheral sympathetic activity [88]. Moreover, intense fear conditioning procedures can induce sustained freezing expression that might reach a ceiling effect in control animals, thus masking potential increase of freezing acquisition following manipulation of CB_1 receptor signaling.

However, the importance of the ECS in extinction of conditioned freezing is well accepted. Indeed, it was first reported in 2002 that constitutive *CB_1*-KO mice failed in adapting their freezing response when exposed to repeated or prolonged CS presentation as compared to their wild-type littermates [35, 76, 79]. Acute injection of SR141716A (3 mg/kg) before extinction training in wild-type mice confirmed that endocannabinoids (eCBs) play a major role in extinction of cued conditioned freezing through CB_1 receptor activation. Importantly, the necessity of the ECS signaling in extinction of conditioned freezing was later demonstrated for contextual fear conditioning [85–88], but also for other fear conditioned responses including fear conditioned analgesia, startle reflex potentiation and inhibitory avoidance [89–92]. Moreover, increasing eCBs availability by administering the inhibitor of eCBs uptake and breakdown AM404, enhances extinction of the conditioned startle reflex in rats [91]. It is important to note that CB_1 receptors signaling is not necessary for extinction of conditioned response to appetitive stimuli, suggesting a specific involvement of the ECS in adaptation of conditioned aversive stimuli [92–94].

The endogenous role of CB_1 receptors has been also evaluated in the two-way active avoidance task. This test is carried-out in a shuttle box composed of two identical compartments separated by a door. Animals learn to flee into the other compartment at the onset of a cue (e.g. a light or tone signal) to avoid a punishment (e.g. an electric footshock). *CB_1*-KO mice were shown to learn this task better than their wild-type littermates [95]. However, Bura et al. [96] were not able to replicate these data using a pharmacological approach.

As mentioned previously, extinction of conditioned freezing can be attributed to both associative (new learning) and non-associative processes (habituation) [33]. In an attempt to verify the role of CB_1 receptors in the non-associative component of extinction, Kamprath et al. (2006) reported that sensitized *CB_1*-KO mice by a strong footshock delivery were impaired in within-session habituation to a neutral tone presented 24 h later, suggesting that the ECS mediates extinction of freezing at least in part through modulating habituation-like processes [35]. Interestingly, the specific deletion of CB_1 receptors on forebrain principal or cortical glutamatergic neurons leads to a similar delayed habituation of the freezing response following sensitization (Table 6.1), suggesting that the CB_1 receptors-mediated control of cortical glutamatergic projection neurons is important for the non-associative component of extinction of conditioned freezing [97].

Neuronal Mechanisms of CB_1-Dependent Modulation of Conditioned Fear Responses

Marsicano et al. (2002) reported that the tissue levels of both the eCBs anandamide (AEA) and 2-arachidonoyl glycerol (2-AG) were increased in the amygdala of fear conditioned wild-type mice following the first CS re-exposure. Moreover, the impaired freezing extinction in mice was associated with the absence of long-term depression (LTD) in the BLA of *CB_1*-KO mice, pointing-out the crucial role of the amygdalar eCB tone in the process of extinction, potentially by regulating both GABAergic and glutamatergic transmission in this brain [98]. While freezing extinction induces the activation of extracellular signal-regulated kinase (ERK1/2) and the phosphatase calcineurin in wild-type animals [99, 100], this effect was strongly reduced in many brain regions of the *CB_1*-KO mice, especially in the BLA and in the prefrontal cortex [79]. Furthermore, it was suggested that the facilitator effect

Table 6.1 Cell-type specific control of CB1 receptors on learned fear responses as assessed in different CB_1-KO mouse lines

Targeted cell types	Mouse lines	Behavioral phenotypes in fear memory paradigms	References
All body cells	CB_1-KO	Sustained freezing to fear-conditioned tone	[76]
		Normal or sustained freezing to conditioned context	[84, 85]
		Sustained freezing to fear-sensitized tone	[35, 97]
		Improved active avoidance	[118]
		Improved passive avoidance	[118]
Forebrain principal neurons	CamKII-CB_1-KO	Sustained freezing to fear-sensitized tone	[97]
Cortical glutamatergic neurons	Glu-CB_1-KO	Sustained freezing to fear-conditioned tone	[117, 118]
		Normal active avoidance	[118]
		Improved passive avoidance	[118]
		Sustained freezing to fear-sensitized tone	[97]
Forebrain GABAergic interneurons	GABA-CB_1-KO	Normal freezing to fear-conditioned tone	[117]
		Decreased freezing to fear-conditioned tone	[118]
		Improved active avoidance	[118]
		Normal passive avoidance	[118]
D1 receptors – expressing cells	D1-CB_1-KO	Sustained freezing to fear-conditioned tone	[115]
		Sustained freezing to conditioned context	
Hypothalamic paraventricular nuclei and mediobasal amygdala	Sim1-CB_1-KO	Decreased freezing to fear-conditioned tone	[116]
		Sustained digging to fear-conditioned tone	
		Normal freezing to fear-conditioned context	
		Normal active avoidance	
Tryptophan hydroxylase 2 –containing cells from the raphe nucleus	TPH2-CB_1-KO	Decreased freezing to fear-conditioned tone	[117]

of the endogenous CB_1 receptor activation involved inhibition of cholecystokinin (CCK) release from GABAergic neurons within the BLA [101]. Recently, a distinct contribution of the CeA and the BLA in the CB_1 receptors-mediated modulation of freezing extinction was proposed [102]. Intra-CeA administration of AM251 in mice impaired within-session extinction of freezing response 24 h following conditioning. Conversely, intra-BLA application of the same drug selectively blocked extinction of conditioned freezing on the subsequent days. These data were associated with a time-dependent facilitation of depolarization-induced suppression of excitation (DSE) and depolarization-induced suppression of inhibition (DSI) in the CeA, suggesting that the ECS acts on both excitatory and inhibitory transmissions within the amygdala to ensure appropriate adaption of fear-conditioned freezing.

Cumulative evidence indicates that the mPFC is a critical site for CB_1 receptors-dependent modulation of conditioned fear circuits. Exposition to an odor previously paired with a footshock strongly increases the bursting activity of a subpopulation of neurons in the mPFC receiving monosynaptic inputs from the BLA [52]. In the same olfactory fear-conditioning procedure, a CB_1 receptor antagonist applied into the mPFC blocked the acquisition of conditioned freezing. This effect was associated with an impairment of the associative firing of single neurons in the mPFC, but also with an altered LTP at BLA-prelimbic cortex synapses [103–105], indicating that CB_1 receptor transmission within the BLA-mPFC pathway is necessary for encoding olfactory fear conditioning. Choi et al. (2012) showed that the levels of CB1 receptors in the mPFC were positively correlated with freezing behavior in classical fear conditioning in mice [106]. However, intra-mPFC administration of AM251 facilitated the acquisition of conditioned freezing and cardiovascular responses in a contextual fear conditioning procedure, and had no effect on conditioned startle reflex, suggesting a complex control of prefrontal CB_1 receptors onto fear acquisition that may differ according to the nature of the conditioned stimuli [107, 108]. Few studies examined the role of prefrontal CB_1 receptors on the extinction process, all indicating a facilitation of freezing extinction after endogenous or exogenous CB_1 receptor activation [108, 109].

The endogenous activity of CB_1 receptors in the hippocampus is necessary for extinction of freezing in contextual fear conditioning experiments [110] and CB_1 receptor blockade impairs the induction of LTP in CA1 pyramidal neurons of hippocampal slices in a $GABA_A$ receptors-dependent manner [111]. These results suggest that activation of CB_1 receptors might mediate extinction of conditioned freezing to context by promoting induction of LTP via a $GABA_A$ receptor-mediated mechanism.

CB_1 receptors are abundant in the dorsal column of the periaqueductal gray (dlPAG) that modulate defensive responses [112]. Enhancing eCBs availability into the dlPAG by local administration of AM404 attenuates the recall of conditioned freezing to context in a CB_1 receptor-dependent manner [113]. Interestingly, activation of CCK1 receptors inhibits GABAergic synaptic transmission via activation of CB_1 receptors [114], suggesting that the ECS modulates conditioned freezing expression by regulating, at least in part, GABAergic transmission within the dlPAG.

The use of conditional mutant mouse lines expressing cell-type specific deletion of CB_1 receptors brought important information about the mechanisms underlying CB_1 receptor-mediated control of learned fear responding (Table 6.1). Interestingly, the endogenous CB_1 receptor activity exerted on specific cells induce different behavioral consequences on fear expression. The bidirectional consequences of CB_1 receptor deletion in cortical glutamatergic neurons and in forebrain GABAergic interneurons are discussed in the following sections. CB_1 receptors located on dopamine D_1 receptors-expressing neurons participate in the attenuation of conditioned freezing expression observed in constitutive CB_1-KO mice [115]. Surprisingly, CB_1 receptors located on serotoninergic neurons from the raphe nucleus and in the paraventricular hypothalamus and mediobasal amygdala exert an opposite effect, suggesting that the ECS in these regions mediates conditioned freezing expression [116, 117].

Taken together, these studies suggest a complex and region-dependent involvement of the endogenous CB_1 receptor signaling in the control of conditioned fear.

A Role for CB_1 Receptors in the Choice of Coping Strategies Towards Threatening Stimuli?

In a recent study performed in our laboratory [118], by allowing animals sufficient time to express different fear-induced behaviors, we described the temporal expression of the freezing response (passive coping) [7, 23] and active coping behaviors (rearing, digging, wall rearing/ sniffing) [7, 8, 39] during classical fear conditioning. Interestingly, the decrease of freezing over tone presentation is associated with a concomitant increase of active coping behaviors, which becomes the dominant response pattern at the end of the session. We then confirmed that the amount of freezing is negatively associated with active avoidance performances in the same animals [67, 70]. Conversely, we observed that the amount of active coping behaviors is positively correlated with active avoidance performances. Thus, the type of response adopted in classical fear conditioning predicts individual active avoidance performances, suggesting that conditioned fear responses are subjected to individual variability.

The constitutive deletion of CB_1 receptors in *CB_1*-KO mice leads to strong freezing expression that prevents the development of active coping behaviors in classical conditioning. However, the same deletion induces both higher passive and active avoidance learning as compared to wild-type littermates, indicating that the relationship between the fear coping strategies adopted in classical fear conditioning and avoidance learning performances is disturbed following constitutive CB_1 receptors inactivation. We found that the dominant freezing response, adopted in classical fear conditioning, and the higher passive avoidance learning displayed by *CB_1*-KO mice are likely accounted by the deletion of CB_1 receptors on cortical glutamatergic neurons (see Fig. 6 **in** [118]). Conversely, the deletion of CB_1 receptors on forebrain GABAergic neurons in GABA-*CB_1*-KO mice leads to the immediate adoption of active coping behaviors in classical fear conditioning and to a facilitation of active avoidance learning (see Fig. 6 **in** [118]). Acute low and high doses of THC have been proposed to preferentially act at CB_1 receptors on glutamatergic neurons and GABAergic neurons, respectively [119]. Conversely, we observed a dose-dependent biphasic effect of THC on the coping styles in wild-type C57BL/6N mice in classical fear conditioning, with low doses favoring active responses and higher doses promoting passive behaviors. Considering that low and high doses of THC have been suggested to act primarily through CB_1 receptors expressed on glutamatergic and GABAergic neurons, respectively [119], these data reveal a cell type-specific control of CB_1 receptors over the behavioral strategy engaged in response to fear conditioned stimuli.

Local re-expression of CB_1 receptors in the amygdala (see Fig. 6 in [118]) of constitutive CB_1-KO restores the temporal relationship between freezing and active coping in classical fear conditioning by decreasing freezing responses, suggesting that CB_1 receptors in the amygdala participate in the adoption of the coping style by animals when faced to aversive stimuli.

As pointed out above, responses to aversive stimulus could be pathological as in the case of anxiety disorders. The group of Beat Lutz in Mainz (Germany) recently showed that there is also a biphasic modulation of anxiety responses by the ECS [120]. In short, the CB_1 receptor-mediated control of glutamatergic transmission is responsible for the anxiolytic-like effect induced by low doses of CB_1 receptor agonists. Conversely, the anxiogenic phenotype observed after injection of high doses of cannabinoids is mediated by action of CB_1 receptors in GABAergic neurons. However, it is still not clear if in such phenomenon is also participating the amygdalar circuits.

Individual Variability of the Fear Responses

Although freezing behavior is a well-established index of conditioned fear, other behaviors may compete with freezing under a variety of conditions. It has been proposed that rodents confronted by potential dangers such as a predator odor or a shock probe can show behaviors oriented toward the threat to facilitate both visual and/ or olfactory detection [121–123]. These behaviors, including defensive burying and rearing, are thus considered as attempts to investigate the potential danger when its source is ambiguous [8, 65, 123–125]. For instance, if sawdust is present in a cage with an aversive encounter, rodents can actively dig to bury the danger source with the bedding [8, 123]. In addition, digging holes is a normal escape strategy for rodents in the wild. Whereas this is obviously impossible in laboratory conditions, the digging response in the sawdust might also represent an innate escape response towards imminent threatens, such as fear conditioned stimuli.

Rearing behavior could be induced in rats by either exposition to novel environments [126] or by the introduction of a cat odor in a familiar place [125]. In this last example, rearing is accompanied with an increase of blood pressure. This is also observed when rats are returned to the previously cat odor-associated context [125]. Moreover, rearing can compete with freezing when rodents are introduced to inescapable contexts previously associated with shocks and rearing can be learned as a conditioned avoidance response [8, 65]. Indeed, as suggested by others and us, freezing is just one, although likely the temporally most immediate and pronounced, of the many behaviors that rodent might exhibit in response to a fear conditioned stimulus.

It has been proposed that freezing and escape behaviors are part of the innate species-specific defense responses of the rodent behavioral repertoire that are engaged by aversive stimuli [13, 65], but the conditions in which a particular behavior is selected are not known. The test situation can determine the type of defensive response adopted by the subjects. For example, classical fear conditioning settings are characterized by the absence of escape routes and it has been proposed that rodents naturally tend to freeze in this condition [23]. An alternative explanation is that the threat proximity determines the type of response adopted [121]. A third hypothesis is that the type of defensive behavior can be determined by stable individual trait-like characteristics [6, 7, 127]. Indeed, even within inbred populations of mice or rats exposed to aversive situations, the type of response primarily adopted can distinguish subpopulations. Several studies reported that some individuals within a population of wild-type, naive rats or mice show very low conditioned freezing response after classical fear conditioning, and some individuals never acquire active avoidance even after an extended training [67, 70, 128]. Moreover, rats have been selected and bred with regard to their defensive behavioral profiles in a variety of experimental situations, so as to generate strains that are characterized, for instance, by their ability to learn active avoidance [129]. Indeed, the Roman High Avoidance (RHA) and Low Avoidance (RLA) rats are characterized by high and low performances in the active avoidance paradigm, respectively. Interestingly, these animals also differ in the amount of conditioned freezing induced by classical fear conditioning, with RHA showing a decreased freezing response and RLA displaying increased freezing response to CS [130, 131].

Consistently, conditioned freezing during the initial stages of two-way active avoidance learning is negatively correlated to the efficiency in the acquisition of the task [70], supporting the idea that freezing tendency to CS runs against the appearance of active avoidance responses. Our results also indicate that active coping behaviors measured in classical fear conditioning are positively correlated to active avoidance performances in wild-type animals. Accordingly, comparative studies have proposed a positive link between defensive digging and two-way active avoidance performances [9].

However, it is important to remind that classical fear conditioning and instrumental conditioning such as avoidance learning involve quite different learning processes. Avoidance conditioning has long been viewed as a two-stage learning process in which animals initially undergoes Pavlovian conditioning to form an association between the shock and the CS in the apparatus [15, 132]. Subsequently, the subjects learn the instrumental response to avoid the shock. Further, the "fear" aroused by the presence of the CS motivates learning of the instrumental response. Following this theory, "fear" reduction is associated with successful avoidance and has been proposed to reinforce avoidance learning. In this case, "fear" is defined as the presence of Pavlovian defensive responses. However, if one considers fear as a central state that can result from exposure to innate or learned stimuli [15] without assumption about the relationship between the type of learning and fear emotional state, fear could be indicated in animals by either classical or instrumental conditioned fear responses. Therefore, the fact that there is a bimodal control of fear coping strategies by the ECS supports the idea that individual variability in defensive strategies might bias the interpretation of performances observed in a conditioned aversive test.

The ECS as a Determinant of the Types of Fear Coping Strategies

The deletion of CB_1 receptors on cortical glutamatergic neurons and in forebrain GABAergic neurons leads to distinct coping strategies in classical fear conditioning, and consistently enhances passive and active avoidance responding, respectively [113]. Additionally, it has been also proposed that the endocannabinoid AEA could have and essential role in the selection of behavioral responses under aversive conditions [133–135]. We thus proposed that CB_1 receptors signaling is an important physiological determinant of fear coping styles and that suppression of the inhibitory control on GABAergic and glutamatergic neurotransmissions likely facilitates the adoption of active or passive fear coping strategies, respectively. Consistently, increasing GABAergic tone by systemic pharmacological administration of $GABA_A$ receptor agonists including benzodiazepines enhances acquisition of shuttle box active avoidance [136–138]. Moreover, it has been shown that the acute stimulation of CB_1 receptors into the CeA impairs consolidation of one-trial dark-light passive avoidance, an effect reversed by co-administration of NMDA [139]. In classical fear conditioning, Lehner et al. [128] measured extracellular GABA concentration by *in vivo* microdialysis in the BLA of two rat groups selected upon their conditioned freezing response intensity, i.e. "high freezer" versus "low freezer". Although GABA levels were identical prior and following the conditioning session in both rat groups, they observed an increase of extracellular GABA during a 10-min CS re-exposure only in "low freezer" animals, suggesting that high GABA release in the BLA might decrease freezing responses. Unfortunately, active coping behaviors were not analyzed in this study. Altogether, these data are in agreement with our proposed bimodal control of CB_1 receptors on glutamatergic and GABAergic transmission in fear coping styles and further support a key role of the amygdala in this mechanism. However, there is also evidence showing that CB_1 receptor-dependent control of other neurotransmitter systems such as catecholamine, dopamine or noradrenaline transmission could play an essential role in the modulation of stress coping behaviors [140]. The ECS may thus either enhance or inhibit responses to aversive stimuli, possibly caused by its modulatory activity on diverse neurotransmitters. The Fig. 6.2 exemplifies the cannabinoid action in amygdalar circuits in the control of fear coping strategies.

Innate tendencies to passively or actively cope with fearful stimuli can be assessed in other behavioral paradigms that allow animals to choose between different patterns of responses efficient for avoiding a harmful stimulus. For instance, in the defensive burying paradigm, animals are confronted with an electrified probe

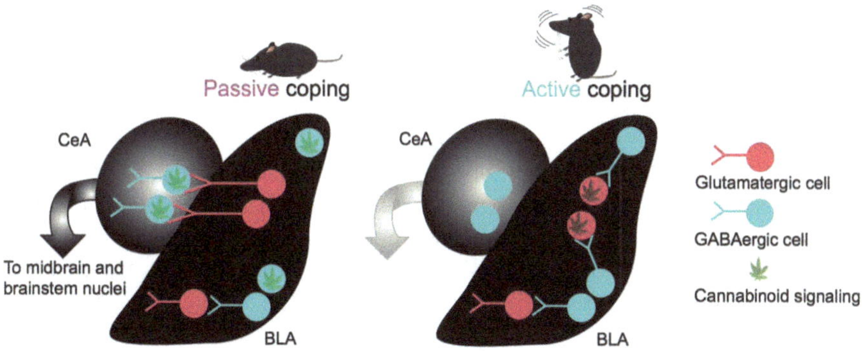

Fig. 6.2 Putative bimodal control of fear expression by cannabinoid signaling in amygdalar circuits. The absence of cannabinoid signaling in glutamatergic cells leads to an increase in passive responses (*left*) and potentially to a hyperactivation of the central amygdala (*CeA*), the main amygdalar output. Conversely, the augmentation of active coping strategies observed in animals lacking CB_1 receptors in GABAergic cells (*right*) could be due to an increase of inhibition of glutamatergic cells in the basolateral amygdala (*BLA*), which, in turn, will result in a reduction of CeA excitation. Thus, amygdalar CB_1 signaling is key in maintaining equilibrium between passive and active coping responses

inserted into their home cage. In response to a brief contact with the probe, animals are observed either to actively bury the probe with the bedding or to display freezing [7, 8, 141]. Both response patterns can be considered as successful coping because they lead to shock avoidance. A positive correlation between shuttle box active avoidance and burying behavior in the defensive burying and a negative correlation between active avoidance and freezing behavior to the probe insertion were observed in the RLA/ RHA rat strains [142]. Furthermore, since both coping behaviors can be expressed within the task, the shock-probe burying paradigm is well suited to describe the neural circuitry of active and passive coping strategies. Indeed, one study assessed the involvement of CB_1 receptor signaling in this test [143]. In the constitutive *CB_1*-KO mice, a decrease of burying time but no change in freezing time was observed, and mutant mice made fewer contacts with the electrified probe. In addition, the acute pharmacological blockade of CB_1 receptors leads to similar consequences only at a middle dose-range (i.e. 3 mg/kg but not 1 and 10 mg/kg), suggesting that the coping strategies adopted in this test might vary as a function of the amount of CB_1 receptors targeted. Interestingly, individual differences in behaviors adopted in the defensive burying test have been associated with distinct patterns of both endocrine reactivity and brain regions activity as assessed by immediate-early genes expression [9]. Thus, exploring the phenotype of the *CB_1*-KO mouse lines in the defensive burying test would be an interesting perspective to further understand the brain mechanisms underlying the control of fear coping strategies by the ECS.

Interestingly, a recent positron emission tomography (PET) study in healthy humans revealed a strong inverse correlation between the brain binding of CB_1 receptors and personality traits linked to novelty-seeking [144], which have been associated with proactive coping [131]. In another study, Rabinak et al. [145] showed that cannabinoid drugs are able to modulate the prefrontal-limbic circuits during fear extinction in humans. Considering that the very large majority of CB_1 protein in the brain is expressed by GABAergic neurons [119, 146–148], the decreased levels of CB_1 binding in "novelty-seekers" observed in the PET study are likely due to reduced expression levels in inhibitory circuits. Thus, the PET data, together with the results using conditional mutant mice [149] suggest that the differential impact of CB_1 receptors on fear coping strategies might be extended to human subjects. Consistently, the therapeutic potential of cannabinoid drugs for the treatment of anxiety disorders is currently under focus [145, 150–153].

Conclusions

The data reviewed above underline the complex relationship between ECS signaling in the brain and fear responses. Differences in specific behavioral traits addressed by different paradigms, but also slight variations in the parameters used in the same tasks can indicate various and even opposite roles of ECS. However, these studies underline that the measure of single behavioral responses might not be sufficient to determine the levels of learning and, thereby, of fear expressed by animals in different experimental conditions. Indeed, in our experiments addressing different "styles" of fear coping behavior in mutant mice lacking CB_1 receptors in specific neuronal populations, we found that summing all the fear-induced behaviors during testing, results in a progressive extinction of global fear responses. However, this extinction is much slower than the classical decrease of freezing and, importantly, is not influenced by genetic manipulations of CB_1 receptors. Thus, ECS signaling might be important to "choose" a specific strategy to cope with fear, but not in determining the proper learning and extinction of the fear memory trace *per se*. This hypothesis will need further studies to be confirmed or rejected. All in all, the impact of ECS signaling in fear processes is far from being fully understood, but its study represents an excellent way to better understand fear memories and their expression.

References

1. Ohman A. Fear and anxiety: evolutionary, cognitive, and clinical perspectives. In: Lewis M & Haviland-Jones JM, editors. Handbook of emotions, 2nd edition. New York: Guilford Publications; 2000. pp. 573–93.
2. American Psychiatric Association Diagnostic and statistical manual of mental disorders : DSM-IV-TR. Washington: American Psychiatric Association; 2000.
3. Gross JJ. Antecedent- and response-focused emotion regulation: divergent consequences for experience, expression, and physiology. J Pers Soc Psychol. 1998;74(1):224–37.
4. Weiss JM. Effects of coping responses on stress. J Comp Physiol Psychol. 1968;65(2):251–60.
5. Lazarus RS, Folkman S. Stress, appraisal, and coping. New York: Springer; 1984.

6. Wechsler B. Coping and coping strategies: a behavioural view. Appl Anim Behav Sci. 1995;43(2):123–34.
7. Koolhaas JM, Korte SM, De Boer SF, Van Der Vegt BJ, Van Reenen CG, Hopster H, De Jong IC, Ruis MAW, Blokhuis HJ. Coping styles in animals: current status in behavior and stress-physiology. Neurosci Biobehav Rev. 1999;23(7):925–35.
8. De Boer SF, Koolhaas JM. Defensive burying in rodents: ethology, neurobiology and psychopharmacology. Eur J Pharmacol. 2003;463(1–3):145–61.
9. Koolhaas JM, De Boer SF, Coppens CM, Buwalda B. Neuroendocrinology of coping styles: Towards understanding the biology of individual variation. Front Neuroendocrinol. 2010;31(3):307–21.
10. Darwin C. The expression of the emotions in man and animals. London: John Murray; 1872.
11. Ekman P. Emotions inside out: 130 years after Darwin's the expression of the emotions in man and animals. New York: New York Academy of Sciences; 2003.
12. LeDoux JE. Evolution of human emotion: a view through fear. Prog Brain Res. 2012;195:431–42.
13. Bolles RC. Species-specific defense reactions and avoidance learning. Psychol Rev. 1970;77(1):32.
14. Blanchard RJ, Blanchard DC. Passive and active reactions to fear-eliciting stimuli. J Comp Physiol Psychol. 1969;68(1):129–35.
15. McAllister WR, McAllister DE, Douglass WK. The inverse relationship between shock intensity and shuttle-box avoidance learning in rats: a reinforcement explanation. J Comp Physiol Psychol. 1971;74(3):426–33.
16. Pavlov IP. Conditioned reflexes. London: Routledge and Kegan Paul; 1927.
17. Fanselow MS. Conditioned and unconditional components of post-shock freezing. Pavlov J Biol Sci. 1980;15(4):177–82.
18. LeDoux JE. Emotion circuits in the brain. Annu Rev Neurosci. 2000;23:155–84.
19. LeDoux JE, Sakaguchi A, Reis DJ. Subcortical efferent projections of the medial geniculate nucleus mediate emotional responses conditioned to acoustic stimuli. J Neurosci. 1984;4(3):683–98.
20. LaBar KS, LeDoux JE, Spencer DD, Phelps EA. Impaired fear conditioning following unilateral temporal lobectomy in humans. J Neurosci. 1995;15(10):6846–55.
21. Blanchard DC, Griebel G, Blanchard RJ. Mouse defensive behaviors: pharmacological and behavioral assays for anxiety and panic. Neurosci Biobehav Rev. 2001;25(3):205–18.
22. Blanchard DC, Griebel G, Blanchard RJ. The mouse defense test battery: pharmacological and behavioral assays for anxiety and panic. Eur J Pharmacol. 2003;463(1–3):97–116.
23. Blanchard RJ, Fukunaga KK, Blanchard DC. Environmental control of defensive reactions to footshock. Bull Psychon Soc. 1976;8(129):40.
24. Maren S. Synaptic transmission and plasticity in the amygdala. An emerging physiology of fear conditioning circuits. Mol Neurobiol. 1996;13(1):1–22.
25. Ehrlich I, Humeau Y, Grenier F, Ciocchi S, Herry C, Lüthi A. Amygdala inhibitory circuits and the control of fear memory. Neuron. 2009;62(6):757–71.
26. Hartley CA, Phelps EA. Changing fear: the neurocircuitry of emotion regulation. Neuropsychopharmacology. 2010;35(1):136–46.
27. Phillips RG, LeDoux JE. Differential contribution of amygdala and hippocampus to cued and contextual fear conditioning. Behav Neurosci. 1992;106(2):274–85.
28. Kim JJ, Fanselow MS. Modality-specific retrograde amnesia of fear. Science. 1992;256(5057):675–7.
29. Calandreau L, Trifilieff P, Mons N, Costes L, Marien M, Marighetto A, Micheau J, Jaffard R, Desmedt A. Extracellular hippocampal acetylcholine level controls amygdala function and promotes adaptive conditioned emotional response. J Neurosci. 2006;26(52):13556–66.
30. Trifilieff P, Herry C, Vanhoutte P, Caboche J, Desmedt A, Riedel G, Mons N, Micheau J. Foreground contextual fear memory consolidation requires two independent phases of hippocampal ERK/CREB activation. Learn Mem. 2006;13(3):349–58.
31. Rescorla RA. Probability of shock in the presence and absence of CS in fear conditioning. J Comp Physiol Psychol. 1968;66(1):1–5.

32. Cain CK, Blouin AM, Barad M. Temporally massed CS presentations generate more fear extinction than spaced presentations. J Exp Psychol Anim Behav Process. 2003;29(4):323–33.
33. Myers KM, Davis M. Mechanisms of fear extinction. Mol Psychiatry. 2007;12(2):120–50.
34. Kamprath K, Wotjak CT. Nonassociative learning processes determine expression and extinction of conditioned fear in mice. Learn Mem. 2004;11(6):770–86.
35. Kamprath K, Marsicano G, Tang J, Monory K, Bisogno T, Di Marzo V, Lutz B, Wotjak CT. Cannabinoid CB1 receptor mediates fear extinction via habituation-like processes. J Neurosci. 2006;26(25):6677–86.
36. Fanselow MS. Shock-induced analgesia on the formalin test: effects of shock severity, naloxone, hypophysectomy, and associative variables. Behav Neurosci. 1984;98(1):79–95.
37. Fanselow MS, Lester LS. A functional behavioristic approach to aversely motivated behavior: predatory imminence as a determinant of the topography of defensive behavior. In: Bolles RC, Beecher MD, editors. Evolution and learning. Hillsdale: Lawrence Erlbaum Associates; 1988. pp. 185–212.
38. LeDoux JE, Cicchetti P, Xagoraris A, Romanski LM. The lateral amygdaloid nucleus: sensory interface of the amygdala in fear conditioning. J Neurosci. 1990;10(4):1062–9.
39. Gozzi A, Jain A, Giovannelli A, Giovanelli A, Bertollini C, Crestan V, Schwarz AJ, Tsetsenis T, Ragozzino D, Gross CT, Bifone A. A neural switch for active and passive fear. Neuron. 2010;67(4):656–66.
40. Kim JJ, Jung MW. Neural circuits and mechanisms involved in Pavlovian fear conditioning: a critical review. Neurosci Biobehav Rev. 2006;30(2):188–202.
41. Quirk GJ, Mueller D. Neural mechanisms of extinction learning and retrieval. Neuropsychopharmacology. 2008;33(1):56–72.
42. Makkar SR, Zhang SQ, Cranney J. Behavioral and neural analysis of GABA in the acquisition, consolidation, reconsolidation, and extinction of fear memory. Neuropsychopharmacology. 2010;35(8):1625–52.
43. Sotres-Bayon F, Quirk GJ. Prefrontal control of fear: more than just extinction. Curr Opin Neurobiol. 2010;20(2):231–5.
44. Johansen JP, Cain CK, Ostroff LE, LeDoux JE. Molecular mechanisms of fear learning and memory. Cell. 2011;147(3):509–24.
45. Amaral DG, Insausti R. Retrograde transport of D-(3H)-aspartate injected into the monkey amygdaloid complex. Exp Brain Res. 1992;88(2):375–88.
46. Rogan MT, LeDoux JE. LTP is accompanied by commensurate enhancement of auditory-evoked responses in a fear conditioning circuit. Neuron. 1995;15(1):127–36.
47. Rogan MT, Stäubli UV, LeDoux JE. Fear conditioning induces associative long-term potentiation in the amygdala. Nature. 1997;390(6660):604–7.
48. Ciocchi S, Herry C, Grenier F, Wolff SBE, Letzkus JJ, Vlachos I, Ehrlich I, Sprengel R, Deisseroth K, Stadler MB, Müller C, Lüthi A. Encoding of conditioned fear in central amygdala inhibitory circuits. Nature. 2010;468(7321):277–82.
49. LeDoux JE, Iwata J, Cicchetti P, Reis DJ. Different projections of the central amygdaloid nucleus mediate autonomic and behavioral correlates of conditioned fear. J Neurosci. 1988;8(7):2517–29.
50. Amorapanth P, LeDoux JE, Nader K. Different lateral amygdala outputs mediate reactions and actions elicited by a fear-arousing stimulus. Nat Neurosci. 2000;3(1):74–9.
51. Nader K, Majidishad P, Amorapanth P, LeDoux JE. Damage to the lateral and central, but not other, amygdaloid nuclei prevents the acquisition of auditory fear conditioning. Learn Mem. 2001;8(3):156–63.
52. Laviolette SR, Lipski WJ, Grace AA. A subpopulation of neurons in the medial prefrontal cortex encodes emotional learning with burst and frequency codes through a dopamine D4 receptor-dependent basolateral amygdala input. J Neurosci. 2005;25(26):6066–75.
53. Herry C, Trifilieff P, Micheau J, Lüthi A, Mons N. Extinction of auditory fear conditioning requires MAPK/ERK activation in the basolateral amygdala. Eur J Neurosci. 2006;24(1):261–9.
54. Kim J, Lee S, Park K, Hong I, Song B, Son G, Park H, Kim WR, Park E, Choe HK, Kim H, Lee C, Sun W, Kim K, Shin KS, Choi S. Amygdala depotentiation and fear extinction. Proc Natl Acad Sci U S A. 2007;104(52):20955–60.

55. Sotres-Bayon F, Bush DEA, LeDoux JE. Acquisition of fear extinction requires activation of NR2B-containing NMDA receptors in the lateral amygdala. Neuropsychopharmacology. 2007;32(9):1929–40.
56. Morgan MA, Romanski LM, LeDoux JE. Extinction of emotional learning: contribution of medial prefrontal cortex. Neurosci Lett. 1993;163(1):109–13.
57. Quirk GJ, Russo GK, Barron JL, Lebron K. The role of ventromedial prefrontal cortex in the recovery of extinguished fear. J Neurosci. 2000;20(16):6225–31.
58. Milad MR, Quirk GJ. Neurons in medial prefrontal cortex signal memory for fear extinction. Nature. 2002;420(6911):70–4.
59. Quirk GJ, Likhtik E, Pelletier JG, Paré D. Stimulation of medial prefrontal cortex decreases the responsiveness of central amygdala output neurons. J Neurosci. 2003;23(25):8800–7.
60. Ji J, Maren S. Electrolytic lesions of the dorsal hippocampus disrupt renewal of conditional fear after extinction. Learn Mem. 2005;12(3):270–6.
61. Herry C, Ciocchi S, Senn V, Demmou L, Müller C, Lüthi A. Switching on and off fear by distinct neuronal circuits. Nature. 2008;454(7204):600–6.
62. Estes WK, Skinner BF. Some quantitative properties of anxiety. J Exp Psychol. 1941;29(5):390–400.
63. Sidman M. Classical avoidance without a warning stimulus. J Exp Anal Behav. 1962;5:97–104.
64. Mineka S. The role of fear in theories of avoidance learning, flooding, and extinction. Psychol Bull. 1979;86(5):985.
65. Cain CK, LeDoux JE. Escape from fear: a detailed behavioral analysis of two atypical responses reinforced by CS termination. J Exp Psychol Anim Behav Process. 2007;33(4):451–63.
66. Choi J-S, Cain CK, LeDoux JE. The role of amygdala nuclei in the expression of auditory signaled two-way active avoidance in rats. Learn Mem. 2010;17(3):139–47.
67. Lázaro-Muñoz G, LeDoux JE, Cain CK. Sidman instrumental avoidance initially depends on lateral and basal amygdala and is constrained by central amygdala-mediated Pavlovian processes. Biol Psychiatry. 2010;67(12):1120–7.
68. Sarter M, Markowitsch HJ. Involvement of the amygdala in learning and memory: a critical review, with emphasis on anatomical relations. Behav Neurosci. 1985;99(2):342–80.
69. Steimer T, la Fleur S, Schulz PE. Neuroendocrine correlates of emotional reactivity and coping in male rats from the Roman high (RHA/Verh)-and low (RLA/Verh)-avoidance lines. Behav Genet. 1997;27(6):503–12.
70. Vicens-Costa E, Martinez-Membrives E, López-Aumatell R, Guitart-Masip M, Canete T, Blázquez G, Tobena A, Fernández-Teruel A. Two-way avoidance acquisition is negatively related to conditioned freezing and positively associated with startle reactions: a dissection of anxiety and fear in genetically heterogeneous rats. Physiol Behav. 2010;103(2):148–56
71. Díaz-Morán S, Palència M, Mont-Cardona C, Cañete T, Blázquez G, Martínez-Membrives E, López-Aumatell R, Tobeña A, Fernández-Teruel A. Coping style and stress hormone responses in genetically heterogeneous rats: comparison with the Roman rat strains. Behav Brain Res. 2012;228(1):203–10.
72. Price JL. Comparative aspects of amygdala connectivity. Ann N Y Acad Sci. 2003;985:50–8.
73. Schiller D, Delgado MR. Overlapping neural systems mediating extinction, reversal and regulation of fear. Trends Cogn Sci. 2010;14(6):268–76.
74. Delgado MR, Jou RL, Ledoux JE, Phelps EA. Avoiding negative outcomes: tracking the mechanisms of avoidance learning in humans during fear conditioning. Front Behav Neurosci. 2009;3:33.
75. Li J, Schiller D, Schoenbaum G, Phelps EA, Daw ND. Differential roles of human striatum and amygdala in associative learning. Nat Neurosci. 2011;14(10):1250–2.
76. Marsicano G, Wotjak CT, Azad SC, Bisogno T, Rammes G, Cascio MG, Hermann H, Tang J, Hofmann C, Zieglgänsberger W, Di Marzo V, Lutz B. The endogenous cannabinoid system controls extinction of aversive memories. Nature. 2002;418(6897):530–4.
77. Medina JF, Repa JC, Mauk MD, LeDoux JE. Parallels between cerebellum- and amygdala-dependent conditioning. Nat Rev Neurosci. 2002;3(2):122–31.

78. Fanselow MS, Poulos AM. The neuroscience of mammalian associative learning. Annu Rev Psychol. 2005;56:207–34.
79. Cannich A, Wotjak CT, Kamprath K, Hermann H, Lutz B, Marsicano G. CB1 cannabinoid receptors modulate kinase and phosphatase activity during extinction of conditioned fear in mice. Learn Mem. 2004;11(5):625–32.
80. Plendl W, Wotjak CT. Dissociation of within- and between-session extinction of conditioned fear. J Neurosci. 2010;30(14):4990–8.
81. Arenos JD, Musty RE, Bucci DJ. Blockade of cannabinoid CB1 receptors alters contextual learning and memory. Eur J Pharmacol. 2006;539(3):177–83.
82. Mikics E, Dombi T, Barsvári B, Varga B, Ledent C, Freund TF, Haller J. The effects of cannabinoids on contextual conditioned fear in CB1 knockout and CD1 mice. Behav Pharmacol. 2006;17(3):223–30.
83. Ledent C, Valverde O, Cossu G, Petitet F, Aubert JF, Beslot F, Böhme GA, Imperato A, Pedrazzini T, Roques BP, Vassart G, Fratta W, Parmentier M. Unresponsiveness to cannabinoids and reduced addictive effects of opiates in CB1 receptor knockout mice. Science. 1999;283(5400):401–4.
84. Jacob W, Marsch R, Marsicano G, Lutz B, Wotjak CT. Cannabinoid CB1 receptor deficiency increases contextual fear memory under highly aversive conditions and long-term potentiation in vivo. Neurobiol Learn Mem. 2012;98(1):47–55.
85. Suzuki A, Josselyn SA, Frankland PW, Masushige S, Silva AJ, Kida S. Memory reconsolidation and extinction have distinct temporal and biochemical signatures. J Neurosci. 2004;24(20):4787–95.
86. Pamplona FA, Prediger RDS, Pandolfo P, Takahashi RN. The cannabinoid receptor agonist WIN 55,212-2 facilitates the extinction of contextual fear memory and spatial memory in rats. Psychopharmacology (Berl). 2006;188(4):641–9.
87. Haller J, Mátyás F, Soproni K, Varga B, Barsy B, Németh B, Mikics E, Freund TF, Hájos N. Correlated species differences in the effects of cannabinoid ligands on anxiety and on GABAergic and glutamatergic synaptic transmission. Eur J Neurosci. 2007;25(8):2445–56.
88. Bellocchio L, Soria-Gómez E, Quarta C, Metna-Laurent M, Cardinal P, Binder E, Cannich A, Delamarre A, Häring M, Martín-Fontecha M, Vega D, Leste-Lasserre T, Bartsch D, Monory K, Lutz B, Chaouloff F, Pagotto U, Guzman M, Cota D, Marsicano G. Activation of the sympathetic nervous system mediates hypophagic and anxiety-like effects of CB_1 receptor blockade. Proc Natl Acad Sci U S A. 2013;110(12):4786–91.
89. Finn DP, Beckett SRG, Richardson D, Kendall DA, Marsden CA, Chapman V. Evidence for differential modulation of conditioned aversion and fear-conditioned analgesia by CB1 receptors. Eur J Neurosci. 2004;20(3):848–52.
90. Finn DP, Jhaveri MD, Beckett SRG, Kendall DA, Marsden CA, Chapman V. Cannabinoids modulate ultrasound-induced aversive responses in rats. Psychopharmacology (Berl). 2004;172(1):41–51.
91. Chhatwal JP, Davis M, Maguschak KA, Ressler KJ. Enhancing cannabinoid neurotransmission augments the extinction of conditioned fear. Neuropsychopharmacology. 2005;30(3):516–24.
92. Niyuhire F, Varvel SA, Thorpe AJ, Stokes RJ, Wiley JL, Lichtman AH. The disruptive effects of the CB1 receptor antagonist rimonabant on extinction learning in mice are task-specific. Psychopharmacology (Berl). 2007;191(2):223–31.
93. Hölter SM, Kallnik M, Wurst W, Marsicano G, Lutz B, Wotjak CT. Cannabinoid CB1 receptor is dispensable for memory extinction in an appetitively-motivated learning task. Eur J Pharmacol. 2005;510(1–2):69–74.
94. Harloe JP, Thorpe AJ, Lichtman AH. Differential endocannabinoid regulation of extinction in appetitive and aversive Barnes maze tasks. Learn Mem. 2008;15(11):806–9.
95. Martin M, Ledent C, Parmentier M, Maldonado R, Valverde O. Involvement of CB1 cannabinoid receptors in emotional behaviour. Psychopharmacology (Berl). 2002;159(4):379–87.
96. Bura SA, Castañé A, Ledent C, Valverde O, Maldonado R. Genetic and pharmacological approaches to evaluate the interaction between the cannabinoid and cholinergic systems in cognitive processes. Br J Pharmacol. 2007;150(6):758–65.

97. Kamprath K, Plendl W, Marsicano G, Deussing JM, Wurst W, Lutz B, Wotjak CT. Endocannabinoids mediate acute fear adaptation via glutamatergic neurons independently of corticotropin-releasing hormone signaling. Genes Brain Behav. 2009;8(2):203–11.
98. Azad SC. Activation of the Cannabinoid Receptor Type 1 Decreases Glutamatergic and GABAergic Synaptic Transmission in the Lateral Amygdala of the Mouse. Learn Mem. 2003;10(2):116–28.
99. Mansuy IM, Mayford M, Jacob B, Kandel ER, Bach ME. Restricted and regulated overexpression reveals calcineurin as a key component in the transition from short-term to long-term memory. Cell. 1998;92(1):39–49.
100. Lin CH, Yeh SH, Lin CH, Lu KT, Leu TH, Chang WC, Gean PW. A role for the PI-3 kinase signaling pathway in fear conditioning and synaptic plasticity in the amygdala. Neuron. 2001;31(5):841–51.
101. Chhatwal JP, Gutman AR, Maguschak KA, Bowser ME, Yang Y, Davis M, Ressler KJ. Functional interactions between endocannabinoid and CCK neurotransmitter systems may be critical for extinction learning. Neuropsychopharmacology. 2009;34(2):509–21.
102. Kamprath K, Romo-Parra H, Häring M, Gaburro S, Doengi M, Lutz B, Pape H-C. Short-term adaptation of conditioned fear responses through endocannabinoid signaling in the central amygdala. Neuropsychopharmacology. 2011;36(3):652–63.
103. Laviolette SR, Grace AA. Cannabinoids potentiate emotional learning plasticity in neurons of the medial prefrontal cortex through basolateral amygdala inputs. J Neurosci. 2006;26(24):6458–68.
104. Tan H, Lauzon NM, Bishop SF, Bechard MA, Laviolette SR. Integrated cannabinoid CB1 receptor transmission within the amygdala-prefrontal cortical pathway modulates neuronal plasticity and emotional memory encoding. Cereb Cortex. 2010;20(6):1486–96.
105. Tan H, Lauzon NM, Bishop SF, Chi N, Bechard M, Laviolette SR. Cannabinoid transmission in the basolateral amygdala modulates fear memory formation via functional inputs to the prelimbic cortex. J Neurosci. 2011;31(14):5300–12.
106. Choi K, Le T, McGuire J, Xing G, Zhang L, Li H, Parker CC, Johnson LR, Ursano RJ. Expression pattern of the cannabinoid receptor genes in the frontal cortex of mood disorder patients and mice selectively bred for high and low fear. J Psychiatr Res. 2012;46(7):882–9.
107. Lisboa SF, Reis DG, da Silva AL, Corrêa FMA, Guimarães FS, Resstel LBM. Cannabinoid CB1 receptors in the medial prefrontal cortex modulate the expression of contextual fear conditioning. Int J Neuropsychopharmacol. 2010;13(9):1163–73.
108. Kuhnert S, Meyer C, Koch M. Involvement of cannabinoid receptors in the amygdala and prefrontal cortex of rats in fear learning, consolidation, retrieval and extinction. Behav Brain Res. 2013;250:274–84.
109. Do Monte FH, Souza RR, Bitencourt RM, Kroon JA, Takahashi RN. Infusion of cannabidiol into infralimbic cortex facilitates fear extinction via CB1 receptors. Behav Brain Res. 2013;250:23–7.
110. De Oliveira Alvares L, Pasqualini Genro B, Diehl F, Molina VA, Quillfeldt JA. Opposite action of hippocampal CB1 receptors in memory reconsolidation and extinction. Neuroscience. 2008;154(4):1648–55.
111. Lin Q-S, Yang Q, Liu D-D, Sun Z, Dang H, Liang J, Wang Y-X, Chen J, Li S-T. Hippocampal endocannabinoids play an important role in induction of long-term potentiation and regulation of contextual fear memory formation. Brain Res Bull. 2011;86(3–4):139–45.
112. Carrive P, Leung P, Harris J, Paxinos G. Conditioned fear to context is associated with increased Fos expression in the caudal ventrolateral region of the midbrain periaqueductal gray. Neuroscience. 1997;78(1):165–77.
113. Resstel LBM, Lisboa SF, Aguiar DC, Corrêa FMA, Guimarães FS. Activation of CB1 cannabinoid receptors in the dorsolateral periaqueductal gray reduces the expression of contextual fear conditioning in rats. Psychopharmacology (Berl). 2008;198(3):405–11.
114. Mitchell VA, Jeong H-J, Drew GM, Vaughan CW. Cholecystokinin exerts an effect via the endocannabinoid system to inhibit GABAergic transmission in midbrain periaqueductal gray. Neuropsychopharmacology. 2011;36(9):1801–10.

115. Terzian AL, Drago F, Wotjak CT, Micale V. The Dopamine and Cannabinoid Interaction in the Modulation of Emotions and Cognition: Assessing the Role of Cannabinoid CB1 Receptor in Neurons Expressing Dopamine D1 Receptors. Front Behav Neurosci. 2011;5:49.
116. Dubreucq S, Kambire S, Conforzi M, Metna-Laurent M, Cannich A, Soria-Gomez E, Richard E, Marsicano G, Chaouloff F. Cannabinoid type 1 receptors located on single-minded 1-expressing neurons control emotional behaviors. Neuroscience. 2012;204:230–44.
117. Dubreucq S, Matias I, Cardinal P, Häring M, Lutz B, Marsicano G, Chaouloff F. Genetic dissection of the role of cannabinoid type-1 receptors in the emotional consequences of repeated social stress in mice. Neuropsychopharmacology. 2012;37(8):1885–900.
118. Metna-Laurent M, Soria-Gómez E, Verrier D, Conforzi M, Jégo P, Lafenêtre P, Marsicano G. Bimodal control of fear-coping strategies by CB_1 cannabinoid receptors. J Neurosci. 2012;32(21):7109–18.
119. Bellocchio L, Lafenêtre P, Cannich A, Cota D, Puente N, Grandes P, Chaouloff F, Piazza PV, Marsicano G. Bimodal control of stimulated food intake by the endocannabinoid system. Nat Neurosci. 2010;13(3):281–3.
120. Rey AA, Purrio M, Viveros M-P, Lutz B. Biphasic effects of cannabinoids in anxiety responses: CB1 and GABA(B) receptors in the balance of GABAergic and glutamatergic neurotransmission. Neuropsychopharmacology. 2012;37(12):2624–34.
121. Blanchard RJ, Blanchard DC. Antipredator defensive behaviors in a visible burrow system. J Comp Psychol. 1989;103(1):70–82.
122. McNaughton N, Corr PJ. A two-dimensional neuropsychology of defense: fear/anxiety and defensive distance. Neurosci Biobehav Rev. 2004;28(3):285–305.
123. Blanchard DC, Griebel G, Pobbe R, Blanchard RJ. Risk assessment as an evolved threat detection and analysis process. Neurosci Biobehav Rev. 2010;35(4):991–8.
124. Blanchard RJ, Blanchard DC, Rodgers J, Weiss SM. The characterization and modelling of antipredator defensive behavior. Neurosci Biobehav Rev. 1990;14(4):463–72.
125. Dielenberg RA, Carrive P, McGregor IS. The cardiovascular and behavioral response to cat odor in rats: unconditioned and conditioned effects. Brain Res. 2001;897(1–2):228–37.
126. Platel A, Porsolt RD. Habituation of exploratory activity in mice: a screening test for memory enhancing drugs. Psychopharmacology (Berl). 1982;78(4):346–52.
127. Ursin H, Olff M. Psychobiology of coping and defence strategies. NeuropsychoBiology. 1993;28(1–2):66–71.
128. Lehner M, Taracha E, Skórzewska A, Turzynska D, Sobolewska A, Maciejak P, Szyndler J, Hamed A, Bidzinski A, Wislowska-Stanek A, others. Expression of c-Fos and CRF in the brains of rats differing in the strength of a fear response. Behav Brain Res. 2008;188(1):154–67.
129. Martin JR, Oettinger R, Driscoll P, Buzzi R, Bättig K. Effects of chlordiazepoxide and imipramine on maze patrolling within two different maze configurations by psychogenetically selected lines of rats. Psychopharmacology (Berl). 1982;78(1):58–62.
130. Aguilar R, Gil L, Flint J, Gray JA, Dawson GR, Driscoll P, Giménez-Llort L, Escorihuela RM, Fernández-Teruel A, Tobeña A. Learned fear, emotional reactivity and fear of heights: a factor analytic map from a large F(2) intercross of Roman rat strains. Brain Res Bull. 2002;57(1):17–26.
131. Steimer T, Driscoll P Divergent stress responses and coping styles in psychogenetically selected Roman high-(RHA) and low-(RLA) avoidance rats: behavioural, neuroendocrine and developmental aspects. Stress Amst Neth. 2003;6(2):87–100.
132. Mowrer OH, Lamoreaux RR. Fear as an intervening variable in avoidance conditioning. J Comp Psychol. 1946;39:29–50.
133. Haller J, Goldberg SR, Pelczer KG, Aliczki M, Panlilio LV. The effects of anandamide signaling enhanced by the FAAH inhibitor URB597 on coping styles in rats. Psychopharmacology (Berl). 2013;230(3):353–62.
134. Haller J, Aliczki M, Pelczer KG, Spitzer K, Balogh Z, Kantor S. Effects of the fatty acid amide hydrolase inhibitor URB597 on coping behavior under challenging conditions in mice. Psychopharmacology (Berl). 2014;231(3):593–601.

135. Gunduz-Cinar O, Hill MN, McEwen BS, Holmes A. Amygdala FAAH and anandamide: mediating protection and recovery from stress. Trends Pharmacol Sci. 2013;34(11):637–44.
136. Lal H, Forster MJ. Flumazenil improves active avoidance performance in aging NZB/BlNJ and C57BL/6NNia mice. Pharmacol Biochem Behav. 1990;35(3):747–50.
137. Fernández-Teruel A, Escorihuela RM, Boix F, Tobena A. Effects of different handling-stimulation procedures and benzodiazepines on two-way active avoidance acquisition in rats. Pharmacol Res. 1991;24(3):273–82.
138. Escorihuela RM, Fernández-Teruel A, Zapata A, Núñez JF, Tobeña A. Flumazenil prevents the anxiolytic effects of diazepam, alprazolam and adinazolam on the early acquisition of two-way active avoidance. Pharmacol Res. 1993;28(1):53–8.
139. Ghiasvand M, Rezayof A, Zarrindast MR, Ahmadi S. Activation of cannabinoid CB1 receptors in the central amygdala impairs inhibitory avoidance memory consolidation via NMDA receptors. Neurobiol Learn Mem. 2011;96(2):333–8.
140. Häring M, Grieb M, Monory K, Lutz B, Moreira FA. Cannabinoid CB_1 receptor in the modulation of stress coping behavior in mice: the role of serotonin and different forebrain neuronal subpopulations. Neuropharmacology. 2013;65:83–9.
141. Beninger RJ, MacLennan AJ, Pinel JP. The use of conditioned defensive burying to test the effects of pimozide on associative learning. Pharmacol Biochem Behav. 1980;12(3):445–8.
142. Boersma GJ, Scheurink AJW, Wielinga PY, Steimer TJ, Benthem L. The passive coping Roman Low Avoidance rat, a non-obese rat model for insulin resistance. Physiol Behav. 2009;97(3–4):353–8.
143. Degroot A, Nomikos GG. Genetic deletion and pharmacological blockade of CB1 receptors modulates anxiety in the shock-probe burying test. Eur J Neurosci. 2004;20(4):1059–64.
144. Van Laere K, Goffin K, Bormans G, Casteels C, Mortelmans L, de Hoon J, Grachev I, Vandenbulcke M, Pieters G. Relationship of type 1 cannabinoid receptor availability in the human brain to novelty-seeking temperament. Arch Gen Psychiatry. 2009;66(2):196–204.
145. Rabinak CA, Angstadt M, Sripada CS, Abelson JL, Liberzon I, Milad MR, Phan KL. Cannabinoid facilitation of fear extinction memory recall in humans. Neuropharmacology. 2013;64:396–402.
146. Katona I, Sperlágh B, Maglóczky Z, Sántha E, Köfalvi A, Czirják S, Mackie K, Vizi ES, Freund TF. GABAergic interneurons are the targets of cannabinoid actions in the human hippocampus. Neuroscience. 2000;100(4):797–804.
147. Kawamura Y, Fukaya M, Maejima T, Yoshida T, Miura E, Watanabe M, Ohno-Shosaku T, Kano M. The CB1 cannabinoid receptor is the major cannabinoid receptor at excitatory presynaptic sites in the hippocampus and cerebellum. J Neurosci. 2006;26(11):2991–3001.
148. Monory K, Massa F, Egertová M, Eder M, Blaudzun H, Westenbroek R, Kelsch W, Jacob W, Marsch R, Ekker M. The Endocannabinoid System Controls Key Epileptogenic Circuits in the Hippocampus. Neuron. 2006;51(4):455–66.
149. Lafenêtre P, Chaouloff F, Marsicano G. Bidirectional regulation of novelty-induced behavioral inhibition by the endocannabinoid system. Neuropharmacology. 2009;57(7–8):715–21.
150. Das RK, Kamboj SK, Ramadas M, Yogan K, Gupta V, Redman E, Curran HV, Morgan CJA. Cannabidiol enhances consolidation of explicit fear extinction in humans. Psychopharmacology (Berl). 2013;226(4):781–92.
151. Klumpers LE, Cole DM, Khalili-Mahani N, Soeter RP, Te Beek ET, Rombouts SARB, van Gerven JMA. Manipulating brain connectivity with δ^9-tetrahydrocannabinol: a pharmacological resting state FMRI study. NeuroImage. 2012;63(3):1701–11.
152. Heitland I, Klumpers F, Oosting RS, Evers DJJ, Leon Kenemans J, Baas JMP. Failure to extinguish fear and genetic variability in the human cannabinoid receptor 1. Transl Psychiatry. 2012;2:e162.
153. Stern CAJ, Gazarini L, Takahashi RN, Guimarães FS, Bertoglio LJ. On disruption of fear memory by reconsolidation blockade: evidence from cannabidiol treatment. Neuropsychopharmacology. 2012;37(9):2132–42.

Part II
Cannabinoid Modulation of Reward and Motivation

Chapter 7
Subjective and Cognitive Effects of Cannabinoids in Marijuana Smokers

Marie R. Ehrler, Erin C. McGlade and Deborah A. Yurgelun-Todd

Abstract This chapter is aimed at summarizing the existing literature with respect to the subjective and objective cognitive profiles associated with marijuana use. Although subjective reports of marijuana use include feelings of relaxation and increased creativity, the altered state of consciousness and reduced inhibitory and executive cognitive functions may be enough to cause impairment in daily functioning. However, a chronically high level of consumption for an extended period of time is likely to result in increased risk of developing impairments such as affective dysregulation, inefficient cognitive control, underachievement, lower estimated intellectual capacity, and increased potential for additional drug use.

Keywords Marijuana · Cognitive impairment · Subjective effects · Cognitive effects

Introduction

The multidimensional summation and interaction of individual differences in genetic determinants, physiology, and psychology are key determinants in what is known as the subjective human experience. Drug induced changes in subjective experience are impacted not only by the pharmacokinetic and pharmacodynamic characteristics of the drug itself, but also by an individual's brain composition in neurotransmitter concentration, the number and sensitivity of neurotransmitter receptors, and the concentration of enzyme available to metabolize the drug [1]. Furthermore, subjective recall and description of individual experience become complicated by co-occurring situationally dependent factors, expectancy, and personal history. Reactions to a drug may even modify subsequent use, thereby changing the recall of

D. A. Yurgelun-Todd (✉) · M. R. Ehrler · E. C. McGlade
Department of Psychiatry, University of Utah School of Medicine,
Salt Lake City, UT 84108, USA
e-mail: deborah.yurgelun-todd@hsc.utah.edu

P. Campolongo, L. Fattore (eds.), *Cannabinoid Modulation of Emotion, Memory, and Motivation,* DOI 10.1007/978-1-4939-2294-9_7

previous experience [2]. Variations in use characteristics such as dose, means of administration, quality and potency, depth of inhalation and environmental setting which modify physiological and psychological reactions, make precise quantification and comparison extremely difficult even under the most controlled conditions. Participant variables such as age of drug initiation, length of use, total lifetime use, recreational versus pathological use, and existence of comorbid mental health and substance use disorders will also affect subjective reports [1, 3]. Thus, examining and incorporating subjective experience into a cognitive model poses numerous methodological challenges.

Neurocognitive assessment provides a more objective approach to the quantification of the effects of marijuana, although conclusive statements regarding these effects are also difficult. The identification and interpretation of cognitive deficits are impacted by the methodological complexities described above. In addition there are difficulties inherent in empirical measurement of abstract concepts of cognitive function.

Despite these limitations, there appears to be some shared subjective experiences that may be relegated to a collective and largely prototypical and generalizable description. With these factors in mind, the goal of this chapter is to highlight the more consistent findings regarding the subjective human experience of cannabis intoxication as well as the short-term and long-term cognitive consequences of cannabis exposure.

Subjective Effects of Exogenous Cannabinoids in Marijuana Smokers

Cannabis has historically been classified as a mild sedative-hypnotic agent, with effects that have been described as similar to alcohol and antianxiety agents in lower doses, while higher doses have been described as inducing a state of relaxed euphoria, heightened sensations, and dissociation of ideas [4].

More than 60 identified cannabinoids have been identified within the marijuana plant and Δ-9-tetrahydrocannabinol (THC) is considered the psychoactive component, producing significantly more subjective and physiological effects than other well known cannabinoids (e.g., cannabidiol or cannabichromene) [5]. Many laboratory studies administer synthetic THC rather than whole plant marijuana for more precise control over dosage and potency; interestingly, the subjective effects between synthetic THC and whole-plant marijuana have been shown to produce some minor differences [6]. Although physiological and behavioral effects appear negligible, marijuana tends to increase ratings on measures of sedation, drowsiness, tiredness, and result in more pronounced reports of intoxication, especially at lower doses. THC in comparison induces a small increase in stimulant effect and ratings of both euphoria and dysphoria [6]. The differences in subjective experiences between synthetic THC and whole-plant marijuana are generally considered to be due to the additional cannabinoids, despite their lack of ability to consistently produce significant intoxication on their own [6–8]. Moreover, means of administration will moderate subjective effects. For example, “edibles,” which are marijuana laced

Table 7.1 Methods of cannabis administration

Blunt/joint	Pinch Hitter/ one Hitter	Glass, chillum, and steamroller pipes	Home-made: apple, beer/ soda can, water bottle, tin foil	Bubbler, bong, and gravity bong
Hookah	Vaporizer	Dabs, 'Wax', Hot Knives	Tinctures, Tonics, Hash Oil	Topical Application

food items, produce a different subjective profile marked by longer onset (3 h) and an increased length of subjective high (4–6 h) [9] when compared with inhalation (0.5 h to peak experience, 1.75–3.5 h duration) [10,11]. While joints are generally rated as more pleasurable than smoking blunts (marijuana mixed with tobacco) [12], pipes tend to be the method of choice when the goal is a larger initial hit, and bongs, water pipes, hookahs, and vaporizers provide larger hits of filtered smoke for smoother delivery. Other methods include topical and sublingual administrations of highly concentrated tonics, oils, and various extractions of extremely high potency 'wax' which are heated and inhaled. These methods of administration are presented in Table 7.1.

Positive effects are frequently endorsed with marijuana intoxication and may include euphoria, heightened sexual pleasure, relaxation, and a general sense of well-being [13]. Acute adverse effects can include anxiety, panic reactions, and changes in visual perception, which are most often reported by naïve users smoking large amounts over a short time span [5, 14, 15]. While marijuana intoxication is generally described in terms of positive and negative effects [1, 2], the range of subjective experience is quite large, spanning multiple neurobiological domains. That is, the range of subjective effects may be further relegated to categories of physiological, emotional, sensory, cognitive, and motoric reactions that under some circumstances may be objectively reflected in neurocognitive assessments and neuroimaging. These are presented in Table 7.2 [1–3, 5, 6, 10–14, 16–28].

Subjective experiences are highest at approximately 0.25–0.5 h after intake of cannabis and remain significantly elevated up to 1.75 h in low concentrations (1.8 %) and up to 3.5 h in higher doses (3.6 %) [10,11]. However, subjective differences in the magnitude of effects between doses has generally appeared to be negligible, with statistically significant changes in subjective experience with virtually any active dose [5, 6, 17, 24, 29–31]. There are a few exceptions, however. For instance with increasing doses, escalation in heart rate, aggression, anxiety and agitation may be observed, which strongly correlate with experiences of depersonalization [6, 11, 16, 17, 24, 32]. This relationship suggests that lower doses of THC/cannabis generally induce pleasant effects including relaxation, whereas the higher doses appear to induce increased levels of brain arousal. Evidence for increased arousal is reflected in both regional cerebral blood flow in paralimbic regions corresponding to changes in emotionality [33] as well as subjective reports of euphoria, tension, anger, or mixtures thereof [6, 11, 16].

Reports of subjective experiences have also described 'cannabis consciousness' as

> a state in which at least a few prejudices and predispositions may be temporarily suspended so that something long-ignored can be seen afresh, as for the first time … this marijuana experience can provide a kind of cognitive training that may subsequently help enlarge and enrich one's outlook in desirable and entirely voluntary ways [34 (p. 95–96)].

Table 7.2 Subjective effects of acute intoxication

Positive		Negative	Neutral
Emotional	Calmness	Aggression	Depersonalization
	Contentment	Anger	
	Self confidence	Anxiety	
	Elation	Depression	
	Friendliness	Dysphoria	
	Happiness	Guilt	
	Mellow	Irritability	
	Jovial/laughter	Misery	
	On top of the world	Decreased motivation	
	Peaceful	Panic	
	Social	Restlessness	
		Suspiciousness	
		Tension	
		Disinterest	
		Social withdrawal	
Sensory	Creativity	Blurred vision	Visual perceptual disturbance
	Enhanced sensations (touch & taste)		Psychotomimetic effects
	Enhanced color sensation/perception		Changes in perceptions of space
	Enhanced sexual pleasure		Hunger/increased appetite
	Appreciation of music		Temporal disorientation
	Decreased muscle pain		
	Decreased stomach upset		
Cognitive	Insightful	Poor attention	Distractibility
	Perceived clarity of thought	Poor concentration	
	Profound ideas	Confusion	
		Impaired thinking	
		Poor memory	
Physiological	Alertness	Dizziness	Chills
	Stimulation	Lightheaded	Tiredness
	Improved sleep	Fatigue	Drowsiness
	Sexual arousal	Dry mouth & throat	
	Sedation	Head ache	
	Relaxed	Red, dry eyes	
		Jitteriness	
		Laziness	
		Panic reactions	
		Respiratory complications	
		Coughing	
		Increased thirst	
Motor		Clumsiness	Inhibition
		Slowed reaction time	Talkativeness

One example of this subjective experience is the enhanced appreciation of music that is frequently cited among marijuana users and musicians alike, who report that this particular state of altered consciousness is valuable to their creativity in creating a sense of time expansion (temporal disintegration) and increased insight [13, 34–36]. While over 50% of marijuana smokers have reported subjective enhancements of creativity in general [13], creativity as a construct has been difficult to both define and measure objectively. It has been hypothesized that creativity may arise from mild dissociative states and/or disinhibition of the frontal cortical functions [27]. Enhanced sexual desire and pleasure are also frequently noted among marijuana users, with heightened sensations of touch and taste particularly endorsed [37]. However, this experience is more likely to be true among young adults and sexual dysfunction may also be observed in some circumstances [38]. It has also been suggested that these effects may be related to cultural mysticism and expectancy rather than to the direct pharmacologic effects of cannabis [39].

Positive Effects Associated with Greater Frequency Use

Scherrer et al. [3] differentiated individuals into four classes of typical response patterns to marijuana intoxication: (1) high responders—those who were characterized by frequent endorsement of subjective effects, (2) positive responders—those who experienced mostly positive effects, (3) mixed/relaxed responders—those who reported low energy and endorsed negative effects, (4) and low responders—those who had few endorsements of positive or negative subjective effects. Not surprisingly, individuals with the greatest duration and frequency of use are likely to report more positive subjective effects and few negative effects, to be younger, to be male, and to meet diagnostic criteria for cannabis abuse and depen-dence [1–3]. Although the acute effects of cannabis are often rated as positive or pleasurable, when heavy marijuana users (former and current) rated the perceived positive and negative effects of their marijuana use, they overwhelmingly reported lower levels of life satisfaction than controls on several measures. Specifically, they reported that marijuana had negatively impacted their social life (70%), physical health (81%), mental health (60%), cognition (91%), memory (91%), and career (79%) [20, 21].

Cognitive Effects of Exogenous Cannabinoids in Marijuana Smokers

There is broad consensus that the acute intoxication effects of marijuana cause mild to moderate impairment in neurocognitive and neurophysiological functioning, as well as various structural abnormalities with chronic use [40–43]. While THC concentration and rate of absorption will vary with method of administration, acute cannabis intoxication is consistently reported to cause impairment in a multitude of different cognitive domains. Because no aspect of the brain exists or functions in isolation, the complexity and interconnectedness of these neurocognitive systems makes it difficult to relegate cannabis associated cognitive deficits to any unitary or isolated domain. In fact, there is substantial evidence that cognitive deficits associated with cannabis use include multiple cognitive domains [44, 45] and affect several neural processing networks [46, 47].

However, while neuropsychological measures can be very useful clinically, not all measures are created equal in terms of their sensitivity to detect deficits. Interpretation of what these tests purport to measure varies from study to study based on study aims. The variety of measures available and lack of standardization across experimental paradigms has led to the use of such a wide variety of measures that direct comparisons of results are somewhat difficult. Further, sample populations are extremely varied in terms of age of onset, frequency and duration of use, quantity of lifetime use, length of abstinence, and dose administered in experimental trials. In addition, the majority of samples are predominantly male, which may be a significant limitation considering men also tend to use cannabis more heavily, meet criteria for abuse and dependence more often [3], report more subjectively positive responses than females [2], and display differential patterns of task performance on some measures [48]. Further, more recent findings suggest women experience increased sensitivity to marijuana intoxication, resulting in more pronounced clinical issues and increased vulnerability to developing cannabis use disorder [49]. Also, a significant portion of the data collected is based on self-report, which can be notably biased. Explanations and conclusions are reported with these factors in mind.

Intelligence

There have been a number of reports of lower intelligence associated with lifetime cannabis use, however several of limitations impede the drawing of causal conclusions. For example, there is a tendency for smokers to receive significantly lower educational attainment and be of lower income status than non-smokers [50]. These are factors known to influence scores on intelligence quotient (IQ) assessments and other cognitive measures. Educational effects have been demonstrated largely on tests of verbal ability [51], but also on less obvious measures such as tests of spatial memory and even simple line drawings [52]. Unavoidable confounding variables and lack of prospective longitudinal studies make it difficult to parse out whether smokers began smoking because of inherently lower cognitive abilities, whether their propensity to smoke led to decreased interest or motivation for advanced education and occupational attainment [50], or perhaps a combination of both. Familial influences such as genetic and environmental circumstances may additionally create a set of vulnerabilities including increased probabilities for first use and early school dropout prior to first use [53]. In addition, cannabis use may impact psychosocial adjustment and development, psychological and physical functioning, interpersonal relationships, employment, and influence progression from abuse to dependence [54]. A cannabis-associated "cultural divergence" in which members of a drug using culture diverge from the mainstream in their interests and skills may also lead to deficits in skills measured by standardized, mainstream measures of neuropsychological ability [55]. These variables likely work synergistically, exacerbating or perpetuating one another, ultimately impacting IQ.

At the opposite end of the spectrum, a higher degree of cognitive reserve or premorbid ability may be a protective factor against the detriments of long-term heavy cannabis use, setting a higher threshold for neurocognitive impairment [56]. In one of the few prospective longitudinal studies available, Fried and colleagues [57] examined IQ scores before, during, and after cessation of regular cannabis use to determine what impact it may have on IQ. The researches followed 70 youth from birth, evaluating their IQ's between the ages of 9–12 and again between the ages of 17–20. While there were no differences in IQ during preteen IQ assessment and prior to any cannabis use, current cannabis use was significantly correlated with a dose related decline in IQ. Heavy current use was defined as at least five joints per week and heavy users demonstrated an average decrease of 4.1 IQ points which was in contrast to gains in IQ points for light current users (less than five joints per week), former users, and non-users. Of note, previous heavy users had smoked a mean of 37 joints per week whereas current heavy users smoked an average of 14 joints per week, yet no negative effects were observed among subjects who had been previously heavy users. These findings seem to highlight the detriments particular to current heavy use. Among the current users, only the number of joints smoked per week was negatively related to change in IQ from preteen to young adult. Duration of marijuana (MJ) use, total quantity of MJ use and former use of MJ were not found to be correlated with change in IQ. Interestingly, although current heavy users experienced a decrease in IQ scores, their scores were still above average. Therefore if they had not been assessed prior to initiation of use, these subjects would not have displayed detectable deficits, as their scores remained within the average range despite declines in functioning.

In a less controlled study, Block and Ghoneim [58] matched participants on intellectual functioning before the onset of drug use utilizing scores from standardized testing administered in the fourth grade. Participants ranged in age from 18–42 and cannabis users were only included if they had used at least weekly for the last 2 years. Adult participants were then administered various subtests from the 12th grade version of the Iowa Tests of Educational Development in addition to other clinical measures. Results indicated overall impairments among heavy cannabis users (seven or greater times per week) relative to non-users. Light and intermediate cannabis use (1–4 times weekly and 5–6 times weekly respectively) was not associated with deficits. Although there are a number of limitations affecting the generalizability of this study, the results are nonetheless intriguing.

Based on current evidence it is unclear whether there is a direct causal relationship between cannabis use and lower IQ; however, heavy current cannabis use does appear to be highly correlated with measures of intellectual capacity.

Attention

The ability to encode, maintain, and properly act upon behaviorally relevant information is critical for successful goal driven cognitive processing [59]. The use of cannabis has been reported to impact a number of processes in the attentional system, which in turn impact performance on neuropsychological domains including learning and memory. The attentional system must utilize aspects of both automatic and controlled processes, which require the ability to engage, disengage, and shift focus for appropriate responsivity to perceptual or cognitive stimuli [52]. Data-driven, bottom up processing is generally considered a rapid and automatic process arising from at least some different cortical circuits than the more effortful top-down processing [60]. The conscious, effortful attention is considered a top-down, conceptually driven process [60], which utilizes volitional control and specific allocation of attentional resources [61]. The former represents lower levels of processing, while the latter represents higher order processing and is considered a frontal executive function. Not surprising, impairment within lower levels inevitably affect higher, more advanced levels of processing.

Basic arousal is a prerequisite for attention and is a physiological state mediated by the brainstem, which prepares the organism for sensory and motor processing [62]. Decreases in arousal or drowsiness can lower the ceiling for how much processing may occur at any one time within the attentional system, which is of limited capacity already [52]. It has been found that lower doses of cannabis (<7 mg) result in physiological arousal such as increased heart rate, which is paradoxically related to subjective feelings of drowsiness, impaired attention, and cognition. With increasing doses, heart rate increases in a dose-dependent manner and drowsiness seems to dissipate, producing more stimulation within the central nervous system associated with subjective reports of agitation, aggression, and anxiety; however, this stimulatory effect is not necessarily associated with improved attention or cognitive performance [16].

Cannabis users have demonstrated impairments on more complex measures of attention such as the ability to select and discriminate relevant stimuli and sustain attention on particular tasks resulting in more incorrect responses [48, 63, 64]. While occasional use of cannabis does not seem to significantly impair the ability to filter irrelevant auditory stimuli during acute intoxication [33], chronic use does. For example, event related potentials (ERPs) are known to be temporal representations of electrical brain activity related to sensory, motor, and cognitive events [65]. In a series of studies done by Solowij and colleagues [64, 66, 67], ERPs were recorded from long-term cannabis users during a complex auditory selective attention task, which were compared with the performance of nonuser controls. Not only did long-term users display dysfunctional allocation of attentional resources and evaluation strategies, increasing duration of cannabis use was associated with progressively impaired mode of information processing, where complex irrelevant information was not properly filtered [64]. In sum, these findings indicate increased distractibility and deficits in selective and sustained auditory attention with chronic cannabis use.

Visual Attention

Speed of visual information processing during simple target detection tasks, may remain intact in the face of cannabis use [68, 69], and in some cases may actually provide functional advantage for target detection because of faulty inhibition [70]. However, with more complex tasks of selective and divided visual attention, cannabis users tend to be significantly slower to detect both central and peripheral visual stimuli [48]. In fact, slower reaction times were predicted by earlier age of onset (prior to age 16), suggesting that cannabis may interfere with the development of visual processing ability [71]. Deficits in visual attention, combined with slower reaction time, may represent one of the more harmful side effects of cannabis use, as it can result in reduced driving ability. In fact, cannabis does have a significant impairing effect on driving [72], resulting in increased risk of responsibility for collisions and driving related fatalities [73].

Executive Function

Among cannabis users, both short-term and long-term frequent heavy use has been associated with impaired executive functioning [44]. The term executive function refers to response choice and execution of adaptive responses such as formulating goals with long-term consequences, generating alternatives, selecting and initiating goal directed behaviors, self-monitoring the adequacy and correctness of one's behavior, correcting and modifying the behavior when conditions change, and persisting in the face of distraction. The cerebral locations responsible for these functions are attributable to the frontal lobes, which include the primary, premotor, and supplementary motor areas, the dorsolateral, orbital and basal areas, as well as anterior cingulate gyri—all of which have numerous cortical and subcortical connections [74, 75]. Frontal lobe functioning is among the last cognitive functions to fully develop and optimal functioning does not occur until after adolescence [74]. Negative influences on prefrontal development include prenatal environment, physical and psychosocial factors, and use of drugs of abuse [76], which likely include cannabis.

Executive functions are among the most complex cognitive processes and are essential to an individual's ability to respond to novel situations in adaptive ways. They are the basis of many cognitive and emotional skills necessary for efficient functioning. For example, planning involves the ability to identify, organize, and carry out an intention. In order to do this, an individual must be able to conceptualize, think ahead, employ effective impulse control, and exercise reasonably intact memory functions. Further, efficient planning entails decision-making and judgment of stimuli that must be held in short-term memory and depends on a complex network of allocated attentional resources distributed by the dorsal prefrontal cortex [52]. The frontal executive system is therefore a top-down, regulatory system in which lower cognitive functions, such as attention, are orchestrated and manipulated through more advanced skill sets such as mental flexibility, impulse control, working memory, and judgment and decision making.

Mental Flexibility

Not surprising, acute cannabis intoxication can result in reduced functional abilities to quickly and fluidly switch between tasks [48] and current users not under the direct influence also show impairments [44]. While light users (< 10 times per month) may not demonstrate blatant impairments in flexibility of thought, heavy users (approximately 5 days per week) have demonstrated impairments in concept formation of simple mental categories and perseverative behavioral responses [74] reflecting an inability to incorporate feedback to guide or change incorrect response selection [56]. These deficits appear to increase with increased use [56].

Flexibility of thought has also been considered one aspect of creativity and increase in subjective creativity is a commonly cited incentive for using cannabis. However, regular users display diminished capacities on measures of mental flexibility as well as other measures of creativity such as divergent thinking, elaboration, fluency, and originality [77].

Impulse Control

Impulsivity is a complex construct; however, impulsive behavior has been associated with decreased inhibition, decreased attention, and decreased motor control. An ability to inhibit inappropriate or otherwise irrelevant responses is a hallmark of frontal executive functioning. Impairment within this domain has been one of the more consistently reported deficits associated with cannabis use despite other relatively intact cognitive functions [78-80]. This effect can be particularly pronounced in individuals exhibiting lower cognitive reserves [56] and conditions creating increased task complexity with increasing demands seem to exacerbate the negative impact on response inhibition [81]. Although early onset and longer duration of use are associated with greater impairments in impulse control [79, 81, 82], this could simply reflect poor impulse control as a contributing factor to the initiation of cannabis use and/or the inability to discontinue use, rather than a consequence of cannabis use per se [79].

Working Memory

Human and animal studies confirm that the frontal lobes play an important role in working memory for both sensory and motor events [60]. Working memory is a temporary and limited capacity memory store for manipulation of information that may be used in more complex cognitive tasks such as language comprehension, learning, and reasoning. Working memory is considered more complex than short-term memory because it requires the simultaneous storage and processing of information and calls upon the central executive system for allocation of attention to both visuospatial and phonological/language based information [83]. It is vulnerable to distraction because of reliance on the higher order attentional systems.

Cannabis intoxication can cause difficulty maintaining a coherent train of thought because of the intrusion of irrelevant information and deficits in both verbal and spatial working memory have been reported following acute intoxication [31, 82]. Among sober frequent users, spatial and verbal working memory may remain intact [68, 84]; however, it is unclear how increased task difficulty might impact these findings.

Decision Making

Impairments in inhibition and performance monitoring associated with cannabis use may translate globally into faulty decision making [80] reflected in both altered neuromodulation of the frontal cortex [85, 86] and overt behavior. Decision-making can be thought of as a two-stage process involving an evaluation of a situation followed by a corresponding behavior. Adaptive decision-making involves properly weighing the potential rewards and punishments associated with a particular behavioral response and responding to the situation accordingly [56]. Impairments of this type are seen in the acutely intoxicated [87] as well as in recent and chronic users abstinent at the time of evaluation [56]. Similarly, while moderate frequency use (defined as 3 times per week) has been related to elevated risky decision-making and impaired executive planning [68], dose related alterations in heavy users seem to persist following weeks of abstinence [56]. During situations in which rewards and punishment must be weighed concurrently, long-term heavy MJ users tend to make more decisions that lead to larger immediate gains despite acknowledgment of more costly losses, make fewer rational decisions overall, and require more attempts to obtain correct planning solutions [68, 88]. Despite the existence of clear deficits, it remains unclear whether the basis for such deficits is directly attributable to cannabis exposure or preexisting genetic and behavioral risk factors. Either way, an inability to properly balance rewards and punishments likely contributes to continued use [88].

Learning and Memory

Important aspects of memory that could differentially affect quantification of a single individual's capacity include that individual's ability to encode new information, consolidate this information to memory, and retrieve that information when necessary. Generally speaking, free recall will be more difficult than recognition. While there are several neuropsychological measures of verbal learning and memory, most assess the ability to learn new information over a series of trials, immediate recall of that new information, as well as delayed memory after a brief interim period, and the ability to recognize the new information.

Among some of the most rigorous studies of cannabis users, deficits in the ability to learn and remember new information have been reported, and heavier, longer-term users have exhibited more difficulty with both verbal and visual information

[44, 56, 58, 89, 90]. Regardless of the type of linguistic information presented [91] impairments have been noted in both immediate and delayed free recall following acute intoxication [17, 44, 92, 93]. Practice and repetition do not appear to facilitate performance [93] and impairments increase with dose [16, 91]. For long-term users, learning curves reflect a less steep gradient, with fewer overall words recalled and a greater proportion of individuals falling within the "very poor" learning range compared to short-term users [90].

Recognition memory, however, appears to remain relatively intact, suggesting that cannabis may have less of an effect on the encoding (acquisition and retention) of new information and cause more detriments to the retrieval of newly acquired information [56, 82, 92, 93]. Although retrieval cues can facilitate recall, intrusion errors are substantially increased and primacy effects suggest interference with the transformation of information to longer-term memory [91–93].

Motor Function

The primary motor, premotor, and supplementary motor areas are all located within the frontal system and the prefrontal cortex receives input connections from the limbic system, which can affect the motor system in multiple ways [74, 75]. As previously discussed, impaired behavioral inhibition or increased disinhibition is a hallmark of executive dysfunction and is often measured in terms of reaction time and errors of commission. Lopez-Larson et al. [94] evaluated activation differences within the motor network utilizing a functional magnetic resonance imaging (fMRI) bilateral finger-tapping task in heavy MJ using adolescents. Findings indicated that healthy controls had significantly greater activation than MJ users within the cingulate gyrus and cerebellum, which also correlated with lifetime MJ smokes. These results suggest that the cerebellum and the cognitive/attentional component of the motor network (cingulate) may be significantly affected by heavy MJ use, which may lead to impairments in motor function, cognition, and mood.

Numerous studies have reported impairments in psychomotor speed and dexterity with cannabis intoxication, and greater impairment with heavier use [17, 40, 56, 82, 95]. Reaction times are slower among heavy users; however, with more complex tasks, heavy and light users appear similarly impaired [56]. Impairment in motoric reaction times seems to be one of the more long-lasting deficits and has been observed in abstinent participants at 12 [96], 24 [44], and 48 h following use [40]. However, deficits are not apparent after 4 weeks of abstinence [63].

In Zuurman et al.'s [16] systematic and comprehensive literature review, motor control and visuomotor control appeared to demonstrate an inverse dose response association where lower doses were associated with more impairment than moderate doses, and higher doses demonstrated the least impairment. However, this review included studies with varying designs including double-blinded (57 %), single blinded (26 %), open design (7 %), and several that were unknown (10 %). It is therefore not possible to attribute these findings to acute versus chronic use or whether or not these were residual symptoms of short- or long-term abstinence within adult or adolescent populations. Interestingly, studies utilizing the Cambridge Neuropsychological Test Automated Battery (CANTAB) have consistently found negative results (absence of impairments) in psychomotor functioning, possibly reflecting test insensitivity [68, 88]. Few others have reported intact psychomotor functioning upon acute intoxication when using other, less utilized measures [33].

Language

Changes in verbal abilities appear to be consistently implicated in long-term heavy cannabis use [44, 55]. These deficits have been observed in both language production, such as phonemic and semantic fluency, as well as in lower overall verbal IQ, deficient verbal learning, and impaired recall of verbal information [44, 55]. However, bias toward verbally based methods of assessment may artificially increase the likelihood of finding verbal, as opposed to non-verbal deficits.

Temporal Disintegration

Another very consistent finding among the acutely intoxicated as well as abstinent short- and long-term chronic users is impaired time estimation, also known as temporal disintegration [48, 81, 96, 97]. Theoretically, this is considered a cognitive phenomenon in which the 'individual has difficulty retaining, coordinating and serially indexing memories, perceptions, and expectations relevant to the goal he is pursuing' [98]. It is thought to relate to memory, temporal order or context, and perhaps working memory [97]. Subjectively, temporal disintegration is experienced as a confusion of past, present, and future and can be clinically related to depersonalization experiences, but has also been associated with increased insight, creativity, and musicality [35]. Objectively, cannabis users tend to misperceive the passing of time evidenced by both under and overestimates of time elapse [48, 81, 96], resulting in spontaneous increases in self-timed behaviors such as counting and tapping [97] and tend to sacrifice accuracy for speed [81]. Disruptions of temporal gauging appear to increase with increased doses [11, 98].

Residual Effects

In several well-controlled studies, a variety of neuropsychological deficits have been reported among long-term users following 12–72 h abstinence [50, 54, 58, 99, 100, 101]. Pope et al. [50] compared current heavy users (>5000 times in lifetime) and former heavy users (>5000 times in lifetime, but fewer than 12 times in the last 3 months) on several measures of neurocognitive performance during baseline and after 1, 7, and 28 days abstinence. At days 0, 1, and 7, the current heavy users scored significantly below former users on recall of word lists, and level of impairment was associated with urinary THC concentrations. However, by day 28, current heavy users showed no differences from former users on ten different neurocognitive measures, rendering the two groups virtually indistinguishable. Authors suggest that the subtle cognitive impairments observed in the active cannabis users during the first week of abstinence may represent actual residual cognitive deficits or abstinence phenomena rather than a persistent loss of functional ability. Furthermore, there were no correlations between cumulative lifetime use of cannabis and cognitive deterioration.

Although there is evidence that heavy users exhibit at least some residual cognitive deficits following discontinuation of use, these deficits seem to be limited to several days or weeks and may represent the persistence of THC in the body or a withdrawal effect rather than directly related residual effects per se. Any cognitive deficits observed within a one-month period of discontinuation seem to remediate after approximately one-month abstinence [47, 50, 82, 102].

Persisting Cognitive Impairments Following Long-Term Use

Despite subjective reports of lasting effects following prolonged abstinence (months and years), deficits appear to be quantifiably minimal or negligible in adult populations [103]. Research findings strongly support cognitive deficits associated with acute cannabis intoxication, and even lingering residual impairments for up to several weeks following use; however, there is less clear evidence that deficits persist longer than 28 days and even less evidence that cannabis causes permanent neurotoxic damage. Some ex-users have reportedly demonstrated deficient filtering of auditory information and alterations of electrophysiology following 2 or more years of abstinence [64, 66, 67], but others report no significant cognitive impairments [104, 105]. Additional research using controlled study approaches is needed before any conclusions may be drawn regarding long-term residual cognitive changes as a consequence of cannabis exposure.

Age of Onset and Cognitive Dysfunction

While the literature on deficits associated with cannabis use in adults has been inconsistent, research findings clearly suggest adolescents are more vulnerable than adults to neurocognitive changes associated with cannabis use. Younger age of onset in particular has been associated with poorer cognitive performance spanning multiple domains as well as negative long-term behavioral and mental health consequences [63, 78, 106–108]. The frontal executive system seems to be particularly vulnerable in this younger population reflected in volumetric abnormalities [109], decreased white matter fiber integrity [79, 108], altered neurocircuitry [94], and compensatory changes in functional connectivity involving cognitive control regions of the frontal cortex [46]. This is of particular concern because optimal frontal lobe functioning does not occur until after adolescence, as it is the last region to fully develop [74]. Chronic cannabis use has been associated with diminished neuronal and axonal integrity in the dorsolateral prefrontal cortex [110] and may be particularly disruptive to prefrontal cortical maturation [111] as well as the underlying glutamatergic and GABAergic functions essential to frontal cortical circuits and inhibitory processes [112].

Earlier age of onset, greater frequency of use, and total lifetime use are associated with poorer cognitive performance, particularly on measures of executive control and inhibitory responses [41]. Earlier age of regular use and greater duration of exposure have been associated with poorer cognitive performance over and above recent use [113]. Problems with frontal executive functions may manifest as intensified novelty-seeking, increased impulsivity and decreased future orientation [109], longer reaction times and more errors of commission [40]. Other deficits include difficulties in selective and sustained attention, learning and memory, working memory, and processing speed [40, 41, 45, 50, 55, 57, 64, 66, 71, 80, 99, 107, 109, 114, 115]. Furthermore, deficits in cognitive functioning are evident at least 6 weeks following abstinence in adolescent users [116], representing a longer duration of residual effects. Lower verbal IQ has also been reported, although this result is likely due to multiple factors as discussed previously [55].

Discussion

Subjective and self-report profiles of the drug response suggest a number of positive experiences may occur after consumption; however, the objective assessment of cognitive outcomes suggests potentially negative effects particularly after chronic heavy use. This is especially true for youth where study findings indicate more negative and longer lasting neurobehavioral and affective consequences. Thus both positive and negative subjective effects are commonly endorsed, often simultaneously, in areas of emotional, sensory, cognitive, physiological, and motor functioning. Nevertheless, particular reactions to smoking appear to be mediated by individual characteristics as well as dose and potency consumed, with the lower doses inducing the more positive emotional reactions and high doses tending to produce discomfort experienced as paranoia, panic, and anxiety. While several studies have evaluated cognition after extended periods of abstinence and overall report no significant long-term deficits, these findings do not discount alterations in brain function or integrity. It is possible that some neuropsychological measures do not have the sensitivity to identify subtle brain malfunctions or may not have a sufficient difficulty level to elicit dysfunctional responding.

Where neuropsychological measures have failed to detect significant differences or otherwise yielded inconclusive or inconsistent results, neuroimaging techniques have found subtle abnormalities in brain circuitry, even with prolonged abstinence [117]. Changes in regional cerebral blood flow during task engagement, but statistically normal neurocognitive performance implies a compensatory mechanism, in which the brain seems to work harder to maintain average functioning [117]. This has important implications for individuals who may have lower cognitive reserve and are easily cognitively fatigued as may occur in individuals with lower intellectual capacity, head injury, stroke, or other neurocognitive complications. For some individuals, cognitive impairments may only be observable and quantifiable in situations of stress, increased task difficulty or when it is critical to inhibit behaviors.

Despite the observed trends of lower estimates of intellectual capacity among cannabis users, it is unclear how the use of cannabis is related to cognitive capacity. The relationship between cannabis use and general intellectual functioning is multifaceted and seems to have a complex relationship with various risk factors for initial use as well as decreased propensity for scholastic achievement. For example, it is not clear to what extent cannabis use impedes age appropriate gains or slows the acquisition of an age appropriate knowledge base.

Early onset use before the age of 16 is particularly concerning as the brain is still developing and is vulnerable to substance induced physiological changes. Not surprisingly, study findings suggest cannabis may have more lasting effects on young users. Complex emotional reactions and executive cognitive functions depend on the integrity of frontal neurocircuitry known to be impacted by early onset cannabis use and alterations to this system have the potential to interfere with successful functional maturation. Interference with this maturation could potentially result in lifelong consequences such as chronic emotional and cognitive dysregulation resembling aspects of ADHD, major depression, anxiety, and schizophrenia. For some individuals, it has been reported that heavy chronic cannabis use may result in lower quality of life, lower life satisfaction, and lower educational attainment. In addition, heavy users are more vulnerable to future drug use, as illustrated by the gateway hypothesis.

According to the gateway hypothesis, initiation and hierarchical sequential progression into harder drugs is fueled by a stage of intermediary marijuana use, which does not happen by opportunistic chance [118]. Age of initiation and intensity of substance use are considered the strongest predictors of subsequent drug use trajectory, including escalation in frequency, severity of pathological use, and severity of drug type [119–122]. According to a 2002 report by the Substance Abuse and Mental Health Services Administration (SAMHSA), adults who were aged 26 or older, had the highest prevalence of heroin, cocaine, and prescription drug use if they initiated marijuana use before the age of 15. Among those who had never used marijuana, less than 1 % had ever used cocaine or heroin and 5 % had used prescription drugs recreationally. In a 25-year longitudinal study (N=1265), Ferguson et al. [123] reported similar trends of decreased use with increasing age and also reported 98 % had used marijuana prior to other illicit drugs. Heavier drug use, in turn, is often related to compounding emotional and cognitive difficulties.

Through numerous studies, this drug use trajectory has been repeatedly demonstrated the world over and under various normative cultural use patterns; however, other explanations exist [112, 124, 125]. For example, it is unclear whether the progressive drug use trend is a direct consequence of the drug itself, increasing drug-seeking behavior through biochemical mechanisms, or due to peer group interactions, social availability, and paradigm shift in beliefs about the harmful consequences of drug use [124, 126, 127]. Likely the result of complex interactions among these factors, the gateway phenomenon is alarming given the relationship between age of first time marijuana use, frequency of use, and significant likelihood of progression into harder drugs of abuse.

There are several limitations to this review. First, the variations in sample compositions among the studies reviewed impede accurate large-scale comparisons. For example, comparisons of adolescents versus adults, naïve users versus chronic users, psychiatrically and emotionally vulnerable populations versus well-educated high functioning individuals will result in inconsistent patterns of subjective and objective clinical assessment. Also, given restrictions on time, money, and resources, cognitive assessment measures are often chosen out of convenience, brevity of assessment, or used as an adjunct measure in an investigation to meet more complex ends.

In summary, the effects of marijuana use have not yet been fully appreciated within the academic community, or society at large. Marijuana has been implicated in psychosocial adjustment and developmental outcomes, psychological and physical well-being, interpersonal relationships, and employment, ultimately lowering overall life satisfaction [50, 54]. Ongoing and future studies of marijuana's effects on the human brain (cognitive and affective) are likely to include more sensitive clinical measures as well as a range of neuroimaging techniques such as magnetic resonance imaging (MRI), functional magnetic resonance imaging (fMRI), magnetic resonance spectroscopy (MRS), and diffusion tension imaging (DTI). These imaging methods are invaluable in identifying subtle brain changes in both normal and abnormal developmental trajectories as well as substance-induced changes that lead to altered behavior. Combining neurocognitive measures with neuroimaging techniques will enhance our understanding of the neural substrates underlying overt behavior and perhaps help elucidate the point at which neurobiological changes induce changes in behavior. Incorporating these techniques within longitudinal paradigms will help distinguish between premorbid conditions that lead to initial use versus marijuana induced changes in structure and function, and the degree to which neuroplasticity can counteract the long-term consequences within various marijuana using populations.

References

1. Lyons MJ, Toomey R, Meyer JM, Green AI, Eisen SA, Goldberg J, True W, Tsuang MT. How do genes influence marijuana use? The role of subjective effects. Addiction. 1997;92(4):409–17.
2. Zeiger JS, Haberstick BC, Corley RP, Ehringer MA, Crowley TJ, Hewitt JK, Hopfer CJ, Stallings MC, Young SE, Rhee SH. Subjective effects to marijuana associated with marijuana use in community and clinical samples. Drug Alcohol Depend. 2010;109:161–6.

3. Scherrer JF, Grant JD, Duncan AE, Sartor CE, Haber JR, Jacob T, Bucholz KK. Subjective effects to cannabis are associated with use, abuse and dependence after adjusting for genetic and environmental influences. Drug Alcohol Depend. 2009;105:76–82.
4. Julien RM. Marijuana: a unique sedative-euphoriant-psychedelic drug. New York: WH Freeman Company; 1998.
5. Ilan AB, Gevins A, Coleman M, ElSoohly MA, de Wit H. Neurophysiological and subjective profile of marijuana with varying concentrations of cannabinoids Behav Pharmacol. 2005;16:487–96.
6. Wachtel SR, ElSohly MA, Ross SA, Ambre J, de Wit H. Comparison of the subjective effects of delta-9-tetrahydorcannibinol and marijuana in humans. Psychopharmacol. 2002;161:331–9.
7. Karniol IG, Shirakawa I, Takahashi RN, Knobel E, Musty RE. Effects of delta-9-tetrahydrocannabinol and cannabinol in man. Pharmacol. 1975;13(6):502–12.
8. Musty RE, Karniol IG, Shirakawa I, Takahashi RN, Knobel E. Interactions of delta-9-tetrahydrocannabinol and cannabinol in man. In: Braude MC, Szara S, editors. New York: Raven; 1976, pp. 559–63.
9. Tashkin DP, Shapiro BJ, Frank I. Acute pulmonary physiologic effects of smoked marijuana and oral delta-9-tetrahydrocannabinol in healthy young men. N Engl J Med. 1973;289:336–41.
10. Fant RV, Heishman SJ, Bunker EB, Pickworth WB. Acute and residual effects of marijuana in humans. Pharmacol Biochem Behav. 1998;60(4):777–84.
11. Mathew RJ, Wilson WH, Humphreys D, Lowe JV, Weithe K.E. Depersonalization after marijuana smoking. Biol Psychiatry. 1993;33:431–41.
12. Cooper ZD, Haney M. Comparison of subjective, pharmacokinetic, and physiological effects of marijuana smoked as joints and blunts. Drug Alcohol Depend. 2009;103:107–13.
13. Green B, Kavanagh D, Young R. Being stoned: a review of self-reported cannabis effects. Drug Alcohol Depend. 2003;22:453–60.
14. Hall W, Degenhardt L. Adverse health effects of non-medical cannabis use. Lancet. 2009;374:1383–91.
15. Tennant FS, Groesbeck J. Psychiatric effects of hashish. Arch Gen Psychiatry. 1972;27:133–6.
16. Zuurman L, Ippel AE, van Gerven JMA. Biomarkers for the effects of cannabis and THC in healthy volunteers. Brit J Clin Pharmacol. 2008;67(1):5–21.
17. Ramesh D, Haney M, Cooper ZD. Marijuana's dose-dependent effects in daily marijuana smokers. Exp Clin Psychopharmacol. 2013;21(4):287–93.
18. Hollister LE. Health aspects of cannabis. Pharmacol Rev. 1986;38(1):1–20.
19. Glass M, Dragunow M, Faull RLM. Cannabinoid receptors in the human brain: a detailed anatomical and quantitative autoradiographic study in the fetal, neonatal, and adult human brain. Neurosci. 1997;77(2):299–318.
20. Gruber AJ, Pope HG, Hudson JI, Yurgelun-Todd D. Attributes of long-term heavy cannabis users: a case control study. Psychol Med. 2003;33:1415–22.
21. Elkashef A, Vocci F, Huestis M, Haney M, Budney A, Gruber A, el-Guebaly N. Marijuana neurobiology and treatment. Subst Abuse. 2008;29(3):17–29.
22. Haertzen CA, Hickey JE. Addiction Research Center Inventory (ARCI): measurement of euphoria and other drug effects. In: Bozareth MA, editor. New York: Springer-Verlag; 1987.
23. Gruber AJ, Pope HG. Marijuana use among adolescents. Pediatr Clin North Am. 2002;49:389–413.
24. Hart CL, Ilan AB, Gevins A, Gunderson EW, Role K, Colley J, Foltin RW. Neurophysiological and cognitive effects of smoked marijuana in frequent users. Pharmacol Biochem Behav. 2010;96:333–41.
25. Hart CL, von Gorp W, Haney M, Foltin R, Fischman MW. Effects of acute smoked marijuana on complex cognitive performance Neuropsychopharmacol. 2001;25(5):757–65.
26. Block RI, Erwin WJ, Farinpour R, Braverman K. Sedative, stimulant, and other subjective effects of marijuana: relationships to smoking technique. Pharmacol Biochem Behav. 1998;59(2):405–12.

27. Schafer J, Brown SA. Marijuana and cocaine effect expectancies and drug use patterns. J Consult Clin Psychol. 1991;59(4):558–65.
28. Metrik J, Kahler CW, McGeary JE, Monti PM, Rohsenow DJ. Acute effects of marijuana smoking on negative and positive affect. J Cogn Psychother. 2011;25(1).
29. Babor T, Mendelson JH, Greenberg I, Keuhnle JC. Marijuana consumption and tolerance to physiological and subjective effects. Arch Gen Psychiatry. 1975;32:1548–52.
30. Heishman SJ, Stitzer ML, Yingling JE. Effects of tetrahydrocannabinol content on marijuana smoking behavior, subjective reports, and performance. Pharmacol Biochem Behav. 1989;34:173–9.
31. Heishman SJ, Stitzer ML, Bigelow GE. Alcohol and marijuana: comparative dose effect profiles in humans. Pharmacol Biochem Behav. 1989;31:649–55.
32. Melges FT, Tinklenberg JR, Hollister LE, Gillespie HK. Temporal disintegration and depersonalization during marijuana intoxication. Arch Gen Psychiat. 1970;23(3):204–10.
33. O'Leary DS, Block RI, Koeppel JA, Flaum M, Schultz SK, Andreasen NC, Ponto LB, Watkins GL, Hurtig RR, Hichwa RD. Effects of smoking marijuana on brain perfusion and cognition. Neuropsychopharmacol. 2002;26(6):802–16.
34. Webster P. Marijuana and music: a speculative exploration. J Cann Ther. 2001;1(2):93–105.
35. Fachner J. The space between the notes: research on cannabis and music perception. Looking Back, Looking Ahead-Popular Music Studies. 2002;20:308–19.
36. Grinspoon L, Bakalar, JB. Marijuana: the forbidden medicine. New Haven: Yale University Press; 1993.
37. Weller RA, Halikas JA. Marijuana and sexual behavior. J Sex Res. 1984;20(2):186–93.
38. Cohen S. Cannabis and sex: multifaceted paradoxes. J Psychoact Drugs. 1982;14(1–2):55–8.
39. Koff WC. Marijuana and sexual activity. J Sex Res 1974;10(3):194–204.
40. Abdullaev Y, Posner MI, Nunnally R, Dishion TJ. Functional MRI evidence for inefficient attentional control in adolescent chronic cannabis abuse. Behav Brain Res. 2010;215:45–57.
41. Gruber AJ, Dahlgren MK, Lukas SE. Age of onset of marijuana use and executive function. Psychol Addic Behav. 2012;26(3):496–506.
42. Metrik J, Kahler CW, McGeary JE, Monti PM, Rohsenow DJ. Acute effects of marijuana smoking on negative and positive affect. J Cogn Psychother. 2011;25(1).
43. Kanayama G, Rogowska J, Pope HG, Gruber S, Yurgelun-Todd D. Spatial working memory in heavy cannabis users: a functional magnetic resonance imaging study. Psychopharmacol. 2004;176:239–47.
44. Messinis L, Kyprianidou A, Malefaki S, Papathanasopoulos P. Neuropsychological deficits in long-term frequent cannabis users. Neurol. 2006;66:737–9.
45. Jacobus J, Bava S, Cohen-Zion M, Mahmood O, Tapert SF. Functional consequences of marijuana use in adolescents. Pharmacol Biochem Behav. 2009;92(4):559–65.
46. Harding IH, Solowij N, Harrison BJ, Takagi M, Lorenzetti V, Lubman DI, Seal ML, Pantelis C, Yucel M. Functional connectivity in brain networks underlying cognitive control in chronic cannabis users. Neuropsychopharmacol. 2012;37:1923–33.
47. Jacobus J, Goldenberg D, Wierenga CE, Tolentino NJ, Liu TT, Tapert SF. Altered cerebral blood flow and neurocognitive correlates in adolescent cannabis users. Psychopharmacol. 2012;222:675–84.
48. Anderson BM, Rizzo M, Block RI, Godfrey D, Pearlson MD, O'Leary DS. Sex, drugs, and cognition: effects of marijuana. J Psychoact Drugs 2010;42(4):413–24.
49. Cooper ZD, Haney M. Investigation of sex-dependent effects of cannabis in daily cannabis smokers. Drug Alcohol Depend 2014;136:85–91.
50. Pope HG, Gruber AJ, Hudson JI, Huestis MA, Yurgelun-Todd D. Neuropsychological performance in long-term cannabis users. Arch Gen Psychiatr. 2001;58:909–15.
51. Lichtenberger EOK, A.S. Essentials of WAIS-IV Assessment. In: Kaufman AS, Kaufman, NL. editors. Hoboken: Wiley; 2009.
52. Lezak MD, Howieson DB, Loring DW. Neuropsychological assessment. 4th ed. New York: Oxford University Press; 2004.

53. Verweij KJH, Huizink AC, Agrawal A, Marin NG, Lynskey MT. Is the relationship between early-onset cannabis use and educational attainment causal or due to common liability? Drug Alcohol Depend 2013;133(2):580–6.
54. Solowij N, Grenyer BFS. Are adverse consequences of cannabis use age dependent? Addiction. 2002; 97:1083–6.
55. Pope HG, Gruber AJ, Hudson JI, Cohane G, Huestis MA, Yurgelun-Todd D. Early onset cannabis use and cognitive deficits: what is the nature of the association. Drug Alcohol Depend. 2003;69:303–10.
56. Bolla KI, Brown K, Eldreth D, Tate K, Cadet JL. Dose-related neurocognitive effects of marijuana use. Neurol. 2002;59:1337–43.
57. Fried P, Watkinson B, James D, Gray R. Current and former marijuana use: preliminary findings of a longitudinal study of effects on IQ in young adults. Can Med Ass J. 2002;166(7):887–91.
58. Block RI, Ghoneim MM. Effects of chronic marijuana use on human cognition. Psychopharmacol. 1993;110:219–28.
59. Awh E, Vogel EK, Oh SH. Interactions between attention and working memory. Neurosci. 2006 139:201–8.
60. Kolb B, Whishaw IQ. Fundamentals of human neuropsychology. 6th ed. New York: Macmillan; 2009.
61. Connor CE, Egeth HE, Yantis S. Visual attention: bottom-up versus top-down. Curr Biol. 2004;14:850–2.
62. Heilman KM, Watson RT, Valenstein E. Neglect and related disorders. Heilman KM, Valenstein E, editors. New York: Oxford University Press;1985.
63. Jacobsen LK, Mencl WE, Westerveld M, Pugh KR. Impact of cannabis use on brain function in adolescents. Ann N Y Acad Sci. 2004;1021:384–90.
64. Solowij N, Michie PT, Gox AM. Effects of long-term cannabis use on selective attention: an event related potential study. Pharmacol Biochem Behav. 1991;40:683–8.
65. Bressler SL, Ding M. Event? Related Potentials. In: Akay M, editor. Wiley encyclopedia of biomedical engineering. 2006.
66. Solowij N, Michie PT, Gox, AM. Differential impairments of selective attention due to frequency and duration of cannabis use. Biol Psychiatr. 1995;37:731–9.
67. Solowij N. Do cognitive impairments recover following cessation of cannabis use? Life Sci. 1995;56(23/24):2119–26.
68. Grant JD, Scherrer JF, Lynskey MT, Agrawal A, Duncan AE, Haber JR, Heath AC, Bucholz KK. Associations of alcohol nicotine, cannabis, and drug use/dependence with educational attainment: evidence from cotwin-control analyses. Alcohol Clin Exp Res. 2012;36:1412–20.
69. Heishman SJ, Huestis MA, Henningfield JE, Cone EJ. Acute and residual effects of marijuana: profiles of plasma THC levels, physiological, subjective, and performance measures. Pharmacol Biochem Behav. 1990; 37:561–5.
70. Vivas AB, Estevez A, Moreno M, Panagis G, Flores P. Use of cannabis enhances attentional inhibition Hum Psychopharmacol. 2012;27:464–9.
71. Ehrenreich H, Rinn T, Kunert HJ, Moeller MR, Poser W, Schilling L, Gigerenzer G, Hoehe MR. Specific attentional dysfunction in adults following early start of cannabis. Psychopharmacol. 1999;142:295–301.
72. O'Kane CJ, Tutt DC, Bauer LA. Cannabis and driving: a new perspective. Emerg Med. 2002;14:296–303.
73. Laumon B, Gadegbeku B, Martin JL, Biecheler MB. Cannabis intoxication and fatal road crashes in France: population based case-control study. Brit Med J. 2005;331(1371):1–6.
74. Malloy C, Jenkins P. Frontal lobe function and dysfunction. In: Snyder NR. editor. Washington, DC: American Psychological Association; 2009.
75. Damasio A. The frontal lobes. In: Valenstein L. editor. New York: Oxford University Press; 2003.
76. Mackey A, Raizada R, Bunge SA. Environmental influences on prefrontal development. In: Stuss D, Knight R, editors. Oxford Handbook of Frontal Lobe Functions. Oxford: Oxford University Press; 2012.

77. Bourassa M, Vaugeois P. Effects of marijuana use on divergent thinking, creativity. Creat Res J. 2001;13(3–4):411–6.
78. Gruber AJ, Dahlgren MK, Sagar KA, Gonenc A, Killgore WDS. Age of onset of marijuana use impacts inhibitory processing. Neurosci Lett. 2012;511:89–94.
79. Gruber AJ, Dahlgren MK, Yurgelun-Todd D. Why so impulsive? White matter alterations are associated with impulsivity in chronic marijuana smokers. Exp Clin Psychopharmacol. 2011;19(3):231–42.
80. Gruber AJ, Yurgelun-Todd D. Neuroimaging of marijuana smokers during inhibitory processing: a pilot investigation. Brain Res Cogn Brain Res. 2005;23:107–18.
81. Solowij N, Stephens RS, Roffman RA, Babor T, Kadden R, Miller M, Christiansen K, McRee B, Vendetti J. Cognitive functioning of long-term heavy cannabis users seeking treatment. JAMA. 2002;287(9):1123–31.
82. Ilan B, Smith ME, Gevins A. Effects of marijuana on neurophysiological signals of working and episodic memory. Psychopharmacol. 2004;176(2):214–22.
83. Baddeley A. Working memory. Sci. 1993;225(5044):556–9.
84. Jager G, Kahn RS, Van Den Brink W, Van Ree JM, Ramsey NF. Long-term effects of frequent cannabis use on working memory and attention: an fMRI study. Psychopharmacol. 2006;185:358–68.
85. De Bellis MD, Wang L, Bergman SR, Yaxley RH, Hooper SR, Huettel SA. Neural mechanisms of risky decision making and reward response in adolescent cannabis use disorder. Drug Alcohol Depend. 2013;133(1):134–45
86. Boggio PS, Zaghi S, Villani AB, Fecteau S, Pascual-Leone A, Fregni F. Modulation of risk-taking in marijuana users by transcranial direct current stimulation (tDCS) of the dorsolateral prefrontal cortex (DLPFC). Drug Alcohol Depend. 2010;112:220–5.
87. Lane SD, Cherek DR, Tcheremissine OV, Lieving LM, Pietras CJ. Acute marijuana effects on human risk taking. Neuropsychopharmacol. 2005;30(4):800–9.
88. Witlow CT, Liguori A, Livengood LB, Hart SL, Mussat-Witlow BJ, Lamborn CM, Laurienti PJ, Porrino LJ. Long-term heavy marijuana users make costly decisions on a gambling task. Drug Alcohol Depend. 2004;76:107–11.
89. Grant I, Gonzalez R, Carey CL, Natarajan L, Wolfson T. Non-acute (residual) neurocognitive effects of cannabis use: a meta-analytic study. J Int Neureopsychol Soc. 2003;9:679–89.
90. Solowij N, Michie PT. Cannabis and cognitive dysfunction: parallels with endophenyo types of schizophrenia? J Psych Neurosci. 2007;32(1):30–52.
91. Miller LL, Cornett TL. Marijuana: dose effects on pulse rate, subjective estimates of intoxication, free recall, and recognition memory. Pharmacol Biochem Behav. 1978;9:573–7.
92. Ranganathan M, D'Souza DC. The acute effects of cannabinoids on memory in humans: a review. Psychopharmacol. 2006;188:225–44.
93. Miller LL, McFarland D, Cornett TL, Brightwell D. Marijuana and memory impairment: effect on free recall and recognition memory. Pharmacol Biochem Behav. 1977;7:99–103.
94. Lopez-Larson MP, Rogowska J, Bogorodzki P, Beuler CE, McGlade EC, Yurgelun-Todd DA. Cortico-cerebellar abnormalities in adolescents with heavy marijuana use. Psychiatr Res. 2012;202:224–32.
95. Wilson WH, Ellinwood EH, Mathew RJ, Johnson K. Effects of marijuana on performance of a computerized cognitive neuromotor test battery. Psychiatr Res. 1993;51:115–25.
96. Chait LD. Subjective and behavioral effects of marijuana the morning after smoking. Psychopharmacol. 1990;100:328–33.
97. O'Leary DS, Block RI, Turner BM, Koeppel J, Magnotta VA, Ponto L, Watkins GL, Hichwa RD, Andreasen NC. Marijuana alters the human cerebellar clock. Neuroreport. 2003 ;14(8):1145–51.
98. Melges FT, Tinklenberg JR, Hollister LE, Gillespie HK. Temporal disintegration and depersonalization during marijuana intoxication. Arch Gen Psychiat. 1970;23(3):204–10.
99. Pope HG, Yurgelun-Todd. The residual cognitive effects of heavy marijuana use in college students. JAMA. 1996;275(7):521–7.

100. Heishman SJ, Huestis MA, Henningfield JE, Cone EJ. Acute and residual effects of marijuana: profiles of plasma THC levels, physiological, subjective, and performance measures. Pharmacol Biochem Behav. 1990;37:561–5.
101. Fletcher JM, Page JB, Francis DJ, Copeland K, Naus MJ, Davis CM, Morris R, Krauskopf D, Satz P. Cognitive correlates of long-term cannabis use in Costa Rican men. Arch Gen Psychiat. 1996;53:1051–7.
102. Iversen L. Long-term effects of exposure to cannabis. Curr Opin Pharmacol. 2005;5:69–72.
103. Schreiner A, Dunn M. Residual effects of cannabis use on neurocognitive performance after prolonged abstinence: a meta-analyis. Exp Clin Psychopharmacol. 2012;20(5):420–9.
104. Lyons JJ, Bar JL, Panizzon MS, Toomey R, Eisen S, Xian H. Neuropsychological consequences of regular marijuana use: a twin study. Psychol Med. 2004;34:1239–50.
105. Eisen SA, Chantarujikapong S, Xian H, Lyons MJ, Toomey R, True WR, Scherrer JF, Goldberg J, Tsuang MT. Does marijuana use have residual adverse effects on self-reported health measures, sociodemographics and quality of life? A monozygotic co-twin control study in men. Addiction. 2002;97:1137–44.
106. Gruber S, Rogowska J, Yurgelun-Todd D. Altered affective response in marijuana smokers: An fMRI study. Drug Alchol Depend. 2009;105(1–2):139–53.
107. Schwartz RH, Gruenewald PJ, Klitzner M, Gedio P. Short-term memory impairment in cannabis dependent adolescents. Am J Dis Child. 1989;143:1214–9.
108. Yucel M, Zalesky A, Takagi MJ, Bora E, Fornito A, Ditchfield M, Egan GF, Pantelis C, Lubman DI. White-matter abnormalities in adolescents with long-term inhalant and cannabis use: a diffusion magnetic resonance imaging study. J Psychiatry Neurosci. 2010;35(6):409–12.
109. Churchwell JC, Lopez-Larson M, Yurgelun-Todd DA. Altered frontal cortical volume and decision making in adolescent cannabis users. Front Psychol. 2010;1:1–8.
110. Hermann D, Sartorius A, Welzel H, Walter S, Skopp G, Ende G, Mann K. Dorsolateral prefrontal cortex n-acetylaspartate/total creatine (NAA/tCr) loss in male recreational cannabis users. Biol Psychiatry. 2007;61:1281–9.
111. Rubino T, Zamberletti E, Parolaro D. Adolescent exposure to cannabis as a risk factor for psychiatric disorders. J Psychopharmacol. 2012;26(2):177–88.
112. Prescot AP, Locatelli AE, Renshaw PF, Yurgelun-Todd D. Neurochemical Alterations in Adolescent Chronic Marijuana Smokers: a Proton MRS Study. Neuroimage. 2011;57(1):69–75.
113. Solowij N, Jones KA, Rozman ME, Davis SM, Ciarroci J, Heaven PCL, Pesa N, Yucel M. Reflection impulsivity in adolescent cannabis users: a comparison with alcohol-using and non-substance-using adolescents. Psychopharmacol. 2012;219:575–86.
114. Stiglick A, Kalant H. Residual effects of chronic cannabis treatment of behvior in mature rats. Psychopharmacol. 1985;85:436–9.
115. Loeber R, Yurgelun-Todd D. Human neuroimaging of acute and chronic marijuana use: implications for frontocerebellar dysfunction. Hum Psychopharmacol. 1999;14:291–304.
116. Schweinsburg AD, Brown SA, Tapert SF. The influence of marijuana use on neurocognitive functioning in adolescents Curr Drug Abuse Rev. 2008;1(1):99–111.
117. Eldreth DA, Matochik JA, Cadet JL, Bolla KI. Abnormal brain activity in prefrontal brain regions in abstinent marijuana users. NeuroImage. 2004;23:914–20.
118. Kandel D. Stages and pathways of drug involvement; examining the gateway hypothesis. Cambridge:Cambridge University Press. 2002;p. 1.
119. Kandel D, Yamaguchi K, Chen K. Stages of progression in drug involvement from adolescence to adulthood: further evidence for the gateway theory. J Stud Alcohol. 1992;53:447–57.
120. Valenzuela E, Fernandez, M. The sequence of drug use: testing the gateway hypothesis in Latin America. J Int Drug Alcohol Tob Res. 2011;1(1):1–8.
121. Agrawal A, Neale M, Prescott C, Kendler KA. A twin study of early cannabis use and subsequent use and abuse/dependence of other illicit drugs. Psychol Med. 2004;34:1227–37).
122. Van Gundy KR, Rebellon CJ. A life-course perspective on the "gateway hypothesis". J Health Soc Behav. 2010;51(3):244–59.

123. Ferguson DM, Boden JM, Norwood LJ. Cannabis use and other illicit drug use: testing the cannabis gateway hypothesis. Addiction. 2006;101:556–69.
124. Lynskey M, Heath AC, Bucholz KK, Slutske WS, Madden PA, Nelson EC, Statham DJ, Martin NG. Escalation of drug use in early-onset cannabis users vs co-twin controls. JAMA. 2003;289(4):427–33.
125. Pudney S. Home Office Research Study 253: The road to ruin? Sequences of initiation into drug use and offending by young people in Britain. Home Office Research, Development, and Statistics Directorate. 2002.
126. Kilmer B, Caulkins J, Pacula R, MacCoun R, Reuter P. Altered state? Assessing how marijuana legalization in California could influence marijuana consumption and public budgets. Drug Policy Research Center, R and Corporation. 2010;42.
127. Hall W, Room R, Bondy S. WHO project on health implications of cannabis use: a comparative appraisal of the health and psychological consequences of alcohol, cannabis, nicotine and opiate use. World Health Organization. 1998.

Chapter 8
Endocannabinoid-Dopamine Interactions Shape Ethologically Relevant Behavior through Computation of Conditioned Stimuli

Erik B. Oleson and Joseph F. Cheer

Abstract Endocannabinoids are theorized to modulate cue-motivated behavior by amplifying dopamine release within the mesolimbic dopamine system. Here, we summarize the neurochemical results that revealed how endocannabinoids, particularly 2-arachidonoylglycerol within the ventral tegmental area, augment cue-evoked dopamine release events into the nucleus accumbens while concurrently facilitating cue-motivated behavior. Data are initially described within the context of the classical role attributed to the mesolimbic system—promoting reward seeking. We then expand our discussion beyond simple reward-directed behaviors by discussing a potential role for endocannabinoid-mesolimbic dopamine interactions in: switching an animal's motivation from feeding to foraging behavior when reward availability is temporally delayed, facilitating the extinction of fear memories by amplifying dopaminergic updates of the association between aversive stimuli and their predictors to the fear network, and promoting the avoidance of harmful stimuli through newly discovered negative reinforcement mechanisms.

Keywords Endocannabinoids · 2-Arachidonoylglycerol · Dopamine · Motivation · Negative reinforcement · Aversion · Fear · Reward · Ventral tegmental area · Nucleus accumbens

Introduction

Introduction to Cue-Motivated Behavior

Survival often demands motivated displays of behavior devoted to obtaining valued commodities from the environment. Early studies on animal behavior noted the existence of distinct behavioral repertoires associated with obtaining such commodities. From these studies it was inferred that the 'appetite' of an animal produces a readiness to act, and out of this altered motivational state arise incipient actions

J. F. Cheer (✉) · E. B. Oleson
University of Maryland School of Medicine, Baltimore, MD, USA
e-mail: jchee001@umaryland.edu

P. Campolongo, L. Fattore (eds.), *Cannabinoid Modulation of Emotion, Memory, and Motivation,* DOI 10.1007/978-1-4939-2294-9_8

devoted to acquiring the desired commodity [1]. These incipient actions are clearly distinct from those proceeding obtainment, such as consumption of the commodity. One ethologically relevant example is sex. The initial component of sexual behavior is often described as being appetitive in nature [2–4]. During a sexual exchange an animal first attempts to find and attract a potential mate—sometimes even invoking some ornate display of behavior in the course of the courting process. The proceeding copulatory, or consummatory response is generally more reflexive in nature, and ultimately results in reducing the animal's motivational drive for the sex. Another biologically important example is feeding. During a feeding event, the appetitive act of seeking out a food source often precedes the consummatory act of placing the food into the mouth and swallowing. Again, these two responses are behaviorally distinct, as the amount of effort required to seek out the commodity fails to influence the more reflexive act of food consumption [1, 5], while feeding ultimately reduces motivation for the specific food that was ingested [6, 7]. A commonality between the appetitive responses of seeking sex and food is that both of these incipient actions require the animal to explore their surroundings to gain access to the desired commodity. During exploration, critical associations with environmental stimuli are formed that, in turn, acquire the ability to recapitulate previously successful behavioral repertoires through the recruitment of motivational processes [8, 9]. This conserved feature of appetitive behavior implies the existence of discrete neural circuitry devoted to guiding incipient actions as animals seek valuable commodities from their environment.

Introduction to the Mesolimbic Dopamine System

The mesolimbic dopamine system, a subcortical neural pathway—highly conserved across vertebrate species [10]—is theorized to promote incipient actions by generating a teaching signal that draws animals toward favorable stimuli and away from harmful ones [11–13]. This dopamine signal is generated by phasic bursts of dopamine neural activity within the midbrain [12] that are heterogeneously transmitted as transient release events throughout terminal regions of the mesolimbic pathway, such as the nucleus accumbens [14, 15]. Using cutting-edge neurochemical techniques, we can measure these transient release events in the nucleus accumbens when animals are presented with motivationally salient stimuli [16–18]. The nucleus accumbens has been fittingly described as a limbic-motor [19] and Pavlovian-instrumental [20] interface, which conveys the important theoretical construct that this particular brain region is critically involved in transforming information from motivational-salient environmental cues into incipient actions devoted to obtaining highly valued commodities. The nucleus accumbens primarily receives dopaminergic afferents from the ventral tegmental area of the midbrain. The ventral tegmental area not only consists of dopamine neurons, rather, this brain region is composed of GABA, glutamate and dopamine neurons that interact to influence accumbal dopamine release. The proportional expression of these different neural subtypes within the ventral tegmental area is currently thought to be approximately: 60% dopaminergic, 25% GABAergic and 15% glutamatergic neurons [21–23].

Introduction to the Endocannabinoid System

While many research groups have confirmed a role for accumbal dopamine concentrations in cue-motivated behavior [24–30], the importance of endocannabinoid modulation of dopamine release in these processes is just beginning to be realized. This is primarily due to the historical fact that Arvid Carlson first identified dopamine to be critically involved in controlling motor behavior in 1957 [31], but endocannabinoids were not discovered by Raphael Mechoulam until the mid-1990s [32, 33]. The endocannabinoid system is comprised of lipid signaling molecules—2-arachidonoylglycerol and anandamide are the best characterized [32, 33], their G protein-coupled receptor targets (CB1 and CB2); although it should be noted that anandamide also exhibits some binding affinity for TRPV1 receptors [34], their synthetic enzymes (diacylglycerol lipase and N-arachidonoyl phosphatidylethanolamine phospholipase D), their hydrolytic enzymes (monoacylglycerol lipase (MAGL) and fatty acid amide hydrolase (FAAH)) [35–37], and a recently characterized transport system [38].

Endocannabinoid-Dopamine Interaction in Cue-Motivated Reward Seeking

Insights from Endocannabinoid Psychopharmacology and Neurochemistry

Advances in our understanding of the endocannabinoid system and rational drug design led to the development of new pharmacological tools allowing investigators, for the first time, to focus on how endocannabinoids might modulate cue-motivated behavior. A number of psychopharmacology studies first investigated the effects of disrupting endocannabinoid signaling on motivated behavior by using cannabinoid receptor antagonists/inverse agonists. These initial studies revealed that disrupting endocannabinoid signaling decreases motivation for both food [39] and drugs of abuse [40–42]. Additional research revealed that disrupting endocannabinoid signaling is particularly effective at reducing environmental influences over motivated behavior [43–45]. The observation that disrupting endocannabinoid signaling is particularly effective at reducing cue-motivated behavior, regardless of the commodity that the cue predicts [44], suggested the existence of a central neural mechanism through which endocannabinoids modulate their effects on motivation. Simultaneously, neurochemists began using cannabinoid receptor antagonists to confirm that exogenous cannabinoids, like the primary psychoactive component of cannabis—delta-9-tetrahydrocannabinol, increase dopamine concentrations in the nucleus accumbens by binding to cannabinoid CB1 receptors in the brain [46–50].

Theoretical Neural Mechanism of Endocannabinoid-Dopamine Interactions

The finding that CB1 receptor activation is required to observe accumbal dopamine release events raised the intriguing possibility that maybe endocannabinoids capably modulate dopamine release events in their own right. However, it was also realized that dopamine cell bodies lack CB1 receptors. Emerging electrophysiological evidence revealed an indirect mechanism through which endocannabinoids might modulate dopamine release during periods of phasic neural activity—a mechanism that has become known as depolarization induced suppression of inhibition/excitation [51]. Endocannabinoids, unlike the majority of vesicular released neurotransmitters, are synthesized post-synaptically and released on demand during periods of high neural activity [52, 53]. Specifically, activation of $G_{q/11}$-coupled metabotropic receptors and/or postsynaptic depolarization is thought to activate voltage gated Ca^{2+} channels, allowing for an influx of intracellular Ca^{2+} that, in turn, activates the synthetic enzymes responsible for producing endocannabinoids [54]. The newly synthesized endocannabinoids are then released from the postsynaptic domain and act retrogradely at presynaptic CB1 receptors. Cannabinoid CB1 receptors are typically coupled to G_i G-proteins which, when activated, ultimately reduce neurotransmitter release through modulation of presynaptic K^+ and Ca^{2+} conductances [55]. While endocannabinoids bind to CB1 receptors on both GABAergic and glutamatergic terminals in the ventral tegmental area [56], an abundance of evidence suggests that a suppression of GABA release onto dopamine neurons ultimately facilitates dopamine neural activity *via* disinhibition of the dopamine cell bodies [57–59]. According to this line of logic, endocannabinoid levels should be highest in the ventral tegmental area after periods of high dopaminergic neural activity, and consequently, this increase in endocannabinoid tone should amplify accumbal dopamine release in a manner that facilitates cue-motivated behavior.

Disrupting Endocannabinoid Signaling Decreases the Neural Mechanisms of Cue-Motivated Reward Seeking

To investigate the role of endocannabinoids in the neural mechanisms of cue-motivated behavior (Fig. 8.1a), we first treated animals with a cannabinoid receptor antagonist/inverse agonist while measuring accumbal dopamine release during a reward seeking task [60]. In this case, animals were trained to respond on a lever for brain stimulation reward—electrical currents delivered to the ventral tegmental area. Brain stimulation reward is a highly reinforcing stimulus, well known to

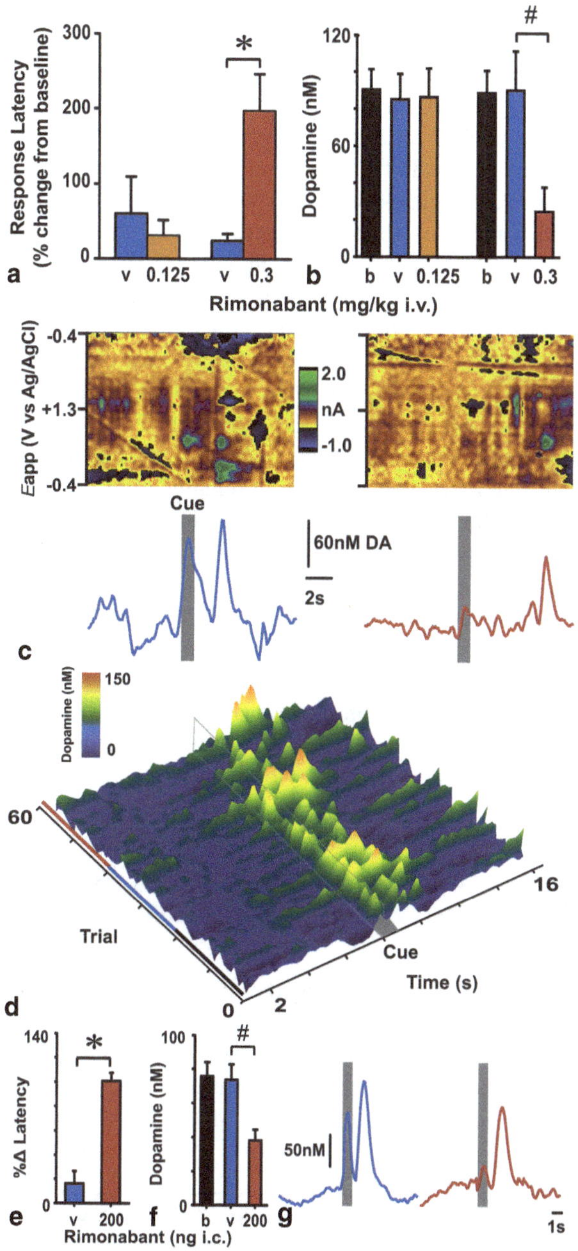

Fig. 8.1 Depiction of all behavioral procedures described herein. **a** In a representative food seeking trial the animal is presented with a cue-light and response lever. Pressing the lever once results in the immediate delivery of a rewarding stimulus (e.g., brain stimulation reward, highly palatable food), retraction of the lever, and a 10 s inter-trial-interval. **b** In a periodic reinforcement trial (i.e., fixed interval) the animal is presented with a lever but delivery of reward is delayed until a response occurs after fixed period of time (e.g., 30 s). Responses occurring prior to culmination

maintain high rates of lever responding during operant behavior. Presentation of a preceding environmental cue (e.g., light) signaled the availability of brain stimulation reward. As occurs when animals are presented with environmental predictors of various rewarding stimuli, including sex [61], food [62, 63], and drugs [64, 65], the conditioned cue evoked a transient dopamine release event in the nucleus accumbens (Fig. 8.2c). Importantly, we determined that a greater magnitude of the dopamine release event corresponded with a decrease in the animal's latency to respond for brain stimulation reward—suggesting that the cue-evoked dopamine release event motivated the incipient action aimed at obtaining the desired stimulus [60]. When the animal was systemically treated with the cannabinoid receptor antagonist/inverse agonist rimonabant, however, cue evoked dopamine concentrations decreased (Fig. 8.2b, c), while the latency to respond for brain stimulation reward increased (Fig. 8.2a). The corresponding decrease in cue-evoked dopamine release and cue-motivated behavior is shown across trials in Fig. 8.2d. While these data appeared to show that endocannabinoids are necessary to observe cue-evoked dopamine release during motivated behavior, it remained unclear whether disrupting endocannabinoids in the ventral tegmental area alone would be sufficient to suppress the neural mechanisms of cue-motivated behavior. To investigate the precise role of endocannabinoids in the ventral tegmental area during cue-motivated behavior, we infused the cannabinoid antagonist/inverse agonist rimonabant directly into the ventral tegmental area. As occurred following systemic administration, disrupting endocannabinoid signaling in the ventral tegmentum concurrently decreased cue-evoked dopamine concentrations and reward seeking behavior (Fig. 8.2e, f, g). These data demonstrated that endocannabinoids are required to amplify cue-evoked dopamine release during motivated behavior.

2-Arachidonoylglycerol, but not Anandamide, Facilitates the Neural Mechanisms of Cue-Motived Reward Seeking

Our research question next shifted to whether the endocannabinoid responsible for amplifying cue-evoked dopamine release during cue-motivated behavior was 2-arachidonoylglycerol and/or anandamide. We began by systemically increasing 2-arachidonoylglycerol or anandamide concentrations through pharmacological antagonism of their respective degradative enzymes, MAGL and FAAH [35–37]. JZL184

---

of the interval are recorded, but produce no scheduled consequence. **c** In a negative reinforcement session the animal is presented with a cue light (i.e., warning signal) and lever. If the lever is pressed within the first 2 s of warning signal presentation the animal prevents the occurrence of footshock by immediately entering an inter-trial period signaled by a tone (i.e., safety period) for 20 s (avoidance response; *red* line). After the 2 s warning period elapses footshocks commence. A lever response will now terminate ongoing footshock and produce the 20 s safety period (escape response; *blue* line). **d** In a fear conditioned session the animal is presented with tone in association with an inescapable footshock (3 presentations, each culminating with footshock). 24 hours later, the animal is presented with the tone alone and fear-induced freezing behavior is assessed

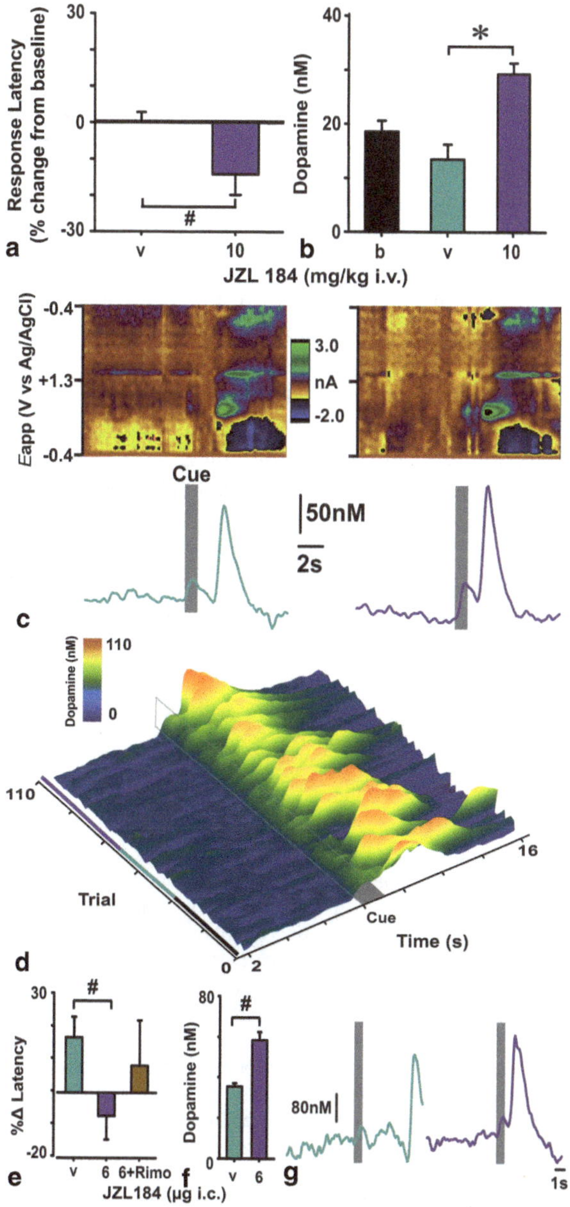

Fig. 8.2 Endocannabinoids sculpt ethologically relevant pattern of dopamine release during food seeking. **a** Response latency (a metric of reward seeking) for brain stimulation reward maintained in the brain stimulation reward task. A high (0.3 mg/kg i.v.; *red bar*) but not low (0.125 mg/kg i.v.; *orange bar*) dose of rimonabant increased the latency to respond for brain stimulation reward in comparison to vehicle (v, *blue bar*). **b** Mean dopamine concentration observed during the first second of cue presentation under baseline (b), vehicle (v), and drug conditions. Rimonabant at a high (0.3 mg/kg i.v.; *red bar*) but not low (0.125 mg/kg i.v.; *orange bar*) dose decreased the concentration of cue-evoked dopamine in comparison to vehicle. **c** Representative color plots (*top*) and

is a known MAGL inhibitor [37] while URB597 is a known FAAH inhibitor [36]. A clear dissociation between the effects of the two endocannabinoids on cue-motivated behavior became readily apparent. Augmenting 2-arachidonoylglycerol levels increased motivation for food and brain stimulation reward, whereas, increasing anandamide levels failed to change behavior. It was also confirmed that JZL184 significantly increased 2-arachydonoylglycerol levels in the ventral tegmental area by assessing midbrain tissue content immediately following brain-stimulation reward seeking sessions. Importantly, we found that systemically augmenting 2-arachydonoylglycerol levels increased cue-evoked dopamine concentrations in the nucleus accumbens while decreasing the latency to respond for brain stimulation reward (Fig. 8.3a, b, c, d). Again we wanted to determine whether increasing 2-arachydonoylglycerol in the ventral tegmental area alone would be sufficient to augment cue-evoked dopamine release and facilitate cue-motivated behavior. In confirmation of a central role for 2-arachydonoylglycerol in amplifying the neural mechanisms of cue-motivated behavior by disinhibiting ventral tegmental area dopamine neurons, intrategmental infusions of JZL184 increased cue-evoked dopamine release into the nucleus accumbens while concurrently decreasing the latency to respond for brain stimulation reward (Fig. 8.3e, f, g). These data suggest that 2-arachydonoylglycerol facilitates dopamine signaling, and consequently guides animals to appetitive stimuli and motivates their incipient actions devoted to obtaining the desired commodities.

dopamine concentration traces (*bottom*) show the effects of rimonabant on cue-evoked dopamine events in individual trials. *Top*: Representative color plots topographically depict the voltammetric data with time on the x axis, applied scan potential (*E*app) on the y axis and background-subtracted faradaic current shown on the z-axis in pseudocolor. Dopamine can be identified by an oxidation peak (*green*) at + 0.6 V and a smaller reduction peak (*yellow*) at −0.2 V. *Bottom*: Corresponding traces show the concentration of dopamine (nM) detected at the time of cue presentation (*gray bar*) following vehicle (*left*; *blue* trace) and rimonabant (*right*; *red* trace) administration. **d** A representative surface-plot shows changes in dopamine concentration (z axis) across trials (y axis) during baseline (*black line*), vehicle (*blue line*), and rimonabant (*red line*) conditions. Data are centered around lever presentation on the *x* axis. **e** Disrupting endocannabinoid signaling within the ventral tegmental area is sufficient to decrease reward seeking. Intrategmental rimonabant (200 ng i.c.; *red bar*) significantly increased response latency in comparison to vehicle (v, *blue bar*). **f** Mean dopamine concentrations observed during first second of cue-presentation under baseline (b), vehicle (v), and drug conditions. Intra-tegmental rimonabant (200 ng i.c.; *red bar*) significantly decreased the concentration of cue-evoked dopamine in comparison to vehicle. **g** Representative dopamine concentration traces from individual trials after vehicle (*left*; *blue* trace) and rimonabant (200 ng i.c.; *right*; *red* trace) treatment. From Oleson EB, Beckert MV, Morra JT, Lansink CS, Cachope R, Abdullah RA, et al. Endocannabinoids shape accumbal encoding of cue-motivated behavior via CB1 receptor activation in the ventral tegmentum. Neuron. 2012;73(2):360–73; used with permission

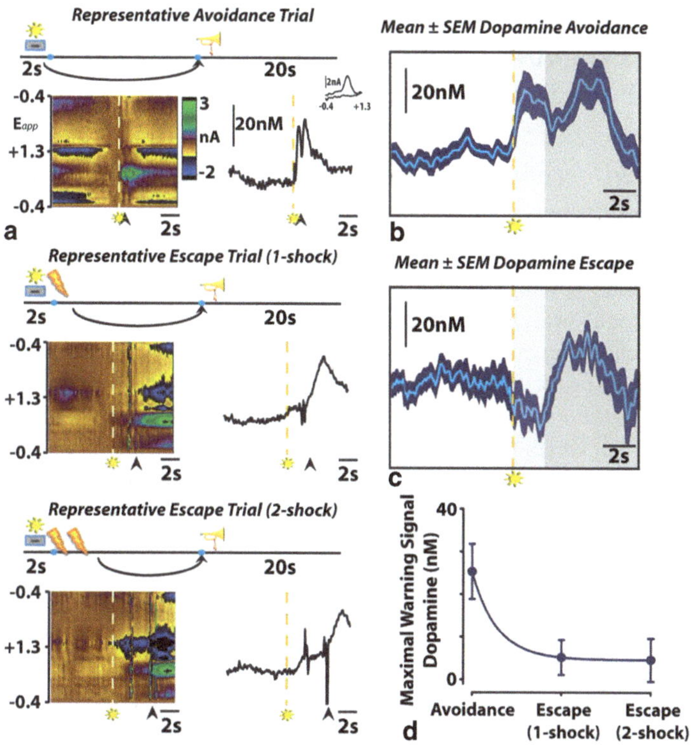

Fig. 8.3 **a** Augmenting 2-arachidonoylglycerol levels facilitate the neural mechanisms of cue-motivated behavior. JZL184 (10 mg/kg i.v., *purple bar*) decreased response latency in comparison to vehicle (v, *blue bar*). **b** Facilitated reward seeking was accompanied by an increase in cue-evoked dopamine concentration. **c** Representative color plots (*top*) and dopamine concentration traces (*bottom*) show the effects of JZL184 (*right*, *purple* trace) in comparison to vehicle (*left*, *green* trace) during individual trials. **d** A representative surface plot illustrates changes in dopamine concentration across trials (y axis) under baseline (*black* line), vehicle (*green* line), and rimonabant (*purple* line) conditions. **e** Augmenting 2-arachidonoylglycerol in the ventral tegmental area is sufficient to facilitate reward seeking. JZL184 (6 mg, ipsilateral, *purple bar*) decreased response latency in comparison to DMSO (*green bar*). Post-treatment with a subthreshold dose of rimonabant (1.25 mg/kg i.v.) reversed the JZL184-induced decrease in reward latency. **f** Facilitated reward seeking occurred simultaneously with an increase in cue-evoked dopamine concentration in comparison to vehicle. **g** Representative traces show the effects of intrategmental vehicle (*left*, *green* trace) and JZL184 (*right*, *purple* trace) on cue-evoked dopamine concentration in individual trials. From Oleson EB, Beckert MV, Morra JT, Lansink CS, Cachope R, Abdullah RA, et al. Endocannabinoids shape accumbal encoding of cue-motivated behavior via CB1 receptor activation in the ventral tegmentum. Neuron. 2012;73(2):360–73; used with permission

To Feed or Forage: Endocannabinoid-Dopamine Interactions Uniquely Modulate Motivated Behavior when Reward is Periodically Available

Periodic Reinforcement Requires Interoceptive Timing Cues to Guide Behavior

Often times in an ethological setting the environment fails to guide incipient actions toward highly valued commodities using a single external, or exteroceptive cue. In some cases, repertoires of animal behavior are forced to adapt to temporal constraints dictating availability of the desired commodity. For example, it might be beneficial for an animal's motivation to switch from feeding to foraging during periods of time in which the primary food source is unavailable. In such instances, the focus of an animal's appetite might rely on interoceptive timing cues encoded by discrete neural circuitry. Within a certain window of time, these interoceptive timing cues might prompt the animal to continue seeking the primary food source, however, after a substantial intermittent delay, the animal might be motivated to forage for alternative options. Previous accounts of interval timing proposed that the brain contains a neural timer that resets to zero following receipt of a desired commodity [66, 67] and that dopamine is critically involved in processing these second-to-second estimations [68–75]. Based on our previous data and the well-documented phenomenon that cannabinoids alter timing behavior [76–80], we speculated that endocannabinoids might modulate dopaminergic encoding of interoceptive timing cues and alter whether an animal is motivated to feed or forage.

Periodic Reinforcement Produces A Unique Behavioral Repertoire

We therefore theorized that cannabinoids might modulate dopamine release and differentially motivate incipient actions during experimental conditions of periodic reinforcement. To address this research question we measured real-time accumbal dopamine release during food-directed behavior maintained under a fixed interval schedule (Fig. 8.1b). On a fixed interval schedule, animals are provided unlimited access to an active lever that dispenses food pellets, but food delivery only occurs after a fixed period of time. Fixed interval procedures produce a characteristic temporal response pattern on the active lever. After food delivery, responding initially slows before accelerating to a maximal rate at the interval terminus [81]. In addition to producing an accelerating temporal response pattern on the active lever, the interval between reinforced responses in fixed interval procedures produces a critical range of intermittency that fosters irrelevant, or adjunctive behavior [82, 83]. In contrast to active lever responses, adjunctive behavior typically increases during the initial part of the interval before declining toward the interval terminus [84]. As such, it was previously suggested that responses on an inactive lever might represent adjunctive behavior in two-lever operant tasks when primary reinforcement is periodically available [85]. One interpretation of this behavioral repertoire during food-maintained responding is that active lever responses reflect the animal seeking a primary food source, whereas, adjunctive behavior reflects, in part, the animal foraging for an alternative option. Thus, in our fixed interval task we attempted to assess how cannabinoids modulate dopamine release, and how these changes in dopamine release might relate to both feeding and foraging behavior.

Mesolimbic Dopamine Release Encodes Interval Time During Periodic Reinforcement

As it remained unknown precisely how dopamine release is related to interval time during conditions of periodic reward availability, we began by characterizing changes in accumbal release over defined periods of time using a fixed interval task. We predicted that dopamine release would increase across the interval duration in parallel with the frequency of active lever responses. Contrary to our hypothesis, dopamine release decreased across the interval, even lawfully resetting to different periods of time, while also signaling the delivery of food reward [86]. These data suggest that accumbal dopamine release encodes the period of elapsed time between occurrences of reward availability, but does so in an inversely proportional manner to interval time.

Cannabinoids Augment Dopamine Release and Accelerate Primary Reward Seeking During Period Reinforcement

To investigate how cannabinoids alter accumbal dopamine release and behavior under conditions in which reward is periodically available, we first treated mice with the synthetic cannabinoid WIN 55,212–2 while they responded for food under a fixed interval schedule. WIN 55,212–2 accelerated the temporal response pattern on the active lever while concurrently increasing accumbal dopamine release through the initial portion of the interval [86]. Pretreatment with a low dose of a cannabinoid CB1 receptor antagonist/inverse agonist blocked the increase in accumbal dopamine release and acceleration of responding on the active lever, thereby demonstrating that these effects required cannabinoid CB1 receptor activation. These data prompted us to further investigate the role of the endocannabinoids 2-arachydonoylglycerol and anandamide in these effects, in addition to characterizing their effects on the complete behavioral repertoire—appetitive seeking of the primary food source versus foraging for alternative options.

2-Arachidonoylglycerol, But Not Anandamide, Accelerates the Temporal Response Pattern of Primary Reward Seeking During Periodic Reinforcement

To assess the specific contributions of the brain's two best characterized endocannabinoids during periodically reinforced reward seeking, we employed a similar approach to that described above. We pharmacologically increased 2-arachidonoylglycerol and anandamide by inhibiting their respective degradative enzymes, MAGL and FAAH [35–37]. In the fixed interval task, elevating 2-arachidonoylglycerol levels significantly accelerated the temporal response on the active lever in a manner resembling WIN 55,212–2, while elevating anandamide levels failed to change behavior [86]. Thus, reward directed behaviors that are guided by either exteroceptive or interoceptive timing cues are under control of the endocannabinoid 2-arachidonoylglycerol, whereas, under multiple conditions of reward availability, anandamide fails to modulate incipient actions aimed at acquiring valued commodities.

Endocannabinoids Promote Both Primary Reward Seeking and Foraging Behavior

As previously noted, after some period of time has elapsed in which the primary commodity of desire remains unavailable, it might be beneficial for an animal's motivation to switch from seeking the primary reward to looking for alternative options. In the experimental laboratory, such behaviors may become manifest as reinforcement-irrelevant, or adjunctive behavior, and be measured by responses on the inactive lever. It might not be surprising, therefore, that increasing 2-arachidonoylglycerol levels with JZL184 suppressed inactive lever responses [86] while active lever responses increased. These data might suggest that competing motivation to seek out alternative options is minimized by heightened dopamine concentrations that serve to motivate the animal to seek the primary food reward. Interestingly, disrupting endocannabinoid signaling by treating animals with a cannabinoid CB1 receptor antagonist/inverse agonist also reduced inactive lever responses [86]. This finding suggests that endocannabinoids might promote behavioral fitness by motivating behaviors directed at both obtaining primary reinforcement and alternative options. An endogenous endocannabinoid tone might be required to motivate the animal to seek out alternative options, or forage, while increased concentrations of 2-arachidonoylglycerol might promote the incentive to continually commit incipient actions aimed at obtaining the primary reward of interest. According to the aforementioned model of endocannabinoid-dopamine interactions, it is likely that heightened phasic activation of dopamine neurons augments basal concentrations of 2-arachidonoylglycerol, which consequently increases dopamine release and motivates the animal to continue seeking the primary food reward. Taken together, this line of logic suggests that endocannabinoid tone is required to promote an animal's propensity to forage for alternative options, but amplified signaling serves to exclusively motivate incipient actions directed at obtaining the desired commodity.

Potential Endocannabinoid-Dopamine Interactions During the Avoidance of Aversion and Aversion Itself

Introduction to Dopaminergic Encoding of Aversive Stimuli

To survive, animals must be concerned about not only seeking reward stimuli in the environment, but also avoiding aversive ones. While it was once thought that the mesolimbic dopamine system is a simple reward circuit, recent evidence demonstrates that midbrain dopamine neurons are critically involved in guiding animals away from potentially harmful outcomes [87–89]. Dopamine depletion impairs learning about aversive events [90], while restoring dopamine to terminal fields overcomes these deficits [91]. An abundance of evidence shows that aversion, in addition to reward, is represented by dopamine neural activity [87, 92–94]. Indeed, electrophysiological studies reveal that anatomically distinct populations of dopamine neurons encode either aversion or reward [87, 92, 94]. As a consequence, real-time accumbal dopamine concentration transients are detected when animals are presented with predictors of aversion and its avoidance [93–95]. These data suggest that dopamine neurons guide incipient actions as animals seek to avoid aversive stimuli in the environment and raise the intriguing possibility that endocannabinoids modulate the processes of negative reinforcement. For more thorough reviews on the role of dopamine in the processing of aversion we refer the reader to: [13, 96, 97].

Dopamine Encodes Warning Stimuli During Conditioned Avoidance: A Potential Role for Endocannabinoid Modulation of Negative Reinforcement

We recently became interested in exploring endocannabinoid-dopamine interactions during events that increase the probability of incipient actions devoted to avoiding the particular event—a psychological concept known as negative reinforcement. This experimental goal required an initial characterization of accumbal dopamine release during conditioned avoidance, a scenario in which an animal must respond to an environmental cue by carrying out some incipient action to prevent the delivery of an aversive stimulus. To characterize dopamine release during conditioned avoidance we measured subsecond accumbal dopamine release events during behavior maintained in an operant conditioned footshock procedure (Fig. 8.1c). During the operant conditioned footshock procedure we implemented, a stimulus light is presented to the animal as a warning signal for 2 s prior to the delivery of

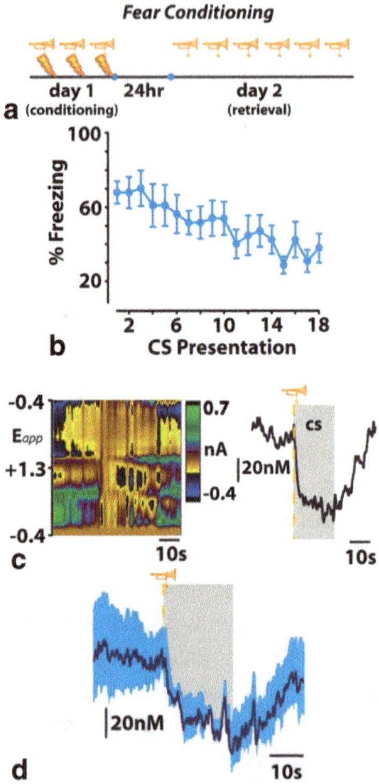

Fig. 8.4 **a** Dopamine encodes conditioned stimuli during negative reinforcement. Representative color plots (*left*) and dopamine concentration traces (*right*) show avoidance (*top*), one-footshock escape (*middle*), and two-footshock escape (*bottom*) responses. Arrows indicate lever responses, lightning bolts indicate footshocks, trumpets indicate safety periods, levers + lights indicate warning signals. *Left*, Voltammetric current (z-axis) plotted against applied scan potential (Eapp; y-axis) and time (x-axis). *Right*, dopamine concentration traces plotted as a function of time. Inset shows cyclic voltammogram for dopamine. **b** Warning signal presentation increases dopamine release when rats successfully avoid footshock. Mean ± SEM traces depict the time course of changes in subsecond dopamine release as animals minimize punishment by avoiding footshock. Dashed lines represent warning stimulus onset, around which mean data are grouped. Color representations: light *gray*, maximum warning stimulus duration; dark *gray*, safety period. **c** Warning signal presentation inhibits dopamine release when rats fail to avoid or escape footshock. Only one-footshock escape trials are included in the mean. **d** Maximal dopamine concentration evoked by warning signal presentation predicts successful punishment avoidance. From Oleson EB, Gentry RN, Chioma VC, Cheer JF. Subsecond dopamine release in the nucleus accumbens predicts conditioned punishment and its successful avoidance. The Journal of neuroscience. 2012;32(42):14804–8; used with permission

recurring footshocks (Fig. 8.4a). A response lever is simultaneously extended into the operant chamber when the warning signal is illuminated. If the animal responds on the lever during the initial 2 s warning period the lever retracts, the light dims, a tone sounds and the footshock delivery is prevented for the duration of a 20 s safety period. If the animal fails to respond to the 2 s warning signal the ensuing footshocks may still be relieved by pressing the lever, thereby terminating ongoing shock and producing a 20 s safety period signaled by a tone. Thus, the animal may either elicit an avoidance response by pressing the lever during the 2 s warning signal or elicit an escape response by pressing the lever after shocks commence. Under these conditions, warning signal evoked accumbal dopamine release was exclusively observed during successful avoidance responses (Fig. 8.4b, c, d). In other words, the occurrence of accumbal dopamine release during warning signal presentation predicted whether or not the animals would successfully avoid foot-shock. These data suggest that accumbal dopamine release encodes cues predicting negative reinforcement and may motivate incipient actions devoted to the avoidance of the stimuli they predict. As this form of dopamine signaling was previously shown to be under modulatory control of endocannabinoid signaling, we speculate that manipulating endocannabinoid transmission will alter the dopaminergic neural mechanisms of cue-motivated avoidance behavior. Future studies will assess whether disruptions in endocannabinoid signaling will reduce dopamine release evoked by cues predicting negative reinforcement, and whether increases in ventral tegmental area 2-arachidonoylglycerol levels will facilitate this neural mechanism of cue-motivated avoidance. These advances in our understanding of the role of endocannabinoids in the neural mechanisms of negative reinforcement may lead to the development of novel treatment approaches for psychopathologies involving negative reinforcement, such as drug addiction [98, 99].

Dopamine Encodes Aversion

Endocannabinoids may also modulate dopaminergic neural encoding of aversion. Precisely how dopamine neurons encode aversive stimuli remains a controversial subject. For the purpose of this chapter, we will focus exclusively on real-time electrochemical measurements of accumbal dopamine release during presentations of aversive stimuli and their conditioned predictors. The Roitman group first demonstrated that, in the freely-moving animal, accumbal dopamine release oppositely encodes rewarding and aversive stimuli [100]. Specifically, they demonstrated that an aversive quinine solution suppresses, whereas an appetitive sucrose solution increases the frequency of dopamine transient events. This observation led to speculation that a decrease in accumbal dopamine release events might also encode conditioned aversive stimuli, or fear memories—powerful stimuli that the endocannabinoid system is known to modulate [100].

Dopamine Encodes Conditioned Fear: A Potential Role for Endocannabinoid-Dopamine Interactions in the Extinction of Fear Memories

Two concurring reports recently characterized how accumbal dopamine release events encode conditioned predictors of fear by employing standard fear condi-

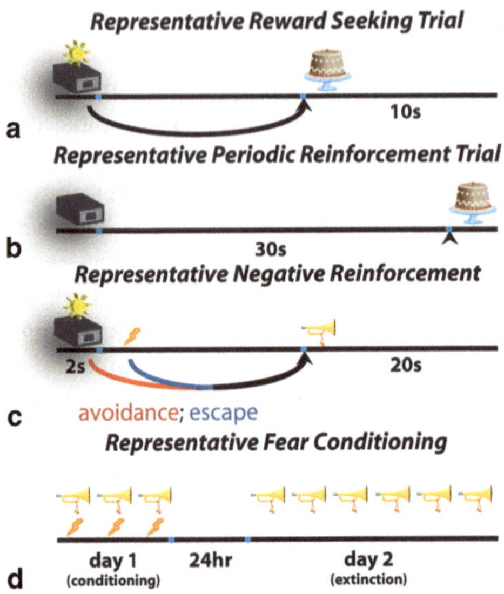

Fig. 8.5 Dopamine encodes conditioned stimuli predicting aversive stimuli. **a, b** Fear conditioning produces freezing behavior that extinguishes across repeated trials of conditioned stimulus (CS) presentation on fear-memory retrieval day. **c** Representative color plot (*left*) and corresponding dopamine concentration trace (*right*) show a CS-induced decrease in dopamine release. **d** Mean ± SEM dopamine concentration trace during presentations of a CS that produces a conditioned freezing response. *Gray* represents CS duration. From Oleson EB, Gentry RN, Chioma VC, Cheer JF. Subsecond dopamine release in the nucleus accumbens predicts conditioned punishment and its successful avoidance. The Journal of neuroscience. 2012;32(42):14804–8; used with permission

tioned methodologies. In a fear-conditioned task, an experimental animal is initially presented with three consecutive tones, each culminating with a potent inescapable foot shock (Fig. 8.5a; Fig. 8.1d). During this initial fear-conditioning phase of the experiment, the animal begins to freeze when the tone in presented. This freezing response is thought to be a behavioral manifestation of a negative emotional state of fear [101–104]. The ability of the tone to produce a freezing response is then assessed 24 h after the initial fear-conditioning period within a novel context. Now, when presented alone the conditioned predictor of fear initially elicits a freezing response, however, the percentage of time spent freezing dissipates over repeated tone presentations (Fig. 8.5b). Theoretically, the magnitude of the freezing response is attenuated over time because the fearful memory that the tone represents extinguishes over repeated pairings. As occurred when animals tasted an aversive quinine solution, tone presentations that produced a freezing response also suppressed the frequency of dopamine release events detected in the core region of the nucleus accumbens (Fig. 8.5c, d) [93, 94]. As we previously demonstrated that 2-arachidonoylglycerol levels in the ventral tegmental area facilitate dopamine release during cue-motivated behavior, it is possible that facilitating endocannabinoid levels may enhance the extinction of fear-memories, in part, by augmenting dopaminergic encoding of the environmental predictor of the aversive stimulus. Enhancing neural representations of the association between a fearful event and its predictive stimulus may strengthen dopaminergic updates to the fear network, which might ultimately facilitate extinction of the cue-induced fear response. Uncovering such a neural mechanism might aid the development of treatments strategies for psychopathologies characterized by conditioned maladaptive fear responses, such as post-traumatic stress disorder.

Conclusion

While we have unambiguously demonstrated that endocannabinoid-dopamine interactions in the ventral tegmental area guide cue-motivated incipient action sequences directed at primary rewards [60], this is just the beginning. It is becoming clear that the mesolimbic dopamine system is more than a mere reward circuit, rather we theorize that endocannabinoid-dopamine interactions within the ventral tegmental area evolved to encode various ethologically-relevant stimuli in a manner that promotes survival. Of note, we believe there is strong potential that endocannabinoid-dopamine interactions are also capable of guiding animals to seek alternative resources when the primary commodity of desire is unavailable, facilitating the extinction of fear memories by amplifying neural signals to the fear network that encode information regarding the association between fearful stimuli and their conditioned predictors, and guiding animals away from potentially harmful stimuli by promoting negative reinforcement.

References

1. Craig W. Appetites and aversions as constituents of instincts. Biol Bull. 1918;34(2):91–107.
2. Beach FA. Sexual attractivity, proceptivity, and receptivity in female mammals. Horm behav. 1976;7(1):105–38.
3. Everitt BJ. Sexual motivation: a neural and behavioural analysis of the mechanisms underlying appetitive and copulatory responses of male rats. Neurosci Biobehav Rev. 1990;14(2):217–32.
4. Balthazart J, Absil P, Gérard M, Appeltants D, Ball GF. Appetitive and consummatory male sexual behavior in Japanese quail are differentially regulated by subregions of the preoptic medial nucleus. J Neurosci. 1998;18(16):6512–27.
5. Kanarek RB, Collier G. Patterns of eating as a function of the cost of the meal. Physiol Behav. 1979;23(1):141–5.
6. Rolls BJ, Rolls ET, Rowe EA, Sweeney K. Sensory specific satiety in man. Physiol Behav. 1981;27(1):137–42.
7. Rolls BJ, Rowe EA, Rolls ET, Kingston B, Megson A, Gunary R. Variety in a meal enhances food intake in man. Physiol Behav. 1981;26(2):215–21.
8. Thorndike EL. Animal intelligence: an experimental study of the associative processes in animals. Psychol Monogr: Gen Appl. 1898;2(4):i-109.
9. Bindra D. A motivational view of learning, performance, and behavior modification. Psychol Rev. 1974;81(3):199.

10. Smeets WJ, Marin O, Gonzalez A. Evolution of the basal ganglia: new perspectives through a comparative approach. J Anat. 2000;196(4):501–17.
11. Redgrave P, Gurney K. The short-latency dopamine signal: a role in discovering novel actions? Nat Rev Neurosci. 2006;7(12):967–75.
12. Schultz W, Dayan P, Montague PR. A neural substrate of prediction and reward. Science. 1997;275(5306):1593–9.
13. Oleson EB, Cheer JF. On the role of subsecond dopamine release in conditioned avoidance. Front Neurosci. 2013;7.
14. Sombers LA, Beyene M, Carelli RM, Wightman RM. Synaptic overflow of dopamine in the nucleus accumbens arises from neuronal activity in the ventral tegmental area. J Neurosci. 2009;29(6):1735–42.
15. Wightman RM, Heien ML, Wassum KM, Sombers LA, Aragona BJ, Khan AS, et al. Dopamine release is heterogeneous within microenvironments of the rat nucleus accumbens. Eur J Neurosci. 2007;26(7):2046–54.
16. Garris PA. Advancing neurochemical monitoring. Nat Methods. 2010;7(2):106–8.
17. Clark JJ, Sandberg SG, Wanat MJ, Gan JO, Horne EA, Hart AS, et al. Chronic microsensors for longitudinal, subsecond dopamine detection in behaving animals. Nat Methods. 2009;7(2):126–9.
18. Wightman RM. Probing cellular chemistry in biological systems with microelectrodes. Science. 2006;311(5767):1570–4.
19. Mogenson GJ, Jones DL, Yim CY. From motivation to action: functional interface between the limbic system and the motor system. Prog Neurobiol. 1980;14(2):69–97.
20. Cardinal RN, Parkinson JA, Hall J, Everitt BJ. Emotion and motivation: the role of the amygdala, ventral striatum, and prefrontal cortex. Neurosci Biobehav Rev. 2002;26(3):321–52.
21. Swanson L. The projections of the ventral tegmental area and adjacent regions: a combined fluorescent retrograde tracer and immunofluorescence study in the rat. Brain Res Bull. 1982;9(1):321–53.
22. Margolis EB, Lock H, Hjelmstad GO, Fields HL. The ventral tegmental area revisited: is there an electrophysiological marker for dopaminergic neurons? J Physiol. 2006;577(3):907–24.
23. Yamaguchi T, Wang H-L, Li X, Ng TH, Morales M. Mesocorticolimbic glutamatergic pathway.J Neurosci. 2011;31(23):8476–90.
24. Bromberg-Martin ES, Matsumoto M, Hikosaka O. Dopamine in motivational control: rewarding, aversive, and alerting. Neuron. 2010;68(5):815–34.
25. Berridge KC, Robinson TE. What is the role of dopamine in reward: hedonic impact, reward learning, or incentive salience? Brain Res Rev. 1998;28(3):309–69.
26. Salamone JD, Correa M. The mysterious motivational functions of mesolimbic dopamine. Neuron. 2012;76(3):470–85.
27. Berridge KC. From prediction error to incentive salience: mesolimbic computation of reward motivation. Eur J Neurosci. 2012;35(7):1124–43.
28. Baldo BA, Kelley AE. Discrete neurochemical coding of distinguishable motivational processes: insights from nucleus accumbens control of feeding. Psychopharmacology. 2007;191(3):439–59.
29. Wise RA. Dopamine, learning and motivation. Nat Rev Neurosci. 2004;5(6):483–94.
30. Saddoris M, Sugam J, Cacciapaglia F, Carelli R. Rapid dopamine dynamics in the accumbens core and shell: learning and action. Front Biosci (Elite edition). 2012;5:273–88.
31. Carlsson A, Lindqvist M, Magnusson T. 3, 4-Dihydroxyphenylalanine and 5-hydroxytryptophan as reserpine antagonists. Nature. 1957;180(4596):1200.
32. Devane WA, Hanus L, Breuer A, Pertwee RG, Stevenson LA, Griffin G, et al. Isolation and structure of a brain constituent that binds to the cannabinoid receptor. Science. 1992;258(5090):1946–9.
33. Mechoulam R, Ben-Shabat S, Hanus L, Ligumsky M, Kaminski NE, Schatz AR, et al. Identification of an endogenous 2-monoglyceride, present in canine gut, that binds to cannabinoid receptors. Biochem Pharmacol. 1995;50(1):83–90.

34. van der Stelt M, Trevisani M, Vellani V, De Petrocellis L, Moriello AS, Campi B, et al. Anandamide acts as an intracellular messenger amplifying Ca2+ influx via TRPV1 channels. EMBO J. 2005;24(17):3026–37.
35. Cravatt BF, Giang DK, Mayfield SP, Boger DL, Lerner RA, Gilula NB. Molecular characterization of an enzyme that degrades neuromodulatory fatty-acid amides. Nature. 1996;384(6604):83–7.
36. Fegley D, Gaetani S, Duranti A, Tontini A, Mor M, Tarzia G, et al. Characterization of the fatty acid amide hydrolase inhibitor cyclohexyl carbamic acid 3′-carbamoyl-biphenyl-3-yl ester (URB597): effects on anandamide and oleoylethanolamide deactivation. J Pharmacol Exp Ther. 2005;313(1):352–8.
37. Long JZ, Li W, Booker L, Burston JJ, Kinsey SG, Schlosburg JE, et al. Selective blockade of 2-arachidonoylglycerol hydrolysis produces cannabinoid behavioral effects. Nat Chem Biol. 2009;5(1):37–44.
38. Fu J, Bottegoni G, Sasso O, Bertorelli R, Rocchia W, Masetti M, et al. A catalytically silent FAAH-1 variant drives anandamide transport in neurons. Nat Neurosci. 2011;15(1):64–9.
39. Ward S, Dykstra L. The role of CB1 receptors in sweet versus fat reinforcement: effect of CB1 receptor deletion, CB1 receptor antagonism (SR141716A) and CB1 receptor agonism (CP-55940). Behav Pharmacol. 2005;16(5–6):381–8.
40. Solinas M, Goldberg SR. Motivational effects of cannabinoids and opioids on food reinforcement depend on simultaneous activation of cannabinoid and opioid systems. Neuropsychopharmacology. 2005;30(11):2035–45.
41. Economidou D, Mattioli L, Cifani C, Perfumi M, Massi M, Cuomo V, et al. Effect of the cannabinoid CB1 receptor antagonist SR-141716A on ethanol self-administration and ethanol-seeking behaviour in rats. Psychopharmacology. 2006;183(4):394–403.
42. Cohen C, Perrault G, Voltz C, Steinberg R, Soubrie P. SR141716, a central cannabinoid (CB1) receptor antagonist, blocks the motivational and dopamine-releasing effects of nicotine in rats. Behav Pharmacol. 2002;13(5–6):451–63.
43. Le Foll B Goldberg SR. Cannabinoid CB1 receptor antagonists as promising new medications for drug dependence. J Pharmacol Exp Ther. 2005;312(3):875–83.
44. De Vries TJD Schoffelmeer AN. Cannabinoid CB1 receptors control conditioned drug seeking. Trends Pharmacol sci. 2005;26(8):420–6.
45. Ward SJ, Walker EA, Dykstra LA. Effect of cannabinoid CB1 receptor antagonist SR141714A and CB1 receptor knockout on cue-induced reinstatement of Ensure® and corn-oil seeking in mice. Neuropsychopharmacology. 2007;32(12):2592–600.
46. Chen J, Paredes W, Li J, Smith D, Lowinson J, Gardner EL. Δ9-Tetrahydrocannabinol produces naloxone-blockable enhancement of presynaptic basal dopamine efflux in nucleus accumbens of conscious, freely-moving rats as measured by intracerebral microdialysis. Psychopharmacology. 1990;102(2):156–62.
47. Chen J, Paredes W, Lowinson JH, Gardner EL. Strain-specific facilitation of dopamine efflux by Δ9-tetrahydrocannabinol in the nucleus accumbens of rat: An in vivo microdialysis study. Neurosci Letters. 1991;129(1):136–40.
48. Tanda G, Pontieri FE, Di Chiara G. Cannabinoid and heroin activation of mesolimbic dopamine transmission by a common μ1 opioid receptor mechanism. Science. 1997;276(5321):2048–50.
49. Cheer JF, Wassum KM, Heien ML, Phillips PE, Wightman RM. Cannabinoids enhance subsecond dopamine release in the nucleus accumbens of awake rats. J Neurosci. 2004;24(18):4393–400.
50. Oleson EB, Cheer JF. A brain on cannabinoids: the role of dopamine release in reward seeking. Cold Spring Harb Perspect Med. 2012;2(8).
51. Alger B, Pitler T. Retrograde signaling at GABAA-receptor synapses in the mammalian CNS. Trends Neurosci. 1995;18(8):333–40.
52. Freund TF, Katona I, Piomelli D. Role of endogenous cannabinoids in synaptic signaling. Physiol Rev. 2003;83(3):1017–66.

53. Melis M, Perra S, Muntoni AL, Pillolla G, Lutz B, Marsicano G, et al. Prefrontal cortex stimulation induces 2-arachidonoyl-glycerol-mediated suppression of excitation in dopamine neurons. J Neurosci. 2004;24(47):10707–15.
54. Wilson RI, Nicoll RA. Endocannabinoid signaling in the brain. Science. 2002;296(5568):678–82.
55. Cachope R. Functional diversity on synaptic plasticity mediated by endocannabinoids. Philos Trans R Soc B: Biol Sci. 2012;367(1607):3242–53.
56. Mátyás F, Urbán GM, Watanabe M, Mackie K, Zimmer A, Freund TF, et al. Identification of the sites of 2-arachidonoylglycerol synthesis and action imply retrograde endocannabinoid signaling at both GABAergic and glutamatergic synapses in the ventral tegmental area. Neuropharmacology. 2008;54(1):95–107.
57. Cheer J, Marsden C, Kendall D, Mason R. Lack of response suppression follows repeated ventral tegmental cannabinoid administration: an in vitro electrophysiological study. Neuroscience. 2000;99(4):661–7.
58. Lupica CR, Riegel AC. Endocannabinoid release from midbrain dopamine neurons: a potential substrate for cannabinoid receptor antagonist treatment of addiction. Neuropharmacology. 2005;48(8):1105–16.
59. Szabo B, Siemes S, Wallmichrath I. (Short communication) Inhibition of GABAergic neurotransmission in the ventral tegmental area by cannabinoids. Eur J Neurosci. 2002;15(12):2057–61.
60. Oleson EB, Beckert MV, Morra JT, Lansink CS, Cachope R, Abdullah RA, et al. Endocannabinoids shape accumbal encoding of cue-motivated behavior via CB1 receptor activation in the ventral tegmentum. Neuron. 2012;73(2):360–73.
61. Robinson DL, Phillips PE, Budygin EA, Trafton BJ, Garris PA, Wightman RM. Subsecond changes in accumbal dopamine during sexual behavior in male rats. Neuroreport. 2001;12(11):2549–52.
62. Roitman MF, Stuber GD, Phillips PE, Wightman RM, Carelli RM. Dopamine operates as a subsecond modulator of food seeking. J Neurosci. 2004;24(6):1265–71.
63. McCutcheon JE, Beeler JA, Roitman MF. Sucrose-predictive cues evoke greater phasic dopamine release than saccharin-predictive cues. Synapse. 2012;66(4):346–51.
64. Willuhn I, Burgeno LM, Everitt BJ, Phillips PE. Hierarchical recruitment of phasic dopamine signaling in the striatum during the progression of cocaine use. Proc Natl Acad Sci. 2012;109(50):20703–8.
65. Phillips PE, Stuber GD, Heien ML, Wightman RM, Carelli RM. Subsecond dopamine release promotes cocaine seeking. Nature. 2003;422(6932):614–8.
66. Gallistel CR, Gibbon J. Time, rate, and conditioning. Psychol Rev. 2000;107(2):289.
67. Gibbon J. Scalar expectancy theory and Weber's law in animal timing. Psychol Rev. 1977;84(3):279.
68. Buhusi CV, Meck WH. What makes us tick? Functional and neural mechanisms of interval timing. Nature Rev Neurosci. 2005;6(10):755–65.
69. Coull JT, Cheng R-K, Meck WH. Neuroanatomical and neurochemical substrates of timing. Neuropsychopharmacology. 2011;36(1):3–25.
70. Matell MS, Meck WH. Neuropsychological mechanisms of interval timing behavior. Bioessays. 2000;22(1):94–103.
71. Meck WH, Penney TB, Pouthas V. Cortico-striatal representation of time in animals and humans. Curr Opin Neurobiol. 2008;18(2):145–52.
72. Coull JT, Hwang HJ, Leyton M, Dagher A. Dopamine precursor depletion impairs timing in healthy volunteers by attenuating activity in putamen and supplementary motor area. J Neurosci. 2012;32(47):16704–15.
73. Bromberg-Martin ES, Matsumoto M, Nakahara H, Hikosaka O. Multiple timescales of memory in lateral habenula and dopamine neurons. Neuron. 2010;67(3):499–510.
74. Narayanan NS, L and BB, Solder JE, Deisseroth K, DiLeone RJ. Prefrontal D1 dopamine signaling is required for temporal control. Proc Natl Acad Sci. 2012;109(50):20726–31.

75. Wittmann M. The inner sense of time: how the brain creates a representation of duration. Nat Rev Neurosci. 2013;14(3):217–23.
76. Lieving LM, Lane SD, Cherek DR, Tcheremissine OV. Effects of marijuana on temporal discriminations in humans. Behav Pharmacol. 2006;17(2):173–83.
77. Anderson BM, Rizzo M, Block RI, Pearlson GD, O'Leary DS. Sex, drugs, and cognition: effects of marijuana. J Psychoactive Drugs. 2010;42(4):413–24.
78. Hicks RE, Gualtieri CT, Mayo Jr J, Perez-Reyes M. Cannabis, atropine, and temporal information processing. Neuropsychobiology. 1984;12(4):229–37.
79. O'Leary DS, Block RI, Turner BM, Koeppel J, Magnotta VA, Ponto LB, et al. Marijuana alters the human cerebellar clock. Neuroreport. 2003;14(8):1145–51.
80. Tart CT. Marijuana intoxication: Common experiences. Nature. 1970;226(5247):701–4.
81. Dews PB. The theory of fixed-interval responding. In: Schoenfeld WN, editor. The theory of reinforcement schedules. New York: Appleton-Century-Crofts, Educational Division, Meredith Corporation, 1970. p. 43–61.
82. Falk JL. The nature and determinants of adjunctive behavior. Physiol Behav. 1971;6(5):577–88.
83. Falk JL. Drug abuse as an adjunctive behavior. Drug Alcohol Depend. 1998;52(2):91–8.
84. Killeen P. On the temporal control of behavior. Psychol Rev. 1975;82(2):89.
85. Olds M. Reinforcing effects of morphine in the nucleus accumbens. Brain Res. 1982;237(2):429–40.
86. Oleson EB, Cachope R, Fitoussi A, Tsutsui K, Wu S, Gallegos J, et al. Cannabinoid receptor activation shifts temporally engendered patterns of dopamine release. Neuropsychopharmacology. 2014;39(6):1441–52.
87. Lammel S, Ion DI, Roeper J, Malenka RC. Projection-specific modulation of dopamine neuron synapses by aversive and rewarding stimuli. Neuron. 2011;70(5):855–62.
88. Matsumoto M, Hikosaka O. Two types of dopamine neuron distinctly convey positive and negative motivational signals. Nature. 2009;459(7248):837–41.
89. Margolis EB, Mitchell JM, Ishikawa J, Hjelmstad GO, Fields HL. Midbrain dopamine neurons: projection target determines action potential duration and dopamine D2 receptor inhibition. J Neurosci. 2008;28(36):8908–13.
90. Zis AP, Fibiger HC, Phillips AG. Reversal by L-dopa of impaired learning due to destruction of the dopaminergic nigro-neostriatal projection. Science. 1974;185(4155):960–2.
91. Darvas M, Fadok JP, Palmiter RD. Requirement of dopamine signaling in the amygdala and striatum for learning and maintenance of a conditioned avoidance response. Learn Mem. 2011;18(3):136–43.
92. Cohen JY, Haesler S, Vong L, Lowell BB, Uchida N. Neuron-type-specific signals for reward and punishment in the ventral tegmental area. Nature. 2012;482(7383):85–8.
93. leson EB, Gentry RN, Chioma VC, Cheer JF. Subsecond dopamine release in the nucleus accumbens predicts conditioned punishment and its successful avoidance. J Neurosci. 2012;32(42):14804–8.
94. Badrinarayan A, Wescott SA, Vander Weele CM, Saunders BT, Couturier BE, Maren S, et al. Aversive stimuli differentially modulate real-time dopamine transmission dynamics within the nucleus accumbens core and shell. J Neurosci. 2012;32(45):15779–90.
95. Navratilova E, Xie JY, Okun A, Qu C, Eyde N, Ci S, et al. Pain relief produces negative reinforcement through activation of mesolimbic reward–valuation circuitry. Proc Natl Acad Sci. 2012;109(50):20709–13.
96. Lammel S, Lim BK, Malenka RC. Reward and aversion in a heterogeneous midbrain dopamine system. Neuropharmacology. 2014;76:351–9.
97. McCutcheon JE, Ebner SR, Loriaux AL, Roitman MF. Encoding of aversion by dopamine and the nucleus accumbens. Front Neurosci. 2012;6:137.
98. Koob GF, Buck CL, Cohen A, Edwards S, Park PE, Schlosburg JE, et al. Addiction as a stress surfeit disorder. Neuropharmacology. 2013;76:370–82.
99. Oleson EB, Cachope R, Fitoussi A, Cheer JF. Tales from the dark side: do neuromodulators of drug withdrawal require changes in endocannabinoid tone? Prog Neuropsychopharmacol Biol Psychiatry. 2014;52:17–23.

100. Roitman MF, Wheeler RA, Wightman RM, Carelli RM. Real-time chemical responses in the nucleus accumbens differentiate rewarding and aversive stimuli. Nat Neurosci. 2008;11(12):1376–7.
101. Maren S. Neurobiology of Pavlovian fear conditioning. Annu Rev Neurosci. 2001;24(1):897–931.
102. Johansen JP, Cain CK, Ostroff LE, LeDoux JE. Molecular mechanisms of fear learning and memory. Cell. 2011;147(3):509–24.
103. LeDoux JE. Emotion circuits in the brain. Annu Rev Neurosci. 2000;23(1):155–84.
104. LeDoux JE. Emotion: clues from the brain. Annu Rev Psychol. 1995;46(1):209–35.

Chapter 9
Synthetic Cannabinoid Effects on Behavior and Motivation

William D. Wessinger, Jeffery H. Moran and Kathryn A. Seely

Abstract The abuse of synthetic cannabinoids is a new phenomenon where little information is known about the behavioral effects of these emerging drugs. In addition, motivation for first time use and chronic abuse is unknown. This chapter presents studies that may explain the motivation of synthetic cannabinoid use by correlating the abuse of synthetic cannabinoids with what is known about marijuana. Although the putative pharmacological targets of these cannabinoid compounds are similar, it seems likely their actions are not identical, which could affect the behavior and motivation of an individual to seek one drug over another, as well as causing them to have differing, yet similar, toxicological effects. Thus, this chapter explores what is known about the similarities and differences in the behavioral and motivational effects of synthetic cannabinoids and marijuana.

Keywords Synthetic cannabinoid · Marijuana · Δ9-THC · Behavior · Mental illness · Addiction · Motivation · Rodent models · Human studies

Introduction

The use of synthetic cannabinoids as drugs of abuse is a relatively new phenomenon with reports of Internet sales beginning around 2004 [1]. Hence, little information is known about the behavioral aspects of the abuse of these substances. In contrast, decades of research and clinical reports of marijuana use have been recorded, which may predict the behavioral effects and motivation for abuse of the synthetic cannabinoid derivatives.

K. A. Seely (✉)
Arkansas Department of Health, Arkansas Public Health Laboratory, Little Rock, AR, USA
e-mail: katie.seely@arkanasas.gov

W. D. Wessinger · J. H. Moran
Department of Pharmacology and Toxicology, College of Medicine, University of Arkansas for Medical Sciences, Little Rock, AR, USA

P. Campolongo, L. Fattore (eds.), *Cannabinoid Modulation of Emotion, Memory, and Motivation,* DOI 10.1007/978-1-4939-2294-9_9

Marijuana

Marijuana, or cannabis, consists of dried leaves and flowers, and sometimes also seeds and stems, from the hemp plant, *Cannabis sativa*. A similar plant derivative, hashish or hash, is made from the dried resinous exudates of the flowering tops. *Cannabis sativa* contains a large and complex mixture of chemical compounds, the best known of which are the cannabinoids. Because these are natural constituents of the cannabis plant, they are sometimes referred to as phytocannabinoids to distinguish them from endogenous cannabinoid receptor ligands (endocannabinoids) or from synthetic cannabinoids [2]. Over a hundred different phytocannabinoids have been identified in marijuana [3] where the most abundant psychoactive cannabinoid in marijuana is Δ9-Tetrahydrocannabinol (Δ9-THC).

The federal government of the United States classifies the marijuana plant as a Schedule I drug (i.e., no recognized medical value and a high abuse potential); however, synthetic Δ9-THC (dronabinol, Marinol®) has recently been reclassified in 2013 from Schedule II to Schedule III and is approved by the U.S. Food and Drug Administration for a few indications. Dronabinol and another synthetic cannabinoid, nabilone (CESAMET®, Schedule II) is approved for the treatment of nausea and vomiting associated with cancer chemotherapy and as an appetite stimulant for AIDS-wasting syndrome. A variety of other medical uses for marijuana or synthetic Δ9-THC have been proposed including treatment of spasticity in multiple sclerosis and chronic pain management for fibromyalgia and other neuropathic pain [1, 4].

Pharmacological Effects of Marijuana

Marijuana produces a number of well-described subjective pharmacological effects consisting of a euphoric dreamy state and altered perceptions, which are presumably the reason for consumption. Effects on perception include visual imagery, an enhanced sense of hearing, increased appetite, and an altered ability to judge distance and the passage of time. Other effects include impaired short-term memory and compromised balance. Complex processes that require perception, attention, and information processing are diminished. These impairments can last 4–8 h [5, 6].

Physiologically, marijuana consumption produces notable cardiovascular effects that include a rapid, dose-dependent increase in heart rate that may be accompanied by an increase in blood pressure [7, 8]. In addition, there is often vasodilation of the scleral and conjunctival vessels resulting in bloodshot eyes.

High doses of marijuana may be associated with less common effects such as anxiety and panic attacks, paranoid feelings, and delusions. Doses high enough to achieve these severe side effects generally occur after oral consumption because more marijuana may be ingested than intended due to the delay in absorption [5].

Marijuana Tolerance

Dramatic tolerance occurs to many effects of marijuana in animals and humans. Within a week of daily intravenous administration, a dose (1.8 mg/kg Δ9-THC) that initially completely suppressed key pecking for food in pigeons, no longer produced an effect. With continued daily administration, McMillan et al. [9] were able to administer doses 20 times the original dose without disrupting behavior. In an extension of these studies [10], the investigators continued increasing the daily dose until 100 times the original dose could be administered without suppressing behavior. Importantly, this 180 mg/kg dose of Δ9-THC was lethal to non-tolerant pigeons. The full extent of tolerance development could not be fully explored due to limited solubility of Δ9-THC. A ten-fold development of tolerance was also demonstrated in rats, along with cross-tolerance to other cannabinoids, but not to morphine [10]. Tolerance has been demonstrated in a wide variety of other laboratory animals (mice, rabbits, dogs and monkeys) against numerous other endpoints including antinociception, ataxia, catalepsy, hypothermia, suppression of locomotor activity, and other effects (see review [11]).

Controlled studies of Δ9-THC tolerance in humans have been infrequent and, as a whole, been inconclusive. For example, early reports in the 1960s and 1970s referred to a phenomenon of "reverse tolerance" in which first time users did not experience the euphoric effects of marijuana [11]. Another study quantitatively examined tolerance in humans by administering escalating doses of oral Δ9-THC every 3.5–6 h (40–120 mg) for 6 days. Tolerance developed to the subjective, euphoric intoxication, but not to the cardiovascular effects [12].

Tolerance to the cardiovascular effects occurs to a lesser degree and bloodshot eyes can still occur in individuals who are tolerant to other effects [12]. A mild withdrawal effect has been noted in heavy long-term users, and many chronic users who suffer negative consequences, such as job loss or court-mandated treatment programs, may maintain their use patterns despite efforts to abstain or reduce use due to withdrawal [13–15].

Development of Synthetic Cannabinoids

The use of marijuana for illicit or medicinal purposes has been occurring for hundreds, if not thousands, of years [1]. Synthetic Δ9-THC has been used medicinally for several decades; however, the new synthetic cannabinoids, which are not derived from Δ9-THC, appeared for sale in on the Internet in 2004 for use in illicit drug markets [1]. Many of the first "K2" or "Spice" synthetic cannabinoids were synthesized by John W. Huffman, a medicinal chemist at Clemson University, for use in hypothesis-driven research to learn more about cannabinoid receptors, and to explore the potential of using cannabinoid compounds to treat a variety of diseases. Years later, clandestine chemists synthesized these compounds and began selling these drugs that were, for all intents and purposes, legal but never approved for medical use due to adverse side effects, like euphoria [16].

How does decades of information reported on marijuana correlate to the new wave of synthetic cannabinoid abuse? Why do people abuse synthetic cannabinoid drugs when marijuana is readily available? What is the motivation to abuse these new drugs? Do individuals chronically abuse synthetic cannabinoids and, if so, why? Generally, chronic abuse of any illicit or addictive substances may be correlated to the inherently rewarding nature of these substances to the human brain, and hence motivation for use. Unfortunately, little is known about the motivation and reward to use synthetic cannabinoids.

Human Motivation and Dependence

No consensus has been reached about the motivation for some individuals to choose to consume illicit drugs, like synthetic cannabinoids. Many hypotheses indicate that a decrease of dopaminergic activity within the ventral striatum, an award center in the brain, may lead to a lack of reward for that individual [17]. A decrease in dopaminergic activity in the ventral straitum has been associated with a propensity to use illicit substances [17, 18] suggesting some users may be motivated to abuse drugs to increase the activity in hypoactive areas of the brain [19]. Hence, drugs are particularly rewarding and addicting for these individuals. In contrast, not everyone that uses drugs, such as alcohol or marijuana, becomes addicted and/or chronic users, but several factors can predict who is likely to abuse. Some factors include genetics, environment, and developmental factors, such as age, stress, and access to drugs [20].

Neurotransmitters and Dependence

The role of dopamine in drug reward and addiction has been extensively studied [20]. As determined by real time PET scans, the euphoria experienced after drug administration is both dose-dependent and time-dependent (e.g., how quickly the drug is administered). Euphoria is correlated with dopamine release within the ventral striatum, specifically within the nucleus accumbens, where the greater the dopamine release, the greater euphoria or "high" [20, 21]. Both Δ9-THC and synthetic cannabinoids, like WIN 55,212–2, induce dopamine release within the nucleus accumbens and increase neural firing of dopaminergic neurons [22]. Also, the extent of dopamine release could be influenced by genetic factors as has been demonstrated in different strains of rats. Some strains of rats actively seek out drug or self-administered drugs more than other strains [23, 24]; thus, some individuals are more susceptible to the rewarding effects of cannabinoids as well as to the potential of addiction. Additionally, cannabinoids are known to modulate the release of other neurotransmitters, like glutamate and GABA, which may play a role in the inherently rewarding nature of these drugs of abuse [25, 26]. For example, an imbalance of glutamate and GABA could be linked to depression, anxiety and suicidal behavior, which could lead to drug-seeking by these individuals to attenuate the severe adverse effects and, in return, potentiate the rewarding nature of drugs of abuse [27].

How do the reward centers of the brain promote the use of synthetic cannabinoid abuse? At this time no one knows. Several case studies indicate that synthetic cannabinoids do cause changes within the brain and the endocannabinoid system because chronic use of synthetic cannabinoids can lead to dependence and withdrawal. Hence, motivation for continued use results from limiting withdrawal or substituting for Δ9-THC [28].

Synthetic Cannabinoid Dependence

Gunderson et al. reported data from three individuals who, at the time of the study, used synthetic cannabinoids and exhibited symptoms of dependence [28]. One individual with a 10-year history of marijuana abuse admitted to switching to synthetic cannabinoids in attempt to limit withdrawal symptoms caused by cessation of marijuana use to comply with mandated drug screens. When using synthetic cannabinoids, this individual experienced no withdrawal symptoms from marijuana, but noticed the synthetic cannabinoids were more potent than marijuana. A second individual with a history of marijuana use switched to synthetic cannabinoids for a new type of "high" and admitted to being attracted to the flashy packaging and multitude of available products. This person reported the synthetic cannabinoids produced an "extreme high", similar to high quality marijuana, but the duration of action was shorter. Interestingly, this individual became concerned about the safety of the synthetic cannabinoids and voluntarily tapered synthetic cannabinoid use. A third individual noted synthetic cannabinoid use increased the severity of withdrawal symptoms and led to continued use of these products. Like the second individual, this third person compared the high of synthetic cannabinoids to marijuana, but with a shorter duration of action [28]. All three cases met the DSM-IV-TR criteria for cannabinoid dependence and were motivated to continue use to limit withdrawal symptoms, such as irritability, cravings, and anxiety [28]. Generally, most individuals report a positive experience and many co-abused with alcohol, marijuana, or tobacco [29].

Another case study [30] described an individual who used the purported synthetic cannabinoid "Spice Gold" daily for 8 months. Chronic abuse led to dependence that became apparent during a shortage of "Spice Gold" in which the individual experienced withdrawal symptoms. The main motivation for continued use was to prevent withdrawal symptoms of nervousness and general unrest.

Motivation to Abuse Synthetic Cannabinoids

As mentioned in one of the above cases, the packaging and various brand names of synthetic cannabinoid products can be very appealing (Fig. 9.1), yet deceptive. The

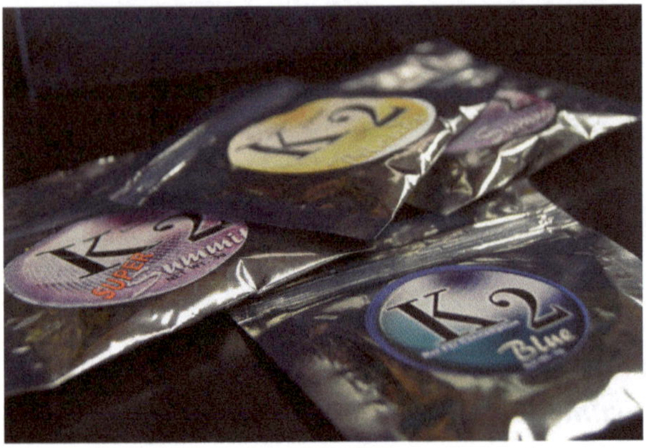

Fig. 9.1 Example of flashy, foil packages containing synthetic cannabinoids. (Photo acknowledgement to Cindy L. Moran, affiliated with the Arkansas State Crime Laboratory)

packaging does not indicate what compound or compounds are within the product, but instead are labeled with brand names like "K2", "Spice", "Aroma", "Genie", and "Dream" [31, 32]. The packages are usually brightly colored foil packets containing dried plant material spiked with synthetic cannabinoids (Fig. 9.1). Sometimes the packages state "incense", "potpourri", "not suitable for children under 18", or "not for human consumption" [32]. Also, many packages try to circumvent existing laws and regulations by indicating that the product does not contain specific compounds, like JWH-018 or JWH-122, which may be illegal or legal, depending on location. Analysis of several of these products detected illicit compounds, even when the same illicit compounds were indicated to not be within the package [31]. Hence, the packages are deceptively labeled, the brand names are made to confuse both consumers and regulators, and are marketed to a group of people that may be first-time drug users or uninformed individuals who believe these products are safe [33].

Other motivations for synthetic cannabinoid abuse are more logical and not necessary neurological. For example, some individuals have chosen to use synthetic cannabinoids because of the "legal" status of these compounds. Also, reports have demonstrated that many like to use synthetic cannabinoids due to the euphoric effects produced, others are curious about these new substances, and the widespread belief that these drugs are safe. An Internet survey of individuals who had used synthetic cannabinoids revealed that many believe these drugs have little risk to health and are not dangerous [29]. Some users have indicated that synthetic cannabinoids have a greater potency than marijuana and are more readily available since synthetic cannabinoids can be bought at local gas stations, on the Internet, and in head shops [31]. The easy availability of the synthetic cannabinoids added to the perception that these drugs are safe. Alarmingly, fourteen percent of respondents to the Internet survey agreed with the statement, "If Spice products were not safe for human use, they would not be marketed and sold in stores" [29].

A significant hazard of using marijuana is loss of employment. Marijuana is metabolized to very long lasting metabolites that are both active and inactive, but are excreted in urine and detectable for days, weeks, or even months after last use. Positive drug tests can result in loss of jobs and/or benefits such a workman's compensation. Many users of synthetic cannabinoids are aware that these substances cannot be detected in normal drug screens [34], which is a motivating factor for not only employees subjected to random drug tests, but also athletes, individuals enrolled in marijuana cessation programs, and those required to pass court-mandated drug screens [29, 31, 35]. Currently, no pharmacokinetic studies have been completed in humans to determine the length of the various metabolites of synthetic cannabinoids remain within the body and what metabolite could be a biomarker, such as with the carboxylated Δ9-THC metabolite that has a long half-life and can be detected weeks after marijuana use.

Marijuana Versus Synthetic Cannabinoids

Although many synthetic cannabinoid users indicate these drugs usually have a similar "high" to marijuana, the duration of action of the synthetic cannabinoids is generally shorterthan marijuana [28, 35]. A study in rhesus monkeys trained to discriminate Δ9-THC from vehicle demonstrated the same phenomenon. In this study, the monkeys were trained to press one lever after Δ9-THC had been administered or press the other lever if saline was administered. When tested after administration of Δ9-THC, the monkeys responded on the Δ9-THC-appropriate lever for over 3 h. When JWH-018 and JWH-073, two popular synthetic cannabinoids, were administered both compounds also elicited responding on the Δ9-THC-appropriate lever indicating the drug effects were similar to Δ9-THC. Interestingly, the duration of action of the synthetic cannabinoids was shorter; responding after administration of JWH-018 lasted 2 h and 20 min and responding after administration of JWH-073 lasted only 1 h and 40 min [35, 36]. The relatively short duration of action of these drugs is disconcerting since, as seen with other shorter-acting illicit drugs, abuse and development of tolerance is increased due to the propensity to re-administered drug more often to maintain the "high". Ginsburg et al., explained:

> Because of a relatively short duration of action, JWH-018 and JWH-073 might be administered more frequently than Δ9-THC to achieve a similar time course of effect as Δ9-THC. Such frequent, repeated use could present additional abuse and dependence liability for these shorter-action drugs by strengthening the association between stimulus and drug effects, thereby leading to more habitual use [36].

Memory and Psychiatric Disorders

Several adverse side effects have been associated with marijuana use, although these effects are usually less serious than with many other abused and illicit substances. As previously mentioned, use of marijuana has been known to cause anxiety and panic attacks. Also, marijuana use has been associated with memory loss [37], and this concern is applied to continued use of synthetic cannabinoids, especially among adolescents, which could lead to memory and learning problems. A significant hazardous side effect is loss of judgment when under the influence of the drug. Such loss of judgment could cause users to place themselves and others in dangerous situations.

Memory Deficiencies

Reports have indicated synthetic cannabinoid abuse, especially chronic abuse, can lead to memory disturbances, lack of motivation, and a relapse of psychiatric disorders in sensitive individuals. Unfortunately, most studies lack an exact identification of the compounds present in the products used by these individuals. Many state use of "K2" or "Spice" or "Aroma", but no toxicological testing of these products was conducted to ascertain if the effects were caused by one or more of many synthetic cannabinoids or other drugs, such as the cathinone-like stimulants (generally referred to as "bath salts") or simply marijuana. This confounding factor complicates most clinical reports and caution should be used with interpreting data when no analysis of the drug has been completed.

A case study of a chronic user of "Spice Gold" indicated he became "listless and [had] problems thinking clearly" after smoking this purported brand of synthetic cannabinoid [30]. This same individual had memory lapses and neglected duties and interests. A different individual smoked "K2" and admitted he only recalled "bits and pieces" of his experience while under the influence of the purported synthetic cannabinoid [38]. According to a survey administered by Castellanos et al., 100 % of adolescents that used synthetic cannabinoids had memory lapses [34]. Interestingly, 82 % of the respondents noted negative effects, such as irritability and anxiety. Also, other studies have noted the propensity for severe anxiety and irritability after using a synthetic cannabinoid or withdrawing from synthetic cannabinoid use [30, 39]. Other negative effects include confusion, sudden depression, paranoia, agitation, and loss of consciousness [1, 31] (Table 9.1 [30, 34, 40–42]).

Antidotal Self-Reports

Since no human studies have been completed on the behavioral effects of synthetic cannabinoids, aside from published clinical reports of extreme cases, the next best information is from antidotal reports of self-use from the website erowid.org. Several individuals reported similar, pleasant "highs" like that with marijuana, but many report bad experiences. Paranoia and intense anxiety is relatively common [43, 44]. Also, extremely dry mouth is noted by most individuals [43, 44]. Another user chronicled his battle with tolerance, dependence, and severe withdrawal indicating the addiction to synthetic cannabinoids was "the worst". This individual lost 20 pounds (nearly 10 kg) in 2 months and his appetite had not fully recovered after 8 weeks of being abstinent [45]. A different chronic abuser complains of

Table 9.1 Central and behavioral effects of synthetic cannabinoids. (Modified from [33], used with permission)

Effect	Example case report
Memory changes	All individuals evaluated reported memory changes [34]
Confusion	Observed in a 27-year-old man after smoking "SpicyXXX" [40]
Sedation	Reported in a 21-year-old man after smoking Spice [40]
Agitation	Reported by 20- to 30-year-old patients following consumption of synthetic cannabinoids [41]
Irritability	Reported by 15- to 19-year-old patients who smoked "Spice" or "K2" [34]
Psychosis	New-onset psychosis in healthy, young men [42]
Anxiety	Reported in 15- to 19-year-old boys after consuming synthetic cannabinoids [34]
Tolerance, dependence, withdrawal	Tolerance to "Spice Gold" reported by a 20-year-old male after smoking daily for 8 months [30]

Each report is an example of a single report of each effect. More case reports are available on each effect

major adverse effects but continues to use synthetic cannabinoids because he has an "unreasonable addiction" to these compounds and cannot stop. His major adverse effects include pain in his lungs, coughing up black mucus, joint pain, and kidney pain [46]. Interestingly, a report published in Centers for Disease Control and Prevention, Morbidity and Mortality Weekly Report, followed 16 cases of acute kidney injury in six U.S. states where the acute kidney injury is associated with the use of XLR-11, a synthetic cannabinoid that emerged in 2012 [47].

Link to Psychosis

Like marijuana, abuse of synthetic cannabinoids can lead to psychotic relapse in individuals predisposed for psychotic disorders. Every-Palmer et al., reported that nearly 70% of patients previously diagnosed and treated for psychotic disorders displayed psychotic symptoms after using JWH-018. Synthetic cannabinoid-induced psychosis is not necessarily related to chronic use, either. The patients with a history of psychosis noted a psychotic relapse within 24 h of using "Aroma", a purported synthetic cannabinoid-containing product, with the psychotic relapse lasting a few days to several weeks. These patients also noted pronounced anxiety, aggression, and paranoia. Although this study presented much needed information about psychosis after synthetic cannabinoid use, the study was subjective due to the methodology of only interviewing patients who were currently seeking treatment for diagnosed mental disorders, and the analytical testing of "Aroma" product was completed in a separate, unrelated study [35].

Interestingly, a different case study included an individual with psychotic symptoms who had no previous history of mental illness or violence [38]. This individual admitted to smoking a large amount of "K2"—"more than usual"—and approximately 15 min later began banging his head on the sidewalk pavement. After the police were called and arrived at the scene, this individual crawled underneath a police car and stated he "didn't deserve to live" [38]. Another case study of acute psychosis featured an individual who had been diagnosed with acute psychotic disorder in the past due to a history of substance abuse, specifically codeine cough syrup and marijuana [48]. This person voluntarily stopped taking cough syrup, but during a 4-week period of abuse of "K2", this individual began exhibiting psychotic behavior again and was hospitalized. After detoxification and treatment, this individual returned to normal with no psychotic symptoms [48].

Generally, individuals susceptible to psychosis may easily relapse after using a synthetic cannabinoid, like with marijuana abuse, but it cannot be determined if synthetic cannabinoid use cause the psychosis or if these individuals abuse synthetic cannabinoids due to psychosis. Unlike marijuana, the synthetic cannabinoids seem to be able to induce psychosis in those without history of mental illness, such as the case study presented above by Thomas et al [31]. Likewise, other case studies have reported individuals with no prior mental illness who have had hallucinations while intoxicated on JWH-018 and JWH-073 [40]. Another case study described three individuals who presented with paranoia, hallucinations, and severe agitation after smoking "Space", another purported brand of synthetic cannabinoid [49]. The severity of these reported symptoms have not been associated with marijuana use and indicate that synthetic cannabinoids are more efficacious than marijuana.

Cannabinoid Pharmacology

Why do the synthetic cannabinoids have distinct effects when compared to classical cannabinoids, like Δ9-THC? The question may be answered by the pharmacology of the cannabinoid receptors.

CB1 Receptors

The cannabinoid type 1 (CB1) receptors are G-protein coupled receptors that function with a basal level of activity. This activity can be modulated up or down depending on the action of the drug. Agonists for CB1 receptors increase the activity, inverse agonists decrease activity, and neutral antagonists do nothing to the basal state of the receptor, but a neutral antagonist can attenuate the activity of both agonists and inverse agonists. Most of the synthetic cannabinoids identified in these illicit products have been characterized as full agonists for the CB1 receptor. On the other hand, Δ9-THC in marijuana is characterized as a partial agonist with a limited ability to activate the CB1 receptor. This important pharmacological difference between these agonistic compounds—either full or partial agonists—may explain some of the distinctive signs and symptoms seen with synthetic cannabinoid use.

Modulation of CB1 receptors from their basal state, resulting in either increased or decreased activity, has been shown to lead to emotional and behavioral changes. These changes may involve modulation of the endocannabinoid system. Other, non-cannabinoid drugs of abuse (e.g., ethanol, nicotine, and cocaine) can change levels of endocannabinoids, specifically anandamide and 2-arachidonylglycerol, and increase dopamine release [22, 50]. Interestingly, when rimonabant, a CB1-receptor antagonist/inverse agonist, was administered, the drug-induced dopamine increase was reversed [22, 50] (see also Chaps. 14, 15 and 19 of this book). Hence, the endocannabinoid system is probably involved in the dopamine reward system and endocannabinoids could impact reward, motivation, and drug-seeking behaviors [22]. Other animal studies have indicated that chronic Δ9-THC, nicotine, and alcohol modulate the endocannabinoid tone within the prefrontal cortex by increasing the concentration of endocannabinoids. This change in endocannabinoid tone could be involved in drug-seeking behavior. Thus, increases in endocannabinoid levels may be linked to drug addiction and motivation [51].

CB1 Receptors and Behavior

In addition to endocannabinoids, CB1 receptors have been shown to be involved in emotion, behavior, and motivation due to their distribution in brain structures such as the amygdala, hippocampus, and prefrontal cortex [52]. Many of these studies utilize knock-out mice, or mice genetically modified to not express a specific receptor or gene (i.e., CB1 receptors), and compare the knock-out animals to wild-type mice, or mice with no genetic modifications. Due to breeding paradigms, knock-out mice and wild-type mice can be from a single litter; hence, littermate wild-type mice are considered the best control for knock-out mice studies.

Changes in receptor activity and regulation have been linked to aggression and depression in knock-out animals indicating involvement in behavior and motivation [52]. For example, studies have demonstrated that CB1 receptor knock-out mice spent more time threatening and attacking other mice in the same cage when compared to the CB1 wild-type mice [53]. In addition to aggression, the same CB1 knock-out mice were more impulsive and more likely to demonstrate anxiolytic behaviors than the littermate wild-type mice [53, 54]. These behaviors may be due to changes in serotonin levels and expression of serotonin receptors within specific brain regions, such as the amygdala and brain stem [53]. In humans, aggression and impulsivity are behaviors usually observed together and have been linked to low serotonin levels. Thus, individuals exhibiting overly aggressive behavior are commonly treated with selective serotonin re-uptake inhibitors (SSRIs) to increase the concentration of serotonin [53].

Cannabidiol: Another Cannabinoid in Marijuana

As mentioned, many studies have been completed with marijuana use in humans. Important to note is that Δ9-THC is only one of over 100 different cannabinoids in marijuana. Another significant cannabinoid component of marijuana is cannabidiol that is an inverse agonist at CB1 receptors [55, 56]. Cannabidiol has been shown to be neuroprotective by attenuating Δ9-THC-induced anxiety and Δ9-THC-induced psychosis [57, 58]. Also, cannabidiol has been shown to reduce withdrawal symptoms and cardiovascular effects of Δ9-THC indicating that cannabidiol, when used concurrently with Δ9-THC in marijuana, may protect against the adverse effects of the cannabinoid agonist [59].

Over the past few decades there is significant evidence that the potency of marijuana has increased, due to increased Δ9-THC concentrations [3, 60–62]. For example, data from law enforcement confiscated samples within the United States showed concentrations of Δ9-THC in marijuana rising from a mean of 3.4 % in 1993 to 8.8 % by 2008, and the prevalence of high potency (>9 %) marijuana samples also greatly increased [3]. Additionally, another study indicated a possible downward trend in cannabidiol concentration in marijuana [62]. If the concentration of the inverse agonist cannabidiol has not increased at the same rate as Δ9-THC [63], more adverse effects may be associated with marijuana than in past years due to the lack of the protection from cannabidiol [62]. Hence, marijuana users may begin to exhibit the severe effects now being associated with synthetic cannabinoid abuse.

Rimonabant

The only long-term study of synthetic cannabinoids in humans was with rimonabant (SR141716). Although this synthetic cannabinoid is an inverse agonist, or decreases basal activity of the CB1 receptors, this study indicated that changing the activity of the CB1 receptor has significant, adverse effects. Many participants in the study noted nausea, depression, anxiety, and suicidal ideation [64–66]. Interestingly, a prominent study—the STRADIVARIUS randomized trial of rimonabant—found that 43.4 % of rimonabant-treated patients experienced psychiatric side effects, which is significantly higher than the number in the placebo-treated patients (only 28.4 %) [66]. Of the rimonabant-treated group, one patient successfully committed suicide whereas none did in the placebo group [66]. Similarly, rimonabant was approved for use in Europe in 2006, but withdrawn from the market in 2009 due to safety concerns, namely five individuals who successfully committed suicide while taking rimonabant [67]. These side effects are similar to case studies of synthetic cannabinoid agonist use (e.g., depression, anxiety, and suicidal ideation), which potentially correlates to modulations of the endocannabinoid system as well as how these effects may be due to a larger, global change within the brain, such as receptor up- or down-regulation.

CB2 Receptors

In addition to the CB1 receptors, the CB2 receptor is also well-characterized and located predominately on immune cells. Although CB2 receptors are mainly found outside of the central nervous system, recent reports have indicated these receptors are located within the brain [68, 69], especially during neuro-inflammatory conditions [70]. Many of the synthetic cannabinoids bind CB2 receptors with high affinity, but no psychoactive effects are tied to these other cannabinoid receptors due to their primary peripheral distribution. Since CB2 receptors are located mainly on immune cells, they are potential drug targets for a variety of autoimmune diseases, like multiple sclerosis, due to immunosuppression [71]. Some people may seek out synthetic cannabinoids as an alternative to medical marijuana, especially if these individuals live somewhere medical marijuana is not legal. Unfortunately, the use of synthetic cannabinoids could be detrimental for the individual due to unintentional immunosuppression as well as the extensive adverse effects associated with synthetic cannabinoids that are not associated with marijuana.

Self-Administration and Drug Discrimination

Very few animal behavioral studies have been published using the emergent synthetic cannabinoid compounds that have recently been detected flashy packaging. However, other synthetic cannabinoids that are full agonists at CB1 receptors, such as CP-55,940, HU-210, and WIN 55,212–2, have been subject to animal studies. Among the commonly employed methods for testing the abuse potential of drugs are drug self-administration and drug discrimination studies.

In *drug self-administration*, rodents and sometimes non-human primates are implanted with an intravenous catheter through which drugs can be administered. Animals make operant responses such as pressing a response lever or using nose-poke holes where a beam of light is broken by a rat or mouse when the rodent pokes its nose into the hole. In each case, correct or appropriate responses produce drug infusion. Drugs that have *reinforcing effects* increase the frequency of the response that produces drug administration. Drugs that are not reinforcing fail to maintain high levels of responding. For example, saline or vehicle infusions typically have very low operant responding levels. Hence, self-administration studies are a common method to evaluate the abuse liability of a drug because, with very few exceptions, drugs that maintain responding for self-administration in animal studies are likely to be abused by humans.

In *drug discrimination studies*, animals are trained to recognize the effect produced by a particular training drug. Most drugs that are active in the brain produce effects, including subjective effects, which are specific and recognizable. These effects recognized by the subject, either animal or human, are called *discriminative stimulus effects*. Laboratory animals can be trained to recognize these effects and make behavioral responses, such as level pressing or nose-poking, after drug is administered. Two-lever drug discrimination is a commonly used experiment and subjects are trained to respond on levers to obtain food or water. One lever corresponds to drug administration and the other corresponds to saline or vehicle administration. A correct response (e.g., if drug or vehicle was delivered) results in the presentation of food or water. Drugs that produce similar subjective effects to the training drug will elicit responding on the same lever; these drugs are said to *substitute* for the training drug. Well-trained subjects typically select the correct lever (training drug- or saline-appropriate lever) with 85–100 % accuracy. Drugs that produce only responding on the saline/vehicle-appropriate lever usually are saline, vehicles, a low dose, or drugs that are dissimilar to the training drug (e.g., opioid agonists tested in Δ9-THC-trained subjects). Generally, similar drug discriminative effects in animals are believed to reflect the subjective effects of drugs in humans and drugs that are similar in drug discrimination studies are likely to have similar abuse potential in humans.

Animal Studies

Rats will dose-dependently self-administer WIN 55,212–2 indicating that the drug serves as a reinforcer [72]. While a low dose of 6.25 μg/kg/injection failed to maintain self-administration level greater than vehicle, higher doses ranging from 12.5 to 50 μg/kg/injection maintained self-administration rates significantly greater than vehicle. Also, the discriminative properties of synthetic cannabinoids have been studied in laboratory animals. In one study, rats were trained to discriminate CP-55,940 from vehicle by reinforcing correct lever pressing with food. Responding on one lever was only reinforced after administration of a dose of CP-55,940 and responding on the other lever was only reinforced after administration of vehicle. After the rats were trained to respond correctly, test sessions were conducted using different doses of CP-55,940 as well as of Δ9-THC, WIN 55,212–2, or cannabinol. All these cannabinoids dose-dependently substituted for CP-55,940. Substitution was specific for cannabinoids since several drugs from other classes (phencyclidine, haloperidol and diazepam) failed to elicit responding on the CP-55,940-appropriate lever [73]. Also, JWH-018 and JWH-073, two compounds detected in synthetic cannabinoid products, dose-dependently substituted for Δ9-THC in monkeys trained to discriminate Δ9-THC from vehicle [36].

In another experiment, a single dose of CP-55,940 was administered in combination with different doses of rimonabant, a CB1 inverse agonist. Rimonabant dose-dependently decreased responding on the CP-55,940-appropriate lever indicating that the stimulus properties of CP-55,940 are mediated by CB1 receptors [73]. In similar experiments, CP-55,940 dose-dependently substituted for Δ9-THC in rats and rhesus monkeys [74]. In addition, rimonabant dose-dependently shifted the dose-response curves for Δ9-THC, JWH-018 and JWH-073 rightward in a parallel manner, indicating rimonabant decreased the potency of the cannabinoid agonists through interactions with the CB1 receptor [36]. The results of these drug discrimination experiments indicate that the discriminative stimulus properties of Δ9-THC and the synthetic cannabinoids tested are mediated via CB1 receptors. Hence, it is highly likely that the subjective effects described in humans after use of synthetic cannabinoids and Δ^9-THC are mediated by actions at CB1 receptors.

Drug-Seeking Behavior

Interviews of synthetic cannabinoid users indicated many individuals co-abuse other substances, such as alcohol, marijuana, and tobacco [29]. So, does abusing synthetic cannabinoids cause the seeking of other substances? One animal study suggests this could be true because low doses of the synthetic cannabinoid WIN 55,212–2 induced mice to increase alcohol binge-drinking behavior [75]. In a variation of self-administration procedures, rats were trained to self-administer intravenous cocaine infusions by nose-poke responses. Cocaine was very reinforcing and maintained high levels of nose-poking behavior. After responding for cocaine stabilized, the cocaine infusions were discontinued for at least 14 days, although the animals continued to be exposed to the same behavioral testing environment. When cocaine was no longer available, nose-poke responding dropped to very low levels. Then, subcutaneous injections of HU-210 dose-dependently reinstated nose-poke responding previously associated with cocaine infusions, which was interpreted as cocaine-seeking behavior. Likewise, cocaine infusions after prolonged discontinued dosing also reinstated nose-poke behavior, but infusions of saline did not. Similarly, exposure to environmental cues (e.g., stimulus lights and tones) previously paired with cocaine infusions, or exposure to intermittent foot-shock, both reinstated nose-poke cocaine-seeking behavior. Rimonabant dose-dependently antagonized cocaine-infusion induced, as well as environmental-cue induced, reinstatement of nose-poke behavior. In contrast, the effect of the foot-shock stressor on drug-seeking behavior was not attenuated by rimonabant [76].

Adverse Effects

Why are there multiple human case studies published demonstrating the adverse side effects of synthetic cannabinoid use? Low doses of synthetic cannabinoids can have pleasant effects, but like most reinforcing effects, these effects seem to be biphasic. Perhaps this is because there are other effects that might also be expressed in a dose-dependent manner. Across a variety of animal models of anxiety such as the social interaction test, elevated plus maze, and hole board exploration test, low to mid-doses of CP-55,940 produced anxiolytic-like effects in rodents [52, 77, 78]. Using different animal models such as analysis of open-field behavior for defensive withdrawal responses and conditioned-place avoidance, high doses of HU-210 and CP-55,940 produced lasting anxiogenic effects [79, 80]. Interestingly, the extent of adverse effects after synthetic cannabinoids use could be due to the anxiety state prior to use. For example, in one study rats were chronically stressed for 21 days by exposure to unpredictable stressors. When these stressed rats were treated with a low or high dose of HU-210 and placed in the elevated plus maze, both dose groups of rats responded in a manner consistent with heightened anxiety. In contrast, in un-stressed control animals, a low dose of HU-210 induced an anxiolytic response [81]. Therefore, the response an individual has while using a synthetic cannabinoid may be related to the dose of the compound, as well as the state of anxiety prior to use.

Predicted Human Effects: Lessons Learned from Animal Studies

Since the synthetic cannabinoids are self-administered in several laboratory animal models, the general conclusion is these drugs are reinforcers. A major difference between synthetic cannabinoids and Δ9-THC in self-administration studies is that Δ9-THC is generally not a reinforcer in rodent self-administration models, and most attempts in other laboratory species have also been unsuccessful. However, in squirrel monkeys, under specific experimental conditions, Δ9-THC is reliably self-administered [82]. Generally, animals reliably discriminate the central nervous system effects of synthetic cannabinoids as being similar to Δ9-THC, suggesting a shared mechanism of action of these discriminative stimulus effects. Experiments with specific CB1 inverse agonists/antagonists suggest these effects are produced by actions at CB1 receptors. In animals, the synthetic cannabinoids produce adverse effects associated with anxiety and can produce drug-seeking behaviors. Thus, predicted human effects are what have been reported in case studies: anxiety, especially with high doses; co-abuse with other substances; and a propensity of continued abuse, which may lead to tolerance and dependence.

Conclusions

Although synthetic cannabinoids and Δ9-THC act at the same receptors and produce many shared effects, these compounds are not necessarily equal. Like Δ9-THC, synthetic cannabinoids produce euphoria, which is appealing to many users. Unlike Δ9-THC, the adverse side effects associated with synthetic cannabinoid use seem to be more severe than after dosing with marijuana (summarized in Table 9.1). Differences in drug potency, receptor specificity, and differing pharmacokinetics (e.g., faster onset, shorter duration of action) may correlate with greater incidence of adverse reactions after abuse of synthetic cannabinoids. Also, the shorter duration of action of the synthetic cannabinoids may lead to continued abuse, tolerance, and dependence. Published clinical case studies and antidotal self-reports seem to support this notion; however, there is little information about the behavioral effects of synthetic cannabinoids in controlled experiments. In general, what is known of the effects of marijuana on behavior and motivation may not predict the same adverse effects after usage of synthetic cannabinoids.

References

1. Seely KA, Prather PL, James LP, Moran JH. Marijuana-based drugs: innovative therapeutics or designer drugs of abuse? Mol Interv. 2011;11(1):36–51.
2. ElSohly MA, Slade D. Chemical constituents of marijuana: the complex mixture of natural cannabinoids. Life Sci. 2005;78(5):539–48.
3. Mehmedic Z, Chandra S, Slade D, Denham H, Foster S, Patel AS, et al. Potency trends of Δ^9-THC and other cannabinoids in confiscated cannabis preparations from 1993 to 2008. J Forensic Sci. 2010;55(5):1209–17.
4. Bostwick JM. Blurred boundaries: the therapeutics and politics of medical marijuana. Mayo Clin Proc. 2012;87(2):172–86.
5. O'Brien CP. Drug addiction. In: Brunton LL, Chabner BA, Knollmann BC, editors. Goodman & Gilman's the pharmacological basis of therapeutics. 12th ed. New York: McGraw-Hill; 2010. (Chap. 24). p. 649–68.
6. Lüscher C. Drugs of abuse. In: Katzung BG, Masters SB, Trevor AJ, editors. Basic and clinical pharmacology. 11th ed. New York: McGraw-Hill; 2009. (Chap. 32). p. 553–68.
7. Malinowska B, Baranowska-Kuczko M, Schlicker E. Triphasic blood pressure responses to cannabinoids: do we understand the mechanism? Br J Pharmacol. 2012;165(7):2073–88.
8. Sidney S. Cardiovascular consequences of marijuana use. J Clin Pharmacol. 2002;42 Suppl 1:64S–70S.
9. McMillan DE, Harris LS, Frankenheim JM, Kennedy JS. l-Δ^9-trans-tetrahydrocannabinol in pigeons: tolerance to the behavioral effects. Science. 1970;169(3944):501–3.
10. McMillan DE, Ford RD, Frankenheim JM, Harris RA, Harris LS. Tolerance to active constituents of marihuana. Arch Int Pharmacodyn Ther. 1972;198(1):132–44.
11. Adams IB, Martin BR. Cannabis: pharmacology and toxicology in animals and humans. Addiction. 1996;91(11):1585–614.
12. Gorelick DA, Goodwin RS, Schwilke E, Schwope DM, Darwin WD, Kelly DL, et al. Tolerance to effects of high-dose oral delta9-tetrahydrocannabinol and plasma cannabinoid concentrations in male daily cannabis smokers. J Anal Toxicol. 2013;37(1):11–6.
13. Budney AJ, Hughes JR, Moore BA, Vandrey R. Review of the validity and significance of cannabis withdrawal syndrome. Am J Psychiatry. 2004;161(11):1967–77.
14. Budney AJ, Moore BA. Development and consequences of cannabis dependence. J Clin Pharmacol. 2002;42 Suppl 11:28S–33S.
15. Budney AJ, Vandrey RG, Hughes JR, Moore BA, Bahrenburg B. Oral delta-9-tetrahydrocannabinol suppresses cannabis withdrawal symptoms. Drug Alcohol Depend. 2007;86(1):22–9.
16. Wiley JL, Marusich JA, Huffman JW, Balster RL, Thomas BF. Hijacking of basic research: the case of synthetic cannabinoids. Methods Rep RTI Press. 2011;2011.
17. Blum K, Braverman ER, Holder JM, Lubar JF, Monastra VJ, Miller D, et al. Reward deficiency syndrome: a biogenetic model for the diagnosis and treatment of impulsive, addictive, and compulsive behaviors. Psychoactive Drugs. 2000;32 Suppl i–iv: 1–112.
18. van Hell HH, Vink M, Ossewaarde L, Jager G, Kahn RS, Ramsey NF. Chronic effects of cannabis use on the human reward system: an fMRI study. Euro Neuropsychopharmacol. 2010;20(3):153–63.
19. Garavan H, Weierstall K. The neurobiology of reward and cognitive control systems and their role in incentivizing health behavior. Prev Med. 2012;55 Suppl:S17–23.
20. Volkow ND, Wang GJ, Fowler JS, Tomasi D, Telang F. Addiction: beyond dopamine reward circuitry. Proc Natl Acad Sci U S A. 2011;108(37):15037–42.
21. Volkow ND, Fowler JS, Wang GJ, Baler R, Telang F. Imaging dopamine's role in drug abuse and addiction. Neuropharmacology. 2009;56 Suppl 1:3–8.
22. Oleson EB, Cheer JF. A brain on cannabinoids: the role of dopamine release in reward seeking. Cold Spring Harb Perspect Med. 2012;2(8).
23. Gardner EL. Addictive potential of cannabinoids: the underlying neurobiology. Chem Phys Lipids. 2002;121(1–2):267–90.

24. Gardner EL. Endocannabinoid signaling system and brain reward: emphasis on dopamine. Pharmacol Biochem Behav. 2005;81(2):263–84.
25. Kalivas PW. The glutamate homeostasis hypothesis of addiction. Nat Rev Neurosci. 2009;10(8):561–72.
26. Koob GF. Neural mechanisms of drug reinforcement. Ann N Y Acad Sci. 1992;654:171–91.
27. Higuera-Matas A, Miguens M, Coria SM, Assis MA, Borcel E, del Olmo N, et al. Sex-specific disturbances of the glutamate/GABA balance in the hippocampus of adult rats subjected to adolescent cannabinoid exposure. Neuropharmacology. 2012;62(5–6):1975–84.
28. Gunderson EW, Haughey HM, Ait-Daoud N, Joshi AS, Hart CL. "Spice" and "K2" herbal highs: a case series and systematic review of the clinical effects and biopsychosocial implications of synthetic cannabinoid use in humans. Am J Addict. 2012;21(4):320–6.
29. Vandrey R, Dunn KE, Fry JA, Girling ER. A survey study to characterize use of Spice products (synthetic cannabinoids). Drug Alcohol Depend. 2012;120(1–3):238–41.
30. Zimmermann US, Winkelmann PR, Pilhatsch M, Nees JA, Spanagel R, Schulz K. Withdrawal phenomena and dependence syndrome after the consumption of "spice gold". Dtsch Arztebl Int. 2009;106(27):464–7.
31. Fattore L, Fratta W. Beyond THC: the new generation of cannabinoid designer drugs. Front Behav Neurosci. 2011;5:60.
32. Spaderna M, Addy PH, D'Souza DC. Spicing things up: synthetic cannabinoids. Psychopharmacology. 2013;228(4):525–40.
33. Seely KA, Lapoint J, Moran JH, Fattore L. Spice drugs are more than harmless herbal blends: a review of the pharmacology and toxicology of synthetic cannabinoids. Prog Neuropsychopharmacol Biol Psychiatry. 2012;39:234–43.
34. Castellanos D, Singh S, Thornton G, Avila M, Moreno A. Synthetic cannabinoid use: a case series of adolescents. J Adolesc Health. 2011;49(4):347–9.
35. Every-Palmer S. Synthetic cannabinoid JWH-018 and psychosis: an explorative study. Drug Alcohol Depend. 2011;117(2–3):152–7.
36. Ginsburg BC, Schulze DR, Hruba L, McMahon LR. JWH-018 and JWH-073: {Delta}9-Tetrahydrocannabinol-like discriminative stimulus effects in monkeys. J Pharmacol Exp Ther. 2012;340(1):37–45.
37. Solowij N, Stephens RS, Roffman RA, Babor T, Kadden R, Miller M, et al. Cognitive functioning of long-term heavy cannabis users seeking treatment. JAMA. 2002;287(9):1123–31.
38. Thomas S, Bliss S, Malik M. Suicidal ideation and self-harm following K2 use. J Okla State Med Assoc. 2012;105(11):430–3.
39. Schneir AB, Cullen J, Ly BT. "Spice" girls: synthetic cannabinoid intoxication. J Emerg Med. 2010;40(3):296–9.
40. Simmons J, Cookman L, Kang C, Skinner C. Three cases of "spice" exposure. Clin Toxicol (Phila). 2011;49(5):431–3.
41. Forrester MB, Kleinschmidt K, Schwarz E, Young A. Synthetic cannabinoid exposures to Texas poison control centers. J Addict Dis. 2011;30(4):351–8.
42. Hurst D, Loeffler G, McLay R. Psychosis associated with synthetic cannabinoid agonists: a case series. Am J Psychiatry. 2011;168(10):1119.
43. PippUK. Spiced out: experience with Spice Products (Various). erowid.org; 2009; Available from: http://www.erowid.org/experiences/exp.php?ID=76319. Accessed 8 May 2013.
44. SirDoughboy. Hell Inside my Own Mind: experience with Products—Spice and Synthetic Cannabinoids (Black Mamba). erowid.org; 2012; Available from: http://www.erowid.org/experiences/exp.php?ID=85174. Accessed 8 May 2013.
45. Riley. Withdrawal from Herbal Incense: experience with 'Gorilla Dro' (JWH-210 & Unknown). erowid.org; 2012; Available from: http://www.erowid.org/experiences/exp.php?ID=96302. Accessed 8 May 2013.
46. SpacedOUTonSPICE. Excessive cannabinoid agonism: experience with Spice & AM-2201. erowid.org; 2012; Available from: http://www.erowid.org/experiences/exp.php?ID=96704. Accessed 8 May 2013.

47. Centers for Disease Control and Prevention (CDC). Acute kidney injury associated with synthetic cannabinoid use—multiple states, 2012. MMWR Morb Mortal Wkly Rep. 2013;62(6):93–8.
48. Tung CK, Chiang TP, Lam M. Acute mental disturbance caused by synthetic cannabinoid: a potential emerging substance of abuse in Hong Kong. East Asian Arch Psychiatry. 2012;22(1):31–3.
49. Bebarta VS, Ramirez S, Varney SM. Spice: a new "legal" herbal mixture abused by young active duty military personnel. Subst Abus. 2012;33(2):191–4.
50. Cheer JF, Wassum KM, Sombers LA, Heien ML, Ariansen JL, Aragona BJ, et al. Phasic dopamine release evoked by abused substances requires cannabinoid receptor activation. J Neurosci. 2007;27(4):791–5.
51. Maldonado R, Valverde O, Berrendero F. Involvement of the endocannabinoid system in drug addiction. Trends Neurosci. 2006;29(4):225–32.
52. Viveros MP, Marco EM, File SE. Endocannabinoid system and stress and anxiety responses. Pharmacol Biochem Behav. 2005;81(2):331–42.
53. Rodriguez-Arias M, Navarrete F, Daza-Losada M, Navarro D, Aguilar MA, Berbel P, et al. CB1 cannabinoid receptor-mediated aggressive behavior. Neuropharmacology. 2013;75C:172–80.
54. Martin M, Ledent C, Parmentier M, Maldonado R, Valverde O. Involvement of CB1 cannabinoid receptors in emotional behaviour. Psychopharmacology. 2002;159(4):379–87.
55. Pertwee RG. Ligands that target cannabinoid receptors in the brain: from THC to anandamide and beyond. Addict Biol. 2008;13(2):147–59.
56. Thomas A, Baillie GL, Phillips AM, Razdan RK, Ross RA, Pertwee RG. Cannabidiol displays unexpectedly high potency as an antagonist of CB1 and CB2 receptor agonists in vitro. Br J Pharmacol. 2007;150(5):613–23.
57. Karniol IG, Shirakawa I, Kasinski N, Pfeferman A, Carlini EA. Cannabidiol interferes with the effects of delta9-tetrahydrocannabinol in man. Eur J Pharmacol. 1974;28(1):172–7.
58. Zuardi AW, Shirakawa I, Finkelfarb E, Karniol IG. Action of cannabidiol on the anxiety and other effects produced by delta9-THC in normal subjects. Psychopharmacology. 1982;76(3):245–50.
59. Oliere S, Joliette-Riopel A, Potvin S, Jutras-Aswad D. Modulation of the endocannabinoid system: vulnerability factor and new treatment target for stimulant addiction. Front Psychiatry. 2013;4:109.
60. Cascini F, Aiello C, Di Tanna G. Increasing delta-9-tetrahydrocannabinol (Δ-9-THC) content in herbal cannabis over time: systematic review and meta-analysis. Curr Drug Abuse Rev. 2012;5(1):32–40.
61. Potter DJ, Clark P, Brown MB. Potency of Δ^9-THC and other cannabinoids in cannabis in England in 2005: implications for psychoactivity and pharmacology. J Forensic Sci. 2008;53(1):90–4.
62. Swift W, Wong A, Li KM, Arnold JC, McGregor IS. Analysis of cannabis seizures in NSW, Australia: cannabis potency and cannabinoid profile. PLoS One. 2013;8(7):e70052.
63. Pijlman FT, Rigter SM, Hoek J, Goldschmidt HM, Niesink RJ. Strong increase in total delta-THC in cannabis preparations sold in Dutch coffee shops. Addict Biol. 2005;10(2):171–80.
64. Addy C, Wright H, Van Laere K, Gantz I, Erondu N, Musser BJ, et al. The acyclic CB1R inverse agonist taranabant mediates weight loss by increasing energy expenditure and decreasing caloric intake. Cell Metab. 2008;7(1):68–78.
65. Christensen R, Kristensen PK, Bartels EM, Bliddal H, Astrup A. Efficacy and safety of the weight-loss drug rimonabant: a meta-analysis of randomised trials. Lancet. 2007;370(9600):1706–13.
66. Nissen SE, Nicholls SJ, Wolski K, Rodes-Cabau J, Cannon CP, Deanfield JE, et al. Effect of rimonabant on progression of atherosclerosis in patients with abdominal obesity and coronary artery disease: the STRADIVARIUS randomized controlled trial. JAMA. 2008;299(13):1547–60.

67. European Medicines Agency. Assessment Report for Zimulti, EMEA/H/C/000691/A20/0012. 2009.
68. Van Sickle MD, Duncan M, Kingsley PJ, Mouihate A, Urbani P, Mackie K, et al. Identification and functional characterization of brainstem cannabinoid CB2 receptors. Science. 2005;310(5746):329–32.
69. Onaivi ES. Neuropsychobiological evidence for the functional presence and expression of cannabinoid CB2 receptors in the brain. Neuropsychobiology. 2006;54(4):231–46.
70. Shoemaker JL, Seely KA, Reed RL, Crow JP, Prather PL. The CB2 cannabinoid agonist AM-1241 prolongs survival in a transgenic mouse model of amyotrophic lateral sclerosis when initiated at symptom onset. J Neurochem. 2007;101(1):87–98.
71. Mackie K. Cannabinoid receptors: where they are and what they do. J Neuroendocrinol. 2008;20 Suppl 1:10–4.
72. Fattore L, Cossu G, Martellotta CM, Fratta W. Intravenous self-administration of the cannabinoid CB1 receptor agonist WIN 55,212–2 in rats. Psychopharmacology. 2001;156(4):410–6.
73. Wiley JL, Barrett RL, Lowe J, Balster RL, Martin BR. Discriminative stimulus effects of CP 55,940 and structurally dissimilar cannabinoids in rats. Neuropharmacology. 1995;34(6):669–76.
74. Gold LH, Balster RL, Barrett RL, Britt DT, Martin BR. A comparison of the discriminative stimulus properties of delta9-tetrahydrocannabinol and CP 55,940 in rats and rhesus monkeys. J Pharmacol Exp Ther. 1992;262(2):479–86.
75. Linsenbardt DN, Boehm SL, 2nd. Agonism of the endocannabinoid system modulates binge-like alcohol intake in male C57BL/6J mice: involvement of the posterior ventral tegmental area. Neuroscience. 2009;164(2):424;–34.
76. De Vries TJ, Shaham Y, Homberg JR, Crombag H, Schuurman K, Dieben J, et al. A cannabinoid mechanism in relapse to cocaine seeking. Nat Med. 2001;7(10):1151–4.
77. Genn RF, Tucci S, Marco EM, Viveros MP, File SE. Unconditioned and conditioned anxiogenic effects of the cannabinoid receptor agonist CP 55,940 in the social interaction test. Pharmacol Biochem Behav. 2004;77(3):567–73.
78. Marco EM, Perez-Alvarez L, Borcel E, Rubio M, Guaza C, Ambrosio E, et al. Involvement of 5-HT1A receptors in behavioural effects of the cannabinoid receptor agonist CP 55,940 in male rats. Behav Pharmacol. 2004;15(1):21–7.
79. McGregor IS, Issakidis CN, Prior G. Aversive effects of the synthetic cannabinoid CP 55,940 in rats. Pharmacol Biochem Behav. 1996;53(3):657–64.
80. Rodriguez de Fonseca F, Rubio P, Menzaghi F, Merlo-Pich E, Rivier J, Koob GF, et al. Corticotropin-releasing factor (CRF) antagonist [D-Phe12,Nle21,38,C alpha MeLeu37]CRF attenuates the acute actions of the highly potent cannabinoid receptor agonist HU-210 on defensive-withdrawal behavior in rats. J Pharmacol Exp Ther. 1996;276(1):56–64.
81. Hill MN, Gorzalka BB. Enhancement of anxiety-like responsiveness to the cannabinoid CB(1) receptor agonist HU-210 following chronic stress. Eur J Pharmacol. 2004;499(3):291–5.
82. Justinova A, Goldberg SR, Heishman SJ, Tanda G. Self-administration of cannabinoids by experimental animals and human marijuana smokers. Pharmacol Biochem Biohav. 2005;81(2):285–99.

Chapter 10
Cannabinoid Modulation of Rodent Ultrasonic Vocalizations in a Social Context: Communicative and Rewarding Properties

Antonia Manduca, Louk J. M. J. Vanderschuren and Viviana Trezza

Abstract Like many other vertebrates, rodents emit ultrasonic vocalizations (USVs) throughout the entire life span in a variety of socially relevant situations, in order to communicate information regarding individual and group identity, status, mood or environmental conditions. High rates of USVs are emitted by pups during the first days of life when removed from the nest, by juveniles engaging in social play behavior, by adult females during social investigation and by adult males when exposed to females or during aggression. The analysis of social USVs therefore offers a translational tool to study the neurobiological mechanisms underlying socio-affective communication.

Considering the major role of the endocannabinoid system in the regulation of emotional states throughout life, it is not surprising that endocannabinoids modulate the emission of USVs. Here, we will summarize the effects of cannabinoid drugs on USVs emitted in different social contexts, such as isolation-induced USVs emitted by pups and USVs emitted by juvenile and adult rodents during social and sexual behaviors. The data outlined here provide evidence for an important role of the endocannabinoid system in social communication in rodents from birth onward. Thus, the endocannabinoid system may be altered in neuropsychiatric disorders characterized by social dysfunction and communicative deficits.

Keywords Endocannabinoid system · Endocannabinoids · Ultrasonic vocalizations · Socio-affective communication · Rodents

V. Trezza (✉) · A. Manduca
Department of Science, Roma Tre University, Rome, Italy
e-mail: viviana.trezza@uniroma3.it

L. J. M. J. Vanderschuren
Department of Animals in Science and Society, Division of Behavioural Neuroscience, Utrecht University, Utrecht, The Netherlands

P. Campolongo, L. Fattore (eds.), *Cannabinoid Modulation of Emotion, Memory, and Motivation,* DOI 10.1007/978-1-4939-2294-9_10

Introduction

Cannabis derivatives, such as hashish and marijuana, have been used for centuries for both recreational and medical purposes. Today, *Cannabis* preparations are among the most widely used illicit drugs in Western Countries, with the first episodes of *Cannabis* use often occurring during adolescence [1–3], which is a critical phase for the development of the central nervous system (see Chap. 12) [4]. Besides adolescence, *Cannabis* derivatives are often consumed by women of child-bearing age and during pregnancy [5, 6]. *Cannabis* consumption during the prenatal period is a major health problem because of its consequences for embryonic/foetal development [7–10]. Indeed, human and animal studies have shown that cannabinoids readily cross the placental barrier [11] and can be transferred to the offspring through breast milk [12], to induce long lasting neurobehavioral alterations [13–15].

The discovery in 1964 of the main psychoactive component of the plant *Cannabis sativa*, Δ9-tetrahydrocannabinol (THC) [16], led to the identification of the endogenous cannabinoid system which includes (1) cannabinoid receptors (CB1 and CB2); (2) endogenous lipid ligands [endocannabinoids (ECs), mainly anandamide (AEA) and 2-arachidonoylglycerol (2-AG)] and (3) enzymes involved in EC synthesis, transport and degradation [17–20]. From then onwards, an increasing number of experimental studies highlighted the role of the endocannabinoid system in the regulation of many physiological processes as well as the long-lasting neurobehavioral effects of developmental exposure to cannabinoid drugs. Animal studies offer the advantage that they allow for precise control over the possible confounding factors that characterize human studies (such as dosage, number of drugs used, timing of drug exposure, or the influence of social problems typically associated with drug use), and for examination of the independent contribution of cannabinoid drugs to adverse neurodevelopmental consequences.

Substantial information about the emotional and motivational state of the animal under study can be obtained by detecting its vocalizations, most of which often occur in the ultrasonic range and hence are called ultrasonic vocalizations (USVs). In rodents, USVs are considered a measure of affective states and a means of communication [21–24]. Although these calls are generally inaudible to humans, they can be studied in laboratory settings using specialized equipment. Like many other vertebrates, rodents emit USVs to communicate information regarding individual and group identity, status or mood (e.g., dominance, submission, fear or aggression), anticipatory behavior (e.g., approach, play, groom or mount) and environmental conditions (e.g., presence of predators, location of food or separation from mother and siblings) [25–28]. This has important implications for preclinical research, since measuring USVs can be considered an important approach to detect communication deficits in rodent models of human neuropsychiatric disorders, such as autism [22].

Three USV categories are typically differentiated in rodents [24, 29, 30]: (1) USVs emitted by pups after being separated from their mother and littermates (isolation-induced 40-kHz USVs); (2) USVs emitted in aversive situations, such as stress, predator exposure and fighting or during drug withdrawal (aversive 22-kHz USVs); (3) USVs emitted in appetitive situations, such as rough-and-tumble play and mating or in response to drugs of abuse (appetitive 50-kHz USVs).

Data on USVs in rodents have been reported since the 1950s, when it was observed that adult laboratory rats emit calls at low frequencies (around 23–28 kHz) when socially isolated [31]. Two years later, Zippelius and Schleidt reported that infant mice produce USVs when separated from their mother and siblings [32]. Since then, several studies have been performed showing that measuring the USVs emitted by laboratory animals under particular physical, environmental or social conditions can provide new insights into their emotional state and into specific aspects of social interaction.

Considering the major role of the endocannabinoid system in the regulation of emotional states from birth to adulthood [13, 33–36] and the functionality of endocannabinoid neurotransmission from early developmental stages [37–40], it is not surprising that endocannabinoids modulate the emission of USVs. Here, we will summarize the effects of cannabinoid drugs on (1) isolation-induced USVs emitted by pups and (2) appetitive and aversive USVs emitted by juvenile and adult rodents, in order to provide evidence that the endocannabinoid system has an important role in rodent ultrasonic communication.

Isolation-Induced USVs Emitted by Rodent Pups

General Aspects

When isolated from their mother and littermates, infant rodents emit characteristic USVs which have a communicative function, acting as a stimulus for maternal search and retrieval, as well as for nest building and maternal grooming [24]. Pup retrieval is a complex type of social interaction involving both the mother and pup; thus, the rate of calling or call characteristics produced by the pup may alter the behavior of the mother [41]. Likewise, individual differences in the mother's behavior, such as more or less exploration to find a pup, may alter the behavior of the pup [42]. The isolation-induced USVs emitted by the pups have a double meaning: (1) they reflect the affective state of the pups because of the separation from their mother and littermates, and (2) they are the product of physiological processes such as thermoregulation [43]. Regarding the emotional state of the pup as a determinant of USV production, it has been shown that the USVs emitted by 14-day-old pups are suppressed by the presence of an unfamiliar (and possibly infanticidal) adult male rat, a response that is seen in combination with freezing behavior [44, 45]. Furthermore, the reduction in body temperature caused by removal of the pup from the nest is another important stimulus for USV emission: as the environmental temperature and the body temperature of the pup decrease, the number of USVs emitted by the pup increases [46].

The strongest argument in favour of the affective hypothesis of pup isolation-induced USVs comes from the evidence that anxiolytic compounds such as benzodiazepines and other positive modulators of GABA-receptors reduce the emission of these calls [47–50]. Accordingly, anxiogenic drugs, such as pentylenetetrazole, increase USV emission in rat pups without producing appreciable hypothermia [51].

From this, it has been suggested that the emotional state of the pup is a crucial determinant of USV production, and that measuring isolation-induced USVs in rodent pups is useful to investigate the ontogeny of emotionality and the potential anxiolytic or anxiogenic effects of pharmacological or genetic manipulations [52–54].

However, not all drugs that reduce USVs in pups are necessarily anxiolytic: thus, drugs with positive reinforcing properties, such as cocaine [55], 3, 4-methylenedioxymethamphetamine (MDMA) [56], and morphine [57] all reduce USVs in pups. Conversely, drugs that have aversive effects in adult rats in the conditioned place preference paradigm, such as the κ-opioid receptor agonist U50,488, increase USVs in pups [58]. In line with their communicative function, isolation-induced USVs have emerged as an innovative tool to evaluate communication deficits in different animal models of neuropsychiatric disorders characterized by a reduction [22, 59–64] or an unusual repertoire [65] of isolation-induced USVs.

Role of the Endocannabinoid System in the Modulation of Isolation-Induced USVs Emitted by Rodent Pups

Preclinical evidence has demonstrated that cannabinoid drugs affect USV emission in pups isolated from their mother and littermates depending on the type and dose of drug used (Table 10.1), confirming the well-known bidirectional effects of cannabinoids on emotional reactivity [35, 66]. Specifically, it has been shown that the acute administration of the potent cannabinoid CB1 receptor agonist CP55,940 produced a dose-dependent reduction in the frequency of USVs in 12-day-old Long-Evans rat pups isolated from their mother and siblings, accompanied to a substantial drug-induced hypothermia [67]. The cannabinoid CB1 receptor antagonist SR141716A reversed the reduction of USVs induced by CP55,940. Interestingly, the number of USVs emitted by pups administered both CP55,940 and SR141716A was increased compared to animals given the agonist alone, which suggests a possible intrinsic effect of SR141716A. Therefore, it is possible that endocannabinoids inhibit pup USV emission under conditions of isolation, and that SR141716A blocks this inhibitory effect to produce a disinhibition of USVs.

Other preclinical evidence of endocannabinoid modulation of isolation-induced USVs in pups has been provided by the use of the AEA hydrolysis inhibitor URB597, which inhibits the enzyme fatty acid amide hydrolase (FAAH) and leads to prolonged AEA signalling [68], and the endocannabinoid transport inhibitor AM404, which inhibits the still controversial high-affinity transport system that removes ECs from the synaptic space [69]. URB597 reduced the number of isolation-induced USVs emitted by rat pups removed from their nest at doses that had no effect on pup motor activity, an effect that was blocked by SR141716A [68]. Similarly, AM404 reduced USVs in 10-day-old rat pups removed from the nest without alteration of axillary temperature or locomotor activity; again, these effects were mediated by activation of cannabinoid CB1 receptors, as they were antagonized by SR141716A [70]. The reduction in pup USV emission induced by drugs that

Table 10.1 Effects of cannabinoid drugs on isolation-induced USVs in rodent pups

Drug	Mechanism of action	Effect	Antagonism	Exposure	Strain	Refs
CP55,940 (0.1–1 mg/kg; *i.p.*)	CB1 cannabinoid receptor agonist	↓	SR141716A (20 mg/kg; *i.p.*)	Postnatal (to pups at PND 12±1)	Long-Evans Hooded rat pups	McGregor et al. 1996
SR141716A (20 mg/kg; *i.p.*)	CB1 cannabinoid receptor antagonist/inverse agonist	↑				
URB597 (0.1 mg/kg; *i.p.*)	AEA hydrolysis inhibitor	↓	SR141716A (2 mg/Kg; *i.p.*)	Postnatal (to pups at PND 10)	Wistar rat pups	Kathuria et al. 2003
AM404 (1–2 mg/kg, *i.p.*)	Endocannabinoid reuptake inhibitor	↓	SR141716A (1 mg/Kg; *i.p.*)	Postnatal (to pups at PND 10)	Wistar rat pups	Bortolato et al. 2006
$CB1^{-/-}$ knockout pups		↓		Postnatal (PND 4–6–15)	Knockout mice pups	Fride et al. 2005
THC (5 mg/kg, *or.*)	CB1 cannabinoid receptor agonist	↑		Perinatal (to mother from GD 15 to PND 9)	Wistar rat pups (PND 12)	Trezza et al. 2008
WIN55,212–2 (0.5 mg/kg; *s.c.*)	CB1 cannabinoid receptor agonist	↓		Prenatal (to mother from GD 5 to GD 20)	Wistar rat pups (PND 10)	Antonelli et al. 2005
SR141716A (10 mg/kg; *i.p.*)	CB1 cannabinoid receptor antagonist/inverse agonist	↓		Postnatal (to mother from PND 1 to PND 8)	Sabra outbred mice pups (PND 6–8)	Schechter et al. 2012

AEA anandamide, *PND* postnatal day, *GD* gestational day, *i.p.* intraperitoneally, *s.c.* subcutaneously, *or.* orally prolong endocannabinoid activity by interfering with endocannabinoid intracellular transport or hydrolysis has been suggested to reflect the anxiolytic-like properties of these compounds. The anxiolytic-like effects induced by URB597 and AM404 were indeed confirmed in adult animals [68, 70]. On the basis of these data, it was proposed that drugs that interfere with endocannabinoid deactivation, thus prolonging local endocannabinoid activity, might represent new therapeutic tools to treat anxiety-related disorders.

To investigate the role of the endocannabinoid system in the modulation of isolation-induced USVs, Fride and colleagues investigated the consequences of cannabinoid CB1 receptor deletion on the ontogeny of emotionality, evaluating the emission of USVs induced by maternal separation in wild-type and CB1 knockout ($CB1^{-/-}$) pups [71]. As expected, wild-type mouse pups reached the highest frequency of USV emission between days 3 and 6 of age until day 10. In contrast, $CB1^{-/-}$ pups showed very low levels of USVs throughout development [71]. Taken together, these results suggest that (1) endocannabinoids regulate USV production in rodent pups, likely via activation of brain cannabinoid CB1 receptors, and (2) endocannabinoid modulation of emotional reactivity appears at early developmental ages.

Exposure to cannabinoid drugs during pregnancy and/or lactation has also been related to changes in USV emissions in rodent pups. For example, 12-day-old pups exposed to THC during the perinatal period displayed an increased rate of USVs when separated from their mother and siblings [72]. However, a reduction of isolation-induced USVs in rat pups prenatally exposed to the synthetic cannabinoid agonist WIN55,212–2 has also been reported [73]. Differences in the type and dose of cannabinoid agonist used and the time window of exposure may account for the apparent discrepancies between these findings.

A recent study reported that in lactating dams, daily injections of SR141716A during *post-partum* days 1–8 led to fewer vocalizations in pups at postnatal day (PND) 6 and PND 8, together with reduced maternal behaviors [74]. It is thought that the number of calls emitted by mouse pups can reflect maternal responsiveness, and when there is no response or there is a delay to respond to the needs of pups, pups lower this mode of connection with the caregiver [75]. Thus, low levels of maternal care in SR141716A-treated dams may have induced a decrease in the rate of pup USVs accompanied with a lower body weight and hypothermia in pups. These results suggest that the endocannabinoid system is crucial for establishment of maternal behavior during the first *post-partum* week, with a long-term impact on the offspring's socio-emotional development.

Interaction Between Opioid, Cannabinoid and Dopaminergic Systems in the Modulation of Isolation-Induced USVs in Pups

There are close functional interactions between the endocannabinoid, opioid and dopaminergic systems in the regulation of reward processes, including drug, food and social rewards [34, 76–82].

Interactions between cannabinoid and opioid receptors have been predicted on the basis of different observations [83–87]: (1) μ-opioid receptors are frequently co-localized with cannabinoid CB1 receptors in several brain areas, (2) they utilize the same pool of G-proteins and (3) these receptors form functional heterodimers. Concerning the formation of heterodimers, *in vitro* studies suggested that cannabinoid CB1 receptors hetero-oligomerize with μ-opioid receptors [88, 89]. Moreover, the interaction between CB1 and μ-opioid receptors has been demonstrated using Fluorescence Resonance Energy Transfer and co-immunoprecipitation experiments in cells exogenously expressing both receptors [85]. Cannabinoid-opioid crosstalk has been also observed through receptor expression levels, as repeated administration of THC regulates μ-opioid receptor density in a time and region-dependent manner [90]. In addition to the possibility of a direct association, cannabinoid-opioid interactions may also occur at the signal transduction level since both receptors signal through the Gi/o alpha subunit, targeting the cAMP pathway [88]. Although the functional relationship between opioid and cannabinoid systems is becoming better understood, the specific mechanisms of their interaction remain to be elucidated (see Chap. 16).

In the literature, many examples of functional interaction between cannabinoid and dopamine receptors have also been reported (see Chap. 19). Functional antagonism was shown in the rat brain, where the cannabinoid CB1 agonist CP55,940 reduced the affinity of D2 dopamine receptor agonist binding sites in the dorsal and ventral striatum. Similarly, *in vivo* administration of CP55,940 inhibited locomotor activity induced by dopamine D2 receptor stimulation [91].

Concerning the interaction between opioid, cannabinoid and dopaminergic systems in the modulation of isolation-induced USVs in pups, it has been demonstrated that the opioid receptor antagonist naloxone and the dopamine D1 receptor antagonist SCH 23390 did not antagonize the reduction in USV emissions in isolated pups treated with CP55,940 [67]. On the basis of these data, it is possible to speculate that cannabinoid-opioid-dopamine interactions, although already functional in adolescent rats engaged in positive social interactions [81, 82], are either not present or not yet mature in pre-weaning pups.

Affective USVs in Rodents

General Aspects on 22-kHz USVs

The 22-kHz USVs are believed to reflect a negative affective state since rodents emit these low frequency USVs in aversive situations, such as predator exposure [92] and fighting [93, 94], or during withdrawal from drugs of abuse, such as alcohol, benzodiazepines, opiates and psychostimulants [95, 96]. In 1991, Blanchard and colleagues suggested that 22-kHz USVs served as alarm calls to warn conspecifics about danger [92]. Thus, no 22-kHz USVs were detected when a rat was exposed to a natural predator such as a cat in absence of conspecifics. There is evidence that 22-kHz USVs can induce anxiety-related behaviors such as freezing, probably inducing neuronal activity in brain areas implicated in fear regulation, like the amygdala [97]. It has also been shown that rats form memory associations between 22-kHz USVs and aversive stimuli [98, 99] and that 22-kHz USVs play an important role in the social transmission of fear [100].

Under stressful conditions, adult rats emit 22-kHz USVs the duration of which can be reduced by anxiolytic drugs [101, 102] or increased by anxiogenic compounds [101, 103]. The production of 22-kHz USVs in aversive situations depends on a wide range of factors [24], including (1) experimental context, (2) genetic influence and (3) environmental factors. Indeed, 22-kHz USV rate increases with the aversiveness of the context [104]. Thus, rats exposed to higher foot shock intensities during fear conditioning displayed more freezing behavior and vocalized more than rats exposed to lower foot shock intensities, whereas rats exposed to tones but not foot shocks did not emit 22-kHz USVs. Moreover, adult rats receiving a single foot shock of 1 s duration and 0.6 mA intensity emit low levels of USVs that can be increased by anxiogenic drugs, whereas rats receiving six foot shocks of the same duration and intensity emit high levels of USVs that can be reduced by the treatment with anxiolytic drugs [105].

In addition to the aversiveness of the context, individual factors also play an important role in 22-kHz USVs [106]. Thus, rats that were characterized as highly anxious in the elevated plus-maze test emitted more 22-kHz USVs during fear conditioning than less anxious animals. Among others, such individual differences in the emission of 22-kHz USVs could be attributable to early environmental factors. For instance, juvenile stress exposure (i.e. forced swim test, exposure to an elevated platform and immobilization) affects 22-kHz USV production at adulthood [107] and induces long-lasting behavioral, physiological and molecular changes [108, 109]. Indeed, rats exposed to such stressors during the pre-pubertal period emitted more 22-kHz USVs in response to fear conditioning at adulthood than unexposed controls [24].

General Aspects on 50-kHz USVs

Anticipation of a wide variety of rewards (i.e. food, mating, play, drugs of abuse and rewarding brain stimulation) increases the emission of 50-kHz USVs, which are positively correlated to the approach latency for the rewarding stimulus [21]. Conversely, aversive stimuli (i.e. bright light, predator odour, social defeat, foot shock, aversive drugs) have been shown to decrease the emission of 50-kHz USVs [21].

In 2007, Panksepp and co-workers reported that juvenile mice produced high rates of 50-kHz USVs during social interactions [27]. As the production of interaction-induced USVs occurred exclusively during social investigation and no aggressive behavior was observed, the authors speculated that 50-kHz USVs in mice reflected a positive affective state. In the same year, they also demonstrated that genetic differences exist in the emission of USVs in mice, since early-adolescent C57BL/6J emitted more USVs than BALB/cJ mice during social conditioned place preference [110].

Panksepp and Burgdorf further demonstrated that it is also possible to induce 50-kHz USVs in rats by hetero-specific play, such as tickling by a human experimenter [26]. Indeed, they suggested that those rats emitting high frequency 50-kHz USVs experienced the tickling procedure as appetitive, as indicated by short latencies to approach the hand of the experimenter [26]. In addition to the tickling itself, even the presentation of cues associated with it, such as the experimenter's hand, elicited 50-kHz USVs. Interestingly, they found large individual differences in the production of 50-kHz USVs in response to tickling. Thus, while most rats emitted 50-kHz USVs, others did not emit and some even emitted 22-kHz USV [26]. Altogether, these studies demonstrate that social context and social stimuli modulate the production of appetitive 50-kHz USVs which display an important pro-social communicative value to establish or maintain social contact [111].

Despite these findings, controversial data exist in the literature about whether 50-kHz USVs may reflect the rewarding value of adolescent social interactions. Knutson and co-workers were the first to suggest that 50-kHz USVs may index play motivation in rats since they found that adolescent Long-Evans rats emitted more 50-kHz USVs while playing together than while alone [112]. Interestingly, rats separated from the play partner through a screen dividing the testing chamber, but given the opportunity to play the day before, vocalized more than rats that played both days. Conversely, separated rat pairs that had not played before vocalized less than rats that played on both days. On the basis of these data, the authors suggested that one trial of play is sufficient to induce a motivational state that evokes 50-kHz USVs in rats, and that general motor activity alone cannot account for the expression of these vocalizations [21, 112]. However, not all data support the use of 50-kHz USVs in adolescent rats as an index of positive affective state during active social interaction or anticipation for social contact. Thus, Willey et al. found that adolescent Sprague–Dawley rats emit fewer 50-kHz USVs than adults in anticipation of a social partner [113] and during social interactions [114], although adolescent rats usually engage in higher levels of social behaviors, especially social play behavior [114–117] and show greater conditioned place preference for a social stimulus than adults [118]. We also recently demonstrated that positive social interactions in adolescent rats do not necessarily correlate with 50-kHz USVs emission and that drugs that affect social play, like morphine, do not affect the rate of USVs emitted during social interactions [119]. Thus, it cannot generally be assumed that 50-kHz USVs index positive affective states during social interactions. In line with this hypothesis, it has also been shown that a dissociation exists between the emission of 50-kHz USVs and behavioral measures of affective responses during drug self-administration or anticipation. Thus, while 50-kHz USVs can be elicited by administration of psychostimulant drugs [120–123], treatment with rewarding doses of morphine, MDMA, and nicotine has been reported not to change or even decrease 50-kHz USV emission [122, 124–126].

In adult mice, high rates of USVs were found in males when courting and copulating with females [127, 128] depending on social factors such as social status [25] and previous heterosexual contact [129]. As shown by Whitney and co-workers, female urine alone, i.e. in the absence of a female, is sufficient to elicit male USVs

[130], whereas no USV response was detected when male mice were exposed to male mouse urine or female urine from rats or humans [130]. As female-induced USVs reflect a positive affective state in mice, abnormalities in their frequency could underlie communication deficits in rodents. Indeed, reduced female-induced USVs were reported in a genetic animal model of nonsyndromic monogenic heritable autism spectrum disorders (ASDs), i.e. mice lacking the murine ortholog of human Neuroligin-4 (NL-4), thus indicating that the NL-4-KO mice are either less responsive to social stimuli or are otherwise inhibited in their propensity to communicate [131, 132].

Role of the Endocannabinoid System in the Modulation of Affective USVs

In adult rats exposed to one foot shock (1 s duration and 0.6 mA intensity), administration of the cannabinoid CB1 receptor antagonist SR141716A induced a bell-shaped dose–response curve: intermediate doses (1.25–5 mg/kg) significantly increased the duration of 22-kHz USVs, while lower and higher doses were ineffective [105]. The anxiogenic effect of intermediate doses of SR141716A was completely abolished by the allosteric inhibitor of the metabotropic glutamate receptor subtype 5 (mGluR5), MTEP, showing that mGluR5 antagonism abolished the anxiety-like state induced by cannabinoid CB1 receptor antagonism. Conversely, MTEP dose-dependently reduced the duration of USVs in rats exposed to six foot shocks of 1 s duration and 0.6 mA intensity. However, this effect was not influenced by SR141716A [105]. In line with the results obtained in rat pups [67], SR141716A displayed anxiogenic-like effect in adult rats exposed to one foot shock in a sound attenuated shocking chamber, with MTEP completely reduced its anxiogenic effects [105].

It has been repeatedly shown that cannabinoid signalling plays an important role in controlling conditioned fear [133–137]. Studies in rats have demonstrated that the basolateral amygdala (BLA), which is implicated in emotional processing [138] and coordination of appropriate responses to conditioned aversive stimuli [139], is a critical brain area underlying endocannabinoid regulation of fear responses (see Chap. 1). Fear-conditioned rats receiving intra-BLA infusion of the cannabinoid CB1 receptor antagonist SR141716A displayed more 22-kHz USVs than vehicle-treated rats [103]. Furthermore, non-fear-conditioned rats displayed little or no contextually induced freezing behavior or 22-kHz USVs emission during the test trial. In contrast, fear conditioning was associated with significant increases in the duration of freezing and 22-kHz USVs upon re-exposure to the context. Recently, Arnold and colleagues found that Wistar rats treated with the cannabinoid CB1 receptor agonist CP55,940 and re-exposed to a chamber in which they had previously received a foot shock, emitted more 18–30-kHz USVs than vehicle-treated rats. Conversely, CP55,940 had no significant effect upon conditioned USVs in Lewis rats [140], highlighting a strain-dependent effect of the cannabinoid receptor agonist on fear-related USV emission. The anandamide hydrolysis inhibitor URB597 significantly reduced 22-kHz USVs emitted during freezing behavior in the fear conditioning paradigm in male Lister-hooded rats [141], thus supporting the possibility that the endocannabinoid AEA has an important role in fear-related USVs. These results also confirm the different behavioral profile induced by direct (by CP55,940) versus indirect (by URB597) activation of cannabinoid CB1 receptors.

To our knowledge, no study has assessed the effects of cannabinoid drugs on 50-kHz USVs and therefore more research in this direction is warranted. However, since previous studies have shown that (1) cannabinoid drugs, including the anandamide hydrolysis inhibitor URB597, modulate maternal separation-induced USVs in rat pups [67, 68, 70] and adult low frequency USVs [141], (2) cannabinoid compounds affect USVs in adult rats depending on strain [140], and (3) the effects of URB597 on social behavior appear to be context-dependent [81], we recently investigated the effects of URB597 on USV emission during active social interactions in adolescent and adult Wistar and Sprague–Dawley rats tested in different environmental conditions. Thus, our data show that URB597 enhances the frequency of 50-kHz USVs emitted during social interaction by adolescent Wistar and adult Sprague–Dawley rats tested under moderate (i.e. unfamiliar test cage/ low light) and high aversive (i.e. unfamiliar test cage/ high light) conditions respectively. These results confirm previous findings by suggesting that the endocannabinoid system modulates the emission of high frequency USVs (50-kHz) during social behavior in a strain- and context-dependent manner [142].

Table 10.2 summarizes the effects of cannabinoid drugs on affective USVs in rodents.

Conclusions

Measuring ultrasonic communication in rodents offers a translational tool for studying the physiological and neurobiological mechanisms underlying socio-affective communication. This may be particularly relevant for rodent models of human neurodevelopmental disorders including autism and schizophrenia, which are characterized by social dysfunctions, including communication deficits.

The endocannabinoid system is actively present and functional since the earliest stages of ontogenetic development, from fertilization and pre-implantation until prenatal and postnatal life [8, 34, 38, 72, 81, 143, 144]. Altogether, the results outlined here suggest that the endocannabinoid system, plausibly via activation of brain cannabinoid CB1 receptors, is essential in the modulation of isolation-induced USVs in rodent pups, playing a crucial role in mother-infant interaction [14, 67, 70, 71]. Preclinical evidence reported here also indicates that cannabinoid signalling plays an important role in controlling conditioned fear at adulthood [105, 140, 141]; thus, it may serve as a promising target for innovative intervention strategies in fear-related disorders, such as post-traumatic stress disorders (PTSD). Conversely, little information is available on endocannabinoid modulation of positive 50-kHz USVs and therefore more research in this direction is warranted.

Table 10.2 Effects of cannabinoid drugs on affective USVs in rodents

Drug	Mechanism of action	Effect	Antagonism	Behavioral task	Strain	Refs
SR141716A (1.25 and 5 mg/kg, *i.p.*) SR141716A (50 μg/ 0.5 μL *intra BLA*) CP55,940 (10, 25 and 50 μg/kg, *i.p.*) URB597 (0.3 mg/kg, *i.p.*) URB597 (0.1 mg/kg, *i.p.*)	CB1 cannabinoid receptor antagonist/ inverse agonist CB1 cannabinoid receptor antagonist/ inverse agonist CB1 cannabinoid receptor agonist AEA hydrolysis inhibitor AEA hydrolysis inhibitor	↑ 22-kHz duration ↑ 22-kHz duration ↑ 18–30-kHz frequency ↓ 22-kHz duration ↑ 50-kHz frequency ↑ 50-kHz frequency	MTEP (3 mg/kg; *i.p.*)	One foot shock exposure (1 s; 0.6 mA) Fear conditioning Conditioned USV model (re-exposition to a chamber after a foot shock) Fear conditioning Social play behavior Social interaction	Hannover Wistar rats Lister-hooded rats Wistar rats Lister-hooded rats Wistar rats (PND 28–35) Sprague-Dawley rats (PND 80–90)	Varga et al. 2012 Roche et al. 2007 Arnold et al. 2010 Butler et al. 2012 Manduca et al. 2014

AEA anandamide, *BLA* basolateral amygdala, *MTEP* (3-3-((2-methyl-4-thiazolyl)ethynyl)pyridine), *PND* postnatal day, *USV* ultrasonic vocalizations, *i.p.* intraperitoneally

References

1. Hall W, Degenhardt L. Adverse health effects of non-medical cannabis use. Lancet. 2009;374(9698):1383–91.
2. NIDA. NIDA Research monograph series: marijuana abuse NIH. Washington, DC; 2005.
3. SAMHSA. Results from the 2009 national survey on drug use and health. Summary of national findings. Rockville, MD:SAMHSA; 2009.
4. Spear LP. The adolescent brain and age-related behavioral manifestations. Neurosci Biobehav Rev. 2000;24(4):417–63.
5. Fried PA, Smith AM. A literature review of the consequences of prenatal marihuana exposure. An emerging theme of a deficiency in aspects of executive function. Neurotoxicol Teratol. 2001;23(1):1–11.
6. Ebrahim SH, Gfroerer J. Pregnancy-related substance use in the United States during 1996–1998. Obstet Gynecol. 2003;101(2):374–9.
7. Fernandez-Ruiz JJ, Berrendero F, Hernandez ML, Romero J, Ramos JA. Role of endocannabinoids in brain development. Life Sci. 1999;65(6–7):725–36.
8. Fernandez-Ruiz J, Berrendero F, Hernandez ML, Ramos JA. The endogenous cannabinoid system and brain development. Trends Neurosci. 2000;23(1):14–20.
9. Maccarrone M, Wenger T. Effects of cannabinoids on hypothalamic and reproductive function. Handb Exp Pharmacol. 2005;168:555–71.
10. Wang H, Xie H, Sun X, Kingsley PJ, Marnett LJ, Cravatt BF, Dey SK. Differential regulation of endocannabinoid synthesis and degradation in the uterus during embryo implantation. Prostaglandins Other Lipid Mediat. 2007;83(1–2):62–74.
11. Hutchings DE, Martin BR, Gamagaris Z, Miller N, Fico T. Plasma concentrations of delta-9-tetrahydrocannabinol in dams and fetuses following acute or multiple prenatal dosing in rats. Life Sci. 1989;44(11):697–701.
12. Dalterio SL. Cannabinoid exposure: effects on development. Neurobehav Toxicol Teratol. 1986;8(4):345–52.
13. Campolongo P, Trezza V, Palmery M, Trabace L, Cuomo V. Developmental exposure to cannabinoids causes subtle and enduring neurofunctional alterations. Int Rev Neurobiol. 2009;85:117–33.
14. Manduca A, Campolongo P, Trezza V. Cannabinoid modulation of mother-infant interaction: is it just about milk? Rev Neurosci. 2012;23(5–6):707–22.
15. Trezza V, Campolongo P, Manduca A, Morena M, Palmery M, Vanderschuren LJ, Cuomo V. Altering endocannabinoid neurotransmission at critical developmental ages: impact on rodent emotionality and cognitive performance. Front Behav Neurosci. 2012;6:2.
16. Gaoni Y, Mechoulam R. Isolation, structure and partial synthesis of an active constituent of hashish. J Am Chem Soc. 1964;86:1646–7.
17. Devane WA, Hanus L, Breuer A, Pertwee RG, Stevenson LA, Griffin G, Gibson D, Mandelbaum A, Etinger A, Mechoulam R. Isolation and structure of a brain constituent that binds to the cannabinoid receptor. Science. 1992;258(5090):1946–9.
18. Freund TF, Katona I, Piomelli D. Role of endogenous cannabinoids in synaptic signaling. Physiol Rev. 2003;83(3):1017–66.
19. Piomelli D. The molecular logic of endocannabinoid signalling. Nat Rev Neurosci. 2003;4(11):873–84.
20. Kogan NM, Mechoulam R. The chemistry of endocannabinoids. J Endocrinol Invest. 2006;29(3 Suppl):3–14.
21. Knutson B, Burgdorf J, Panksepp J. Ultrasonic vocalizations as indices of affective states in rats. Psychol Bull. 2002;128(6):961–77.
22. Scattoni ML, Crawley J, Ricceri L. Ultrasonic vocalizations: a tool for behavioural phenotyping of mouse models of neurodevelopmental disorders. Neurosci Biobehav Rev. 2009;33(4):508–15.

23. Takahashi N, Kashino M, Hironaka N. Structure of rat ultrasonic vocalizations and its relevance to behavior. PLoS One. 2010;5(11):e14115.
24. Wohr M, Schwarting RK. Affective communication in rodents: ultrasonic vocalizations as a tool for research on emotion and motivation. Cell Tissue Res. 2013;354(1):81–97.
25. D'Amato FR. Courtship ultrasonic vocalizations and social status in mice. Anim Behav. 1991;41:875–85.
26. Panksepp J, Burgdorf J. 50-kHz chirping (laughter?) in response to conditioned and unconditioned tickle-induced reward in rats: effects of social housing and genetic variables. Behav Brain Res. 2000;115(1):25–38.
27. Panksepp JB, Jochman KA, Kim JU, Koy JJ, Wilson ED, Chen Q, Wilson CR, Lahvis GP. Affiliative behavior, ultrasonic communication and social reward are influenced by genetic variation in adolescent mice. PLoS One. 2007;2(4):e351.
28. Portfors CV. Types and functions of ultrasonic vocalizations in laboratory rats and mice. J Am Assoc Lab Anim Sci. 2007;46(1):28–34.
29. Schwarting RK, Wohr M. On the relationships between ultrasonic calling and anxiety-related behavior in rats. Braz J Med Biol Res. 2012;45(4):337–48.
30. Burgdorf J, Panksepp J, Moskal JR. Frequency-modulated 50 kHz ultrasonic vocalizations: a tool for uncovering the molecular substrates of positive affect. Neurosci Biobehav Rev. 2011;35(9):1831–6.
31. Anderson JW. The production of ultrasonic sounds by laboratory rats and other mammals. Science. 1954;119(3101):808–9.
32. Zippelius HM, Schleidt WM. Ultraschall-Laute bei jungen Mäusen. Naturwissenschaften. 1956;46:502.
33. Morena M, Campolongo P. The endocannabinoid system: an emotional buffer in the modulation of memory function. Neurobiol Learn Mem. 2014;112:30–43.
34. Trezza V, Cuomo V, Vanderschuren LJMJ. Cannabis and the developing brain: insights from behavior. Eur J Pharmacol. 2008;585(2–3):441–52.
35. Moreira FA, Lutz B. The endocannabinoid system: emotion, learning and addiction. Addict Biol. 2008;13(2):196–212.
36. Rubino T, Parolaro D. Sexually dimorphic effects of cannabinoid compounds on emotion and cognition. Front Behav Neurosci. 2011;5:64.
37. Fried PA. Conceptual issues in behavioral teratology and their application in determining long-term sequelae of prenatal marihuana exposure. J Child Psychol Psychiatry. 2002;43(1):81–102.
38. Harkany T, Guzman M, Galve-Roperh I, Berghuis P, Devi LA, Mackie K. The emerging functions of endocannabinoid signaling during CNS development. Trends Pharmacol Sci. 2007;28(2):83–92.
39. Lee TT, Gorzalka BB. Timing is everything: evidence for a role of corticolimbic endocannabinoids in modulating hypothalamic-pituitary-adrenal axis activity across developmental periods. Neuroscience. 2012;204:17–30.
40. Marco EM, Laviola G. The endocannabinoid system in the regulation of emotions throughout lifespan: a discussion on therapeutic perspectives. J Psychopharmacol. 2012;26(1):150–63.
41. Brudzynski SM. Principles of rat communication: quantitative parameters of ultrasonic calls in rats. Behav Genet. 2005;35(1):85–92.
42. Hahn ME, Lavooy MJ. A review of the methods of studies on infant ultrasound production and maternal retrieval in small rodents. Behav Genet. 2005;35(1):31–52.
43. Blumberg MS, Sokoloff G. Do infant rats cry? Psychol Rev. 2001;108(1):83–95.
44. Takahashi LK. Ontogeny of behavioral inhibition induced by unfamiliar adult male conspecifics in preweanling rats. Physiol Behav. 1992;52(3):493–8.
45. Takahashi LK. Stimulus control of behavioral inhibition in the preweanling rat. Physiol Behav. 1994;55(4):717–21.
46. Oswalt GL, Meier GW. Olfactory, thermal, and tactual influences on infantile ultrasonic vocalization in rats. Dev Psychobiol. 1975;8(2):129–35.
47. Benton D, Nastiti K. The influence of psychotropic drugs on the ultrasonic calling of mouse pups. Psychopharmacology (Berl). 1988;95(1):99–102.

48. Fish EW, Sekinda M, Ferrari PF, Dirks A, Miczek KA. Distress vocalizations in maternally separated mouse pups: modulation via 5-HT(1A), 5-HT(1B) and GABA(A) receptors. Psychopharmacology (Berl). 2000;149(3):277–85.
49. Nastiti K, Benton D, Brain PF. The effects of compounds acting at the benzodiazepine receptor complex on the ultrasonic calling of mouse pups. Behav Pharmacol. 1991;2(2):121–8.
50. Takahashi A, Yap JJ, Bohager DZ, Faccidomo S, Clayton T, Cook JM, Miczek KA. Glutamatergic and GABAergic modulations of ultrasonic vocalizations during maternal separation distress in mouse pups. Psychopharmacology (Berl). 2009;204(1):61–71.
51. Carden SE, Bortot AT, Hofer MA. Ultrasonic vocalizations are elicited from rat pups in the home cage by pentylenetetrazol and U50,488, but not naltrexone. Behav Neurosci. 1993;107(5):851–9.
52. Branchi I, Santucci D, Alleva E. Ultrasonic vocalisation emitted by infant rodents: a tool for assessment of neurobehavioural development. Behav Brain Res. 2001;125(1–2):49–56.
53. Cuomo V, De Salvia MA, Maselli MA, Santo L, Cagiano R. Ultrasonic calling in rodents: a new experimental approach in behavioural toxicology. Neurotoxicol Teratol. 1987;9(2):157–60.
54. Insel TR, Hill JL, Mayor RB. Rat pup ultrasonic isolation calls: possible mediation by the benzodiazepine receptor complex. Pharmacol Biochem Behav. 1986;24(5):1263–7.
55. Barr GA, Wang S. Behavioral effects of chronic cocaine treatment in the week-old rat pup. Eur J Pharmacol. 1993;233(1):143–9.
56. Winslow JT, Insel TR. Serotonergic and catecholaminergic reuptake inhibitors have opposite effects on the ultrasonic isolation calls of rat pups. Neuropsychopharmacology. 1990;3(1):51–9.
57. Carden SE, Hofer MA. The effects of opioid and benzodiazepine antagonists on dam-induced reductions in rat pup isolation distress. Dev Psychobiol. 1990;23(8):797–808.
58. Carden SE, Davachi L, Hofer MA. U50,488 increases ultrasonic vocalizations in 3-, 10-, and 18-day-old rat pups in isolation and the home cage. Dev Psychobiol. 1994;27(1):65–83.
59. Silverman JL, Yang M, Lord C, Crawley JN. Behavioural phenotyping assays for mouse models of autism. Nat Rev Neurosci. 2010;11(7):490–502.
60. Chadman KK, Gong S, Scattoni ML, Boltuck SE, Gandhy SU, Heintz N, Crawley JN. Minimal aberrant behavioral phenotypes of neuroligin-3 R451C knockin mice. Autism Res. 2008;1(3):147–58.
61. Fujita E, Tanabe Y, Shiota A, Ueda M, Suwa K, Momoi MY, Momoi T. Ultrasonic vocalization impairment of Foxp2 (R552H) knockin mice related to speech-language disorder and abnormality of Purkinje cells. Proc Natl Acad Sci U S A. 2008;105(8):3117–22.
62. Gaub S, Groszer M, Fisher SE, Ehret G. The structure of innate vocalizations in Foxp2-deficient mouse pups. Genes Brain Behav. 2010;9(4):390–401.
63. Nakatani J, Tamada K, Hatanaka F, Ise S, Ohta H, Inoue K, Tomonaga S, Watanabe Y, Chung YJ, Banerjee R, Iwamoto K, Kato T, Okazawa M, Yamauchi K, Tanda K, Takao K, Miyakawa T, Bradley A, Takumi T. Abnormal behavior in a chromosome-engineered mouse model for human 15q11–13 duplication seen in autism. Cell. 2009;137(7):1235–46.
64. Shu W, Cho JY, Jiang Y, Zhang M, Weisz D, Elder GA, Schmeidler J, De Gasperi R, Sosa MA, Rabidou D, Santucci AC, Perl D, Morrisey E, Buxbaum JD. Altered ultrasonic vocalization in mice with a disruption in the Foxp2 gene. Proc Natl Acad Sci U S A. 2005;102(27):9643–8.
65. Scattoni ML, Gandhy SU, Ricceri L, Crawley JN. Unusual repertoire of vocalizations in the BTBR T + tf/J mouse model of autism. PLoS One. 2008;3(8):e3067.
66. Viveros MP, Marco EM, File SE. Endocannabinoid system and stress and anxiety responses. Pharmacol Biochem Behav. 2005;81(2):331–42.
67. McGregor IS, Dastur FN, McLellan RA, Brown RE. Cannabinoid modulation of rat pup ultrasonic vocalizations. Eur J Pharmacol. 1996;313(1–2):43–9.
68. Kathuria S, Gaetani S, Fegley D, Valino F, Duranti A, Tontini A, Mor M, Tarzia G, La Rana G, Calignano A, Giustino A, Tattoli M, Palmery M, Cuomo V, Piomelli D. Modulation of anxiety through blockade of anandamide hydrolysis. Nat Med. 2003;9(1):76–81.
69. Fowler CJ. Anandamide uptake explained? Trends Pharmacol Sci. 2012;33(4):181–5.

70. Bortolato M, Campolongo P, Mangieri RA, Scattoni ML, Frau R, Trezza V, La Rana G, Russo R, Calignano A, Gessa GL, Cuomo V, Piomelli D. Anxiolytic-like properties of the anandamide transport inhibitor AM404. Neuropsychopharmacology. 2006;31(12):2652–9.
71. Fride E, Suris R, Weidenfeld J, Mechoulam R. Differential response to acute and repeated stress in cannabinoid CB1 receptor knockout newborn and adult mice. Behav Pharmacol. 2005;16(5–6):431–40.
72. Trezza V, Campolongo P, Cassano T, Macheda T, Dipasquale P, Carratu MR, Gaetani S, Cuomo V. Effects of perinatal exposure to delta-9-tetrahydrocannabinol on the emotional reactivity of the offspring: a longitudinal behavioral study in Wistar rats. Psychopharmacology (Berl). 2008;198(4):529–37.
73. Antonelli T, Tomasini MC, Tattoli M, Cassano T, Tanganelli S, Finetti S, Mazzoni E, Trabace L, Steardo L, Cuomo V, Ferraro L. Prenatal exposure to the CB1 receptor agonist WIN 55,212–2 causes learning disruption associated with impaired cortical NMDA receptor function and emotional reactivity changes in rat offspring. Cereb Cortex. 2005;15(12):2013–20.
74. Schechter M, Pinhasov A, Weller A, Fride E. Blocking the postpartum mouse dam's CB1 receptors impairs maternal behavior as well as offspring development and their adult social-emotional behavior. Behav Brain Res. 2012;226(2):481–92.
75. D'Amato FR, Scalera E, Sarli C, Moles A. Pups call, mothers rush: does maternal responsiveness affect the amount of ultrasonic vocalizations in mouse pups? Behav Genet. 2005;35(1):103–12.
76. Solinas M, Goldberg SR, Piomelli D. The endocannabinoid system in brain reward processes. Br J Pharmacol. 2008;154(2):369–83.
77. Pierce RC, Kumaresan V. The mesolimbic dopamine system: the final common pathway for the reinforcing effect of drugs of abuse? Neurosci Biobehav Rev. 2006;30(2):215–38.
78. Fattore L, Deiana S, Spano SM, Cossu G, Fadda P, Scherma M, Fratta W. Endocannabinoid system and opioid addiction: behavioural aspects. Pharmacol Biochem Behav. 2005;81(2):343–59.
79. Solinas M, Goldberg SR. Motivational effects of cannabinoids and opioids on food reinforcement depend on simultaneous activation of cannabinoid and opioid systems. Neuropsychopharmacology. 2005;30(11):2035–45.
80. Maldonado R, Valverde O, Berrendero F. Involvement of the endocannabinoid system in drug addiction. Trends Neurosci. 2006;29(4):225–32.
81. Trezza V, Vanderschuren LJMJ. Cannabinoid and opioid modulation of social play behavior in adolescent rats: differential behavioral mechanisms. Eur Neuropsychopharmacol. 2008;18(7):519–30.
82. Trezza V, Vanderschuren LJMJ. Divergent effects of anandamide transporter inhibitors with different target selectivity on social play behavior in adolescent rats. J Pharmacol Exp Ther. 2009;328(1):343–50.
83. Spano MS, Fadda P, Fratta W, Fattore L. Cannabinoid-opioid interactions in drug discrimination and self-administration: effect of maternal, postnatal, adolescent and adult exposure to the drugs. Curr Drug Targets. 2010;11(4):450–61.
84. Shapira M, Vogel Z, Sarne Y. Opioid and cannabinoid receptors share a common pool of GTP-binding proteins in cotransfected cells, but not in cells which endogenously coexpress the receptors. Cell Mol Neurobiol. 2000;20(3):291–304.
85. Hojo M, Sudo Y, Ando Y, Minami K, Takada M, Matsubara T, Kanaide M, Taniyama K, Sumikawa K, Uezono Y. mu-Opioid receptor forms a functional heterodimer with cannabinoid CB1 receptor: electrophysiological and FRET assay analysis. J Pharmacol Sci. 2008;108(3):308–19.
86. Salio C, Fischer J, Franzoni MF, Mackie K, Kaneko T, Conrath M. CB1-cannabinoid and mu-opioid receptor co-localization on postsynaptic target in the rat dorsal horn. Neuroreport. 2001;12(17):3689–92.
87. Fattore L, Spano M, Melis V, Fadda P, Fratta W. Differential effect of opioid and cannabinoid receptor blockade on heroin-seeking reinstatement and cannabinoid substitution in heroin-abstinent rats. Br J Pharmacol. 2011;163(7):1550–62.

88. Scavone JL, Sterling RC, Van Bockstaele EJ. Cannabinoid and opioid interactions: implications for opiate dependence and withdrawal. Neuroscience. 2013;248:637–54.
89. Christie MJ. Opioid and cannabinoid receptors: friends with benefits or just close friends? Br J Pharmacol. 2006;148(4):385–6.
90. Corchero J, Oliva JM, Garcia-Lecumberri C, Martin S, Ambrosio E, Manzanares J. Repeated administration with Delta9-tetrahydrocannabinol regulates mu-opioid receptor density in the rat brain. J Psychopharmacol. 2004;18(1):54–8.
91. Marcellino D, Carriba P, Filip M, Borgkvist A, Frankowska M, Bellido I, Tanganelli S, Müller CE, Fisone G, Lluis C, Agnati LF, Franco R, Fuxe K. Antagonistic cannabinoid CB1/dopamine D2 receptor interactions in striatal CB1/D2 heteromers. A combined neurochemical and behavioral analysis. Neuropharmacology. 2008;54(5):815–23.
92. Blanchard RJ, Blanchard DC, Agullana R, Weiss SM. Twenty-two kHz alarm cries to presentation of a predator, by laboratory rats living in visible burrow systems. Physiol Behav. 1991;50(5):967–72.
93. Kaltwasser MT. Acoustic signaling in the black rat (Rattus rattus). J Comp Psychol. 1990;104(3):227–32.
94. Sales GD. Ultrasound and aggressive behaviour in rats and other small mammals. Anim Behav. 1972;20(1):88–100.
95. Covington HE, 3rd, Miczek KA. Vocalizations during withdrawal from opiates and cocaine: possible expressions of affective distress. Eur J Pharmacol. 2003;467(1–3):1–13.
96. Vivian JA, Farrell WJ, Sapperstein SB, Miczek KA. Diazepam withdrawal: effects of diazepam and gepirone on acoustic startle-induced 22 kHz ultrasonic vocalizations. Psychopharmacology (Berl). 1994;114(1):101–8.
97. Sadananda M, Wohr M, Schwarting RK. Playback of 22-kHz and 50-kHz ultrasonic vocalizations induces differential c-fos expression in rat brain. Neurosci Lett. 2008;435(1):17–23.
98. Bang SJ, Allen TA, Jones LK, Boguszewski P, Brown TH. Asymmetrical stimulus generalization following differential fear conditioning. Neurobiol Learn Mem. 2008;90(1):200–16.
99. Endres T, Widmann K, Fendt M. Are rats predisposed to learn 22 kHz calls as danger-predicting signals? Behav Brain Res. 2007;185:69–75.
100. Kim EJ, Kim ES, Covey E, Kim JJ. Social transmission of fear in rats: the role of 22-kHz ultrasonic distress vocalization. PLoS One. 2010;5(12):e15077.
101. Jelen P, Soltysik S, Zagrodzka J. 22-kHz ultrasonic vocalization in rats as an index of anxiety but not fear: behavioral and pharmacological modulation of affective state. Behav Brain Res. 2003;141(1):63–72.
102. Sanchez C. R-citalopram attenuates anxiolytic effects of escitalopram in a rat ultrasonic vocalisation model. Eur J Pharmacol. 2003;464(2–3):155–8.
103. Roche M, O'Connor E, Diskin C, Finn DP. The effect of CB(1) receptor antagonism in the right basolateral amygdala on conditioned fear and associated analgesia in rats. Eur J Neurosci. 2007;26(9):2643–53.
104. Wohr M, Borta A, Schwarting RK. Overt behavior and ultrasonic vocalization in a fear conditioning paradigm: a dose-response study in the rat. Neurobiol Learn Mem. 2005;84(3):228–40.
105. Varga B, Kassai F, Gyertyan I. Interactions of CB1 and mGlu5 receptor antagonists in food intake, anxiety and memory models in rats. Pharmacol Biochem Behav. 2012;103(2):425–30.
106. Borta A, Wohr M, Schwarting RK. Rat ultrasonic vocalization in aversively motivated situations and the role of individual differences in anxiety-related behavior. Behav Brain Res. 2006;166(2):271–80.
107. Yee N, Schwarting RK, Fuchs E, Wohr M. Juvenile stress potentiates aversive 22-kHz ultrasonic vocalizations and freezing during auditory fear conditioning in adult male rats. Stress. 2012;15(5):533–44.
108. Tsoory M, Cohen H, Richter-Levin G. Juvenile stress induces a predisposition to either anxiety or depressive-like symptoms following stress in adulthood. Eur Neuropsychopharmacol. 2007;17(4):245–56.

109. Tsoory M, Guterman A, Richter-Levin G. Exposure to stressors during juvenility disrupts development-related alterations in the PSA-NCAM to NCAM expression ratio: potential relevance for mood and anxiety disorders. Neuropsychopharmacology. 2008;33(2):378–93.
110. Panksepp JB, Lahvis GP. Social reward among juvenile mice. Genes Brain Behav. 2007;6(7):661–71.
111. Schwarting RK, Jegan N, Wohr M. Situational factors, conditions and individual variables which can determine ultrasonic vocalizations in male adult Wistar rats. Behav Brain Res. 2007;182(2):208–22.
112. Knutson B, Burgdorf J, Panksepp J. Anticipation of play elicits high-frequency ultrasonic vocalizations in young rats. J Comp Psychol. 1998;112(1):65–73.
113. Willey AR, Spear LP. Development of anticipatory 50 kHz USV production to a social stimuli in adolescent and adult male Sprague-Dawley rats. Behav Brain Res. 2012;226(2):613–8.
114. Willey AR, Varlinskaya EI, Spear LP. Social interactions and 50 kHz ultrasonic vocalizations in adolescent and adult rats. Behav Brain Res. 2009;202(1):122–9.
115. Willey AR, Spear LP. The effects of pre-test social deprivation on a natural reward incentive test and concomitant 50 kHz ultrasonic vocalization production in adolescent and adult male Sprague-Dawley rats. Behav Brain Res. 2013;245:107–12.
116. Varlinskaya EI, Spear LP. Acute effects of ethanol on social behavior of adolescent and adult rats: role of familiarity of the test situation. Alcohol Clin Exp Res. 2002;26(10):1502–11.
117. Varlinskaya EI, Spear LP. Social interactions in adolescent and adult Sprague-Dawley rats: impact of social deprivation and test context familiarity. Behav Brain Res. 2008;188(2):398–405.
118. Douglas LA, Varlinskaya EI, Spear LP. Rewarding properties of social interactions in adolescent and adult male and female rats: impact of social versus isolate housing of subjects and partners. Dev Psychobiol. 2004;45(3):153–62.
119. Manduca A, Campolongo P, Palmery M, Vanderschuren LJMJ, Cuomo V, Trezza V. Social play behavior, ultrasonic vocalizations and their modulation by morphine and amphetamine in Wistar and Sprague-Dawley rats. Psychopharmacology (Berl). 2014;231:1661–73.
120. Barker DJ, Root DH, Ma S, Jha S, Megehee L, Pawlak AP, West MO. Dose-dependent differences in short ultrasonic vocalizations emitted by rats during cocaine self-administration. Psychopharmacology (Berl). 2010;211(4):435–42.
121. Thompson B, Leonard KC, Brudzynski SM. Amphetamine-induced 50 kHz calls from rat nucleus accumbens: a quantitative mapping study and acoustic analysis. Behav Brain Res. 2006;168(1):64–73.
122. Simola N, Fenu S, Costa G, Pinna A, Plumitallo A, Morelli M. Pharmacological characterization of 50-kHz ultrasonic vocalizations in rats: comparison of the effects of different psychoactive drugs and relevance in drug-induced reward. Neuropharmacology. 2012;63(2):224–34.
123. Maier EY, Ma ST, Ahrens A, Schallert TJ, Duvauchelle CL. Assessment of ultrasonic vocalizations during drug self-administration in rats. J Vis Exp. 2010;22(41):2041, p. ii. doi: 10.3791/2041.
124. Hamed A, Taracha E, Szyndler J, Krzascik P, Lehner M, Maciejak P, Skórzewska A, Płaźnik A. The effects of morphine and morphine conditioned context on 50 kHz ultrasonic vocalisation in rats. Behav Brain Res. 2012;229(2):447–50.
125. Sadananda M, Natusch C, Karrenbauer B, Schwarting RK. 50-kHz calls in rats: effects of MDMA and the 5-HT(1A) receptor agonist 8-OH-DPAT. Pharmacol Biochem Behav. 2012;101(2):258–64.
126. Wright JM, Deng L, Clarke PB. Failure of rewarding and locomotor stimulant doses of morphine to promote adult rat 50-kHz ultrasonic vocalizations. Psychopharmacology (Berl). 2012;224(4):477–87.
127. Wang H, Liang S, Burgdorf J, Wess J, Yeomans J. Ultrasonic vocalizations induced by sex and amphetamine in M2, M4, M5 muscarinic and D2 dopamine receptor knockout mice. PLoS One. 2008;3(4):e1893.
128. Sewell GD. Ultrasound in adult rodents. Nature. 1967;215(5100):512.

129. Roullet FI, Wohr M, Crawley JN. Female urine-induced male mice ultrasonic vocalizations, but not scent-marking, is modulated by social experience. Behav Brain Res. 2011;216(1):19–28.
130. Whitney G, Alpern M, Dizinno G, Horowitz G. Female odors evoke ultrasounds from male mice. Anim Learn Behav. 1974;2(1):13–8.
131. Jamain S, Radyushkin K, Hammerschmidt K, Granon S, Boretius S, Varoqueaux F, Ramanantsoa N, Gallego J, Ronnenberg A, Winter D, Frahm J, Fischer J, Bourgeron T, Ehrenreich H, Brose N. Reduced social interaction and ultrasonic communication in a mouse model of monogenic heritable autism. Proc Natl Acad Sci U S A. 2008;105(5):1710–5.
132. Radyushkin K, Hammerschmidt K, Boretius S, Varoqueaux F, El-Kordi A, Ronnenberg A, Winter D, Frahm J, Fischer J, Brose N, Ehrenreich H. Neuroligin-3-deficient mice: model of a monogenic heritable form of autism with an olfactory deficit. Genes Brain Behav. 2009;8(4):416–25.
133. Trezza V, Campolongo P. The endocannabinoid system as a possible target to treat both the cognitive and emotional features of post-traumatic stress disorder (PTSD). Front Behav Neurosci. 2013;7:100.
134. Marsicano G, Wotjak CT, Azad SC, Bisogno T, Rammes G, Cascio MG, Hermann H, Tang J, Hofmann C, Zieglgänsberger W, Di Marzo V, Lutz B. The endogenous cannabinoid system controls extinction of aversive memories. Nature. 2002;418(6897):530–4.
135. Chhatwal JP, Ressler KJ. Modulation of fear and anxiety by the endogenous cannabinoid system. CNS Spectr. 2007;12(3):211–20.
136. Chhatwal JP, Davis M, Maguschak KA, Ressler KJ. Enhancing cannabinoid neurotransmission augments the extinction of conditioned fear. Neuropsychopharmacology. 2005;30(3):516–24.
137. Mikics E, Dombi T, Barsvari B, Varga B, Ledent C, Freund TF, Haller J. The effects of cannabinoids on contextual conditioned fear in CB1 knockout and CD1 mice. Behav Pharmacol. 2006;17(3):223–30.
138. Campolongo P, Roozendaal B, Trezza V, Hauer D, Schelling G, McGaugh JL, Cuomo V. Endocannabinoids in the rat basolateral amygdala enhance memory consolidation and enable glucocorticoid modulation of memory. Proc Natl Acad Sci U S A. 2009;106(12):4888–93.
139. Davis M, Whalen PJ. The amygdala: vigilance and emotion. Mol Psychiatry. 2001;6(1):13–34.
140. Arnold JC, Dielenberg RA, McGregor IS. Cannabinoids increase conditioned ultrasonic vocalisations and cat odour avoidance in rats: strain differences in drug-induced anxiety. Life Sci. 2010;87(17–18):572–8.
141. Butler RK, Ford GK, Hogan M, Roche M, Doyle KM, Kelly JP, Kendall DA, Chapman V, Finn DP. Fear-induced suppression of nociceptive behaviour and activation of Akt signalling in the rat periaqueductal grey: role of fatty acid amide hydrolase. J Psychopharmacol. 2012;26(1):83–91.
142. Manduca A, Servadio M, Campolongo P, Palmery M, Trabace L, Vanderschuren LJMJ, Cuomo V, Trezza V. Strain- and context-dependent effects of the anandamide hydrolysis inhibitor URB597 on social behavior in rats. Eur Neuropsychopharmacol. 2014;24(8):1337–48.
143. Fride E, Gobshtis N, Dahan H, Weller A, Giuffrida A, Ben-Shabat S. The endocannabinoid system during development: emphasis on perinatal events and delayed effects. Vitam Horm. 2009;81:139–58.
144. Schneider M. Cannabis use in pregnancy and early life and its consequences: animal models. Eur Arch Psychiatry Clin Neurosci. 2009;259(7):383–93.

Chapter 11
Age-Dependent Effects of Cannabinoids on Neurophysiological, Emotional, and Motivational States

María-Paz Viveros and Eva María Marco

Abstract *Cannabis sativa* preparations are among the illicit drugs most commonly used by young people, including pregnant women. The endocannabinoid (eCB) system, which is involved in the regulation of emotional and motivational homeostasis, synaptic plasticity and cognitive functions, also plays a critical role in diverse phases of brain development. Both perinatal and periadolescent periods are critical for brain eCB system development. Thus, interference of endocannabinoid signalling by cannabis exposure may contribute to explain the enduring negative impact of cannabis on neurodevelopmental processes and the resulting psycho-physio-pathological consequences. In the present chapter we describe and discuss published data dealing with the long-term neurobehavioural effects of cannabis exposure during the prenatal and adolescent periods. Human studies have demonstrated that marijuana consumption by pregnant women critically affects the neurobehavioural development of their children. Investigations using animal models provide useful information for a better understanding of the long-lasting deleterious consequences of cannabis exposure during pregnancy and lactation. Increasing use of cannabis among adolescents is a matter of great public concern that has led to a parallel increase in research on appropriate animal models. Chronic administration of cannabinoid agonists during the periadolescent period causes persistent behavioural alterations related to cognitive deficits, increased risk of psychosis, mood disorders and addiction to cannabis and other drugs of abuse. The underlying mechanisms by which cannabis use may lead to these disorders, including genetic vulnerability and the increasing content of the main psychoactive ingredient in cannabis preparations, delta-9-tetrahydrocannabinol (THC), will be discussed. To conclude, prevention and therapeutic strategies based on scientific knowledge will be proposed.

Keywords Cannabis · Critical age periods · Perinatal · Adolescence · Development · Endocannabinoid system

---

M.-P. Viveros (✉) · E. M. Marco
Departamento de Fisiología (Fisiología Animal II), Facultad de Biología,
Universidad Complutense de Madrid (UCM), Madrid, Spain
e-mail: pazviver@bio.ucm.es

P. Campolongo, L. Fattore (eds.), *Cannabinoid Modulation of Emotion, Memory, and Motivation,* DOI 10.1007/978-1-4939-2294-9_11

Introduction

Cannabis contains psychoactive components, mainly Δ9-Tetrahydrocannabinol (THC), which interfere with the brain's endogenous cannabinoid system (endocannabinoid, eCB, system) through the activation of the cannabinoid type 1 (CB1) and type 2 (CB2) receptors. The eCB system plays a relevant regulatory role in a wide variety of functions; eCB signalling is critically involved not only in processes of synaptic plasticity but also in cognitive functions, motivation, and regulation of emotional homeostasis [1–7]. The eCB system plays a crucial role in diverse phases of brain development [8–11]. Interference of eCB signalling by cannabis exposure during the perinatal and the adolescent periods may contribute to explain the enduring negative impact of cannabis on neurodevelopmental processes and the resulting psycho-physio-pathological consequences [3, 12–19].

The main feature of the recreational use of cannabis is a euphoric effect. This "high" can be accompanied by decreased anxiety and increased sociability. However, acute aversive emotional reactions such as feelings of anxiety, panic and paranoia have also been reported [15, 20]. Dependence on cannabis consumption has been reported and an associated withdrawal syndrome has been described [21–23]. Cannabis withdrawal syndrome includes anxiety and nervousness, craving, decreased appetite and weight loss, restlessness, sleep difficulties, strange dreams, chills, depressed mood, stomach pain, physical discomfort, shakiness, and sweating [21, 24, 25]. The syndrome has a transient course after cessation of cannabis use and is pharmacologically specific. Cannabis withdrawal is reported by up to one-third of regular cannabis users in the general population, and by 50–95 % of heavy users in treatment. The clinical relevance of cannabis withdrawal is demonstrated by the use of cannabis or other substances to relieve its symptoms, by the reports of difficulty in quitting, and by the worsening of treatment outcomes in association with greater withdrawal severity.

Marijuana has been associated with disrupted functioning in a variety of cognitive and performance tasks, and chronic marijuana smoking has been reported to cause persistent memory deficiencies [26, 27]. In addition, pharmacological studies have shown that cannabinoids can induce a full range of transient positive, negative, and cognitive symptoms in healthy individuals that are similar to those seen in schizophrenia. Despite most of the current research has focused on the effects of cannabis on psychosis and schizophrenia, there is also increasing evidence indicating a close relationship between cannabis consumption and an increased risk for depression, anxiety disorders, and drug addiction [13, 15, 28–30].

Cannabis preparations are the illicit drugs most widely used by young people, peaking between 15 and 30 years of age, although a trend has been reported for continued cannabis use in people aged 30–40. Growing evidence from human and animal studies suggests a differential effect of cannabis exposure depending on the age of exposure [31]. In this chapter, we will pay special attention to two periods that appear to be of special vulnerability, i.e. the perinatal and the adolescent period. Substance use by pregnant women poses significant risks to the unborn child. Accumulating evidence from both human and preclinical studies indicates that maternal substance use during pregnancy can affect foetal development, birth weight and infant outcomes. Thus, the prenatal period can be regarded as an important sensitive period of development [16, 32]. Actually, cannabis is the most commonly used illicit substance among pregnant women, and given the lipophilic nature of THC, it is estimated that one-third of THC in the plasma crosses the foetus-placental barrier [33, 34]. Moreover, THC is secreted through the breast milk [35]. Therefore, it is plausible that THC can easily reach the developing foetal brain. In fact, human epidemiological and animal studies have found that prenatal/perinatal cannabis exposure influences brain development and can have long-lasting impacts on cognitive functions and other behavioural aspects, notably reward and emotional responses [12, 16, 36–39].

In this chapter we will consider adolescence as the gradual period of transition from childhood to adulthood, including pubertal maturation. Adolescence represents a developmental period of unique plasticity during which the brain is particularly sensitive to environmental insults such as stress and drugs of abuse. There is evidence indicating that during this sensitive period exposure to drugs may have a greater impact on neurocognition compared to adult exposure [40]. A "window of vulnerability" appears to exist during the adolescent period regarding the onset of certain neuropsychiatric disorders such as schizophrenia and the effects of drugs of abuse [14, 17, 18, 41, 42]. In particular, both human and animal studies indicate that cannabis use during adolescence may produce cognition impairments [26, 27] and depressive symptoms, and may increase the risk to develop psychiatric and substance abuse disorders [3, 18, 39, 43]. We will also discuss a number of factors of vulnerability to the harmful effects of cannabis such as the age of starting to use cannabis, the degree of cannabis exposure and genetic susceptibility, as well as the composition of the cannabis plant consumed.

Developmental Aspects of the Endocannabinoid System

During early phases of neuronal development, eCB signalling is integral for an array of processes including proliferation and differentiation of progenitor cells, neuronal migration, axonal guidance, fasciculation, positioning of cortical interneurons, neurite outgrowth and morphogenesis. At early developmental stages, the eCB system seems to both influence the appearance of key cellular signals and modify the expression of genes that are relevant for neural development [8–12, 44]. Both CB1 receptors and eCB ligands can be detected in the rat [45, 46] and human [47] brain during early developmental periods. Moreover, stimulation of [^{35}S]GTP gamma-S binding by cannabinoid agonists suggests that embryonic CB1 receptors are already functional [48]. During the perinatal period, a common atypical pattern of CB1 receptor expression has been found both in rodents and humans, with high densities of CB1 receptors observed in fibre-enriched areas that are practically devoid of them in the adult brain. This transient pattern of CB1 receptor localization in white matter areas during the prenatal stages suggests a specific role of the eCB system in neural development, which may be important for guidance processes that result in the establishment of cortical-subcortical connections [45–47]. Gaffuri et al. [44] have recently reviewed current knowledge about the effects of CB1 receptor signalling during different phases of brain development, i.e. migration and differentiation of progenitor cells, neurite outgrowth, axonal path finding and synaptogenesis. Authors highlighted the eCB signalling as dependent upon the diacylglycerol lipases (DAGLs), the enzymes responsible for the synthesis of the endocannabinoid 2-arachidonoylglycerol (2-AG). DAGL-dependent eCB signalling regulates axonal growth and guidance during development, and is required for the generation and migration of new neurons in the adult brain. It is now clear that DAGLs and CB1 receptors can modulate growth cone dynamics in vitro, and that they are expressed in advancing growth cones during development likely playing a key role in axonal growth and guidance in vivo. In the same growth cone, 2-AG acts upon CB1 receptors to promote motility [49]. The importance of the eCB system during early developmental periods is further supported by the aberrations that occur following disruption of normal eCB signalling during ontogenetic phases. For example, pharmacological blockade of the CB1 receptor in mid-to-late gestational periods impairs progenitor proliferation in the subventricular zone, disrupts axonal path finding and results in cortical delamination [50]. In turn, *in utero* exposure to THC hampers appropriate interneuron positioning during corticogenesis and results in increased density of cholecystokinin-positive (CCK^+) interneurons in the hippocampus [51].

Development of the eCB system continues during adolescence. In humans, expression patterns of CB1 receptors have been found to increase dramatically from infancy to young adulthood, in regions such as the frontal cortex, striatum and hippocampus [47]. Rodent studies have provided further time- and region-specific data. Ontogeny of cannabinoid receptors in rat striatum, limbic forebrain and ventral mesencephalon is relatively similar, exhibiting a progressive increase that peaks on postnatal days 30 or 40 and then subsequently decrease to adult values [46]. In animal models, the content of the endocannabinoid N-arachidonoylethanolamine (anandamide, AEA) has been observed to gradually increase during early postnatal stages, reaching its maximum in the adolescent brain [8]. Similarly, in rat brain CB1 receptors exhibit a largely postnatal pattern of development, reaching maximal densities during adolescence which later drop to adult expression levels, as detected in the dorsal striatum [45, 46]. Whereas most data available in the literature refer to expression of protein or mRNA for brain CB1 receptors, it would be extremely interesting to examine the developmental changes of CB1 receptor functional activity throughout these critical developmental periods. In the female rat hypothalamus, AEA levels are seen to peak at the onset of puberty and then decline into adulthood [52]. More recent studies have revealed clear developmental fluctuations throughout adolescence in eCB levels in diverse brain regions involved in reward, motivation, and cognition. The most profound alteration was the continuous increase in prefrontal cortex (PFC) of AEA levels throughout the adolescent period; concentrations were almost three times higher in late than early adolescence [53]. However, 2-AG concentrations were lower in the PFC in the later phases than in the beginning of the adolescent period, a finding paralleled within the nucleus accumbens (NAc).

In addition, CB1 receptors were found to vary in the PFC and NAc core during the different phases of adolescence, although the alterations were less marked than for eCB levels. These findings emphasize dynamic alterations in eCB function in mesocorticolimbic regions of the adolescent brain that are relevant to reward and, to a greater extent, to cognition and emotional learning, and underscore the specific association of the eCB system with neurodevelopment, not only for the perinatal period but also during adolescence [53]. Lee et al. [54] have further characterized temporal changes in N-acylethanolamine (NAE) content and fatty acid amide hydrolase (FAAH) activity across the periadolescent period, in PFC, amygdala, hippocampus, and hypothalamus. Four developmental points were analysed, specifically postnatal days (pnd) 25, 35, 45, and 70, representing respectively pre-adolescence, early- to mid-adolescence, late adolescence, and adulthood. The observed age-dependent patterns of NAE content and FAAH activity further demonstrate temporal specificity in the development of the system that could contribute to alterations in stress sensitivity, emotionality, and executive functions which also fluctuate during this developmental period.

Another aspect that deserves further investigation is the possible existence of sex differences in developmental patterns. In the developmental study quoted above Rodriguez de Fonseca et al. [46], found subtle sexual dimorphisms in the rat striatum and ventral mesencephalon but not the limbic forebrain. At pnd 43, subtle differences in the expression of hippocampal CB1 receptors were found, with female rats showing lower cannabinoid CB1 receptor density when compared with males [55]. Moreover, clear sex differences in the expression and functionality of hippocampal CB1 receptors are also evidenced in adult rats. Male rats show higher levels of hippocampal CB1 receptor expression than females [56], which in turn exhibit a pattern of higher CB1 receptor-mediated G protein activation in hippocampus when compared to males [57]. Thus, it seems likely that sexual differences in CB1 receptor expression (at least in certain regions such as the hippocampus) are established beyond pnd 40. Interestingly, however, diverse kinds of stress exert differential effects on hippocampal CB1 receptor expression of male and female rats in both adult [56] and 13-day-old neonate animals [58], suggesting a role for organizational effects of gonadal steroids during the perinatal period. The sexual dimorphism observed in the eCB system may contribute to explain the sex differences observed in cannabinoid-induced behavioural alterations (see Chap. 13).

Despite CB2 receptor was initially claimed as a peripheral cannabinoid receptor, it has been detected in a diversity of brain regions including cerebral cortex, hippocampus, amygdala, hypothalamus, and cerebellum, thus suggesting a role for CB2 receptors in emotional and cognitive function [15]. There is also evidence supporting a role of CB2 receptor in neural development [59, 60]. It would be highly interesting to characterize the developmental pattern of CB2 receptors expression and functionality, as well as to investigate on possible interactions between CB1 and CB2 receptors during brain development.

To sum up, in both the rodent and the human foetal brain, cannabinoid receptors are present from early developmental stages onwards. Moreover, there is evidence that the eCB system has a central signalling role in brain development of rodents.

Endocannabinoid signalling modulates fundamental developmental processes such as cell proliferation, neurogenesis, migration and axonal path finding, and undergoes important changes and fluctuations through the perinatal and the adolescent periods. Therefore, it is plausible that exposure to exogenous cannabinoids during brain development and/or adolescence may impact the normal developmental course, and lead to adverse outcomes [12, 16, 31].

Long-term Effects of Chronic Cannabinoid Exposure During the Perinatal Period

Studies on the effects of cannabinoids in humans have demonstrated that the consumption of marijuana by women during pregnancy affects the neurobehavioural development of their children. While human studies on long-term neurobehavioural effects of drugs of abuse usually include a number of confounding factors that do not allow to control for potentially important environmental factors, preclinical studies allow a tight control of environmental variables and provides insights about potential mechanisms through which prenatal cannabinoid exposure may exert its impact on the developing foetus. Perinatal exposure to THC or synthetic cannabinoid agonists has been shown to induce long-term effects on diverse parameters (see Table 11.1).

Cannabis and Cognitive Deficit

In a recent review on longitudinal cohort studies, Wu et al. [12] reported that cannabis consumption during pregnancy has profound but variable effects on the offspring in several areas of cognitive development, and suggested an association between maternal cannabis use and impaired high-order cognitive function in the offspring. Maternal cannabis use during pregnancy has also been associated with growth restriction in mid and late pregnancy, and with lower body weight at birth, while similar associations were not found for paternal cannabis use during the reproductive period, demonstrating a direct biological effect of maternal intrauterine exposure to cannabis on foetal growth.

Executive functions refer to higher-order cognitive functions such as cognitive flexibility, sustained and focused attention, planning and working memory: prenatal marijuana exposure exerts a negative effect on these functions [31]. For instance, several reports by Fried and co-workers indicate that cannabis has a negative effect on self-regulatory abilities, including tasks that require impulse control, and is associated with deficits in sustained attention and visual memory, analysis and integration [61–64]. By using functional magnetic resonance imaging (fMRI), Smith et al. [65] investigated the long-lasting neurophysiological effects of prenatal marijuana exposure on visuospatial working memory in 18–22 years old young adults. The study revealed that prenatal marijuana exposure alters neural functioning during

Table 11.1 Long-term effects of perinatal cannabinoid exposure on adolescent and adult emotional behaviour. Data extracted from [37–39]

Agonist	Treatment age window	Animals tested	Test	Effect	Reference
Locomotion and exploration					
THC (5 mg/kg, p.o.)	G5–PND 24	Male and oestrous female adult (>PND 70) Wistar rats	Actimeter	Psychomotor activation in adult males and adult oestrous females	[71]
THC (5 mg/kg, p.o.)	G5–PND 24	Male and oestrous female adult (>PND 70) Wistar rats	locomotor activity	Not affected	[73]
THC (0.1, 0.5, 2 mg/kg, p.o.)	GD 5–PND 24	Male and female—ovariecto-mized—adult (>PND 70) Wistar rats	OF	Reduced activity. However, the effects of the highest dose were more accentuated in males but disappeared in females	[72]
WIN55,212–2 (0.5 mg/kg, sc)	G5–G20	Male infant (PND 12), adolescent (PND 40) and adult (PND 80) Wistar rats	Motor activity	Hyperactive behaviour during early postnatal life and adolescence, not at adulthood	[66]
WIN55,212–2 (1.2 mg/kg, i.p.)	PND 15–PND 40	Male adult (>PND 75) Wistar rats	OF	Reduced rearing frequency, and decreased the time spent in the centre of the arena, i.e. increased emotionality	[76]
WIN55,212–2 (0.1 mg/kg, s.c.)	PD 10	Male and females adolescent (PND 30–34) Wistar rats	HB	Reduced locomotor activity	[75]
THC (2 mg/kg, s.c.)	GD 1–GD 22 PD 2–PND 10	Male adult (PND 90) Long Evans rats	OF	No changes in motor activity, but a decrease in activity specific to the inner section, i.e. increased emotionality	[74]
Anxiety-like responses, emotional effects and social behaviour					
THC (5 mg/kg, p.o.)	G5–PND24	Male and oestrous female adult (>PND 70) Wistar rats	EPM	Increased exploratory behaviour of the open arms, only in males	[71]
THC (5 mg/kg, p.o.)	G5–PND24	Male and oestrous female adult (>PND 70) Wistar rats	SI	Increased active social interaction in males. Males exhibited an altered exploration pattern in the socio-sexual approach	[73]
			LD	No effects	

Table 11.1 (continued)

Agonist	Treatment age window	Animals tested	Test	Effect	Reference
CP 55,940 (increasing doses: 0.15, 0.20 and 0.30 mg/kg)	PND 4–PND 25	Male young adult (PND 53–62) Wistar rats	SI	Reduced social interaction	[70]
WIN55,212–2 (0.1 mg/kg, s.c.)	PND 10	Male and females adolescent (PND 30–34) Wistar rats	EPM	Anxiogenic-like effects among male animals	[75]
			FST	Depressive-like behaviour in females (a similar trend observed among males)	
THC (2.5 and 5 mg, p.o.)	GD 15–PND 9	Male adolescent (PND 35) and adult (PND 80) Wistar rats	SI	The highest THC dose inhibited adolescent social interaction and play behaviour	[77]
			EPM	The highest THC dose induced an anxiogenic-like profile at adulthood	
THC (2 mg/kg, s.c.)	GD 1–GD 22 PD 2–PND 10	Male adult (PND 90) Long Evans rats	SI	Subtle effects, only an increase in social sniffing, i.e. investigation	[74]
			FST	No effects	
Cognitive function					
WIN55,212–2 (0.5 mg/kg, sc)	G5–G20	Male adolescent (PND40) and adult (PND80) Wistar rats	Passive avoidance	Disruption in memory retention at adolescence and adulthood	[66]
THC (5 mg/kg, s.c.)	PND 4–PND 14	Male adult (>PND 56, for 5 weeks) Wistar rats	Y-maze test	no effects in the spatial discrimination task (spatial memory)	[69]
				Impairments in the delayed alternation task (working memory)	
WIN55,212–2 (0.5 mg/kg, s.c.)	GD 5–GD 20	Male adult (>PND 80) Wistar rats	Active avoidance	Impairment in acquisition	[67]

Table 11.1 (continued)

Agonist	Treatment age window	Animals tested	Test	Effect	Reference
WIN55,212–2 (1.2 mg/kg, i.p.)	PND 15–PND 40	Male adult (>PND 75) Wistar rats	PPI	Sensorimotor gating deficits	[76]
			NOR	No effect on object recognition memory	
CP 55,940 (increasing doses: 0.15, 0.20 and 0.30 mg/kg)	PND 4–PND 25	Male young adult (PND 53–62) Wistar rats	NOR	Impaired working memory	[70]
THC (5 mg/kg, p.o.)	GD 15–PND 9	Male adult (>PND 80) Wistar rats	Inhibitory avoidance	Decrease ability to remember the task, indicative of long-term memory impairment	[68]
			Social discrimination task	Reduced investigation of a novel partner, indicative of a disruption of the short-term olfactory memory	
Motivation and drug abuse					
THC (5 mg/kg, p.o.)	G5–PND24	Male and oestrous female adult (>PND 70) Wistar rats	CPP Morphine	Enhanced sensitivity to the rewarding properties of morphine	[71]
THC (5 mg/kg, p.o.)	G5–PND24	Male and females adult (>PND 70) Wistar rats	SA-FR1 Morphine	Increase in intravenous morphine self-administration in females (subtle trend among males), suggestive of a higher susceptibility to opiate reinforcing effects	[78]
THC (5 mg/kg, p.o.)	G5–PND24	Male and female adult (>PND 70) Wistar rats	SA-PR Morphine/Food	No effect on morphine or food self-administration, suggesting no changes in the reinforcing efficacy of these stimuli	[79]
THC (5 mg/kg, p.o.)	GD 15–PND 9	Male adult (>PND 60) Wistar rats	SA Ethanol	No effects on ethanol self-administration	[80]

Cannabinoid agonists: THC delta-9-tetrahydrocannabinol, G gestational day, PND postnatal day, OF open field, HB Holeboard, EPM elevated plus maze, SI social interaction, LD light-dark, FST forced swim test, PPI prepulse inhibition, NOR novel object recognition, CPP conditioned place preference, SA self-administration, FR fixed ratio, PR progressive ratio visuospatial working memory processing in young adulthood suggesting that deficits in executive functions induced by prenatal cannabis exposure are long-lasting.

In rodents, prenatal exposure to cannabinoid agonists has been reported to induce notable impairments in cognitive function (see Table 11.1), among which a disruption in memory retention in the passive avoidance task [66], an impairment in the active avoidance task [67], and a long-term impairment in the inhibitory avoidance and in a social discrimination task [68]. Similarly, postnatal administration of cannabinoid compounds induces impairments in working memory and object recognition [69, 70], confirming that the gestational and the perinatal age windows are critical periods for the adverse consequences of cannabis on cognition.

Emotional Long-Term Adverse Effects of Cannabis

Findings from animal studies are often controversial given the diversity of behavioural paradigms employed, the time windows investigated and the drug and dose range employed (see Table 11.1). However, in general, gestational exposure to cannabinoid agonists induces motor activation in rodents [66, 71], or no motor effects [72–74] whereas late postnatal exposure reduces locomotor and exploratory activity [75, 76]. The emotional consequences of perinatal cannabis exposure strictly depend upon the time of cannabinoid exposure. If cannabinoid agonists are administered during either the gestational or the early postnatal period, animals exhibit increased anxiety-related behaviour [74, 77] and inhibited social interaction and play behaviour at adolescence [77], as well as an increased exploratory behaviour in the elevated plus maze [71]. If cannabinoid agonist administration is prolonged until weaning, animals exhibit an increment in exploratory behaviour both in the elevated plus maze [71] and the social interaction test [73]. If cannabinoid agonists are administered during early postnatal life, animals are more anxious and prone to exhibit a depressive-like behaviour, consequences that seem to depend upon the sex of the animals [74, 75].

Early Cannabis Consumption and the Risk of Addiction to Other Drugs of Abuse

Regarding associated risk for drug addiction, studies in rodents have reported that perinatal exposure to THC induced in adult females, but not in adult males, an increase in the amount of morphine consumed in the self-administration paradigm under a fixed-ratio (FR-1) schedule of reinforcement [78]. However, perinatal THC exposure does not affect the reinforcing efficacy of morphine in a progressive ratio (PR) schedule of reinforcement [79]. Taken as a whole, these findings suggest that morphine is particularly preferred by adult females that had been exposed perinatally to THC, but that this vulnerability to morphine may disappear when animals are submitted to a higher requirement to obtain the drug. The possibility that perinatal

THC exposure induces sensitization to opiates has also been addressed by evaluating morphine place preference conditioning in the adult offspring. In this case, the results indicated that THC-exposed offspring of both sexes exhibited enhanced sensitivity to the rewarding effects of morphine [71]. Moreover, these changes in motivation for drugs seem to be specific for opioid consumption since no changes were observed when alcohol was self administered [80].

Neurobiological Mechanisms Underlying Perinatal Cannabinoid Exposure

Preclinical studies provide also insights about potential neurobiological mechanisms underlying perinatal cannabinoid exposure. For example, cannabinoid exposure in pregnant rats can affect the expression of key genes (e.g. related to the neural adhesion molecule L1) for foetal neural development, possibly resulting in neurotransmitter and behavioural disturbances [34]. The dopaminergic and the opioid systems appear to be markedly affected by perinatal cannabinoid administration. The effects on dopaminergic transmission have been widely studied [81]. It has been shown that perinatal exposure to THC affects the functionality of dopaminergic autoreceptors, inducing a greater sensitivity to the presynaptic actions of dopamine D2 receptor agonists [72]. With respect to the endogenous opioid system, perinatal treatment with THC induces a decrease in pain sensitivity and an increase in the tolerance to the analgesic effect of morphine in males [82]. Baseline opioid activity may be affected since females perinatally exposed to THC showed a decrease in proenkephalin gene expression in the caudate-putamen in adulthood [83]. This result may be related with the sexual dimorphism observed in morphine self-administration following perinatal THC exposure, i.e. females self-administer a higher amount of the drug [78].

The glutamatergic system has also been studied in both neurons and glial cells. Developmental THC exposure induces a decrease in the expression of glutamate receptors, which could lead to functional alterations through the inhibition of glutamatergic neurotransmission [84]. Prenatal exposure to WIN 55,212–2 (WIN) induces a remarkable memory impairment that is correlated with alterations in both long-term potentiation (LTP) and glutamate release in the hippocampus. The decrease in hippocampal glutamate outflow appears to be the cause of LTP disruption, which in turn might underlie, at least in part, the long-lasting impairment of cognitive functions caused by the gestational exposure to WIN [66]. Similarly, in a more recent study, Ferraro et al. [85] showed that the cognitive deficit induced by gestational exposure to cannabinoids is associated with alterations of cortical and hippocampal glutamate outflow, cortical neuron morphology and hippocampal long-term potentiation. As a whole, these data support the view that altered glutamate transmission might underlie, at least in part, some of the cognitive deficits affecting the offspring of marijuana users. Last but not least, prenatal THC exposure also affects cerebellar astroglial cells. Both glial fibrillary acidic protein (GFAP) and glutamine synthetase are decreased in astroglial cells not only during THC exposure but also at adult ages. Thus, cannabinoids may exert developmental toxicity not only on neurons but also on astroglial cells, which could contribute to foetal brain growth retardation [86]. In this respect, it is important to note that glial cells also express components of the cannabinoid signalling system and that marijuana-derived compounds act at cannabinoid receptors expressed on glial cells, affecting their functions [87].

Long-Term Effects of Chronic Cannabinoid Exposure During Adolescence

Adolescence is a period of intense growth, reshaping and maturation of grey and white matters in the human brain. This period involves neurocognitive, hormonal and psychosocial changes with considerable modifications in cognition, mood, arousal, motivation, sleeping patterns, personality, social interactions, behaviour and affection. In humans, the ages associated with adolescence are commonly considered to be approximately 12 to 20–25 years of age, whereas in rodents adolescence is considered within the time frame of 28 to 42 pnd. During this period, the brain undergoes radical functional alterations that are associated with a high degree of plastic structural remodelling. A key finding from structural MRI studies is that the volume of grey matter, which contains brain cell bodies and synapses, changes between childhood and adulthood. In the prefrontal cortex, grey matter volume increases during childhood, peaks in early adolescence, and then declines in late adolescence and throughout the twenties. The loss of grey matter during adolescence is thought to be due, at least partly, to synaptic pruning—the process by which excessive synapses are eliminated. This process of synaptic pruning that sculpts neuronal circuitry during critical periods of brain development is sensitive to environmental factors, including exposure to drugs of abuse [88]. Different brain regions have different peaks of maturation, and changes include modifications in the volume of grey and white matter [19].

Cannabis and Cognitive Deficit

The maturational processes that occur during adolescence are likely to confer a higher risk for suffering from adverse consequences of cannabinoid exposure [89]. Persistent cannabis use has been associated with important deficits in cognitive functions. One of the most important study to date on this topic examined the impact of regular marijuana use on intelligence quotient (IQ) and neuropsychological functioning in a longitudinal sample of 1,037 individuals followed from birth to age 38 [27]. Neuropsychological testing was conducted at 13 years old, before initiation of cannabis use, and again at age 38, after a pattern of persistent cannabis use had developed. Results indicated that persistent cannabis use is related to a broad neuropsychological decline across domains of functioning. Indeed, the statistically significant decline in cognitive ability was present even after controlling for years of education. The more persistent the cannabis use, the greater the cognitive decline. Remarkably, the association between persistent cannabis use and cognitive decline was significantly greater for early marijuana onset, i.e. people who began using cannabis before 18 years old. Converging lines of evidence suggest that regular use of marijuana starting before 18 years old is associated with poorer attention, increased deficits in visual search, reduced overall or verbal IQ, and executive functioning [40]. Moreover, if cannabis use started before 18 years, the cognitive deficit remained significant when people had stopped using for at least 1 year before testing. In line with these results Pope et al. [90] has reported that early onset cannabis users, i.e. people who began smoking before age 17, exhibit poorer cognitive performance, especially in verbal IQ, than late-onset users, i.e. people who began smoking at age ≥ 17 or later, or control subjects.

Cannabis and Psychiatric Disorders—A Focus on Schizophrenia

There is now evidence demonstrating an association between increased rates of cannabis use and new cases of schizophrenia. Epidemiological studies suggest a high incidence of schizophrenia within marijuana smokers, and long-term users of cannabis exhibit cognitive deficits similar to those seen in schizophrenia. A series of longitudinal studies in the general population have investigated the role of cannabis as a risk factor for schizophrenia. Overall, it has been found that cannabis-use approximately doubles the odds of developing schizophrenia [28]. Importantly, there appears to be a dose-response relationship, so that the more extensive the use of cannabis the higher the risk. For example Zammit et al. [91] reported that heavy cannabis users were six times more likely than non-users to subsequently receive a diagnosis of schizophrenia, while DiForti et al. [92] found a clear relationship between the frequency of cannabis use and development of a psychotic illness. Importantly, cannabis has been considered a risk factor for development or worsening of schizophrenia, and there is evidence indicating that young people at genetic high risk of schizophrenia are particularly vulnerable to mental health problems associated with cannabis use. Cannabis use has been associated with a decrease in age of onset of schizophrenia, frequently related with a poorer outcome. Moreover, cannabis-using patients experience more positive symptoms and frequency of relapse and hospitalization and respond poorly to antipsychotic medication [3, 13–15, 28, 92]. However, the ultimate proof of a causal relationship between cannabis use and psychotic illness later in life would come from studies in which healthy young people were exposed to THC and followed-up until adulthood. Obviously, for practical and ethical reasons, such an approach is impossible.

Among many other important health risks, it is well known that cannabis induces harmful effects on cognitive function. While any animal model cannot represent the full phenotypic spectrum of a psychiatric disorder, such as schizophrenia or depression, specific phenotypic components of disorders can be used to construct adequate animal models that may be useful to investigate disease mechanisms and that may allow testing novel interventions. Such studies can be performed in animals under well-controlled conditions and allow pharmacological manipulation that may contribute to unravel causative links.

The most common protocols involve treating rats or mice with THC or synthetic cannabinoids during adolescence and then during adulthood, i.e., after a relatively long wash out period, analyzing a series of behavioural responses that are considered to reflect psychotic-like symptoms. One of the most used and accepted paradigms is the so called pre-pulse inhibition (PPI) of the startle response, a measure of sensorimotor gating that reflects the ability of an organism to attain information and process it correctly. Loss of normal PPI is widely accepted as an endophenotype of schizophrenia with high translational validity, and it can be assessed in both animals and humans. Another usual paradigm that is frequently used is the social interaction test, since individuals suffering from schizophrenia often exhibit impaired social interaction (in form of social withdrawal), which is considered a negative symptom of the disorder. Measurements of social behaviour in rats are relatively easy, as they show a well-structured stable degree of social behaviour. Several cognitive tests, including the analysis of working memory, are also employed. Cognitive symptoms associated with schizophrenia include deficits in attention and working memory that lead to an inability to organize one's life and to work effectively [39].

The most relevant results obtained from preclinical studies on long-term effects of adolescent cannabinoid exposure are presented in Table 11.2. Chronic pubertal treatment with the cannabinoid agonist WIN resulted in impaired memory in adulthood as well as in a disrupted PPI of the acoustic startle response [93]. These behavioural alterations resemble schizophrenic like-symptoms since PPI deficit, object recognition memory impairment, and anhedonia are among the endophenotypes of schizophrenia. Importantly, Schneider and Koch [94] also showed that if the chronic treatment with the drug occurs during adulthood, it does not lead to behavioural changes. In another study, a 21-day treatment with the cannabinoid receptor agonist CP in 30-day-old rats resulted in a lasting impairment of working memory [95] and, again, these later behavioural changes are observed in adolescent but not adult treated rats. A more recent study performed in male rats has shown that pubertal, but not adult, chronic WIN administration induced persistent disturbances in object and social recognition memory (indicating impairments in working memory and social memory, respectively) and led to social withdrawal and alterations in social behaviour [93]. Furthermore, acute administration of WIN induces more severe behavioural effects in pubertal than in adult rats [93]. Exposure of male rats to chronic THC causes greater lasting memory deficit and hippocampal alterations in adolescent than adult rats [96]. On the other hand, O'Shea et al. [70] found that chronic exposure to the cannabinoid agonist CP during perinatal, adolescent or early adult-hood induced similar long-term memory impairments in male rats. To explain the different results with respect to their previous study performed in female rats [95], authors claimed that adult males might be more vulnerable than adult females to some detrimental effects of cannabinoids, such as cognitive impairment. In line with this proposal, we have recently shown that, in the novel object recognition test, males are more vulnerable than females to the detrimental effects of chronic adolescent administration of CP [57]. Our results also indicate that in the object location task, only the females showed a significantly impaired performance in response to adolescent (pnd 28–43) cannabinoid exposure, suggesting that diverse aspects of memory function may be differentially affected in each sex [57].

Table 11.2 Long-term effects of adolescent cannabinoid exposure on adult emotional behaviour. Data extracted from [37–39]

Agonist (dose)	Treatment age window	Animals tested	Test	Effect	Reference
Locomotion and exploration					
CP 55,940 (0.4 mg/kg)	PND 35–45	Male and female adult (>PND 75) Wistar rats	HB	Reduced exploration in males, but decreased locomotion and higher exploration in females	[100]
WIN55,212–2 (1.2 mg/kg, 20 injections, ip)	PND 40–65	Male adult (PND 80) Wistar rats	OF	Not affected	[94]
THC (increasing doses: 2.5, 5 and 10 mg/kg, twice a day, ip)	PND 35–45	Male and female adult (>PND 75) Sprague–Dawley rats	Activity cage	No effects on spontaneous locomotor activity	[101]
			OF	Not affected.	
WIN55,212–2 (1.2 mg/kg, 20 injections, ip)	PND 40–65	Male adult (PND 80/105) Wistar rats	OF	More active and increased rearing behaviour; more time in the centre of the arena	[138]
CP 55,940 (0.4 mg/kg)	PND 28–42	Male and female adult Wistar rats	HB	Males not affected, but an increase in exploration of holes and inner region of the arena was observed among females	[139]
THC (increasing doses:2.5, 5 and 10 mg/kg, ip)	PND 28–45	Male and female adolescent (PND 46) Wistar rats	HB	Decrease exploration of the central region of the arena	[140]
		Male and female adult (PND 71) Wistar rats	OF	Decrease time in centre of the arena	
Anxiety-like responses, emotional and social behaviour					
CP 55,940 (0.4 mg/kg)	PND 35–45	Male and female adult (>PND 75) Wistar rats	OF	Trend for an increase in the exploration of the internal area of the maze	[100]
			EPM	More time spent in the open arms of the maze, indicating an anxiolytic-like effect	
WIN55,212–2 (1.2 mg/kg, 20 injections, ip)	PND 40–65	Male adult (PND 75–80) Wistar rats	PR	Lower break point, related to anhedonia	[94]

Table 11.2 (continued)

Agonist (dose)	Treatment age window	Animals tested	Test	Effect	Reference
CP 55,940 (increasing doses: 0.15, 0.20 and 0.30 mg/kg, ip)	PND 30–51	Female adult (>PND 72) Wistar rats	SI	Less social interaction, suggestive of residual anxiogenesis	[95]
CP 55,940 (increasing doses: 0.15, 0.20 and 0.30 mg/kg for 7 days at each dose, ip)	PND 30–51	Male adult (>PND 86) Wistar rats	SI	Reduced time spent in social interaction	[70]
CP 55,940 (increasing doses: 0.15, 0.20 and 0.30 mg/kg for 7 days at each dose, ip)	PND 30–51	Male adult (>PND 88) Wistar rats	LD	No differences but for a small (albeit significant) reduction in time spent in the hide box	
THC (priming dose: 1 mg/kg, two consecutive days; then 5 mg/kg, on alternative days on a CPP schedule, ip)	PND 32–51	Male adult (PND 70) Wistar rats	SI	Reduced social interaction	[96]
WIN55,212–2 (1.2 mg/kg, ip)	PND 40–65	Male adult (PND 66 and 80) Wistar rats	SI	Inadequate increase in social play and self-grooming behaviour	[93]
THC (increasing doses: 2.5, 5 and 10 mg/kg, twice a day, ip)	PND 35–45	Male and female adult (>PND 75) Sprague–Dawley rats	EPM	Not affected	[101]
			Sucrose preference	Decreased sucrose preference in THC male and female animals	

Table 11.2 (continued)

Agonist (dose)	Treatment age window	Animals tested	Test	Effect	Reference
			FST	THC treated females spent more time immobile, i.e. depressive-like phenotype, as well as reduced climbing	
WIN55,212–2 (1.2 mg/kg, 20 injections, ip)	PND 40–65	Male adult (PND 90) Wistar rats	EPM	Increased residence on the open arms i.e. decreased anxiety	[138]
CP 55,940 (0.4 mg/kg, i.p)	PND 28–38	Male and female (>PND 97) Wistar rats	EPM	Less time in the close arms of the maze, and decreased grooming frequency	[141]
THC (increasing doses:2.5, 5 and 10 mg/kg, ip, twice a day)	PND 35–45	Female adult (>PND 75) Sprague–Dawley rats	FST	Increase in immobility in the first session of the test, resembling a passive avoidance coping strategy	[120]
			Palatable food preference	Reduced the consumption of palatable food, suggesting the induction of an anhedonic state	
			SI	Decrease in the duration of social contact with a conspecific	
WIN55,212–2 (1.2 mg/kg, ip)	PND 45–60	Male adult (PND 61, 70, 90 and 135) Sprague–Dawley rats	Sucrose consumption	No effects on sucrose consumption, thus no changes in anhedonia	[142]
WIN55,212–2 (2.5 mg/kg, sc)	PND 32–52	adult (>PND 73) C57BL6 mice	SI	Decreased sociability and disrupted social novelty preference	[143]
			LD	Not affected	
CP 55,940 (0.4 mg/kg)	PND 28–42	Male and female adult Wistar rats	EPM	No effects	[139]
THC (increasing doses:2.5, 5 and 10 mg/kg, ip)	PND 28–45	Male and female adult (PND 46) Wistar rats	EPM	Not affected	[140]

Table 11.2 (continued)

Agonist (dose)	Treatment age window	Animals tested	Test	Effect	Reference
Cognitive function					
WIN55,212–2 (1.2 mg/kg, 20 injections, ip)	PND 40–65	Male adult (PND 85–90) Wistar rats	NOR	Deficit in recognition memory	[94]
CP 55,940 (increasing doses: 0.15, 0.20 and 0.30 mg/kg, ip)	PND 30–51	Female adult (>PND 72) Wistar rats	NOR	Reduced preference for a novel object over a familiar object relative to control animals, suggesting impairment in working memory	[95]
CP 55,940 (increasing doses: 0.15, 0.20 and 0.30 mg/kg for 7 days at each dose, ip)	PND 30–51	Male adult (>PND 79) Wistar rats	NOR	Impairment of working memory	[70]
THC (5 mg/kg, ip)	PND 30–51	Male and female adult (>PND 79) Sprague–Dawley rats	Water maze Spatial version	No spatial learning deficit	[144]
WIN55,212–2 (1.2 mg/kg, ip)	PND 40–65	Male adult (PND 66 and 80) Wistar rats	NOR	Persistent deficit in object recognition	[93]
			SR	Decreased capability for social discrimination and social recognition	
THC (priming dose: 1 mg/kg, two consecutive days; then 5 mg/kg, on alternative days on a CPP schedule, ip)	PND 32–51	Male adult (PND 67) Wistar rats	NOR	Less time investigating the novel object, i.e. cognitive deficit	[96]
CP 55,940 (0.4 mg/kg, i.p)	PND 28–38	Male and female (>PND 97) Wistar rats	NOR	Not affected	[141]

Table 11.2 (continued)

Agonist (dose)	Treatment age window	Animals tested	Test	Effect	Reference
			Water maze	Not affected	
THC (increasing doses: 2.5, 5 and 10 mg/kg, twice a day, ip)	PND 35–45	Male adult (>PND 75) Sprague–Dawley rats	Passive avoidance	Not altered	[118]
			Radial maze	Slight but significant impairment in learning behaviour	
THC (increasing doses: 2.5, 5 and 10 mg/kg, twice a day, ip)	PND 35–45	Female adult (>PND 75) Sprague–Dawley rats	Passive avoidance	Not affected	[121]
			Radial maze	Poorer performance than vehicle administered animals, i.e. impaired spatial working memory	
THC (increasing doses:2.5, 5 and 10 mg/kg, ip, twice a day)	PND 35–45	Female adult (>PND 75) Sprague–Dawley rats	NOR	Significant cognitive deficit	[120]
WIN55,212–2 (1.2 mg/kg, ip)	PND 45–60	Male adult (PND 61, 70, 90 and 135) Sprague–Dawley rats	Water maze	Acquisition in the spatial task impaired only in the short-term, not in the long-term	[142]
			NOR	Long-lasting impairment in the spatial version of NOR, while no affection was observed in the long-term in the non-spatial version of the task	
CP 55,940 (increasing doses: 0.15, 0.20 and 0.30 mg/kg)	PND 29–50	Male adult (>PND 77) Wistar and Lister Hooded rats	NOR	Times spent exploring familiar and novel objects were similar, i.e. cognitive deficit	[145]

Table 11.2 (continued)

Agonist (dose)	Treatment age window	Animals tested	Test	Effect	Reference
			NOL	No preference for the object in the novel location over the familiar location dependent; the delay depended upon the rat strain employed	
THC (increasing doses:2.5, 5 and 10 mg/kg, ip)	PND 28–45	Male and female adult (PND 71–75) Wistar rats	NOR	Impairmet in recognition memory, exclusively among females	[140]
Psychotic-like behaviour					
WIN55,212–2 (1.2 mg/kg, 20 injections, ip)	PND 40–65	Male adult (>PND 85) Wistar rats	PPI	Disrupted PPI	[94]
WIN55,212–2 (1.2 mg/kg, 20 injections, ip)	PND 40–65	Male adult (PND 80/105) Wistar rats	PPI	PPI deficit at both ages	[138]
CP 55,940 (0.4 mg/kg)	PND 28–42	Male and female adult Wistar rats	PPI	No effects among males, but a decreased PPI with the higher intensity (80 dB) prepulse	[139]
WIN55,212–2 (2.5 mg/kg, sc)	PND 32–52	adult (>PND 73) C57BL6 mice	PPI	Reduced overall startle amplitude; but PPI response was not altered	[143]
THC (increasing doses:2.5, 5 and 10 mg/kg, ip)	PND 28–45	Male and female adult (PND 84) Wistar rats	PPI	No effects	[140]

Cannabinoid agonists: THC delta-9-tetrahydrocannabinol, PND postnatal day, HB holeboard, OF open field, EPM elevated plus maze, PR progressive ratio operant behaviour task, SI social interaction, LD light-dark test, FST forced swim test, NOR novel object recognition, SR social Recognition, NOL novel object location, PPI prepulse inhibition

Harte and Dow-Edwards [97] examined the effects of THC administered daily during juvenile or early adolescence (pnd 22–40) or late adolescence (pnd 41–60) on locomotor activity, development of tolerance, and acquisition/retention of spatial avoidance in adulthood. THC causes locomotor depression in both male and female animals treated during early adolescence but only in females treated during late adolescence. Evidence of reverse tolerance to THC is seen in early adolescent treated animals only. In the active place avoidance test, male and female animals administered THC during early adolescence made more errors on the reversal trial requiring flexibility in learning, but in animals treated during late adolescence there are no significant sex or treatment differences. The results of the locomotor activity study suggest that females may be more sensitive to the effects of THC than males, while results of both locomotor activity and active place avoidance studies suggest that early adolescent animals are more vulnerable to these effects than late adolescents/young adults. As a whole, these animal studies indicate that the nature of at least certain long-term residual effects of adolescent cannabinoid exposure may be gender- and task-dependent, and that different time intervals of specific vulnerabilities may exist throughout the periadolescent period. The duration and onset of the treatments are also important factors that may affect outcomes, but it seems clear that chronic adolescent cannabinoid treatments induce deleterious effects on cognitive function that can be observed after a long wash-out period.

Some few data from human studies also suggest the existence of gender differences as regards cannabis-induced residual effects at least in certain aspects of cognitive function in young people [90]. However, there is very scarce information regarding gender differences in residual effects of cannabis in humans, even because the vast majority of human and animal studies typically focus on males and do not recognize the importance of sex [98]. In order to gain further insights into this particular aspect, it is important to highlight the necessity of analyzing the two sexes separately.

All together, the data described above indicate that chronic pubertal cannabinoid treatment in rats results in long-lasting behavioural alterations that reflect certain characteristics of schizophrenia symptomatology, such as deficits in sensorimotor gating, impaired memory, reduced motivation and inappropriate and scarce social behaviour. Acute injections of the typical antipsychotic haloperidol are able to restore sensorimotor gating deficits, while the atypical antipsychotic quetiapine is able to acutely restore deficits in social behaviour induced by developmental cannabinoid exposure, and even exerts some persistent beneficial effects. All these data provide support for using pubertal cannabinoid administration as an animal model for investigating aspects of psychosis and schizophrenia [18].

Emotional Long-Term Adverse Effects of Cannabis

In addition to psychotic-like signs, adolescent cannabis use has been shown to induce other types of psychiatric disorders. Longitudinal research suggests that cannabis use predicts the development of anxiety disorders, depression, suicidal ideation, certain personality disorders, and interpersonal violence. Stronger associations have been found in adolescents relative to adults, and younger age of initiation increases the risk of developing mental health disorders [30, 99]. Preclinical studies specifically focused at analyzing depressive and anxiety responses to cannabinoid exposure agree with human observations (see Table 11.2). For instance, adult rats exposed to CP during the juvenile period (pnd 35–45) show anxiolytic-like responses in adulthood, as measured in the elevated plus-maze and the illuminated open field test, as well as sex-dependent effects regarding locomotion and exploration [100]. However, the effects on anxiety-related responses appear to be dependent on the duration of the pharmacological treatment, and likely the test employed, since a 21-day treatment with CP in 30-day-old rats results in increased anxiety in the social interaction test [95]. Moreover, in this latter study, the behavioural test was performed 23 days after the end of the pharmacological treatment, whereas in our case, the animals were tested approximately 37 days after the end of the treatment [100]. As for other types of emotional response, Rubino et al. [101] demonstrated that chronic administration of THC in adolescent rats induced subtle but lasting alterations in the emotional circuit ending in depressive-like behaviour in adulthood, and that this effect is observed in female but not male rats. These animal findings resemble certain observations in humans showing that frequent cannabis use in teenage girls predicts later depression and anxiety, with daily users carrying the highest risk [102].

Early Cannabis Consumption and the Risk of Addiction to Other Drugs of Abuse

Clinical and epidemiological studies have documented a significant link between repeated early cannabis exposure and an increased risk of other illicit drug use [17]. According to the phenotypic causation—"gateway model"—early initiation of cannabis use might be a risk factor for the consumption of other drugs of abuse [103], though the alternative "correlated liabilities model" proposes that cannabis use and other illicit drug use is influenced by correlated genetic and environmental factors [104]. Ferguson et al. [105] examined the associations between the frequency of cannabis use and the use of other illicit drugs in a 25-year longitudinal study of a birth cohort of 1,265 New Zealand children. They obtained annual assessments of the frequency of cannabis use for the period 14–25 years, together with measures of the use of other illicit drugs from the same time period. Regular or heavy cannabis use was associated with an increased risk of using other illicit drugs, abusing or becoming dependent upon other illicit drugs, and using a wider variety of other illicit drugs [105]. This association was particularly strong during adolescence but declined with increasing age. The findings may support a general causal model but they do not clarify the actual underlying mechanisms and the extent to which these causal mechanisms are direct or indirect. Lynskey et al. [103] have further analysed whether the association between early cannabis use and subsequent progression to use of other drugs and drug abuse/dependence persists after controlling for genetic and shared environmental influences. They found that individuals who used cannabis by age 17 years have odds of other drug use, alcohol dependence, and drug abuse/dependence that were 2.1–5.2 times higher than those of their co-twin, who did not use cannabis before age 17 years. Controlling for known risk factors (early-onset alcohol or tobacco use, parental conflict/separation, childhood sexual abuse, conduct disorder, major depression, and social anxiety) had only negligible effects on these results, and the associations do not differ significantly between monozygotic and dizygotic twins. In view of these data, it seems that associations between early cannabis use and later drug use and abuse/dependence cannot solely be explained by common predisposing genetic or shared environmental factors.

An important limitation of human studies is the difficulty of demonstrating a causal relationship between adolescent cannabis use and the use and/or dependence of other substances. However, animal studies suggest that the association may reflect neurobiological disturbances caused by early cannabis exposure that make individuals more vulnerable to the reinforcing effects of other drugs. In fact, there is evidence suggesting a causal relationship between early cannabis exposure and use or abuse of other addictive substances later in life [17, 36]. Ellgren et al. [106], in a study performed on male rats demonstrated that exposure to THC in adolescent animals produced an increase in heroin self-administration, preproenkephalin mRNA expression and functionality of μ-opioid receptors in adulthood. Accordingly, we found that chronic periadolescent exposure to CP altered morphine self-administration and the opioid system in adult rats in a sex-dependent manner. In particular, CP increases the acquisition of morphine self-administration and decreases μ-opioid receptor functionality in the nucleus accumbens shell in males but not female animals [107]. In line with our results, decreased μ-opioid-coupled G-protein activity was found in the nucleus accumbens shell of male rats exposed prenatally to THC, with no changes in the nucleus accumbens core or caudate putamen [108]. Together, these data suggest that cannabinoid exposure in early stages of development and adolescence produces perdurable changes in μ-opioid receptor functionality that are specific to the nucleus accumbens shell, which is one of the brain regions most closely related to natural and drug-induced reward. Other authors have reported that a chronic treatment with CP during adolescence resulted in a higher rate of cocaine self-administration during the acquisition phase in adult females, whereas no effect was found in males [109]. Thus, the direction of sex differences regarding long-lasting effects of adolescent cannabinoid exposure on self-administration of other drugs of abuse may depend of the specific nature of the drug. Tomasiewicz et al. [110] have shown that over-expression of the pro-enkephalin gene in the nucleus accumbens shell enhances heroin self-administration and heroin-seeking behaviour in animals naïve to THC, whereas knocking down the pro-enkephalin gene in THC-exposed rats reduces heroin intake. Given the well-known interactions between the endocannabinoid and the opioid system and the involvement of the two systems in the brain reward mechanisms, it is likely that exposure to THC during adolescence induces alterations in the opioid system that likely contribute to the development of opiate abuse in adults [17].

The mesolimbic dopaminergic system, which is related to the mechanisms mediating natural and drug-induced reward and the neuropathology of psychoses, is a relevant possible target that might be affected by cannabinoid exposure during puberty. The effects of repeated cannabinoid administration on meso-accumbens dopaminergic neuronal functions and responses to drugs of abuse have been analysed. Animals were pre-treated during adolescence or adulthood, for 3 days, with WIN or vehicle and allowed a 2-week interval. In WIN administered rats dopaminergic neurons were significantly less responsive to the stimulating action of the cannabinoid, regardless of the age of pre-treatment. However, in the adolescent group, but not in the adults, long-lasting cross-tolerance developed to morphine, cocaine and amphetamine [111]. These results suggest that cannabis exposure at a young age may induce long-term neuronal adaptations in the mesolimbic dopaminergic system and hence affect the responses to drugs of abuse. Hurd and co-workers [17] showed that in their model of adolescent THC exposure, reduced levels of Drd2 mRNA, which encodes dopamine D2 receptor, are observed within the nucleus accumbens of adult animals. In addition to adolescent THC exposure, prenatal THC also leads to dysregulation of the Drd2 gene in adulthood. Since a reduced D2 receptor level has long been a characteristic neurobiological feature of addiction vulnerability, that developmental THC exposure reduces Drd2 mRNA expression in the striatum and affects related behavioural traits supports the hypothesis that developmental cannabis may induce a neurobiological state of addiction vulnerability [17].

Neurobiological Mechanisms Underlying Adolescent Cannabinoid Exposure

Several studies provided interesting data suggesting possible neurobiological mechanisms, including molecular and cellular alterations, which may underlie behavioural alterations and psychiatric disorders induced by adolescent cannabinoid exposure, although much more work is necessary to this respect [39]. CP has been reported to impair not only PPI in rats but also auditory gating and neuronal synchrony in limbic areas such as the hippocampus and entorhinal cortex, as evaluated through theta field potential oscillations [112]. It seems clear that, at least in rats, cannabinoid agonists impair auditory gating function in the limbic circuitry, supporting a connection between cannabis abuse and schizophrenia as evaluated through this animal model. More recently, Raver et al. [113] have shown that chronic adolescent, but not adult, cannabinoid exposure in mice suppresses pharmacologically evoked cortical oscillations, that are integral for cognitive processes and are abnormal in patients with schizophrenia, and impairs working memory performance in adults. These data further support a link between chronic adolescent cannabinoid exposure and alterations in adult cortical network activities that underlie cognitive processes. Mice exposed to WIN during adolescence that exhibit in adulthood deficits in PPI and fear conditioning, also show a reduction of hippocampal metabotropic glutamate receptors type 5 (mGluR5) and increased levels of monoacylglycerol lipase

(MAGL) and FAAH, indicative of increases in endocannabinoid uptake and degradation [114]. These data further support the idea that cannabis use during adolescence may be a contributory causal factor in the development of at least certain features of schizophrenia probably in relation to altered endocannabinoid signalling in the hippocampus. Page et al. [115] demonstrated that, in adult rats, repeated administration of WIN induces transient anxiety-like behaviours that correlate with increases in catecholamine synthesizing enzyme expression in the locus coeruleus and in norepinephrine efflux in response to a challenge injection of the same drug. Bambico et al. [116] have recently shown that chronic adolescent, but not adult, exposure to low (0.2 mg/kg) and high (1 mg/kg) doses of WIN leads to depression-like behaviour, while the high dose also induces anxiety-like responses in rats. Electrophysiological recordings revealed that both doses attenuate serotonergic activity, while the high dose also leads to a hyperactivity of noradrenergic neurons only after adolescent exposure. These results suggest that the anxiety-like and depression-like behaviour shown by adult rats exposed to the cannabinoid agonist in the adolescent period might be a result of serotonergic hypoactivity and noradrenergic hyperactivity.

Morphological changes in the hippocampus have been observed following chronic administration of cannabinoids [117, 118]. Two-dimensional gel electrophoresis proteomic analysis conducted on THC-treated hippocampal samples revealed several proteins showing long-lasting alterations in response to THC administration. The greater number of differentially expressed protein spots in adolescent THC-pre-treated rats compared with adult THC-pre-treated rats suggests a greater vulnerability to lasting effects of THC in the former group. Differentially expressed proteins in adolescent THC exposed rats include cytoskeletal and other structural proteins, including transgelin-3 (NP25), α and β tubulin and myelin basic protein [96]. This may be linked to structural changes or remodelling occurring after THC exposure in adolescents and is consistent with observations of cytoarchitectural changes occurring with cannabinoid treatment [117]. As a whole, differentially expressed proteins in the hippocampus of THC pre-exposed adolescents have a variety of functions broadly related to oxidative stress, mitochondrial and metabolic function and regulation of the cytoskeleton and signalling. Reductions in dendrite length and complexity and in the number of dendritic spines in the dentate gyrus of the hippocampus have been also found in these animals [118]. Moreover, recent findings suggest that adolescent cannabinoid exposure may induce long-term alterations in astrocytes [119]. These latter results highlight the potential functional importance of astrocytes and their interaction with the eCB system in relation to long-term consequences of adolescent cannabis exposure.

Most of these studies have been carried out in male animals. When female rats were used, adolescent THC exposure induced a significant reduction in cell proliferation in the dentate gyrus of the hippocampus [120] as well as less synaptic density and/or efficiency throughout the prefrontal cortex [121]. Further studies analyzing both sexes are urgently needed to get a clearer picture of possible differential vulnerabilities in both genders.

MRI studies further suggest that heavy cannabis use may modify brain structure. Just to mention some of them, Yucel et al. [122] showed that heavy cannabis users had bilaterally reduced hippocampal and amygdala volumes with greater effect in the former. Left hemisphere hippocampal volume was inversely associated with cumulative exposure to cannabis and with sub-threshold positive psychotic symptoms. Interestingly, hippocampal abnormalities in schizophrenia are more prominent in the left hemisphere. In another imaging study, long-term use of cannabis during adolescence was associated with gyrification abnormalities in the cortex, suggesting that early cannabis use affected normal neurodevelopment [123]. Arnone et al. [124] used an MRI technique sensitive to the structural integrity of brain tissue that combines with a white matter mapping tractography to investigate structural changes in the corpus callosum. Mean diffusivity, which measures structural integrity, was significantly increased in marijuana users relative to controls in the region of the corpus callosum where white matter passes between the prefrontal lobes. Moreover, there was a trend towards a positive correlation between mean diffusivity and length of use, which suggests the possibility of a cumulative effect of marijuana over time and that a younger age at onset of use may predispose individuals to structural white matter damage.

More recently, Zalesky et al. [125] have found that axonal connectivity is impaired in the right fimbria of the hippocampus (fornix), splenium of the corpus callosum and commissural fibres, suggesting that long-term cannabis use is hazardous to white matter in the developing brain. Adolescent onset marijuana use has also been linked with increased prefrontal cortex white matter diffusivity and increased impulsivity compared to later onset in a sample of well-matched adolescent onset marijuana users [126]. A recent review by Lorenzetti et al. [127] has examined evidence from structural neuroimaging investigations of regular cannabis users. This review supports the notion that regular cannabis use is associated with alterations of brain morphology, specifically medial temporal, frontal and cerebellar brain regions. Greater brain morphological alterations are evident among samples that used higher doses for longer periods. To sum up, structural abnormalities, disturbed brain connectivity and altered brain activation patterns may underlie cognitive impairment, behavioural alterations and vulnerability to certain psychiatric disorders that are observed in long-term heavy cannabis users.

Cannabis Plant Composition

The composition of the cannabis plant ("cannabis brands") has a critical influence on its possible long-term effects. Until recently, the main types of cannabis available on the "street" were marijuana (grass) and resin (hash), but in recent years a more potent variant termed sinsemilla or skunk has become available in many countries. Marijuana and resin have traditionally contained about 4 % THC, but the concentration of THC in skunk in countries such as England and the Netherlands has increased to about 16 and 20 % respectively, partly due to the use of intensive indoor cultivation methods [28]. The content of THC in confiscated cannabis preparations has substantially increased over the past 20 years. Recent data showed an upward trend in the mean THC content, which increased from 3.4 % in 1993 to 8.8 % in 2008 [128]. Thus, cannabis consumption nowadays implies exposure to very high amounts of THC, especially if sinsemilla (skunk) or synthetic cannabinoids are consumed. The risk associated with use of these stronger forms of cannabis needs to be further and deeply evaluated, since it is plausible that there are greater health risks than thought. The risk of psychosis is much greater among people who are frequent cannabis users, and among those using sinsemilla (skunk) rather than traditional hash [92]. It is not surprising that those who use skunk daily are at the highest risk, and public education about the risks of heavy use of high-potency cannabis is therefore urgently needed.

There is growing public health concern about the increasing use of a new generation of synthetic cannabinoid agonists marketed as natural herbal incense mixtures comprised under the "Spice" name. "Spice" refers to a wide variety of herbal mixtures that produce experiences similar to marijuana and that are marketed as "safe", legal alternatives to cannabis. Sold under many names, including K2, fake weed, Yucatan Fire, Skunk, Moon Rocks, and others—and labelled "not for human consumption"—these products contain dried, shredded plant material and chemical additives that are responsible for their psychoactive (mind-altering) effects. Spice products do contain dried plant material, but chemical analyses show that their active ingredients are synthetic cannabinoid compounds, e.g. JWH-018, CP-47,497. Spice users report experiences similar to those produced by marijuana—elevated mood, relaxation, and altered perception—and, in some cases, the effects are even stronger than those of marijuana [129]. Some users report psychotic effects like extreme anxiety, paranoia, and hallucinations. Spice can also raise blood pressure, can induce myocardial ischemia, and, in a few cases, it has been associated with heart attacks. Regular users may experience craving and withdrawal symptoms [130].

THC and cannabidiol (CBD), the two main ingredients of the *Cannabis sativa* plant have distinct symptomatic and behavioural effects. CBD has been demonstrated to have low affinity for both cannabinoid CB1 and CB2 receptors, but it can behave as a CB2 receptor inverse agonist [131]. Recent data suggest that THC and CBD can have opposite effects on regional brain function, which may underlie their different symptomatic and behavioural effects, and the potential ability of CBD to somehow 'buffer' the detrimental consequences of THC [132]. Notably, the ratio of CBD and THC seems to have changed in an unfavourable manner in the last years, and this fact may underlay the increased risk for adverse, and long-lasting detrimental consequences of marijuana consumption during adolescence. Nevertheless, more information is urgently needed in order to further clarify the potential therapeutic effect of CBD and the extent to which it is able to diminish the detrimental effects of THC [15]. Future studies are needed regarding the investigation of the long-term effects of chronic CBD administration alone or in combination with THC, and animal models would be a very useful tool for this purpose.

Genetics Factors of Vulnerability

In spite of the fact that cannabis is the most widely used drug in the world, only a relatively small proportion of users develop psychotic illness, suggesting the relevance of individual genetic factors in the susceptibility to the psychotic-inducing potential of cannabis. To date, most research has focused on the catechol-O-methyltransferase (COMT) gene. COMT is a key enzyme involved in the metabolism of dopamine that is highly expressed in the prefrontal cortex. Caspi et al. [133] showed that a functional polymorphism in the COMT gene moderates the influence of adolescent cannabis use on developing adult psychosis. Homozygous carriers of the COMT valine158 allele (Val/Val) are most likely to exhibit psychotic symptoms and to later develop schizophrenia-like disorders if they have used cannabis during adolescence (relative risk: 10.9). Heterozygous individuals with the valine/methionine (Val/Met) genotype who used cannabis during adolescence show an intermediate risk, while those homozygous for the methionine allele (Met/Met) show the lowest risk (relative risk: 1.1). A subsequent study by Henquet et al. [134] showed that carriers of the Val allele (Val/Val) are more sensitive to THC-induced memory and attention impairments compared to carriers of the Met allele (Met/Met), and are most sensitive to THC-induced psychotic experiences only in the presence of prior evidence of psychometric psychosis liability.

Taken together, it seems that the effects of THC on cognition and psychosis are moderated by COMT genotype, although partially conditioned to the presence of pre-existing psychosis liability. Notably, negative results have also been reported in this regard since Zammit et al. [135] did not report differential effects of cannabis use on schizophrenia due to COMT variations.

More recently, Van Winkel et al. [136] have examined the interactions between cannabis use and 152 single-nucleotide polymorphisms in 42 genes in 740 unaffected siblings of 801 patients with psychosis. Authors showed that genetic variation in AKT1 may mediate the effects on psychosis expression associated with cannabis use. AKT1 is a serine/threonine kinase central in many signal-transduction pathways. Cannabinoids are able to activate the AKT1 pathway through the activation of CB1 and CB2 receptors. Polymorphisms in the AKT1 gene may be involved in cannabis induced psychosis through a mechanism of cannabinoid-regulated AKY1/GSK-3 signalling downstream of the dopamine D2 receptor [28]. However, individual responses to cannabis use might be modulated by several genes rather than by a single polymorphism. Future research is needed to gain insights into genetic vulnerability to the harmful effects of cannabis.

Final Remarks

There is still scarce research available to determine whether sustained abstinence from cannabis results in recovery of cognitive functions. Though certain preliminary findings seem to be hopeful, further research is needed to learn whether cannabis-induced impairments in the brain are reversible. It has been proposed that "interventions geared toward lowering alcohol and drug exposure in teens and young adults that have shown evidence of efficacy need to be implemented more aggressively in schools and college campuses to not only reduce symptoms of drug abuse and dependence, but delay the onset of regular use from early teen years to early adult years in order to prevent long-term neuronal damage and ensure optimal brain health and cognitive functioning in youth" [40]. Yet, besides delaying the onset of use, it should be important to also promote abstinence. In addition to the age of onset, other factors such as genetic vulnerability, dosing, personality traits and amount of THC present in the drug are also important factors that may influence the impact of drug use. For example, genetic background might be a crucial factor in terms of vulnerability, but we are still far from having a clear knowledge about the nature of the genes implicated and from predicting and controlling these risk factors. It is also worth noticing that though cannabis use is most prevalent among adolescents and young adults, it is by no means restricted to this age group and increasing recognition is currently given to cannabis users in older age groups, including individuals who initiate cannabis use at a later age [137]. As pointed out by Agrawal and Lynskey [137], while later onsets are rare, their impact may be fairly profound, and attempts to identify correlates of new onsets and of persistence of cannabis use through adulthood seem to be relevant. Not to forget, the importance of sex differences regarding not only the prevalence of cannabis use but also the possible differential effects of the drug on males and females and the different underlying motivation to consume it (see Chap. 13). This approach may result in better prevention and treatment strategies. A fluent interaction between basic researchers, clinicians and epidemiologists together with a clear message to the society about the detrimental effects of cannabis are urgently needed. Giving healthy alternatives to young people, promoting exercise and considering that availability of the drug are all important aspects to establish efficacious treatment and prevention campaigns.

References

1. Marco EM, Romero-Zerbo SY, Viveros MP, Bermudez-Silva FJ. The role of the endocannabinoid system in eating disorders: pharmacological implications. Behav Pharmacol. 2012;23(5–6):526–36.
2. Bermudez-Silva FJ, Viveros MP, McPartland JM, Rodriguez de Fonseca F. The endocannabinoid system, eating behavior and energy homeostasis: the end or a new beginning? Pharmacol Biochem Behav. 2010;95(4):375–82.
3. Marco EM, Viveros MP. The critical role of the endocannabinoid system in emotional homeostasis: avoiding excess and deficiencies. Mini Rev Med Chem. 2009;9(12):1407–15.
4. Castillo PE, Younts TJ, Chavez AE, Hashimotodani Y. Endocannabinoid signaling and synaptic function. Neuron. 2012;76(1):70–81.
5. Katona I, Freund TF. Multiple functions of endocannabinoid signaling in the brain. Annu Rev Neurosci. 2012;35:529–58.
6. Oleson EB, Cheer JF. A brain on cannabinoids: the role of dopamine release in reward seeking. Cold Spring Harb Perspect Med. 2012;2(8).

7. Gardner EL. Addiction and brain reward and antireward pathways. Adv Psychosom Med. 2011;30:22–60.
8. Harkany T, Guzman M, Galve-Roperh I, Berghuis P, Devi LA, Mackie K. The emerging functions of endocannabinoid signaling during CNS development. Trends Pharmacol Sci. 2007;28(2):83–92.
9. Galve-Roperh I, Palazuelos J, Aguado T, Guzman M. The endocannabinoid system and the regulation of neural development: potential implications in psychiatric disorders. Eur Arch Psychiatry Clin Neurosci. 2009;259(7):371–82.
10. Keimpema E, Alpar A, Howell F, Malenczyk K, Hobbs C, Hurd YL, et al. Diacylglycerol lipase alpha manipulation reveals developmental roles for intercellular endocannabinoid signaling. Sci Rep. 2013;3:2093.
11. Keimpema E, Mackie K, Harkany T. Molecular model of cannabis sensitivity in developing neuronal circuits. Trends Pharmacol Sci. 2011;32(9):551–61.
12. Wu CS, Jew CP, Lu HC. Lasting impacts of prenatal cannabis exposure and the role of endogenous cannabinoids in the developing brain. Future Neurol. 2011;6(4):459–80.
13. Burns JK. Pathways from cannabis to psychosis: a review of the Evidence. Front Psychiatry. 2013;4:128.
14. Fernandez-Espejo E, Viveros MP, Nunez L, Ellenbroek BA, Rodriguez de Fonseca F. Role of cannabis and endocannabinoids in the genesis of schizophrenia. Psychopharmacology (Berl). 2009;206(4):531–49.
15. Marco EM, Garcia-Gutierrez MS, Bermudez-Silva FJ, Moreira FA, Guimaraes F, Manzanares J, et al. Endocannabinoid system and psychiatry: in search of a neurobiological basis for detrimental and potential therapeutic effects. Front Behav Neurosci. 2011;5:63.
16. Huizink AC. Prenatal cannabis exposure and infant outcomes: overview of studies. Prog Neuropsychopharmacol Biol Psychiatry. 2013.
17. Hurd YL, Michaelides M, Miller ML, Jutras-Aswad D. Trajectory of adolescent cannabis use on addiction vulnerability. Neuropharmacology. 2014;76 Pt B:416–24.
18. Viveros MP, Llorente R, Suarez J, Llorente-Berzal A, Lopez-Gallardo M, de Fonseca FR. The endocannabinoid system in critical neurodevelopmental periods: sex differences and neuropsychiatric implications. J Psychopharmacol. 2012;26(1):164–76.
19. Viveros MP, Marco EM, Lopez-Gallardo M, Garcia-Segura LM, Wagner EJ. Framework for sex differences in adolescent neurobiology: a focus on cannabinoids. Neurosci Biobehav Rev. 2011;35(8):1740–51.
20. Crippa JA, Zuardi AW, Martin-Santos R, Bhattacharyya S, Atakan Z, McGuire P, et al. Cannabis and anxiety: a critical review of the evidence. Hum Psychopharmacol. 2009;24(7):515–23.
21. Budney AJ, Hughes JR. The cannabis withdrawal syndrome. Curr Opin Psychiatry. 2006;19(3):233–8.
22. Fattore L, Fadda P, Spano MS, Pistis M, Fratta W. Neurobiological mechanisms of cannabinoid addiction. Mol Cell Endocrinol. 2008;286(1–2 Suppl 1):S97–S107.
23. Hasin DS, O'Brien CP, Auriacombe M, Borges G, Bucholz K, Budney A, et al. DSM-5 criteria for substance use disorders: recommendations and rationale. Am J Psychiatry. 2013;170(8):834–51.
24. Hesse M, Thylstrup B. Time-course of the DSM-5 cannabis withdrawal symptoms in poly-substance abusers. BMC Psychiatry. 2013;13:258.
25. Gorelick DA, Levin KH, Copersino ML, Heishman SJ, Liu F, Boggs DL, et al. Diagnostic criteria for cannabis withdrawal syndrome. Drug Alcohol Depend. 2012;123(1–3):141–7.
26. Solowij N, Pesa N. Cannabis and cognition: short and long-term effects. In: Castle D, Murray R, D'Souza DC, editors. Marijuana and madness. 2nd ed. Cambridge: Cambridge University Press; 2012. p. 91–102.
27. Meier MH, Caspi A, Ambler A, Harrington H, Houts R, Keefe RS, et al. Persistent cannabis users show neuropsychological decline from childhood to midlife. Proc Natl Acad Sci U S A. 2012;109(40):E2657–64.
28. Casadio P, Fernandes C, Murray RM, Di Forti M. Cannabis use in young people: the risk for schizophrenia. Neurosci Biobehav Rev. 2011;35(8):1779–87.

29. Hurd YL, Michaelides M, Miller ML, Jutras-Aswad D. Trajectory of adolescent cannabis use on addiction vulnerability. Neuropharmacology. 2014;76PB:416–24.
30. Copeland J, Rooke S, Swift W. Changes in cannabis use among young people: impact on mental health. Curr Opin Psychiatry. 2013;26(4):325–9.
31. Trezza V, Campolongo P, Manduca A, Morena M, Palmery M, Vanderschuren LJ, et al. Altering endocannabinoid neurotransmission at critical developmental ages: impact on rodent emotionality and cognitive performance. Front Behav Neurosci. 2012;6:2.
32. Passey ME, Sanson-Fisher RW, D'Este CA, Stirling JM. Tobacco, alcohol and cannabis use during pregnancy: Clustering of risks. Drug Alcohol Depend. 2013.
33. Hutchings DE, Martin BR, Gamagaris Z, Miller N, Fico T. Plasma concentrations of delta-9-tetrahydrocannabinol in dams and fetuses following acute or multiple prenatal dosing in rats. Life Sci. 1989;44(11):697–701.
34. Gomez M, Hernandez M, Johansson B, de Miguel R, Ramos JA, Fernandez-Ruiz J. Prenatal cannabinoid and gene expression for neural adhesion molecule L1 in the fetal rat brain. Brain Res Dev Brain Res. 2003;147(1–2):201–7.
35. Perez-Reyes M, Wall ME. Presence of delta9-tetrahydrocannabinol in human milk. N Engl J Med. 1982;307(13):819–20.
36. DiNieri JA, Hurd YL. Rat Models of prenatal and adolescent cannabis exposure. In: Kobeissy FH, editor. Psychiatric disorders: methods and protocols: Springer protocols. USA: Humana Press; 2012. p. 231–42.
37. Viveros MP, Llorente R, Moreno E, Marco EM. Behavioural and neuroendocrine effects of cannabinoids in critical developmental periods. Behav Pharmacol. 2005 Sep;16(5–6):353–62.
38. Schneider M. Cannabis use in pregnancy and early life and its consequences: animal models. Eur Arch Psychiatry Clin Neurosci. 2009;259(7):383–93.
39. Rubino T, Parolaro D. Cannabis abuse in adolescence and the risk of psychosis: a brief review of the preclinical evidence. Prog Neuropsychopharmacol Biol Psychiatry. 2014;52:41–4.
40. Lisdahl KM, Gilbart ER, Wright NE, Shollenbarger S. Dare to delay? The impacts of adolescent alcohol and marijuana use onset on cognition, brain structure, and function. Front Psychiatry. 2013;4:53.
41. Adriani W, Laviola G. Windows of vulnerability to psychopathology and therapeutic strategy in the adolescent rodent model. Behav Pharmacol. 2004;15(5–6):341–52.
42. Schneider M. Adolescence as a vulnerable period to alter rodent behavior. Cell Tissue Res. 2013;354(1):99–106.
43. Viveros MP, Marco EM, File SE. Endocannabinoid system and stress and anxiety responses. Pharmacol Biochem Behav. 2005;81(2):331–42.
44. Gaffuri AL, Ladarre D, Lenkei Z. Type-1 cannabinoid receptor signaling in neuronal development. Pharmacology 2012;90(1–2):19–39.
45. Belue RC, Howlett AC, Westlake TM, Hutchings DE. The ontogeny of cannabinoid receptors in the brain of postnatal and aging rats. Neurotoxicol Teratol. 1995;17(1):25–30.
46. Rodriguez de Fonseca F Ramos JA Bonnin A Fernandez-Ruiz JJ. Presence of cannabinoid binding sites in the brain from early postnatal ages. Neuroreport. 1993;4(2):135–8.
47. Mato S, Del Olmo E, Pazos A. Ontogenetic development of cannabinoid receptor expression and signal transduction functionality in the human brain. Eur J Neurosci. 2003;17(9):1747–54.
48. Berrendero F, Garcia-Gil L, Hernandez ML, Romero J, Cebeira M, de Miguel R, et al. Localization of mRNA expression and activation of signal transduction mechanisms for cannabinoid receptor in rat brain during fetal development. Development. 1998;125(16):3179–88.
49. Reisenberg M, Singh PK, Williams G, Doherty P. The diacylglycerol lipases: structure, regulation and roles in and beyond endocannabinoid signalling. Philos Trans R Soc Lond B Biol Sci. 2012;367(1607):3264–75.
50. Mulder J, Aguado T, Keimpema E, Barabas K, Ballester Rosado CJ, Nguyen L, et al. Endocannabinoid signaling controls pyramidal cell specification and long-range axon patterning. Proc Natl Acad Sci U S A. 2008;105(25):8760–5.
51. Berghuis P, Dobszay MB, Wang X, Spano S, Ledda F, Sousa KM, et al. Endocannabinoids regulate interneuron migration and morphogenesis by transactivating the TrkB receptor. Proc Natl Acad Sci U S A. 2005;102(52):19115–20.

52. Wenger T, Gerendai I, Fezza F, Gonzalez S, Bisogno T, Fernandez-Ruiz J, et al. The hypothalamic levels of the endocannabinoid, anandamide, peak immediately before the onset of puberty in female rats. Life Sci. 2002;70(12):1407–14.
53. Ellgren M, Artmann A, Tkalych O, Gupta A, Hansen HS, Hansen SH, et al. Dynamic changes of the endogenous cannabinoid and opioid mesocorticolimbic systems during adolescence: THC effects. Eur Neuropsychopharmacol. 2008;18(11):826–34.
54. Lee TT, Hill MN, Hillard CJ, Gorzalka BB. Temporal changes in N-acylethanolamine content and metabolism throughout the peri-adolescent period. Synapse. 2013;67(1):4–10.
55. Marco EM, Adriani W, Canese R, Podo F, Viveros MP, Laviola G. Enhancement of endocannabinoid signalling during adolescence: Modulation of impulsivity and long-term consequences on metabolic brain parameters in early maternally deprived rats. Pharmacol Biochem Behav. 2007;86(2):334–45.
56. Reich CG, Taylor ME, McCarthy MM. Differential effects of chronic unpredictable stress on hippocampal CB1 receptors in male and female rats. Behav Brain Res. 2009;203(2):264–9.
57. Mateos B, Borcel E, Loriga R, Luesu W, Bini V, Llorente R, et al. Adolescent exposure to nicotine and/or the cannabinoid agonist CP 55,940 induces gender-dependent long-lasting memory impairments and changes in brain nicotinic and CB(1) cannabinoid receptors. J Psychopharmacol. 2011;25(12):1676–90.
58. Suarez J, Llorente R, Romero-Zerbo SY, Mateos B, Bermudez-Silva FJ, de Fonseca FR, et al. Early maternal deprivation induces gender-dependent changes on the expression of hippocampal CB(1) and CB(2) cannabinoid receptors of neonatal rats. Hippocampus. 2009;19(7):623–32.
59. Duff G, Argaw A, Cecyre B, Cherif H, Tea N, Zabouri N, et al. Cannabinoid receptor CB2 modulates axon guidance. PLoS ONE. 2013;8(8):e70849.
60. Palazuelos J, Ortega Z, Diaz-Alonso J, Guzman M, Galve-Roperh I. CB2 cannabinoid receptors promote neural progenitor cell proliferation via mTORC1 signaling. J Biol Chem. 2012;287(2):1198–209.
61. Fried PA, Watkinson B. Differential effects on facets of attention in adolescents prenatally exposed to cigarettes and marihuana. Neurotoxicol Teratol. 2001;23(5):421–30.
62. Fried PA, Watkinson B, Gray R. A follow-up study of attentional behavior in 6-year-old children exposed prenatally to marihuana, cigarettes, and alcohol. Neurotoxicol Teratol. 1992;14(5):299–311.
63. Fried PA, Watkinson B, Gray R. Differential effects on cognitive functioning in 9–12-year olds prenatally exposed to cigarettes and marihuana. Neurotoxicol Teratol. 1998;20(3):293–306.
64. Fried PA, Watkinson B, Gray R. Differential effects on cognitive functioning in 13–16-year-olds prenatally exposed to cigarettes and marihuana. Neurotoxicol Teratol. 2003;25(4):427–36.
65. Smith AM, Fried PA, Hogan MJ, Cameron I. Effects of prenatal marijuana on visuospatial working memory: an fMRI study in young adults. Neurotoxicol Teratol. 2006;28(2):286–95.
66. Mereu G, Fa M, Ferraro L, Cagiano R, Antonelli T, Tattoli M, et al. Prenatal exposure to a cannabinoid agonist produces memory deficits linked to dysfunction in hippocampal long-term potentiation and glutamate release. Proc Natl Acad Sci U S A. 2003;100(8):4915–20.
67. Antonelli T, Tomasini MC, Tattoli M, Cassano T, Tanganelli S, Finetti S, et al. Prenatal exposure to the CB1 receptor agonist WIN 55,212–2 causes learning disruption associated with impaired cortical NMDA receptor function and emotional reactivity changes in rat offspring. Cereb Cortex. 2005;15(12):2013–20.
68. Campolongo P, Trezza V, Cassano T, Gaetani S, Morgese MG, Ubaldi M, et al. Perinatal exposure to delta-9-tetrahydrocannabinol causes enduring cognitive deficits associated with alteration of cortical gene expression and neurotransmission in rats. Addict Biol. 2007;12(3–4):485–95.
69. O'Shea M, Mallet PE. Impaired learning in adulthood following neonatal delta9-THC exposure. Behav Pharmacol. 2005;16(5–6):455–61.
70. O'Shea M, McGregor IS, Mallet PE. Repeated cannabinoid exposure during perinatal, adolescent or early adult ages produces similar longlasting deficits in object recognition and reduced social interaction in rats. J Psychopharmacol. 2006;20(5):611–21.

71. Rubio P, Rodriguez de Fonseca F, Munoz RM, Ariznavarreta C, Martin-Calderon JL, Navarro M. Long-term behavioral effects of perinatal exposure to delta9-tetrahydrocannabinol in rats: possible role of pituitary-adrenal axis. Life Sci. 1995;56(23–24):2169–76.
72. Moreno M, Trigo JM, Escuredo L, Rodriguez de Fonseca F, Navarro M. Perinatal exposure to delta9-tetrahydrocannabinol increases presynaptic dopamine D2 receptor sensitivity: a behavioral study in rats. Pharmacol Biochem Behav. 2003;75(3):565–75.
73. Navarro M, de Miguel R, Rodriguez de Fonseca F, Ramos JA, Fernandez-Ruiz JJ. Perinatal cannabinoid exposure modifies the sociosexual approach behavior and the mesolimbic dopaminergic activity of adult male rats. Behav Brain Res. 1996;75(1–2):91–8.
74. Newsom RJ, Kelly SJ. Perinatal delta-9-tetrahydrocannabinol exposure disrupts social and open field behavior in adult male rats. Neurotoxicol Teratol. 2008;30(3):213–9.
75. Llorente R, Arranz L, Marco EM, Moreno E, Puerto M, Guaza C, et al. Early maternal deprivation and neonatal single administration with a cannabinoid agonist induce long-term sex-dependent psychoimmunoendocrine effects in adolescent rats. Psychoneuroendocrinology. 2007;32(6):636–50.
76. Schneider M, Drews E, Koch M. Behavioral effects in adult rats of chronic prepubertal treatment with the cannabinoid receptor agonist WIN 55,212–2. Behav Pharmacol. 2005;16(5–6):447–54.
77. Trezza V, Campolongo P, Cassano T, Macheda T, Dipasquale P, Carratu MR, et al. Effects of perinatal exposure to delta-9-tetrahydrocannabinol on the emotional reactivity of the offspring: a longitudinal behavioral study in Wistar rats. Psychopharmacology (Berl). 2008;198(4):529–37.
78. Vela G, Martin S, Garcia-Gil L, Crespo JA, Ruiz-Gayo M, Fernandez-Ruiz JJ, et al. Maternal exposure to delta9-tetrahydrocannabinol facilitates morphine self-administration behavior and changes regional binding to central mu opioid receptors in adult offspring female rats. Brain Res. 1998;807(1–2):101–9.
79. Gonzalez B, de Miguel R, Martin S, Perez-Rosado A, Romero J, Garcia-Lecumberri C, et al. Effects of perinatal exposure to delta9-tetrahydrocannabinol on operant morphine-reinforced behavior. Pharmacol Biochem Behav. 2003;75(3):577–84.
80. Economidou D, Mattioli L, Ubaldi M, Lourdusamy A, Soverchia L, Hardiman G, et al. Role of cannabinoidergic mechanisms in ethanol self-administration and ethanol seeking in rat adult offspring following perinatal exposure to Delta9-tetrahydrocannabinol. Toxicol Appl Pharmacol. 2007;223(1):73–85.
81. Fernandez-Ruiz JJ, Berrendero F, Hernandez ML, Romero J, Ramos JA. Role of endocannabinoids in brain development. Life Sci 1999;65(6–7):725–36.
82. Vela G, Fuentes JA, Bonnin A, Fernandez-Ruiz J, Ruiz-Gayo M. Perinatal exposure to delta 9-tetrahydrocannabinol (delta9-THC) leads to changes in opioid-related behavioral patterns in rats. Brain Res. 1995;680(1–2):142–7.
83. Corchero J, Garcia-Gil L, Manzanares J, Fernandez-Ruiz JJ, Fuentes JA, Ramos JA. Perinatal delta9-tetrahydrocannabinol exposure reduces proenkephalin gene expression in the caudate-putamen of adult female rats. Life Sci. 1998;63(10):843–50.
84. Suarez I, Bodega G, Fernandez-Ruiz J, Ramos JA, Rubio M, Fernandez B. Down-regulation of the AMPA glutamate receptor subunits GluR1 and GluR2/3 in the rat cerebellum following pre- and perinatal delta9-tetrahydrocannabinol exposure. Cerebellum. 2004;3(2):66–74.
85. Ferraro L, Tomasini MC, Beggiato S, Gaetani S, Cassano T, Cuomo V, et al. Short- and long-term consequences of prenatal exposure to the cannabinoid agonist WIN55,212–2 on rat glutamate transmission and cognitive functions. J Neural Transm. 2009;116(8):1017–27.
86. Suarez I, Bodega G, Fernandez-Ruiz JJ, Ramos JA, Rubio M, Fernandez B. Reduced glial fibrillary acidic protein and glutamine synthetase expression in astrocytes and Bergmann glial cells in the rat cerebellum caused by delta(9)-tetrahydrocannabinol administration during development. Dev Neurosci. 2002;24(4):300–12.
87. Stella N. Cannabinoid signaling in glial cells. Glia. 2004;48(4):267–77.
88. Blakemore SJ. Teenage kicks: cannabis and the adolescent brain. Lancet. 2013;381(9870):888–9.

89. SAMHSA. Results from the 2008 National Survey on Drug Use and Health National Findings: U.S. Department of Health and Human Services, Substance Abuse and Mental Health Services Administration, Office of Applied Studies, North Carolina. 2008.
90. Pope HG, Jr., Gruber AJ, Hudson JI, Cohane G, Huestis MA, Yurgelun-Todd D. Early-onset cannabis use and cognitive deficits: what is the nature of the association? Drug Alcohol Depend. 2003;69(3):303–10.
91. Zammit S, Allebeck P, Andreasson S, Lundberg I, Lewis G. Self reported cannabis use as a risk factor for schizophrenia in Swedish conscripts of 1969: historical cohort study. BMJ. 2002;325(7374):1199.
92. Di Forti M, Morgan C, Dazzan P, Pariante C, Mondelli V, Marques TR, et al. High-potency cannabis and the risk of psychosis. Br J Psychiatry. 2009;195(6):488–91.
93. Schneider M, Schomig E, Leweke FM. Acute and chronic cannabinoid treatment differentially affects recognition memory and social behavior in pubertal and adult rats. Addict Biol. 2008;13(3–4):345–57.
94. Schneider M, Koch M. Chronic pubertal, but not adult chronic cannabinoid treatment impairs sensorimotor gating, recognition memory, and the performance in a progressive ratio task in adult rats. Neuropsychopharmacology. 2003;28(10):1760–9.
95. O'Shea M, Singh ME, McGregor IS, Mallet PE. Chronic cannabinoid exposure produces lasting memory impairment and increased anxiety in adolescent but not adult rats. J Psychopharmacol. 2004;18(4):502–8.
96. Quinn HR, Matsumoto I, Callaghan PD, Long LE, Arnold JC, Gunasekaran N, et al. Adolescent rats find repeated Delta(9)-THC less aversive than adult rats but display greater residual cognitive deficits and changes in hippocampal protein expression following exposure. Neuropsychopharmacology. 2008;33(5):1113–26.
97. Harte LC, Dow-Edwards D. Sexually dimorphic alterations in locomotion and reversal learning after adolescent tetrahydrocannabinol exposure in the rat. Neurotoxicol Teratol. 2010;32(5):515–24.
98. Viveros MP, Mendrek A, Paus T, Lopez-Rodriguez AB, Marco EM, Yehuda R, et al. A comparative, developmental, and clinical perspective of neurobehavioral sexual dimorphisms. Front Neurosci. 2012;6:84.
99. Chadwick B, Miller ML, Hurd YL. Cannabis Use during Adolescent Development: Susceptibility to Psychiatric Illness. Front Psychiatry. 2013;4:129.
100. Biscaia M, Marin S, Fernandez B, Marco EM, Rubio M, Guaza C, et al. Chronic treatment with CP 55,940 during the peri-adolescent period differentially affects the behavioural responses of male and female rats in adulthood. Psychopharmacology (Berl). 2003;170(3):301–8.
101. Rubino T, Vigano D, Realini N, Guidali C, Braida D, Capurro V, et al. Chronic delta9-tetrahydrocannabinol during adolescence provokes sex-dependent changes in the emotional profile in adult rats: behavioral and biochemical correlates. Neuropsychopharmacology. 2008;33(11):2760–71.
102. Patton GC, Coffey C, Carlin JB, Degenhardt L, Lynskey M, Hall W. Cannabis use and mental health in young people: cohort study. BMJ. 2002;325(7374):1195–8.
103. Lynskey MT, Heath AC, Bucholz KK, Slutske WS, Madden PA, Nelson EC, et al. Escalation of drug use in early-onset cannabis users vs co-twin controls. JAMA. 2003 22–29;289(4):427–33.
104. Agrawal A, Neale MC, Prescott CA, Kendler KS. Cannabis and other illicit drugs: comorbid use and abuse/dependence in males and females. Behav Genet. 2004;34(3):217–28.
105. Fergusson DM, Boden JM, Horwood LJ. Cannabis use and other illicit drug use: testing the cannabis gateway hypothesis. Addiction. 2006;101(4):556–69.
106. Ellgren M, Spano SM, Hurd YL. Adolescent cannabis exposure alters opiate intake and opioid limbic neuronal populations in adult rats. Neuropsychopharmacology. 2007;32(3):607–15.
107. Biscaia M, Fernandez B, Higuera-Matas A, Miguens M, Viveros MP, Garcia-Lecumberri C, et al. Sex-dependent effects of periadolescent exposure to the cannabinoid agonist CP-55,940 on morphine self-administration behaviour and the endogenous opioid system. Neuropharmacology. 2008;54(5):863–73.

108. Spano MS, Ellgren M, Wang X, Hurd YL. Prenatal cannabis exposure increases heroin seeking with allostatic changes in limbic enkephalin systems in adulthood. Biol Psychiatry. 2007;61(4):554–63.
109. Higuera-Matas A, Soto-Montenegro ML, del Olmo N, Miguens M, Torres I, Vaquero JJ, et al. Augmented acquisition of cocaine self-administration and altered brain glucose metabolism in adult female but not male rats exposed to a cannabinoid agonist during adolescence. Neuropsychopharmacology. 2008;33(4):806–13.
110. Tomasiewicz HC, Jacobs MM, Wilkinson MB, Wilson SP, Nestler EJ, Hurd YL. Proenkephalin mediates the enduring effects of adolescent cannabis exposure associated with adult opiate vulnerability. Biol Psychiatry. 2012;72(10):803–10.
111. Pistis M, Perra S, Pillolla G, Melis M, Muntoni AL, Gessa GL. Adolescent exposure to cannabinoids induces long-lasting changes in the response to drugs of abuse of rat midbrain dopamine neurons. Biol Psychiatry 2004.;56(2):86–94.
112. Hajos M, Hoffmann WE, Kocsis B. Activation of cannabinoid-1 receptors disrupts sensory gating and neuronal oscillation: relevance to schizophrenia. Biol Psychiatry. 2008;63(11):1075–83.
113. Raver SM, Haughwout SP, Keller A. Adolescent cannabinoid exposure permanently suppresses cortical oscillations in adult mice. Neuropsychopharmacology. 2013;38(12):2338–47.
114. Gleason KA, Birnbaum SG, Shukla A, Ghose S. Susceptibility of the adolescent brain to cannabinoids: long-term hippocampal effects and relevance to schizophrenia. Transl Psychiatry. 2012;2:e199.
115. Page ME, Oropeza VC, Sparks SE, Qian Y, Menko AS, Van Bockstaele EJ. Repeated cannabinoid administration increases indices of noradrenergic activity in rats. Pharmacol Biochem Behav. 2007;86(1):162–8.
116. Bambico FR, Nguyen NT, Katz N, Gobbi G. Chronic exposure to cannabinoids during adolescence but not during adulthood impairs emotional behaviour and monoaminergic neurotransmission. Neurobiol Dis. 2010;37(3):641–55.
117. Tagliaferro P, Javier Ramos A, Onaivi ES, Evrard SG, Lujilde J, Brusco A. Neuronal cytoskeleton and synaptic densities are altered after a chronic treatment with the cannabinoid receptor agonist WIN 55,212–2. Brain Res. 2006;1085(1):163–76.
118. Rubino T, Realini N, Braida D, Guidi S, Capurro V, Vigano D, et al. Changes in hippocampal morphology and neuroplasticity induced by adolescent THC treatment are associated with cognitive impairment in adulthood. Hippocampus. 2009;19(8):763–72.
119. Lopez-Gallardo M, Lopez-Rodriguez AB, Llorente-Berzal A, Rotllant D, Mackie K, Armario A, et al. Maternal deprivation and adolescent cannabinoid exposure impact hippocampal astrocytes, CB1 receptors and brain-derived neurotrophic factor in a sexually dimorphic fashion. Neuroscience. 2012;204:90–103.
120. Realini N, Vigano D, Guidali C, Zamberletti E, Rubino T, Parolaro D. Chronic URB597 treatment at adulthood reverted most depressive-like symptoms induced by adolescent exposure to THC in female rats. Neuropharmacology. 2011;60(2–3):235–43.
121. Rubino T, Realini N, Braida D, Alberio T, Capurro V, Vigano D, et al. The depressive phenotype induced in adult female rats by adolescent exposure to THC is associated with cognitive impairment and altered neuroplasticity in the prefrontal cortex. Neurotox Res. 2009;15(4):291–302.
122. Yucel M, Solowij N, Respondek C, Whittle S, Fornito A, Pantelis C, et al. Regional brain abnormalities associated with long-term heavy cannabis use. Arch Gen Psychiatry. 2008;65(6):694–701.
123. Mata I, Perez-Iglesias R, Roiz-Santianez R, Tordesillas-Gutierrez D, Pazos A, Gutierrez A, et al. Gyrification brain abnormalities associated with adolescence and early-adulthood cannabis use. Brain Res. 2010;1317:297–304.
124. Arnone D, Barrick TR, Chengappa S, Mackay CE, Clark CA, Abou-Saleh MT. Corpus callosum damage in heavy marijuana use: preliminary evidence from diffusion tensor tractography and tract-based spatial statistics. Neuroimage. 2008;41(3):1067–74.

125. Zalesky A, Solowij N, Yucel M, Lubman DI, Takagi M, Harding IH, et al. Effect of long-term cannabis use on axonal fibre connectivity. Brain. 2012;135(Pt 7):2245–55.
126. Gruber SA, Silveri MM, Dahlgren MK, Yurgelun-Todd D. Why so impulsive? White matter alterations are associated with impulsivity in chronic marijuana smokers. Exp Clin Psychopharmacol. 2011;19(3):231–42.
127. Lorenzetti V, Solowij N, Fornito A, Lubman DI, Yucel M. The Association between Regular Cannabis Exposure and Alterations of Human Brain Morphology: An Updated Review of the Literature. Curr Pharm Des. 2014;20(13):2138–67.
128. Mehmedic Z, Chandra S, Slade D, Denham H, Foster S, Patel AS, et al. Potency trends of Delta9-THC and other cannabinoids in confiscated cannabis preparations from 1993 to 2008. J Forensic Sci. 2010;55(5):1209–17.
129. Seely KA, Lapoint J, Moran JH, Fattore L. Spice drugs are more than harmless herbal blends: a review of the pharmacology and toxicology of synthetic cannabinoids. Prog Neuropsychopharmacol Biol Psychiatry. 2012;39(2):234–43.
130. EMCDDA. Thematic paper- Understanding the 'Spice' phenomenon. Luxembourg: Office for Official Publications of the European Communities 2009 Contract No.: ISBN 978–92-9168–411–3, doi:10.2810/27063.
131. Thomas A, Baillie GL, Phillips AM, Razdan RK, Ross RA, Pertwee RG. Cannabidiol displays unexpectedly high potency as an antagonist of CB1 and CB2 receptor agonists in vitro. Br J Pharmacol. 2007;150(5):613–23.
132. Bhattacharyya S, Morrison PD, Fusar-Poli P, Martin-Santos R, Borgwardt S, Winton-Brown T, et al. Opposite effects of delta-9-tetrahydrocannabinol and cannabidiol on human brain function and psychopathology. Neuropsychopharmacology. 2010;35(3):764–74.
133. Caspi A, Moffitt TE, Cannon M, McClay J, Murray R, Harrington H, et al. Moderation of the effect of adolescent-onset cannabis use on adult psychosis by a functional polymorphism in the catechol-O-methyltransferase gene: longitudinal evidence of a gene X environment interaction. Biol Psychiatry. 2005;57(10):1117–27.
134. Henquet C, Rosa A, Krabbendam L, Papiol S, Fananas L, Drukker M, et al. An experimental study of catechol-o-methyltransferase Val158Met moderation of delta-9-tetrahydrocannabinol-induced effects on psychosis and cognition. Neuropsychopharmacology. 2006;31(12):2748–57.
135. Zammit S, Spurlock G, Williams H, Norton N, Williams N, O'Donovan MC, et al. Genotype effects of CHRNA7, CNR1 and COMT in schizophrenia: interactions with tobacco and cannabis use. Br J Psychiatry. 2007;191:402–7.
136. van Winkel R van Beveren NJ Simons C. AKT1 moderation of cannabis-induced cognitive alterations in psychotic disorder. Neuropsychopharmacology. 2011;36(12):2529–37.
137. Agrawal A, Lynskey MT. Correlates of later-onset cannabis use in the National Epidemiological Survey on Alcohol and Related Conditions (NESARC). Drug Alcohol Depend. 2009;105(1–2):71–5.
138. Wegener N, Koch M. Neurobiology and systems physiology of the endocannabinoid system. Pharmacopsychiatry. 2009;42 Suppl 1:S79–86.
139. Llorente-Berzal A, Fuentes S, Gagliano H, Lopez-Gallardo M, Armario A, Viveros MP, et al. Sex-dependent effects of maternal deprivation and adolescent cannabinoid treatment on adult rat behaviour. Addict Biol. 2011;16(4):624–37.
140. Llorente-Berzal A, Puighermanal E, Burokas A, Ozaita A, Maldonado R, Marco EM, et al. Sex-dependent psychoneuroendocrine effects of THC and MDMA in an animal model of adolescent drug consumption. PLoS ONE 2013;8(11):e78386.
141. Higuera-Matas A, Botreau F, Miguens M, Del Olmo N, Borcel E, Perez-Alvarez L, et al. Chronic periadolescent cannabinoid treatment enhances adult hippocampal PSA-NCAM expression in male Wistar rats but only has marginal effects on anxiety, learning and memory. Pharmacol Biochem Behav. 2009;93(4):482–90.
142. Abush H, Akirav I. Short- and long-term cognitive effects of chronic cannabinoids administration in late-adolescence rats. PLoS ONE. 2012;7(2):e31731.

143. O'Tuathaigh CM, Clarke G, Walsh J, Desbonnet L, Petit E, O'Leary C, et al. Genetic vs. pharmacological inactivation of COMT influences cannabinoid-induced expression of schizophrenia-related phenotypes. Int J Neuropsychopharmacol. 2012;15(9):1331–42.
144. Cha YM, Jones KH, Kuhn CM, Wilson WA, Swartzwelder HS. Sex differences in the effects of delta9-tetrahydrocannabinol on spatial learning in adolescent and adult rats. Behav Pharmacol. 2007;18(5–6):563–9.
145. Renard J, Krebs MO, Jay TM, Le Pen G. Long-term cognitive impairments induced by chronic cannabinoid exposure during adolescence in rats: a strain comparison. Psychopharmacology (Berl). 2013;225(4):781–90.

Chapter 12
Gender Differences in Cannabis Addiction and Dependence

Caroline Davis and Liana Fattore

Abstract In humans as in animals, males and females are dissimilar in their genetic and hormonally driven behaviour; they process information differently, perceive experiences and emotions in different ways, and display diverse attitudes. In the human population, gender differences in the frequency and patterns of cannabis use have been identified in clinical studies and in anecdotal observations, although the nature of these differences is still poorly explored. The motivations for smoking cannabis are also different between sexes, especially in adolescents. A number of potential factors which could provide a neurobiological basis for gender-based differences in cannabinoid addiction have been identified, among which are organizational and activational effects of gonadal hormones, socio-cultural factors, different stress responsiveness and impulse-control ability as well as different cannabinoid pharmacodynamic and pharmacokinetic in males and females. In this chapter we will review both clinical and laboratory-based research evidence revealing important sex-related differences in cannabinoid-induced effects on reward and motivation.

Keywords Sex-dependent differences · Gender differences · Cannabinoid-taking · Cannabinoid-seeking · Sex hormones · Marijuana smoking · Drug addiction · Self-administration

L. Fattore (✉)
Institute of Neuroscience-Cagliari, National Research Council (CNR), Monserrato, Italy
e-mail: lfattore@in.cnr.it

CNR Institute of Neuroscience-Cagliari, National Research Council-Italy @ Department of Biomedical Sciences, University of Cagliari, Cittadella Universitaria di Monserrato, Monserrato, Cagliari, Italy

C. Davis
Department of Kinesiology and Health Sciences, York University, Toronto, Ontario, Canada

P. Campolongo, L. Fattore (eds.), *Cannabinoid Modulation of Emotion, Memory, and Motivation*, DOI 10.1007/978-1-4939-2294-9_12

Introduction

It is now well-established that strong male-female bias exists in the prevalence of many brain disorders. For instance, while women are more likely to suffer from depression and anxiety, and to experience more eating-related disturbances, most alcoholics, drug addicts, and compulsive gamblers are men [1]. Undeniably, different social attitudes and expectancies for men and women impact their respective risk for, and expression of, mental problems. However, these predispositions have persisted across time and are relatively uniform in most cultures, also suggesting an important role for sex-specific biological causes [2]. Indeed, in the past two decades, there has been a growing recognition of the importance of taking account of both *sex* and *gender* differences in all relevant aspects of health-related research. Accordingly, a number of institutions, funding agencies, and research bodies worldwide have adopted legislation and guidelines for the inclusion of "gender- and sex-based analysis" in the interpretation of data [3]. Regrettably, however, many researchers continue to ignore one or both concepts in their studies, or use the terms interchangeably and therefore inaccurately [4].

Historically, medical science focused largely on *sex* differences—that is, the biological attributes and characteristics associated with the terms 'male' and 'female'—while social-science health research has mainly directed its attention to *gender* effects, which describe the differences between men and women due to socio-cultural and socio-political influences. Human addiction research has also conflated sex and gender terminology ever since this field of research became prominent in the 1980's. Especially in the earlier years, the preponderance of studies examined *gender-related* individual differences concerning emotional, cognitive, and other psychological mechanisms related to the use and abuse of addictive behaviours. Psychosocial factors focusing on family dynamics, history of trauma, ethnic diversity, and gender stereotypes were also the topic of many investigations. Only in recent years—especially the last decade or so—has the study of *sex-related* differences in addiction research burgeoned, although the most robust evidence in this field still derives largely from well-controlled and replicated preclinical studies.

Many human behaviours are not entirely learned, but instead are driven by evolved instincts and urges. A fundamental tenet of evolutionary science is that overtime the human psyche has been shaped by two processes—*natural selection*, which produces adaptations that bestow a survival advantage to the species, and *sexual selection* which fosters traits that confer a species-specific mating advantage [5]. In other words, humans are born with "innate cognitive blueprints" that are imperative for their ability to prosper and to reproduce. Importantly, we also have an evolved capacity to experience considerable pleasure and happiness from these key adaptive pursuits, like eating, drinking, and creating protective shelter, as well as mating and having children. Those experiences in our lives that we find transcendent—whether they are socially-sanctioned and essential to our existence, or illicit vices (as will be discussed later)—all activate an anatomical and neurochemically-defined "pleasure circuit" in the brain [6]. And not surprisingly, given their highly differentiated roles, especially in reproduction, there is strong evidence that sexually-dimorphic forces shape the pronounced preferences, emotions, and behavioural responses that are observable globally between males and females, and that cannot be accounted for by gender-specific parenting styles, cultural attributes, or patterns of socialization.

Gender and Sex Differences in Addictive Behaviours

Behavioural addictions are complex disorders with a multi-faceted etiology including environmental factors, comorbid influences, personality traits, and stress responsivity. Several behaviours, besides the ingestion of psychoactive substances, produce immediate reward that may engender persistent behaviour, despite knowledge of adverse consequences and diminished control over the behaviour. Although the term "addiction" is traditionally linked to the out-of-control use of legal and illegal substances, numerous studies demonstrate that engaging in non-drug-related activities, especially those involving 'natural' rewards, can also result in addictive behaviours to the extent that they act on the brain "pleasure circuit". Thus, the definition of a "behavioural addiction" currently describes any behaviour characterized by a sense of nervousness or awakening before committing the act followed by satisfaction and/or relief once the act has been committed, as well as the inability to resist an urge or drive in the face of the negative consequences that may affect themselves, family, friends, or work. For example, behavioural addictions include compulsive overeating and sexual activity, pathological gambling and internet use, kleptomania and pyromania, compulsive spending, and excessive exercising or working. All these behaviours—often referred to as "impulse control disorders"—share common features such as compulsiveness, impulsivity, impaired decision-making, craving, tolerance, withdrawal, and high rates of relapse. Gender-related differences in the abovementioned addictive behaviours indicate that men typically self-report more frequent problems with overuse of exercise, gambling, and sex, while women are more likely to engage in compulsive shopping and food binging reviewed in [7]. Figure 12.1 summarizes addictions in which gender- and sex-related differences have been described.

Natural Rewards

Reproductive Behaviours

Historically and universally, there is great diversity between men and women in their mating preferences. Men tend to be much more attracted to youth and beauty than are women, who, by contrast, show a stronger preference for high social status and plentiful resources when choosing a parenting partner. While both men and

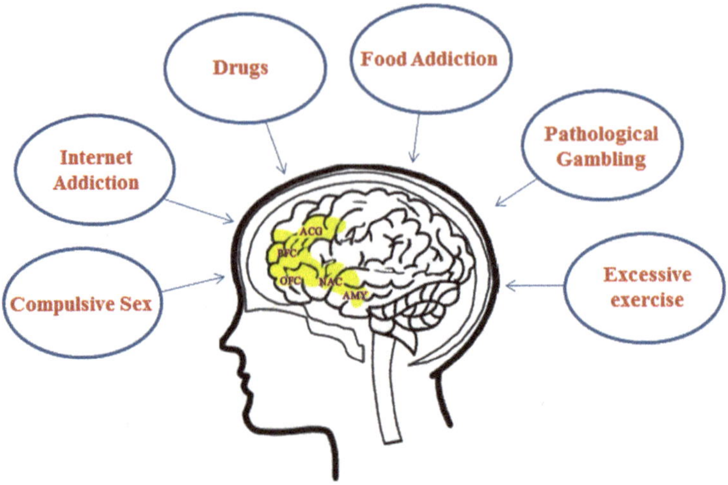

Fig. 12.1 Addictive behaviours that share the ability to affect the brain mesocorticolimbic system and extended amygdala. *ACG* Anterior Cingulate Gyrus, *AMY* Amygdala, *NAC* Nucleus Accumbens, *OFC* Orbito-Frontal Cortex, *PFC* Pre-Frontal Cortex

women engage in explicit 'sexual signaling' when trying to attract a mate, they do so quite differently—each by enhancing the personal cues and attributes favoured by the opposite sex. Another powerful example of differences in reproductive strategies is the expression of romantic jealousy. Although both men and women are equally apt to experience this emotion, men are more likely to react to *sexual* infidelity while women are more distressed by *emotional* infidelity [8]. A widely accepted explanation for such disparity is that the uncertainty of paternity is the greatest threat to men's genetic interests, while for women the peril concerns the potential loss of their partner and his resources [9].

A related reproductive behaviour, also showing pronounced sex differences which emerge early in life, is parental responsiveness. Females typically spend much more time than males nurturing their offspring and engaging in caregiving activities [10]. These differences appear to be largely influenced by *in utero* exposure to sex-specific gonadal hormones. As seen in preclinical studies, testosterone appears to have an inhibitory effect on parental responsiveness via the oxytocin neural system [11]. Congruently, women with lower salivary testosterone reported wanting more children, and at an earlier age, than those with higher levels [12]. Similarly, men with children had lower testosterone levels than childless men [13], although among the fathers, testosterone levels correlated positively with the number of children they had at the end of their life. These findings add to a growing body of research confirming that among men pair bonding and parental care are associated with lower testosterone levels, while the pursuit and acquisition of sex partners is associated with higher hormone levels.

By contrast, estrogen appears to 'feminize' the human brain. Individual differences in trait estrogen levels have been linked, for instance, to more feminine body shapes and more feminine looking faces [14]. It also positively enhances intimacy and attachment motivation [15], and parental responsiveness in women as indicated by their reported ideal number of children [16]. It was also found that the facial attractiveness of women who desired many children was rated as more feminine than those desiring fewer children [16].

Eating Behaviours

Just as *sexual selection* appears to have influenced mate preferences, *natural selection* is likely to have shaped our food preferences and our ability to evaluate the cost-free value of these substances [17]. A sexual dimorphism in fondness for certain tastes would also be expected given the historical differences in acquisition abilities and roles between men and women. As evidence, women tend to display a greater preference for sweet foods compared to their male counterparts [18]—similar to the preclinical differences found in male and female rats [19]—and more likely to experience food cravings [20]. In recent years, we have also learned that binge eating is significantly greater in women than in men [21], and that the most pronounced increases in morbid obesity have occurred in women [22]. For these reasons, the majority of animal models commonly used to study eating disorders use females [23].

In addition, there appear to be sex-specific genetic effects on the intake of certain substances. For instance, the frequency of chocolate consumption was found to have a significantly higher heritability in women than in men [24]. Relatedly, women with a family history of alcoholism preferred significantly higher sucrose concentrations and craved sweets more often than their non-affected counterparts, and compared to the differences found in men [25]. Perhaps this is not surprising since alcohol and sugar are biochemically congruent—alcohol is simply the fermentation of sugar [26]. These findings mesh with the well-established evidence that alcohol-preferring rodents consume greater quantities of a sweet solution than non-alcohol-preferring animals [27, 28], and that the reverse is also the case. Animals bred for high-saccharin consumption also showed elevated rates of drug acquisition and escalation of intake compared to those bred for low-saccharin consumption—and in both groups female rates exceeded those of males [29].

A modern-day study of a still-existent, full-time, hunter-gatherer tribe (the Hadza of Tanzania) is very relevant to this discussion because of their exposure to the paleo-diet, from which our ancestral genetic endowment is believed to originate [17]. This society displayed pronounced sex differences in their preference for certain foods. Men and women equally preferred the most available calorically-dense food (honey). Women, however, ranked berries as the second more preferred food despite their relatively low-caloric value, with meat as the 4th choice, while men chose meat as the 2nd and berries as the 4th. The authors speculate that these differences reflect an evolved disparity in taste preferences because of differences in the dietary requirements of men and women. Specifically, in men there is a greater need for the allocation of energy resources to muscle, while in women it is to fat.

Drug Rewards

In a recent study, it was estimated that "addiction and harmful use" of substances (e.g. alcohol, tobacco, illicit drugs) contributes to the greatest economic burden and cost-of-illness of all "disorders of the brain" [30]. Historically, drug addiction has been more prevalent in men than in women [31]. Recently, however, the gap appears to be narrowing, suggesting that earlier disparities may simply reflect variation in opportunity and gender-role expectations rather than differences in vulnerability [32, 33]. Indeed, many addiction risk factors appear to be more pronounced in women than in men. For instance, women tend to increase their rate of drug consumption more quickly than men, are more likely to relapse, and to have longer periods of drug use before their next attempt to quit [34, 35]. Women with addictions also report more pronounced cravings and subjective drug effects than their male counterparts [36]. In general, pattern of sex differences seems to be similar for most drugs of abuse [37]. In addition, while the genetic and environmental risk factors for addiction are roughly equivalent in proportion, they are mediated both through factors common to all substances, and by substance-specific effects.

Stimulant Drugs: Cocaine, Amphetamines and Methylphenidate

Although there are higher rates of cocaine use and abuse among men, women appear to have a heightened response to some aspects of psychomotor stimulant use—specifically to a phenomenon known as *telescoping*, which describes an accelerated progression from the initiation of drug use to the development of dependence, and to first treatment admission [38]. Moreover, women with these addictions tend to present with a more severe clinical profile despite having used less of the substance, and for a shorter duration, than their male counterparts. Currently, little is known about the biological basis for this augmented time-course to addiction in women. What information is available has only been gathered since the recent development of progressive addiction paradigms in animals, which capture characteristics more relevant to human drug addiction—viz. excessive use and its role in the development of enhanced motivation for drug acquisition [39]. Findings from such studies confirm that females have a heightened vulnerability to develop an addicted phenotype, and that estradiol is critically involved in its development—indeed it may be a requirement [39]. However, once the addiction has developed, there appears to be little difference between females and males in the course of the condition.

While the experimental evidence for sex differences in *behavioural* response to psychomotor stimulant drugs is relatively robust in preclinical studies—with females showing a greater reaction than males [37]—the results of human laboratory studies are more equivocal. For instance, concerning the reinforcing effects of these drugs, some research demonstrates that women are more sensitive than men, while other studies have reported null findings [40]. One moderating factor may relate to drug-dosage. To wit, a recent study found that women demonstrated a greater reinforcement to low-dose *d*-amphetamine using a progressive-ratio reinforcement task, while men found a higher dose more reinforcing [40]. Another consideration appears to be the route of administration. Studies in human and non-human primates indicated minimal sex differences in response to intranasal cocaine. By contrast, women were more responsive than men to smoked cocaine, but the differences were only significant if comparisons were made when women were in the luteal phase of their menstrual cycle [35].

Methylphenidate is a close relative to illicit psychomotor stimulants like cocaine and amphetamine with structural and pharmacological similarities to both these drugs. Commonly known as Ritalin®, methylphenidate has mainly been used therapeutically, and is especially heavily prescribed for the treatment of attention deficit/hyperactivity disorder (ADHD) in children, adolescents, and an increasing number of adults. Regrettably, however, there has been increasing evidence of the non-medical (ab)use of this drug in schools, universities, and the work place to combat fatigue, improve concentration, and sustain productivity [41]. Understanding sex-specific effects of methylphenidate is especially relevant given that ADHD is much more frequently diagnosed in males (2- to 9-fold greater prevalence), although affected females seem to have a more severe symptomatology and a higher genetic loading for the disorder [42]. There is also some evidence—albeit based mostly on animal studies—that exposure to methylphenidate can exert long-term brain and behavioural effects, especially in females [43]. Recent animal studies also indicate that female animals show a greater behavioural response to methylphenidate and increased sensitization to the drug compared to male animals [44], although these effects are not entirely consistent among various strains of rats [45]. Unfortunately, there is a dearth of well-controlled studies of sex differences in the human condition, but what is available appears to be relatively consistent with the animal research. For instance, in a study of the pharmacodynamic response to long-acting methylphenidate in children with ADHD, girls showed a superior and more rapid response following drug administration than boys, but also a more rapid decline as time from dosing increased. Given, however, that female sex hormones have been shown to play an important role in brain dopamine function and response to stimulants, these findings cannot be extrapolated to medication effects in adults taking methylphenidate [46].

Depressant Drugs: Alcohol and Opiates

Similar to the psychostimulants described above, there is a higher rate of alcoholism among men than women. In rodent models using free-drinking paradigms, however, female animals tend to consume more alcohol than their male counterparts [47], and intake is significantly higher in adolescent compared to adult mice [48]. Therefore, it is perhaps not surprising that women tend to develop more alcohol-related medical complications, and at a lower drinking threshold than men [49]. There is also evidence that alcoholic women have more depressive symptoms than men, and use drinking more often to ameliorate negative emotions [49]. Similar negative mood-related sex differences were also found in heavy drinking male and female adolescents [50].

It is therefore of interest that a recent study found greater vulnerability to, and severity of, alcohol-induced damage in certain brain areas, including the hippocampus, in alcoholic females compared to their male counterparts [2]. Since the results of a 14-year follow-up study also indicated that annualized death rates were significantly higher in alcohol-dependent women than in men, and compared to the age- and sex-specific general population—viz 4.6-fold and 1.9-fold, respectively [51]—a better understanding of sexually dimorphic responses to chronic alcohol intake is an important goal for targeted treatment interventions. Despite the widely accepted view that both social and hormonal factors influence the prevalence and severity of alcoholism-related behaviours, it is becoming apparent that established sex differences are also influenced by genetic factors. In other words, these male-female differences are regulated both by gonadal sex and by chromosomal sex [52].

While most of the information on the nature of opiate addiction comes from heroin abusers, it is highly relevant that those seeking treatment for dependence on prescription painkillers currently outnumber individuals with heroin addiction—a factor that may limit the generalizability of earlier research [53]. General observations suggest that adult men are still more likely than women to use *illicit* opiate drugs, but that women are at higher risk of abusing opioids through initial prescription painkiller use [54]. Opiate-dependent women also report stronger cravings for opiates, have higher "addiction severity index" and psychiatric severity scores [36] than men, and more frequently use opioids to cope with anxiety and tension [53]. In summary—and similar to findings with other addictive substances—although the prevalence of addiction to prescription painkillers is roughly similar between men and women, the functional impairment of their disorders appears to be more severe for women.

Nicotine

Historically, adolescent boys were significantly more likely to smoke cigarettes than girls. In recent years, however, there has been a substantial narrowing of the gender gap [55]. Currently, the school appears to be the main setting for initiation among boys, but more girls begin smoking in the home [56]. While these studies suggest gender-specific modeling influences, other research points to possible biological causes for the increasing prevalence of smoking among girls. The results of a sex- and gender-based analytic review of studies over a 20 year period from 1989–2009, indicated that girls are more at risk of smoking than boys when their mothers smoked prenatally [57]. The authors propose that *in utero* exposure to nicotine may increase the sensitivity of the dopamine motivational system to the effects of nicotine—a process that is further enhanced by estrogen production during female puberty.

It is also well-established that individuals with a history of substance use disorders and/or a psychiatric disorder have a higher rate of nicotine dependence than the general population [58, 59], and that a large majority of individuals in drug-treatment programs concurrently use tobacco [60]. However, male-female differences appear to moderate these co-morbidities. For instance, a recent study of patients in a smoking cessation program reported that male smokers were more likely to have a history of alcohol, cocaine, or marijuana abuse/dependence, whereas history of a psychiatric disorder (e.g. anxiety or depression) was significantly associated with being female [61]. These results mesh with the clinical findings that women smokers show greater abstinence-induced increases in negative affective states such as anger, anxiety, and depression [62, 63], and that negative mood induces stronger cravings in women than in men [64]. In a controlled experimental study, women also smoked sooner than men following an implicit negative mood-induction using music [65].

While some research has shown that men are more likely to quit smoking than women, other information, based on longitudinal data from a large national sample in the US, indicated no differences in smoking cessation rates [66]. What emerged, however, was that women have more difficulty quitting in the longer term due to higher smoking-relapse and smoking-reinitiation rates [66]. A review of two decades of smoking cessation treatment research also found that depression had a greater impact on treatment outcomes for women than for men [67].

One of the most common adverse consequences of smoking cessation is weight gain. Typically, women gain more weight than men and exhibit significantly higher levels of weight concerns and body dissatisfaction as a consequence [68]. Among the novel pharmacologic treatments developed to reduce weight gain, the *mu*-opioid receptor antagonist naltrexone has shown some promise for producing long-term reductions in abstinent smokers—albeit this effect was only found in women [69]. King et al. [69] surmised, based on the demonstrated links between brain reward activation and eating behaviours [20, 70], that women may be more sensitive to the opioidergic and dopaminergic processes underlying palatable food consumption, which could thereby be altered more effectively by the *mu* opioid receptor antagonism produced by naltrexone.

Cannabis

Cannabis is by far the most widely used illicit drug, consumed by millions of people worldwide. Although cannabis withdrawal lacks clinical significance and is therefore not recognized in the *Diagnostic and Statistical Manual of Mental Disorders, Fifth Edition* (DSM-V), marijuana users often display a withdrawal syndrome, and attempts to relieve these symptoms facilitate relapse to drug use during cessation attempts [71]. According to United Nation Office on Drugs and Crime, the use of legal and illegal substances is more widespread (i) among males than among females,

(ii) in metropolitan than in rural areas and (iii) among the unemployed than among working population [72]. Male-female differences in cannabis use have been identified in several clinical studies and in anecdotal observations, although the nature of the differences has not been well explored. The most consistent finding arising from the surveys conducted so far on this topic is that boys are more likely to be 'heavy users' than girls [73]. For example, according to the Ontario Student Drug Use Survey conducted between 1999 and 2003, daily marijuana smoking appears to be more common among boys (6.2 %) than girls (2.2 %) [74]. Similarly, in Europe, more males than females use cannabis and attend drug treatment services [75, 73].

Despite differences in prevalence rates, the risk factors associated with marijuana smoking by boys and girls use are not well described. It is possible that cannabis smoking is associated with different emotional or mental states in males than in females, since anxiety disorders tend to be more frequent in the latter [76]. Nevertheless, there is good evidence that males are more than twice as likely to use synthetic cannabinoids as females [72]. Male marijuana smokers also experience greater cardiovascular and subjective effects [77], more evident withdrawal symptoms [78], and have higher circulating levels of THC [79] than female smokers. Male high-school students who smoke marijuana report poor family relationships and problems at school more often than age-matched female students [80], are less likely to be cannabis-only users (i.e. are poly-substance users), and display a higher prevalence of panic attacks and personality disorders [81]. Consistent with these differences, among non-marijuana smokers, men are more sensitive to the subjective effects of THC alone than women [82]. On the contrary, although there are no apparent differences in intoxication or plasma THC levels between men and women after smoking marijuana [83], women report significantly more dizziness than men, and are more susceptible to cannabinoid-induced hemodynamic changes and visuospatial memory impairment [84], smoke marijuana mainly when they feel anxious [85], and show higher sub-type 1 cannabinoid receptor (CB1) protein expression than men, as measured in blood samples [86]. However, when cannabis smokers are matched for use, although men and women significantly differed in body weight, ratings of cannabis' subjective effects that are associated with abuse liability are higher in women than in men [87].

Factors determining vulnerability to cannabis dependence have proven difficult to untangle in human studies. Notably, sexually-dimorphic effects of THC on anxiety-related behaviours and locomotor activity have been described in adolescent rats [88]. In humans, several genes have shown to have critical roles in determining the risk for cannabis use [89–92]. Marijuana withdrawal after abstinence and cue-elicited craving has been associated with two single nucleotide polymorphisms (SNPs) on two genes involved in regulating the endocannabinoid system, cannabinoid receptor 1 (CNR1) and fatty acid amide hydrolase (FAAH) [93]. More specifically, several mutations have been found in CNR1 and FAAH genes, which lead to altered mRNA stability and transcription rate, or a reduction of the activity of the encoded protein [92]. Importantly, these functional mutations are associated not only with marijuana dependence, but also with cocaine, alcohol, heroin and nicotine dependence [94]. Stress relief is the most commonly reported benefit from smoking marijuana, and stress-related factors such as negative life events or trauma have been positively associated with marijuana use. Thus, a different response to traumatic episodes could account, at least partly, for the observed discrepancies in the amount and frequency of marijuana use between men and women [95].

Mental health status and emotional traits such as depression- and anxiety-proneness, may also represent important risk factors for cannabis use. Accordingly, animals with a depressive-like phenotype tend to self-administer higher amount of cannabinoid than controls [96]. In humans, female adolescents reporting relatively poor mental health are also more at risk than boys for frequent and heavy cannabis use [97]. Moreover, girls with marijuana dependence exhibit higher anxiety compared to boys with marijuana dependence [98]. It should, however, be noted that for girls, the drug is typically obtained through their social relationships with boys, supporting the idea that girls smoke marijuana to impress boys, whereas boys 'get high' for the sake of the experience [99].

Finally, cannabinoid pharmacodynamics and pharmacokinetics are likely crucial factors, as sex differences in cannabinoid effects might also be due to differences in muscle mass and fat tissue distribution between males and females. Cannabinoids are lipophilic, and a high concentration is sequestered in fat tissue. Women have a higher percentage of body fat than men, suggesting that women experience weaker effects because more THC is retained by fat cells. Accordingly, a recent pharmacological resting-state FMRI study reported higher average plasma concentration for females compared to males [100]. In addition, cannabinoids may be differentially metabolized to active and inactive metabolites in men and women. In contrast to human males, male rodents have a higher percentage of body fat, which could account, at least partly, for the inconsistent results reported from human and animal studies. For example, female rodents express greater amounts of hepatic cytochrome P-450 isozymes and aldehydeoxygenase activity that may facilitate conversion of THC to potent bioactive metabolites such as 11-hydroxy-THC [101]. Consistent with this, levels of THC metabolites in brain tissue, including 11-hydroxy-delta9-THC, are higher in females than in males, likely contributing to the greater behavioural effects of THC in female compared to male rats [102]. On the other hand, several molecular changes occur differently, and in a region-dependent manner, in the male and female brain after prolonged exposure to THC [103], which are schematically illustrated in Fig. 12.2.

Hormones and the Brain

Gonadal hormones such as androgens (testosterone), estrogens, and progesterone and its metabolites—allopregnanolone and pregnanolone (progestins)—have considerable influence on sexually-differentiated brain development, and these effects begin to occur soon after conception. Prenatal hormone exposure appears to be especially critical for the early *organization* of the brain, while the effects of hormonal changes at puberty are thought to be mostly *activational,* and a matter of

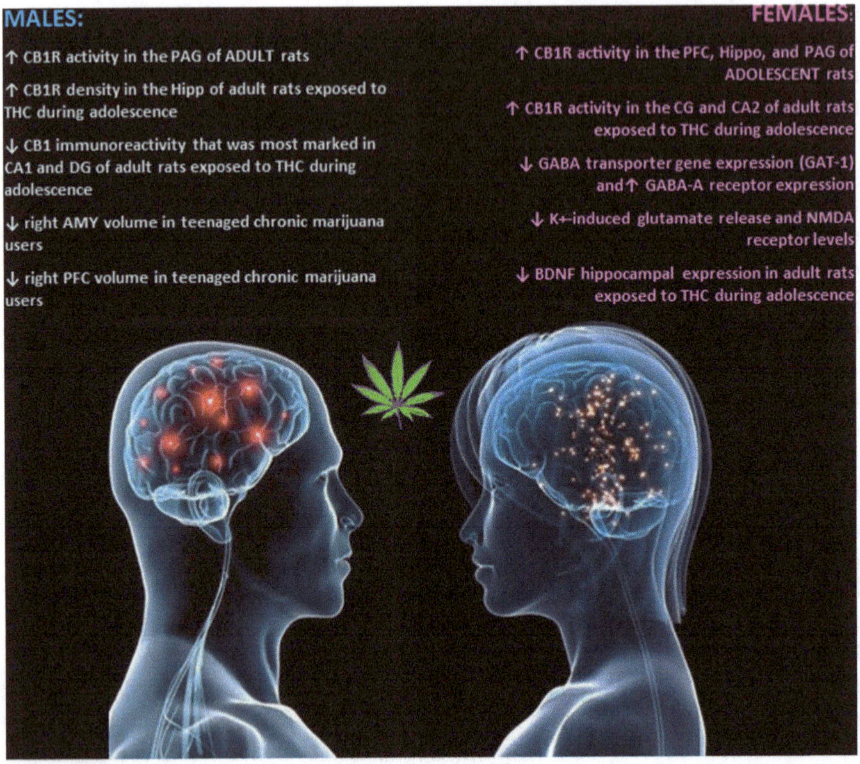

Fig. 12.2 Sex-dependent differences after chronic THC. Schematic list of the molecular changes occurring differently in a region-dependent manner in the male and female brains after prolonged exposure to THC. *AMY* Amygdala, *CA1, CA2* CA1, CA2 fields of the hippocampus, *CG* Cingulate Gyrus, *DG* Dentate Gyrus, *Hipp* Hippocampus, *PAG* Periaqueductal Gray, *PFC* Prefrontal Cortex

'fine-tuning' the earlier pattern of neural growth and maturity [104]. In particular, circulating *in utero* sex-hormone levels can have life-long effects on the personality, and the social and emotional behaviours, of individuals, including their relative risk for a range of psychopathologies like addiction disorders [105]. Compelling evidence has also demonstrated that hormones may interact with stress and reward systems to affect drug-taking and drug-seeking behaviours.

Testosterone

Testosterone has been extensively studied, not only for its role in sexual behaviour but also for its 'masculinizing' effects on psychosocial and psycho-behavioural development. For example, female animals exposed to excess androgen during gestation typically show more masculinized and defeminized behaviours than their non-exposed counterparts [106]. However, because of ethical imperatives against the manipulation of hormones, studies have perforce relied mainly on natural-setting paradigms to investigate the consequences of abnormal hormone levels and variation across individuals in human research. For instance, polycystic ovary syndrome occurs in about 7 % of women of reproductive age and is characterized by several health complications including hyper-androgenism [107]. In a recent longitudinal study comparing the female offspring of women with polycystic ovary syndrome to those of healthy controls, it was found that the former had higher autism-spectrum symptom scores—in keeping with the 'extreme male brain' theory of autism–and lower empathy scores than controls [108]. Even in the control participants, prenatal testosterone levels in amniotic fluid correlated with more masculinized behaviours suggesting that even within the normal range, testosterone makes some individuals more male-sex typical than others. These results mesh with longitudinal data showing conclusively that higher prenatal testosterone exposure contributes to greater masculinization of behaviour in both sexes after birth [104].

Another natural experiment used full life-history data from four generations of male and female lines to test the hypothesis that maternal parity may be a proxy for foetal testosterone exposure because *in utero* levels of testosterone are known to decrease across parities. The results were supportive of the prediction, indicating that low parity sons and high parity daughters had the greatest reproductive potential. Findings endorsed the view that maternal parity has an effect on offspring fitness via varying levels of testosterone across pregnancies—an effect which is moderated by the sex of the child [109]. Twin studies also provide a natural experiment for the role of testosterone on sex-typed behaviours based on the evidence that testosterone is transferred to a female foetus from a male foetus located adjacent to it in the womb, via the mother's bloodstream or the amniotic fluid [105]. Indeed, there is considerable evidence that females from an opposite-sex (OS) twin pair display more masculine behaviours and more male morphological characteristics relative to single females or same-sex (SS) twin females [105]. These *in utero* factors may also increase the relative risk for various disorders and diseases in twins. For example, the prevalence of anorexia and bulimia nervosa has a strong sex-bias with females outnumbering males by a ratio of about 10 to 1. Using (dizygotic) twin-registry data and an outcome measure of disordered-eating symptoms, a significant linear trend in disordered eating was seen with SS female twins showing the highest scores, followed by OS female twins, then OS male twins, and finally SS male twins showing the lowest scores [110]. The OS female twins also had lower disordered-eating scores compared to age-equivalent single females raised with at least one brother.

In animals, testosterone was found to act in a similar way to drugs of abuse, in that it increases the rates of bar pressing for electrical brain stimulation [111], induces a conditioned place preference in rats [112, 113] and mice [114], and sustains oral [115, 116], intravenous and intra-cerebro-ventricular self-administration in male rats and hamsters [117], which are all hallmarks of a drug's rewarding effects.

Regarding the relationship between testosterone levels and cannabis exposure, human studies have shown that neither oral THC nor marijuana cigarettes alter testosterone plasma levels in healthy male volunteers [118, 119], nor in in healthy male marijuana smokers [120], although a depressant effect has also been described [121]. Discrepant data also exist on testosterone levels and cannabis consumption in women, since chronic marijuana use was found to induce no significant effect on testosterone concentration in healthy women [122], while women who use cannabis frequently and for extended periods of time display significantly higher levels of plasma testosterone [123]. However, the absence of a diminution in testosterone level was ascribed to an inadequate or excessive temporal lag between the last exposure to cannabis and testosterone measurement. Indeed, other studies strengthened the idea that cannabis use may temporarily decrease testosterone production [124].

Estrogen and Progesterone

For many years the prevailing view was that feminization was a passive biological process that occurred in the absence of high levels of androgens. More recently, however, there has been a growing recognition that other hormones are necessary for the complete sex-specific development of the female brain [105, 125]. It is now well- established that the rapidly rising levels of estrogen, which occur in females during puberty, remodel the female brain by fostering, for example, alterations in food intake in response to changes in energy availability [126]. Post-pubertal females are also significantly more responsive to the presence of offspring than males. Moreover, these behaviours are induced more rapidly after the initial pregnancy suggesting that the establishment of maternal responses at childbirth fosters long-term changes in the brain [127].

There is also good evidence that estrogen has a role in maintaining dopaminergic cell numbers both in the substantia nigra and the ventral tegmental area (VTA), primarily through its action on estrogen beta receptors [128]. These mechanisms are likely to account for—at least in part—the fact that women are significantly less likely to develop Parkinson's disease, and to do so at a later age [129]. Studies have also shown that anatomical brain changes associated with aging are sexually dimorphic with males showing greater deficits related to prefrontal cortical function than females. These findings may reflect the protective effects of estrogen which continues to be secreted at low levels after the reproductive years [130]. Females also show consistently higher striatal dopamine release in response to psychomotor stimulant drugs [128], which may account for their tendency to develop addictions more rapidly. Notably, imaging studies revealed an augmented reactivity of the reward system in women during the mid-follicular phase when estrogen is unopposed by progesterone [131]. Neurofunctional modulation of the reward system by gonadal steroid hormones in humans is corroborated by the finding of a positive correlation between activation of the amygdalo-hippocampal complex and estradiol level [131].

Addiction research has paid less attention to progesterone than estrogen, but recent findings indicate that progesterone, and its metabolites allopregnanolone and pregnanolone (collectively termed progestins), may also be important in altering the effects of drugs of abuse [132]. Intriguingly, pregnanolone was recently found to suppress cannabinoid self-administration in mice [133].

While estrogen enhances drug-induced effects in both animals and humans, the same effects are attenuated by progesterone. Clinical studies demonstrated that progesterone diminishes the positive subjective effects of cocaine and nicotine in women, and is associated with reduced cocaine craving. That is, the subjective effects of stimulants vary across the menstrual cycle [134], with the greatest subjective effects observed when estrogen levels are high and relatively unopposed by progesterone [135]. In rodents, progesterone also attenuates cocaine conditioned place preference and self-administration [35, 136, 137], while in nonhuman primates progesterone decreases nicotine self-administration [138].

In rhesus monkeys, THC produces a significant decrease in the level of progesterone [139]. On the other hand, estradiol, a key estrogen produced as a neurosteroid in the brains of both males and females, was found to stimulate post-synaptic mobilization of anandamide, which in turn retrogradely suppresses GABA release from CB1 receptor-containing inhibitory synaptic inputs. Most pertinent to the aim of this chapter, is that the estradiol modulation of endocannabinoid tone, and its consequent suppression of inhibition, is sex-specific, occurring in female but not in male rats [140]. Estradiol may also elicit changes in emotional behaviour through an endocannabinoid mechanism, which appears more evident in females than in males [141].

When administered chronically during adolescence in female rats, THC alters their sensitivity to THC during adulthood and causes long-term effects on operant learning and performance tasks that are influenced by the ovarian hormones [142]. In turn, ovarian hormones induce enhancing effects on cannabinoid self-administration and cannabinoid-seeking reinstatement [143, 144]. In addition, human studies have shown that peak plasma anandamide occurs at ovulation and is positively correlated with estradiol [145].

Oxytocin

Estrogen is also known to interact with cortisol and oxytocin in a way that increases women's sensitivity to interpersonal stressors, particularly at puberty [146]. Oxytocin is a neuropeptide produced in the mammalian hypothalamus, and influences a broad range of complex central and peripheral physiological functions. Indeed, in recent years it has become one of the most studied peptides of the human neuroendocrine system [147]. The specific effects of oxytocin vary across individuals and some of the diversity can be attributed to variations in the level of oxytocinergic functioning or its signalling strength [148]. Originally oxytocin was recognized for its role in parturition, lactation, and maternal care, but later research identified its function as a potent regulator of human social behaviour. In particular, it enhances the formation of pair bonding and social attachment [149]. It has also been shown to increase trust and empathy in social situations [104] and to reduce social anxiety and avoidance behaviours [149]. There appear, however, to be sex-specific differences in oxytocin-dependent threat processing, which is generally enhanced in females. For instance, women treated with this hormone displayed more avoidance and safety-seeking behaviours in response to menacing scenes and faces, while the opposite response was seen in men [150]. It is important to note that oxytocin levels are generally higher in women and that estrogen is known to stimulate oxytocin release from the hypothalamus and promote oxytocin gene expression as well as receptor binding in the amygdala [150].

In conjunction with dopamine, oxytocin appears to promote a broad range of reward processes and behaviours such as appetite, sexual behaviour, and parental sensitivity to their offspring [149]. A large body of converging research shows that oxytocin also alters many addiction-related behaviours and responses. For instance, it decreases the severity of alcohol withdrawal and attenuates tolerance to, and consumption of, a range of substances like heroin and cocaine [151]. Moreover, popular party drugs such as MDMA (methylenedioxy-n-methylamphetamine, ecstasy) and GHB (gamma-hydroxybutyric acid) may activate the brain's oxytocin system to produce their characteristic prosocial and prosexual effects [152]. Importantly, oxytocin is down-regulated in the mesolimbic system by chronic cannabinoid exposure [153], and it has been proposed that an estrogen and oxytocin-linked component may play a critical role in cannabis-withdrawal symptoms [154], and in modulating food and water intake [155].

Oxytocin is also higher during early amorous relationships and predicts which couples will remain together after 6 months of courtship [156]. Indeed, there is compelling evidence that romantic attachment is a kind of 'behavioural addiction' since virtually every neurochemical pathway implicated in conventional addiction disorders—especially the oxytocin system—also participates in the development and maintenance of human love [157].

Activational and Organizational Effects of Sex Hormones

The *organization-activation theory* posits that during early brain development the exposure to sex hormones fosters long-lasting (i.e. "organizational") effects that influence neuronal activity and behaviour throughout life. By contrast, the transient actions of sex hormones on behaviour during adulthood are conceptualized as "activational" effects [158]. These actions are particularly manifest during late adolescence, when distinct sex hormone-driven neurodevelopmental changes occur, which may influence behaviours such as initiation to drug use. Sex-dependent differences in mammalian behaviour, and in drug effects on behaviour, have generally been found to rely on both intrinsic sex differences in structural and functional brain organization, and on activational effects of circulating gonadal steroid hormones [159–161].

Sex differences in cannabinoid-induced behavioural effects in rats seem to depend on the activational effects of testosterone in males and/or estradiol in females since, for example, cycling females are more sensitive to THC-induced effects when tested in estrous (i.e. in a high-estradiol state) than in diestrous (i.e. in a low-estradiol state) [162]. Consistent with this, female rats also displayed a faster acquisition, higher maintenance, and later extinction of cannabinoid self-administration behaviour when compared to their male counterparts [143]. This sex dichotomy relies on circulating sex hormones since it was abolished following ovary ablation [143]. Similarly, other activities such as spontaneous locomotion, social behaviour, and sensorimotor gating are more sensitive to the activational effects of sex hormones [163].

Given that steroid receptors can also be activated by neurotransmitters such as dopamine, and that dopamine neurons express steroid receptors [164–166], it can be argued that some of activational effects of sex hormones on behaviour are the result of converging signal downstream activation of these receptors [167]. Accordingly, estrogen is produced at high levels within the brain and acts at synapses where, by changing neuronal excitability, it affects synaptic transmission and activity-dependent plasticity [168]. Estrogen exerts profound effects on mood and memory by acting on both monoamine and neuropeptide transmitter mechanisms in the brain. Indeed, low levels of estrogen in women are commonly associated with the premenstrual syndrome, and with post-natal and post-menopausal depression [169].

In female rats, estradiol engages the endocannabinoid system, potentially through an FAAH-related mechanism, to modulate emotional behaviour [170]. Furthermore, a reciprocal interplay between endocannabinoids and estrogen in acute modulation of inhibitory synapses has been reported to be sex specific [140]. In particular, estradiol acutely suppresses synaptic inhibition in the hippocampus via a sex-specific tonic mobilization of the endocannabinoid anandamide. Remarkably, no evidence of such a tonic mobilization of anandamide was previously found in male rats [171]. Estradiol is also known to regulate CB1 receptor density [172], transcription [173], and signal transduction [174] in some areas of the adult rodent brain, suggesting that CB1 receptor function may be sexually dimorphic. That is, in some regions of the brain, the endocannabinoid levels [175], and the CB1 receptor density and affinity [163, 176], fluctuate as a function of sex, the hormonal cycle, and the presence/absence of the ovarian hormones, supporting the hypothesis of possible sex hormone-dependent differences in the sensitivity of certain neuronal processes triggered by cannabinoid treatment. It is noteworthy that CB1 receptors were decreased in both the prefrontal cortex and the amygdala—two regions critically involved in decision-making and learning processes underlying goal-directed behaviour—in both rodents and human beings [177, 178].

However, the organizational effects of sex hormones in cannabinoid-induced behaviour cannot be disregarded any longer. Sex differences in cannabinoid sensitivity are still observed in guinea pigs gonadectomised as adults [179], and a few sex differences in cannabinoid effects have been observed in adolescent rats [180, 181], suggesting that cannabinoid systems may develop in a sexually dimorphic manner early in life. Intriguingly, females appear to have a higher expression of dopamine D2 receptors and larger concentrations of dopamine in the extracellular milieu in the striatum [127, 182]. These effects appear to be organizational given that similar sex differences in striatal dopamine concentrations were found when comparisons were made between castrated male rats and ovariectomized female rats [182, 183]. Thus, together with the notion that increased striatal dopamine D2 receptor availability is a protective factor for vulnerability to drug abuse and dependence [184, 185], organizational effects of sex hormones might account for higher striatal dopamine and dopamine transporter availability in women than in men [186, 187]. Notably, sex differences were found in a form of endocannabinoid-mediated short-term plasticity at inhibitory synapses at dopamine neurons within the VTA before onset of puberty [188]. Thus, sex specific endocannabinoid signalling at dopamine neurons might contribute to regulate responses to aversive intrinsic properties to cannabinoids, thereby resulting in faster acquisition/initiation of cannabinoid-taking in female rats.

A better understanding of the contribution of both organizational and activational effects of sex hormones to cannabis smoking and to drug addiction in general, may represent a valuable tool in development of preventive and therapeutic strategies.

Sex Differences in Behavioural Traits Predisposing to Cannabinoid Addiction and Relapse

Not only is adolescence a time of pronounced physical changes, but also a watershed occasion for remodelling brain neurotransmitter and hormone systems in preparation for sexually-differentiated roles in reproduction. Both these processes have profound effects on the emerging emotional and behavioural differences between the sexes. Furthermore, and consequent on these neurobiological alternations, adolescence is a developmental period typified by poor impulse-control, risky decision-making, and a heightened reactivity to stress, especially during a time in the life cycle of unprecedented social challenges [189]. It is not surprising, therefore, that adolescence is also a time when most people begin to use and abuse addictive substances.

Stress is a major vulnerability factor in drug addiction; yet, several phenotypes, which include preference for sweets, novelty reactivity, and impulsivity, as well as some environmental conditions (e.g. rearing conditions) are associated with drug-abuse vulnerability. Moreover, there is good evidence that risk for most addictions is driven by environmental and by genetic factors, and may be conferred through heritable impulsive tendencies—effects that appears to be stronger in men [190].

Stress Responsiveness

Stress is strongly implicated in the development and perpetuation of drug and alcohol use and abuse, and has a critical role in the risk for relapse in addicted individuals [191, 192]. Indeed, dysregulation of stress reactivity is a major adverse consequence of long-term drug exposure that plays a primary role in relapse to drug use. A recent National survey found that high-stress teenagers were twice as likely to smoke, drink, and use illegal drugs compared to their low-stress counterparts [193]. Considerable evidence also suggests that women are more susceptible to stress-related disorders like depression than men, and that interactions between the Hypothalamic-Pituitary-Gonadal (HPG) axis and the Hypothalamic-Pituitary-Adrenal (HPA) axis may underlie women's enhanced susceptibility to these conditions [194]. A recent preclinical study demonstrated, for example, that a chronic mix-modality stressor presented to animals during adolescence produced sustained changes in depression-related behaviours in adulthood. Compared to males who showed no significant behavioural changes, the females displayed several anhedonic and anxiety indicators such as decreased sucrose consumptions, decreased activity in the forced swim test, and blunted cortisol response [195].

Sex differences in the psychobiology of stress are also well-established with females producing a stronger HPA response than males. Individual differences in progesterone levels in women appear to be one factor mediating this relationship. For instance, high progesterone levels have been associated with lower stress and cue-induced craving, anxiety, and cardiovascular reactions [196]. This effect may occur because progesterone is a potent positive enhancer of GABA, which is an inhibitory neurotransmitter that diminishes dopamine availability and reduces the stress responses. In addition, these processes are known to weaken drug reward and drug cravings [196]. Preclinical evidence indicates that in a trace-conditioning paradigm, females outperformed males in neutral conditions, but their performance was impaired in response to stress while that of males was enhanced [197]. In other related research, men demonstrated greater activation in brain regions known to regulate emotions during stress, congruent with the evidence that they display more pronounced physiological changes in response to a stressful situation than women do [191]. By contrast, women showed greater neural activation in brain regions associated with high-level cognitive processing and language, consistent with the evidence that they tend to express their emotions verbally and to use verbal coping strategies more than men do.

Stress during adolescence appears to have long-lasting effects on brain development, especially in areas involved in learning, memory, and emotional regulation [44]. In particular, stress has considerable negative impact on one's decision-making capacity [193]. Again, there is evidence that these effects are moderated by sex, with males showing poor decision-making and greater risk taking in stressful situations while females show the opposite pattern of responding [198].

The endocannabinoid system has a homeostatic role in limiting HPA axis activation [199, 200], and CB1 receptors are densely represented in brain regions involved in regulating stress responsivity, including midbrain monoaminergic nuclei like the locus coeruleus and dorsal raphe. In humans, social stress increases circulating levels of anandamide in healthy men but not in women, while levels of 2-arachidonyl glycerol (2-AG) was not affected in either sex [201]. This latter finding is in contrast to previous findings showing that in women circulating level of 2-AG was higher immediately following social stress exposure than before exposure [202]. However, despite contrasting results, both studies suggest a protective role for the endocannabinoid system in anxiety, in line with the notion that dysfunctional endocannabinoid signalling is associated with increased anxiety and depression (see Chapters 5 and 6 of this book). Accordingly, reductions in circulating endocannabinoids have been documented in women with major depression [203].

Phenotypes Associated with Enhanced Vulnerability toCannabinoid Addiction

Poor response inhibition and impulsivity—defined as the tendency to act without thinking and without consideration of future consequences—have been associated with addictive behaviours. Impulsivity is commonly believed to consist of two distinct components: the impulsive action, which involves serious difficulty in inhibiting or controlling a behaviour, and the impulsive choice, which refers to the tendency to choose a smaller but immediate reward over a larger but delayed reward. Gender differences in both behavioural measures have been found, although the direction and magnitude of the differences may considerably vary [204]. In laboratory animals, impulsive action is typically greater in males than females, whereas impulsive choice is typically greater in females. In humans, for example, female heavy-drinkers show higher deficits in response inhibition than their light-drinking counterparts and corresponding male groups [205]. Enhanced impulsive choice predict the rate of acquisition of cocaine self-administration behaviour [206] and are associated with differences in cocaine-seeking behaviour in male and female rats [207], suggesting that impulsive choice, in addition to sex, represents a vulnerability factor to drug use. Very recently, important sex-dependent differences have been reported in the impulsive traits between male and female pre-pubertal rats, with males displaying significantly higher impulsive choices than females [208]. Notably, these sex differences are ascribed by the authors to the organizing actions of testosterone during the neonatal period, and are a consequence of both androgenic and estrogenic actions [208].

Trait impulsivity has been positively associated with greater marijuana use [209, 210] and marijuana-related problems [211, 212]. However, data on sex-dependent differences in impulsivity parameters among marijuana users are still very scarce, and so far have not revealed significant differences between sexes [213].

Use of drugs has also been associated with preference for sweets and for foods with a high sucrose concentration in both human and animal studies, and a genetic contribution has been hypothesised for this positive relationship. In other words, the hedonic response to sweet taste may predict the risk for having drug-related problems. To date, human studies have confirmed an association between sweet preference and a genetic predisposition to alcoholism [214, 215]. Higher consumption of sweets and eagerness to consume sweet foods have also been found in patients under long-term methadone treatment, compared to controls [216]. Notably, preference for sweet taste is significantly enhanced in cocaine-exposed newborns [217] and significantly attenuated in morphine-withdrawn rats [218]. It is still not clear however, whether sweet preference is positively associated with marijuana smoking, and if so, whether it differs between males and females.

Other Risk and Protective Factors for Cannabinoid Addiction

An important feature of drug addiction in humans is the emergence of negative states, like dysphoria, irritability or anhedonia, which are thought to play a critical role in drug craving and relapse. A different sensitivity to the aversive properties of drugs (i.e. drug withdrawal) has been described in animal studies, with males being more sensitive than females to drug withdrawal [136]. Interestingly, progesterone and progesterone-related neurosteroids may ameliorate some of these effects, perhaps through anxiolytic actions mediated via the HPA axis. In the human condition, women are more likely than men to report marijuana withdrawal symptoms, due mostly to their greater incidence of upset stomach [219], including nausea and vomiting [78, 220]. Since withdrawal symptoms can serve as negative reinforcement for relapse, these findings suggest that females are more prone than males to marijuana-use relapse. Environmental stimuli associated with drug taking are also among the factors that have been shown to elicit drug craving in humans and increase the likelihood of relapse, and to reinstate drug-seeking and drug-taking in laboratory animals. Human studies typically use differential-cue paradigms to investigate cue-induced drug craving, such as drug-related imagery scripts and drug-related paraphernalia. These studies tend to report different cue-sensitivity between men and women. For example, female cocaine addicts are more likely to report increased craving in response to cues than males [221], while heroin-dependent women showed stronger heroin cravings and sadness than men [222]. Craving reactivity to smoking cues is also higher in female than in male smokers [223, 224], although discrepant results have been reported [225]. Moreover, the cue of drinking low-alcohol beer increased alcohol craving in men but not in women [226],

Concerning cannabinoids, animal studies have demonstrated that female rats are more vulnerable than males to both visual (light) and auditory (tone) cues previously associated with the delivery of cannabinoid (i.e. during cannabinoid self-administration), and promptly reinstate cannabinoid-seeking behaviour after exposure to an acute cue priming to a greater extent than males [144]. The enhanced reactivity to cannabinoid-associated cues in females seems to be under the control of the ovarian hormones, since their rate of responding for the cannabinoid was no different from that of males after ovariectomy [144].

In a recent study, cannabis-dependent and cannabis-naïve participants were exposed to neutral and marijuana-related cues [227]. When assessing subsequent changes in mood, self-reported craving, and physiologic function, the cannabis-dependent individuals reported increased cravings in response to marijuana-related compared to neutral cue exposure.

Some environmental factors have been found to influence the sensitivity to the rewarding effects of drugs and vulnerability to develop addiction. That is, negative environmental factors may exacerbate use of amphetamine, cocaine, and heroin [228] while positive environmental manipulations, such as housing animals in environmental enrichment during early stages of life, may reduce their rewarding effects [229]. Importantly, early exposure to environmental enrichment induces differential modifications in the expression of the CB1 receptors, FAAH, and monoacylglycerol lipase (MAGL) enzymes in brain regions involved in drug addiction [230]. Unfortunately, whether or not the effect of the environmental enrichment on the endocannabinoid system is sexually dimorphic is still unclear. Yet, the recent finding that environmental enrichment reduces the activating effects of nicotine on the adrenocorticotropic hormone (ACTH) and corticosterone (CORT) in both males (ACTH) and females (ACTH and CORT) with different dose sensitivities [231] is an important step forward in such investigations.

Sex Differences in Cannabinoid-Induced Reward

Studies on interactions between drug rewards and natural rewards have led us to a better understanding of motivation in general. The addictive potential of drugs is believed to be based on the strength of its motivational properties [232]. Drug motivation significantly varies between sexes, and in this sense behavioural animal studies have provided in part a rational basis for how males and females differ in efforts made to gain a reward. More specifically, male and female animals showed different propensity to abuse drugs, different vulnerability to develop drug dependence, and different susceptibility to relapse when abstinent [233, 234]. Female laboratory animals typically self-administer more caffeine [235], cocaine [236, 237], heroin [238], morphine [239, 240], and fentanyl [241] than males. The heightened response to stimulants shown by female rodents has been attributed to the dopamine-enhancing properties of estrogen [242, 243], implying that females are inherently more sensitive than males to the rewarding properties of drugs of abuse, and therefore more biologically susceptible to developing drug addiction and dependence. Ethanol consumption is also greater in female rats [244], mice [245], and vervet monkeys [246] than in males. Such sex differences in ethanol intake are the opposite of those found in humans, where daily average ethanol intake by men is about double that for women after adjusting for body weight and body water. Thus, while men drink alcohol and smoke marijuana more often than women, female animals consistently appear more vulnerable than males to positive reinforcing effects of alcohol and cannabinoids and more motivated to obtain them. Socio-cultural factors are likely to contribute to this divergence. For example, there is a higher social disapproval of smoking marijuana for girls and women [99]. This does not necessarily imply that females find cannabis less rewarding than males, but it can explain why female animals (unaffected by socio-cultural factors) usually display higher drug intake.

The brain structure that assesses whether an experience is pleasurable or adverse, and that connects the experience with its consequences, is the amygdala, a very small area in charge of the affective and emotional processing. Men and women differ in this critical brain area, as cannabinoid CB1 receptors density and efficacy differ between the two sexes, suggesting that male and female users smoke marijuana motivated by different reasons, in an attempt to achieve different effects. Not only brain anatomy but also brain neurochemistry and physiology may differ between males and females. For example, dopamine, serotonin, and GABA neurotransmission systems exhibit significant sex differences in their metabolism and activity [247]. Therefore, it is not surprising that THC, acting in these areas and through mechanisms involving these neurotransmitters, could trigger different responses in males and females. If males and females use different neural paths to reach the same behavioural endpoint, THC exposure may have different neuronal consequences on the male and female brain.

Sexual Dimorphism of the Brain Endocannabinoid System

Many of the gender differences observed in drug use and misuse are determined by sex differences in the brain. Though macroscopically very similar, male and female adult brains are different not only in size (roughly 1.5 and 1.2 kg, respectively) but also in terms of functional and neuroanatomical organization. For example, adolescent female marijuana users exhibit larger right amygdala volumes relative to female controls, and report increased depression and anxiety symptomatology, while male users have similar volumes as male controls [248]. Moreover, abstinent adolescent female and male marijuana users demonstrate, respectively, larger and smaller prefrontal cortex volumes compared to the same-gender controls [249]. Intriguingly, marijuana-related cues increase self-reported cravings and activate the reward brain pathway including the VTA, thalamus, anterior cingulate, insula, and amygdala [250].

Within the central nervous system, cannabinoid CB1 receptors are differentially expressed in males and females in areas of the brain, including forebrain and midbrain structures, that work together to sustain the multifaceted addictive behaviour. Importantly, both human and animal studies have widely demonstrated that limbic brain regions, often referred to as the 'emotional brain', are particularly vulnerable to chronic marijuana use [92, 251]. The limbic brain has evolved to respond to natural rewards, but drugs of abuse also affect these same circuits. To date, CB1 receptor density and function in the male and female brain have been analyzed in post-mortem studies on psychiatric or alcoholic patients [252–254], and in rodents exposed to cannabinoids prenatally [255], during adolescence [256, 257] and adulthood [258], or after sub-chronic treatment with antipsychotics [256]. Besides post-mortem and animal data, few studies have been conducted thus far to assess the in vivo cerebral CB1 receptor distribution and its variation with healthy aging and sex, which reported a region-dependent and gender-related up-regulation of CB1 receptors [259]. Specifically, by using positron emission tomography and a high-affinity, subtype-selective radioligand, it has been shown that binding to CB1 receptors increases with ageing in the basal ganglia, lateral temporal cortex, and limbic system of women, whereas men show higher binding in clusters of the limbic system and cortico-striato-thalamic-cortical circuit [259]. Surprisingly, only a limited number of animal studies have conducted a systematic comparison in adult drug-naïve males and females to investigate possible sex differences in brain CB1 receptor density and function [163, 173, 176, 260].

Recently, evidence has been provided for sex-dependent differences in the density of cannabinoid CB1 receptors in the rat prefrontal cortex (Cg1 and Cg3) and amygdala, brain areas in which estradiol seems to be the major factor responsible for the decreased number of CB1 receptors [163]. The anterior cingulate cortex (Cg1,2) and the prelimbic cortex (Cg3) are part of the neural network that mediates executive control, governing behavioural inhibition, implementation of control, and decision making [261], while the amygdala critically modulates various types of fear and anxiety responses [262, 263]. Notably, a neural circuit between the amygdala and the prefrontal cortex is activated in response to novel and emotionally arousing events [264, 265]. Thus, reduced CB1 receptor density in brain areas involved in cognition and emotional processing in females might account for their higher propensity to cannabinoid addiction. Male and female rats also differed in their spontaneous motor activity and basal level of social anxiety [163], two predisposing traits enhancing susceptibility to addiction, and both known to enhance vulnerability to initiate drug self-administration in animals and accelerate the rate of acquisition and increase the frequency of cannabis smoking in humans [266–269]. Notably, females consistently showed a more vulnerable phenotype to cannabinoid addiction, in line with animal and human studied showing that females display more cocaine-induced locomotor activation and stronger behavioural sensitization to psychostimulants than males [35, 270, 271].

Sexually Dimorphic Effects of Exogenous Cannabinoids

In addition to sex differences observed in effects related to cannabis abuse and dependence, cannabinoids have been shown to influence other aspects of physiology and behaviour such as food intake and energy balance (more evident in males), or anxiety and depression (more evident in females). Notably, the influence of cannabis intake on sexual behaviour and arousability appears to be dose-dependent in both men and women, although only women report facilitatory effects [272]. No gender differences have been observed in the effects of THC on impulsivity [273]. Similar sex differences in responses to cannabinoids have been found in preclinical studies. For example, males are more sensitive to the hyperphagic and hypophagic effects of the CB1 receptor agonists and antagonists, respectively [179], as well as to their hypothermic and hyperthermic effects [273]. Conversely, cannabinoids elicit comparatively greater catalepsy, antinociception and locomotor effects in females than in males [102], and decrease both exploratory behaviour and emotionality/anxiety levels in female, but not male, rodents [274].

Importantly, perinatal exposure to THC decreases proenkephalin gene expression in the caudate-putamen of female but not male rats [275], while female, but not male rats, exposed perinatally to THC self-administer more morphine once they are adults [276]. More recently, it has been reported that repeated exposure to THC produces greater desensitization and down-regulation of CB1 receptors in the brain of adolescent female than male rats [258], demonstrating differential central actions of cannabinoids between the sexes. In support of the role of sex in the effects of THC on CB1 receptor level and CB1/G-protein coupling, it has been shown that THC exposure during adolescence significantly reduces CB1 receptor density and function in the amygdala, VTA and nucleus accumbens of female rats, whereas in male rats it causes significant alterations in the amygdala and hippocampal formation [277].

Chronic intravenous self-administration of the CB1 receptor agonist WIN 55,212-2 alters, and in most cases increases, density and coupling of CB1 receptors in the reward related brain of male Lister Hooded rats [278], who typically exhibit slower acquisition of cannabinoid self-administration and less drug intake than females [143], and a lower response rate for cannabinoids when exposed to acute drug and cue primings after extinction [144]. Intriguingly, in this strain of rats, although the basal spontaneous activity of the dopaminergic neurons of the VTA is not different between sexes, their response to WIN 55,212-2 is differently regulated in male and female Lister Hooded rats [188]. Indeed, the stimulating properties of the CB1 receptor agonist on VTA DA cells are evident in males but blunted in females (see Fig. 12.3).

Sex-dependent differences in the perception of the rewarding effects of cannabinoids can also be explained by the presence, and action in the brain, of estrogens. Indeed, estrogen has powerful anxiolytic effects in rats that involve endocannabinoids [141] and, therefore, it may minimize the aversive effects of cannabinoids, unmasking euphorigenic effects. The estrogenic modulation of the reinforcing effects of cannabinoids mirrors that consistently observed with stimulant drugs [279]. Moreover, since CB1 and estrogen receptors interact in the mesolimbic dopamine system [280, 281], THC exposure increases dopamine activity in the mesolimbic system in an estrogen-sensitive way [282]. Hormonal influences may also account for the finding that adolescent rats perceive cannabinoids less aversive than adult rats [283].

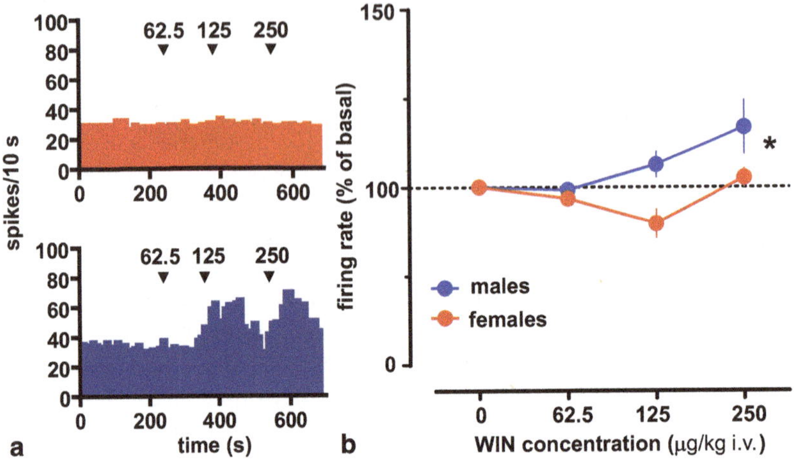

Fig. 12.3 Sex specific effect of the cannabinoid receptor agonist WIN55212–2 on VTA dopamine neuron firing rate. **a** Effect produced by cumulative doses of WIN on a VTA dopamine (DA) neuron recorded from a female (*top*) and a male (*bottom*) rat. *Arrows* indicate the time of injection. Numbers above arrows indicate the dosages expressed in μg/kg i.v. Note the lack of stimulating effect of WIN in a female rat VTA DA neuron. **b** Dose-response curves depicting the sex specific stimulating effect of cumulative doses of WIN ($n=8$ for both sexes) on the firing rate of VTA DA neurons. Data (mean ± SEM) are expressed as percentages of basal firing rate. *$P<0.05$ with respect to pre-drug level. Figure as originally published in Melis et al. (2013) Sex-specific tonic 2-arachidonoylglycerol signaling at inhibitory inputs onto dopamine neurons of Lister Hooded rats. Front. Integr. Neurosci. 2013 Dec 19;7:93. doi:10.3389/fnint.2013.00093

Sex Differences in Cannabinoid-Taking Behaviour

Self-administration protocols provide an advanced animal model of human drug use with high face and predictive validity. The avid self-administration of alcohol, nicotine, heroin, cocaine, and methamphetamine by rodents has been of great utility in unravelling the complex neurobiology of addictive behaviours. However, cannabis self-administration has proven notoriously difficult to obtain in laboratory animals. The breakthrough discovery of intravenous THC self-administration in rhesus monkeys in 2000 [284] followed by that of chronic self-administration of the synthetic cannabinoid agonist WIN 55,212-2 in rats [285] has allowed new studies of the neural, genetic and environmental determinants of cannabis use. Notably, in both studies the critical factor appeared to be the use of very low intravenous doses of CB1 receptor agonists, in line with the notion that at low and high doses cannabinoids exert pleasurable and aversive effects, respectively.

It soon became clear that genes play a critical role in shaping the proclivity towards cannabis consumption. That is, earlier studies involving intracranial self-stimulation and place-preference procedures revealed that different genetic strains of rats had different motivational responses to cannabinoids [286, 287]. Similarly, strain differences were also observed in cannabinoid-induced brain activation and mesolimbic dopamine release [288], as well as in cannabinoid intravenous self-administration [289]. A genetic predisposition to cannabis use is also hinted at in human studies. An appealing early study conducted on identical twins in New Zealand showed that the extent to which cannabis smoking is experienced as pleasurable or aversive was, at least partly, genetically determined [89]. More recently, single nucleotide polymorphisms in the cannabinoid CB1 gene have been associated with susceptibility towards cannabis dependence in adolescents that smoked marijuana [91]. Therefore, the notion that sex also determines the vulnerability to cannabis use was not a completely unexpected finding. In 2007, it was demonstrated for the first time that female rats intravenously self-administer more cannabinoid than males [143 and exhibit higher cannabinoid-seeking reinstatement when abstinent [144]. Notably, cannabinoid self-administration was not only dependent upon sex (intact female rats being more sensitive than males to the reinforcing properties of cannabinoids) but also upon estrous cycle, as ovariectomized female animals were less responsive than cycling females [143, 144]. In this regard, cannabis acts as other drugs of abuse, since even when cocaine or heroin infusions are made contingent upon increasingly higher numbers of bar presses, female rats make substantially more presses than males, and their level of cocaine self-administration varies as a function of estrus cycle [290]. Importantly, in an early human study, increased marijuana use was associated with premenstrual dysphoria and impaired social functioning [291]. However, cannabis intake in women seemed not to vary across phases of the menstrual cycle [292].

Sex Differences in Cannabinoid-Seeking Behaviour

Drug addiction is associated with a loss of 'executive' inhibitory control over maladaptive drug-seeking and drug-taking habits [293]. Maladaptive memories are crucial in persistent drug-seeking, as the brain process by which retrieved memories are consolidated—often referred to as "unrestrained reconsolidation"—may reiterate and strengthen drug memories over long periods of time, which in turn may contribute to compulsive drug-seeking [294]. Sex-related differences have been observed in long-term memory consolidation processes, with females typically having a weaker memory trace which may be more susceptible to disruption [295–297]. Moreover, memory storage for emotional material in humans involves the amygdala, which exhibits a sex-dependent hemispheric asymmetry [298]. Specifically, activity of the left, but not right, amygdala related significantly to enhanced long-term memory for a series of emotionally arousing films in women, while the opposite was observed in men. The greater participation of left-hemisphere amygdala processing of memory for emotionally arousing materials in women meshes with the evidence of left hemisphere amygdala hyperactivation in clinically depressed women [299].

In animal studies, drug-seeking behaviour is extinguished by interrupting the contingency between drug-seeking (i.e. the operant response) and delivery of the drug reward. Female rats took longer than males to extinguish cannabinoid-seeking behaviour, in that they persist in responding for the cannabinoid even in the absence of the contingent reward presentation [143]. This finding corroborates the idea that females are more motivated and, hence, more willing to look for cannabinoid drugs.

Reduction in drug-seeking behaviour following extinction training is a robust but not a permanent condition, since drug-seeking can be reinstated following presentations of a drug prime, a drug-associated stimulus, or a stressor. After either drug or cue priming, intact female rats reinstate responding for the cannabinoid at higher level than males [144], showing a higher propensity to resume cannabinoid-seeking during abstinence. Notably, as during cannabinoid self-administration and extinction training, ovariectomy dampens operant responding for the cannabinoid, revealing a pivotal role for the ovarian hormones not only in regulating cannabinoid-taking but also cannabinoid-seeking behaviour [143, 144].

Extinction of drug-seeking behaviours is believed to depend on cortical-striatal-hypothalamic and cortical-hypothalamic-thalamic pathways which interface with the neural circuits controlling the reinstatement of drug-seeking [300]. When considering sex-dependent levels of the endocannabinoids and the density of the CB1 receptors within these circuits, it is not surprising that males and females display different patterns of extinction and reinstatement of cannabinoid-seeking. For example, in the hypothalamus, the content of endocannabinoids (i.e. anandamide) is higher in females than in males, and fluctuates during the ovarian cycle [173], while in the cortico-striato-thalamic-cortical circuit females show lower density of the CB1 receptor then males [259]. Finally, in the striatum females have larger concentrations of dopamine and a higher expression of dopamine D2 receptors [127, 182], which strictly interact with CB1 receptors in modulating cannabinoid-induced reward (see also Chap. 17 of this book).

Conclusions

Women and men appear to have different needs in maintaining health, coping with diseases and responding to treatment protocols and drugs. Women seem to be more responsive than men to treatment for drug abuse as well [301]. While men report more self-justification after initial use of drugs, women report more help-seeking. Before relapse into drug re-use, women also reported more unpleasant affect and interpersonal problems than men, suggesting that women might benefit more from techniques that enable them to deal more effectively with unpleasant emotions and interpersonal problems.

Recently, several differences in the use and abuse of cannabis have been found between men and women, and animal studies have also revealed that females are more vulnerable to take cannabinoids and to relapse to cannabinoid-seeking while in a drug-free state. Within the brain, cannabinoid CB1 receptors are differently expressed between males and females, and gonadal hormones have been shown to strongly influence cannabinoid-induced rewarding effects. Sex-dependent differences have also been reported in many vulnerability factors that contribute to individual variation in the risk of cannabis addiction, ranging from social and cultural characteristics to genetic susceptibility to brain morphology and neurotransmission. In this "translational era" for research, future challenges will be the deeper knowledge of the underlying reasons for gender differences in vulnerability to cannabis addiction as well as the optimization of sex-tailored preventing strategies and treatment approaches.

References

1. Holden C. Sex and the suffering brain. Science. 2005;308:1574–7.
2. Wiren KM. Males and females are just different: sexually dimorphic responses to chronic ethanol exposure in hippocampal slice cultures. Neurosci Lett. 2013;550:1–5.
3. Greyson DL, Becu ARE, Morgan SG. Sex, drugs and gender roles: mapping the use of sex and gender based analysis in pharmaceutical policy research. Int J Equity Health. 2010;9:26.
4. Johnson JL, Greaves L, Repta R. Better science with sex and gender: facilitating the use of a sex and gender-based analysis in health research. Int J Equity Health. 2009;8:14.
5. Saad G. The consuming instinct. What Darwinian consumption reveals about human nature? Politics Life Sci. 2013 Spring;32(1):58–72.
6. Linden DJ. The Compass of Pleasure. New York: Penguin Group; 2011.
7. Fattore L, Melis M, Fadda P, Fratta W. Sex differences in addictive disorders. Front Neuroendocrinol. 2014;35(3):272–84.
8. Buss DM, Larsen RJ, Westen D, Semmelroth J. Sex differences in jealousy: evolution, physiology, and psychology. Psychol Sci. 1992;3:251–5.
9. Sagarin BJ, Martin AL, Coutinho SA, Edlund JE, Patel L, Skowronski JJ, Zengel B. Sex differences in jealousy: a meta-analytic examination. Evol Hum Behav. 2012;33:595–614.
10. Berman PW. Children caring for babies: age and sex differences in response to infant signals and to the social context. In: Woodhead M, Carr R, Light P, editors. Becoming a person: child development in social context, vol. 1. London: Routledge; 1991
11. Okabe S, Kitano K, Nagasawa M, Mogi K, Kikusui T. Testosterone inhibits facilitating effects of parenting experience on parental behavior and the oxytocin neural system in mice. Physiol Behav. 2013;118:159–64.
12. Deady DK, Law-Smith MJ, Sharp MA, Al-Dujaili EAS. Maternal personality and reproductive ambition in women is associated with salivary testosterone levels. Biol Psychol. 2006;71:29–32.
13. Pollet TV, Cobey KD, van der Meij L. Testosterone levels are negatively associated with childlessness in males but positively related to offspring count in fathers. PLOS One. 2013;4:e60018.
14. Law Smith MJ, Perrett DI, Jones BC, Cornwell RE, Moore FR, Feinberg DR, Boothroyd LG, Durrani S, Stirrat MR, Whiten S, Pitman RM, Hillier SG. Facial appearance is a cue to oestrogen levels in woman. Proc R Soc. 2006;98:111–26.
15. Edelstein RS, Stanton SJ, Henderson MM, Sanders MR. Endogenous estradiol levels are associated with attachment avoidance and implicit intimacy motivation. Horm Behav. 2010;31:145–58.
16. Law Smith MJ, Deady DK, Moore FR, Jones BC, Cornwell RE, Stirrat MR, Lawson JF, Feinberg DR, Perrett DI. Maternal tendencies in women are associated with estrogen levels and facial femininity. Horm Behav. 2012;61:12–6.

17. Berbesque JC. Sex differences in food preferences of Hadza hunter-gatherers. Evol Psychol. 2009;7:601–16.
18. Xu X, Tan L, Himi T, Sadamatsu M, Tsutsumi S, Akaike M, Kato N. Change preference for sweet taste in adulthood induced by perinatal exposure to bisphenol A—a probable link to overweight and obesity. Neurotox Teratol. 2011;33:458–63.
19. Cason AM, Grigson PS. Prior access to a sweet is more protective against cocaine self-administration in female rats than in male rats. Physiol Behav. 2013;112–113:96–103.
20. Davis C, Fattore L, Kaplan AS, Carter JC, Levitan RD, Kennedy JL. The suppression of appetite and food consumption by methylphenidate: the moderating effects of gender and weight status in healthy adults. Int J Neuropsychopharmacol. 2012;15(2):181–7.
21. Javaras KN, Laird NM, Reichborn-Kjennerud T, Bulik CM, Pope HG Jr, Hudson JI. Familiality and heritability of binge eating disorder: results of a case-control family study and a twin study. Int J Eat Disord. 2008;41:174–9.
22. Popkin BM. Recent dynamics suggest selected countries catching up to US obesity. Am J Clin Nutr. 2010;91:284–8.
23. Scherma M, Fattore L, Castelli MP, Fratta W, Fadda P. The role of the endocannabinoid system in eating disorders: neurochemical and behavioural preclinical evidence. Curr Pharm Des. 2014;20(13):2089–99.
24. Keskitalo K, Silventoinen K, Tuorila H, Perola M, Pietilainen KH, Rissanen A, Kaprio J. Genetic and environmental contributions to food use patterns in young adult twins. Physiol Behav. 2008;93:235–42.
25. Pepino MY, Mennella JA. Effects of cigarette smoking and family history of alcoholism on sweet taste preference and food cravings in women. Alcohol-Clin Exper Res. 2007;31:1891–9.
26. Lustig RH. Fructose: metabolic, hedonic, and societal parallels with ethanol. J Am Diet Assoc. 2010;110:1307–21.
27. Belknap JK, Crabbe JC, Young ER. Voluntary consumption of ethanol in 15 inbred mouse strains. Psychopharmacol (Berl). 1993;112:503–10.
28. Li TK, Lumeng L, Doolittle DP. Selective breeding for alcohol preference and associated responses. Behav Genet. 1993;23:163–70.
29. Carroll ME, Morgan AD, Anker JJ, Perry JL, Dess NK. Selective breeding for differential saccharin intake as an animal model of drug abuse. Behav Pharmacol. 2008;19:435–60.
30. Effertz T, Mann K. The burden and cost of disorders of the brain in Europe with the inclusion of harmful alcohol use and nicotine addiction. Eur Neuropsychopharmacol. 2013;23:742–8.
31. Wittchen HU, Jacobi F, Rehm J, Gustavsson A, Svensson M, Jonsson B, Olesen J, Allgulander C, Alonso J, Faravelli C, Fratiglioni L, Jennum P, Lieb R, Maercker A, van Os J, Preisig M, Salvador-Carulla L, Simon R, Steinhausen HC. The size and burden of mental disorders and other disorders of the brain in Europe 2010. Eur Neuropsychopharmacol. 2011;21:655–79.
32. Becker JB. Sexual differentiation of motivation: a novel mechanism? Horm Behav. 2009;55:646–54.
33. Colell D, Sanchez-Niubo A, Domingo-Salvany A. Sex differences in the cumulative incidence of substance use by birth cohort. Int J Drug Policy. 2013;24:319–25.
34. Elman I, Karlsgodt KH, Gastfriend DR. Gender differences in cocaine craving among non-treatment-seeking individuals with cocaine dependence. Am J Drug Alcohol Abuse. 2001;27:193–202.
35. Evans SM, Foltin RW. Does the response to cocaine differ as a function of sex or hormonal status in human and non-human primates? Horm Behav. 2010;58(1):13–21.
36. Back SE, Payne RL, Wahlquist AH, Carter RE, Stroud Z, Haynes L, Hillhouse M, Brady KT, Ling W. Comparative profiles of men and women with opioid dependence: results from a national multisite effectiveness trial. Am J Drug Alcohol Abuse. 2011;37:313–23.
37. Becker JB, Ming H. Sex differences in drug use. Frontiers Neuroendocrinol. 2008;29:36–47.
38. Greenfield SF, Back SE, Lawson K, Brady KT. Substance use in women. Psychiatr Clin North Am. 2010;33:339–55.

39. Ramoa CP, Doyle SE, Naim DW, Lynch WJ. Estradiol as a mechanism for sex differences in the development of an addicted phenotype following extended access cocaine self-administration. Neuropsychopharmacol. 2013;38:698–705.
40. Vansickel AR, Stoops WW, Rush CR. Human sex differences in *d*-amphetamine self-administration. Addiction. 2009;727–31.
41. Wilens TE, Adler LA, Adams J, Sgambati S, Rotrosen J, Sawtelle R, Utzinger L, Fusillo S. Misuse and diversion of stimulants prescribed for ADHD: a systematic review of the literature. J Am Acad Child Adolesc Psychiatry. 2008;47(1):21–31.
42. Dafny N, Yang PB. The role of age, genotype, sex, and route of acute and chronic administration of methylphenidate: a review of its locomotor effects. Brain Res Bull. 2006;393–405.
43. Lee MJ, Yang PB, Wilcox VT, Burau KD, Swann AC, Dafny N. Repetitive methylphenidate administration modulates the diurnal behavioral activity pattern of adult female SD rats. Biol Child Adolesc Psychiatry. 2011;118:285–98.
44. Brown GR, Spencer KA. Steroid hormones, stress and the adolescent brain: a comparative perspective. Neurosci. 2013;249:115–28.
45. Chelaru MI, Yang PB, Dafny N. Sex differences in the behavioral response to methylphenidate in three adolescent rat strains (WKY, SHR, SD). Behav Brain Res. 2012;226:8–17.
46. Sonuga-Barke EJS, Coghill D, Markowitz JS, Swanson JM, Vandenberghe M, Hatch SJ. Sex differences in the response of children with ADHD to once-daily formulations of methylphenidate. J Am Acad Child Adolesc Psychiatry. 2007;46:701–10.
47. Middaugh LD, Kelley BM, Bandy AL, McGroarty KK. Ethanol consumption by C57BL/6 mice: influence of gender and procedural variables. Alcohol. 1999;17:175–83.
48. Strong MN, Yoneyama N, Fretwell AM, Snelling C, Tanchuck MA, Finn DA. "Binge" drinking experience in adolescent mice shows sex differences and elevated ethanol intake in adulthood. Horm Behav. 2010;58:82–90.
49. Abulseoud OA, Karpyak VM, Schneekloth T, Hall-Flavin DK, Louklanova LL, Geske JR, Biernacka JM, Mrazek DA, Frye MA. A retrospective study of gender differences in depressive symptoms and risk of relapse in patients with alcohol dependence. Am J Addict. 2013;22:437–42.
50. Bekman NM, Winward JL, Lau LL, Wagner CC, Brown SA. The impact of adolescent binge drinking and sustained abstinence on affective state. Alcohol Clin Exper Res. 2013;37:1432–9.
51. John U, Rumpf H-J, Bischof G, Hapke U, Hanke M, Meyer C. Excess mortality of alcohol-dependent individuals after 14 years and mortality predictors based on treatment participation and severity of alcohol dependence. Alcohol Clin Exper Res. 2013;37:156–63.
52. Barker JM, Torregrossa MM, Arnold AP, Taylor JR. Dissociation of genetic and hormonal influences on sex differences in alcoholism-related behaviors. J Neurosci. 2010;30:9140–4.
53. McHugh RK, DeVito EE, Dodd D, Carroll KM, Potter JS, Greenfield SF, Connery HS, Weiss RD. Gender differences in a clinical trial for prescription opioid dependence. J Subst Abuse Treat. 2013;45:38–43.
54. Lee CW-S, Ho I-K. Sex differences in opioid analgesia and addiction: interactions among opioid receptors and estrogen receptors. Mol Pain. 2013;9:45.
55. Elliot R. Global efforts to combat smoking: an evaluation of smoking control policies. Soc Hist Med. 2008;21:583–5.
56. Okali C, Greaves L, Fagyas V. Sex differences in smoking initiation among children and adolescents. Pub Health. 2013;127:3–10.
57. Sullivan KM, Bottorff J, Reid C. Does mother's smoking influence girls' smoking more than boys' smoking? A 20-year review of the literature using a sex- and gender-based analysis. Subst Use Misuse. 2011;46:656–68.
58. Morisano D, Bacher L, Audrain-McGowan J, George TP. Mechanisms underlying the comorbidity of tobacco use in mental health and addictive disorders. Can J Psychiatry. 2009;54:356–67.
59. de Leon J, Diaz FJ. A meta-analysis of worldwide studies demonstrates an association between schizophrenia and tobacco smoking behaviors. Schiz Res. 2005;76:135–57.

60. Stein MD, Weinstock MC, Herman DS, Anderson BJ. A smoking cessation intervention for the methadone-maintained. Addiction. 2006;101:599–607.
61. Okoli CT, Torchalla I, Khara M. Sex differences in nicotine dependence among addictions clients accessing a smoking cessation programme in Vancouver, British Columbia, Canada. J Psychiatr Ment Health Nurs. 2012;19(9):776–84.
62. Pergadia ML, Agrawal A, Heath AC, Martin NG, Bucholz KK, Madden PA. Nicotine withdrawal symptoms in adolescent and adult twins. Twin Res Hum Genet. 2010;13:359–69.
63. Pang RD, Leventhal AM. Sex differences in negative affect and lapse behavior during acute tobacco abstinence: A laboratory study. Exper Clin Psychopharmacol. 2013;21:269–76.
64. Perkins KA, Karelitz JL, Giedgowd GE, Conklin CA. Negative mood effects on craving to smoke in women versus men. Addict Behav. 2013;38:1527–31.
65. Weinberger AH, McKee SA. Gender differences in smoking following an implicit mood induction. Nicotine Tob Res. 2012;14:621–5.
66. Weinberger AH, Pilver CE, Mazure CM, McKee SA. Do women really have a harder time quitting smoking? J Women Health. 2013;22:42–3.
67. Weinberger AH, Mazure CM, Morlett A, McKee SA. Two decades of smoking cessation treatment research on smokers with depression: 1990–2010. Nicotine Tob Res. 2013;15:1014–31.
68. Sanchez-Johnsen LAP, Carpentier MR, King AC. Race and sex associations to weigh concerns among urban African American and Caucasian smokers. Addict Behav. 2011;36(1–2):14–7.
69. King AC, Cao D, Zhang L, O'Malley SS. Naltrexone reduction of long-term smoking cessation weight gain in women but not men: a randomized controlled trial. Biol Psychiatry. 2013;73(9):924–30.
70. Kenny PJ. Reward mechanisms in obesity: new insights and future directions. Neuron. 2011;69:664–79.
71. Levin KH, Copersino ML, Heishman SJ, Liu F, Kelly DL, Boggs DL, Gorelick DA. Cannabis withdrawal symptoms in non-treatment-seeking adult cannabis smokers. Drug Alcohol Depend. 2010;111(1–2):120–7.
72. United Nation Office on Drugs and Crime (UNODC). World Drug Report 2013;933:77.
73. Kohn L, Kittel F, Piette D. Peer, family integration and other determinants of cannabis use among teenagers. Int J Adolesc Med Health. 2004;16(4):359–70.
74. Adlaf EM, Mann RE, Paglia A. Drinking, cannabis use and driving among Ontario students. CMAJ. 2003;168(5):565–66.
75. European Monitoring Centre for Drugs and Drug Addiction (EMCDDA). Differences in patterns of drug use between women and men 2005. Lisbon: EMCDDA; 2005. pp. 2–4.
76. Donner NC, Lowry CA. Sex differences in anxiety and emotional behavior. Biochem Behav. 2000;66(4):887–92. (Pflugers Arch. 2013;465(5):601–26).
77. Leatherdale ST, Hammond DG, Kaiserman M, Ahmed R. Marijuana and tobacco use among young adults in Canada: are they smoking what we think they are smoking? Cancer Causes Control. 2007;18(4):391–7.
78. Crowley TJ, Macdonald MJ, Whitmore EA, Mikulich SK. Cannabis dependence, withdrawal, and reinforcing effects among adolescents with conduct symptoms and substance use disorders. Drug Alcohol Depend. 1998;50(1):27–37.
79. Jones AW, Holmgren A, Kugelberg FC. Driving under the influence of cannabis: a 10-year study of age and gender differences in the concentrations of tetrahydrocannabinol in blood. Addiction. 2008;103(3):452–61.
80. Butters JE. Promoting healthy choices: the importance of differentiating between ordinary and high risk cannabis use among high-school students. Subst. Use Misuse. 2005;40(6):845–55.
81. Hasin DS, Keyes KM, Alderson D, Wang S, Aharonovich E, Grant BF. Cannabis withdrawal in the United States: results from NESARC. J Clin Psychiatry. 2008;69(9):1354–63.
82. Haney M. Opioid antagonism of cannabinoid effects: differences between marijuana smokers and non-marijuana smokers. Neuropsychopharmacology. 2007;32(6):1391–403.
83. Wall ME, Sadler BM, Brine D, Taylor H, Perez-Reyes M. Metabolism, disposition, and kinetics of delta-9-tetrahydrocannabinol in men and women. Clin Pharmacol Ther. 1983;34(3):352–63.

84. Mathew RJ, Wilson WH, Davis R. Postural syncope after marijuana: a transcranial Doppler study of the hemodynamics. Pharmacol Biochem Behav. 2003;75(2):309–18.
85. Patton GC, Coffey C, Carlin JB, Degenhardt L, Lynskey M, Hall W. Cannabis use and mental health in young people: cohort study. BMJ. 2002;325(7374):1195–8.
86. Onaivi ES, Chaudhuri G, Abaci AS, Parker M, Manier DH, Martin PR, Hubbard JR. Expression of cannabinoid receptors and their gene transcripts in human blood cells. Prog Neuropsychopharmacol Biol Psychiatry. 1999;23(6):1063–77.
87. Cooper ZD, Haney M. Investigation of sex-dependent effects of cannabis in daily cannabis smokers. Drug Alcohol Depend. 2014;136:85–91.
88. Harte-Hargrove LC, Dow-Edwards DL. Withdrawal from THC during adolescence: sex differences in locomotor activity and anxiety. Behav Brain Res. 2012;231(1):48–59.
89. Lyons MJ, Toomey R, Meyer JM, Green AI, Eisen SA, Goldberg J, True WR, Tsuang MT. How do genes influence marijuana use? The role of subjective effects. Addiction. 1997;92(4):409–17.
90. Bierut LJ, Dinwiddie SH, Begleiter H, Crowe RR, Hesselbrock V, Nurnberger JI Jr, Porjesz B, Schuckit MA, Reich T. Familial transmission of substance dependence: alcohol, marijuana, cocaine, and habitual smoking: a report from the Collaborative Study on the Genetics of Alcoholism. Arch Gen Psychiatry. 1998;55(11):982–8.
91. Hopfer CJ, Young SE, Purcell S, Crowley TJ, Stallings MC, Corley RP, Rhee SH, Smolen A, Krauter K, Hewitt JK, Ehringer MA. Cannabis receptor haplotype associated with fewer cannabis dependence symptoms in adolescents. Am J Med Genet B Neuropsychiatr Genet. 2006;141B(8):895–901.
92. Filbey FM, Schacht JP, Myers US, Chavez RS, Hutchison KE. Individual and additive effects of the CNR1 and FAAH genes on brain response to marijuana cues. Neuropsychopharmacol. 2010;35(4):967–75.
93. Haughey HM, Marshall E, Schacht JP, Louis A, Hutchison KE. Marijuana withdrawal and craving: influence of the cannabinoid receptor 1 (CNR1) and fatty acid amide hydrolase (FAAH) genes. Addiction. 2008;103(10):1678–86.
94. López-Moreno JA, Echeverry-Alzate V, Bühler KM. The genetic basis of the endocannabinoid system and drug addiction in humans. J Psychopharmacol. 2012;26(1):133–43.
95. Fattore L. Considering gender in cannabinoid research: a step towards personalized treatment of marijuana addicts. Drug Test Anal. 2013;5(1):57–61.
96. Amchova P, Kucerova J, Giugliano V, Babinska Z, Zanda MT, Scherma M, Dusek L, Fadda P, Micale V, Sulcova A, Fratta W, Fattore L. Enhanced self-administration of the CB1 receptor agonist WIN55,212-2 in olfactory bulbectomized rats: evaluation of possible serotonergic and dopaminergic underlying mechanisms. Front Neuropsychopharmacology. 2014;5:44.
97. Tu AW, Ratner PA, Johnson JL. Gender differences in the correlates of adolescents' cannabis use. Subst Use Misuse. 2008;43(10):1438–63.
98. Chabrol H, Ducongé E, Casas C, Roura C, Carey KB. Relations between cannabis use and dependence, motives for cannabis use and anxious, depressive and borderline symptomatology. Addict Behav. 2005;30(4):829–40.
99. Haines RJ, Johnson JL, Carter CI, Arora K. "I couldn't say, I'm not a girl"—adolescents talk about gender and marijuana use. Soc Sci Med. 2009;68(11):2029–36.
100. Klumpers LE, Cole DM, Khalili-Mahani N, Soeter RP, Te Beek ET, Rombouts SA, van Gerven JM. Manipulating brain connectivity with δ^9-tetrahydrocannabinol: a pharmacological resting state FMRI study. Neuroimage. 2012;63(3):1701–11.
101. Narimatsu S, Watanabe K, Yamamoto I, Yoshimura H. Sex difference in the oxidative metabolism of delta9-tetrahydrocannabinol in the rat. Biochem Pharmacol. 1991;41(8):1187–94.
102. Tseng AH, Craft RM. CB(1) receptor mediation of cannabinoid behavioral effects in male and female rats. Psychopharmacology (Berl). 2004;172(1):25–30.
103. Fratta W, Fattore L. Molecular mechanisms of cannabinoid addiction. Curr Opin Neurobiol. 2013;23(4):487–92.

104. Auyeung B, Lombardo MV, Baron-Cohen S. Prenatal and postnatal hormone effects on the human brain and cognition. Pflugers Arch—Eur J Physiol. 2013;465:557–71.
105 Cohen-Bendahan CCC, van de Beek C, Berenbaum SA. Prenatal sex hormone effects on child and adult sex-typed behavior: methods and findings. Neurosci Biobehav Rev. 2005;29:353–84.
106. Abbout DH, Barnett DK, Bruns CM, Dumesic DA. Androgen excess fetal programming of female reproduction: a developmental aetiology for polycystic ovary syndrome? Hum Reprod Update. 2005;11:357–74.
107. Norman RJ, Dewailly D, Legro RS, Hickey TE. Polycystic ovary syndrome. Lancet. 2007;370:9588.
108. Palomba S, Marotta R, Di Cello A, Russo T, Falbo A, Orio F, et al. Pervasive developmental disorders in children of hyperandrogenic women with polycystic ovary syndrome: a longitudinal case-control study. Clin Endocrinol. 2012;77:898–904.
109. Skjaervø GR, Røskaft E. Early conditions and fitness: effect of maternal parity on human life-history traits. Behav Ecol. 2013;24:334–41.
110. Culbert KM, Auyeung SM, Burt A, Klump KL. Prenatal hormone exposure and risk for eating disorders. Arch Gen Psychiatry. 2008;65:329–36.
111. Campbell HJ. The effect of steroid hormones on self-stimulation, central and peripheral. Steridologia 1970;1(1):8–24.
112. de Beun R Jansen E Slangen JL Van de Poll NE. Testosterone as appetitive and discriminative stimulus in rats: sex- and dose-dependent effects. Physiol Behav. 1992;52(4):629–34.
113. Alexander G, Packard M, Hines M. Testosterone has rewarding affective properties in male rats: implications for the biological basis of sexual motivation. Behav Neurosci. 1994;108(2):424–8.
114. Arnedo MT, Salvador A, Martinez-Sanchis S, Gonzalez-Bono E. Rewarding properties of testosterone in intact male mice: a pilot study. Parmacol Biochem Behav. 2000;65(2):327–32.
115. Johnson LR, Wood RI. Oral testosterone self-administration in male hamsters. Neuroendocrinology. 2001;73(4):285–92.
116. Wood RI. Oral testosterone self-administration in male hamsters: dose-response, voluntary exercise, and individual differences. Horm Behav. 2002;41(3):247–58.
117. Wood RI, Johnson LR, Chu L, Schad C, Self DW. Testosterone reinforcement: intravenous and intracerebroventricular self-administration in male rats and hamsters. Psychopharmacology (Berl). 2004;171(3):298–305.
118. Cone EJ, Johnson RE, Moore JD, Roache JD. Acute effects of smoking marijuana on hormones, subjective effects and performance in male human subjects. Pharmacol Biochem Behav. 1986;24(6):1749–54.
119. Dax EM, Pilotte NS, Adler WH, Nagel JE, Lange WR. The effects of 9-ene-tetrahydrocannabinol on hormone release and immune function. J Steroid Biochem. 1989;34(1–6):263–70.
120. Cushman P Jr. Plasma testosterone levels in healthy male marijuana smokers. Am J Drug Alcohol Abuse. 1975;2(2):269–75.
121. Barnett G, Chiang CW, Licko V. Effects of marijuana on testosterone in male subjects. J Theor Biol. 1983;104(4):685–92.
122. Block RI, Farinpour R, Schlechte JA. Effects of chronic marijuana use on testosterone, luteinizing hormone, follicle stimulating hormone, prolactin and cortisol in men and women. Drug Alcohol Depend. 1991;28(2):121–8.
123. Kolodny RC, Masters WH, Kolodner RM, Toro G. Depression of plasma testosterone levels after chronic intensive marihuana use. N Engl J Med. 1974;290(16):872–4.
124. Cohen S. The 94-day cannabis study. Ann N Y Acad Sci. 1976;282:211–20.
125. Juraska JM, Sisk CL, DonCarlos LL. Sexual differentiation of the adolescent rodent brain: hormonal influences and developmental mechanisms. Horm Behav. 2013;64:203–10.
126. Swithers SE, McCurley M, Hamilton E, Doerflinger A. Influence of ovarian hormones on development of ingestive responding to alterations in fatty acid oxidation in female rats. Horm Behav. 2008;54:471–7.

127. Becker JB. Gender differences in dopaminergic function in striatum and nucleus accumbens. Pharmacol Biochem Behav. 1999;64(4):803–12.
128. Kuhn C, Johnson M, Thomae A, Luo B, Simon SA, Zhou G, Walker QD. The emergence of gonadal hormone influences on dopaminergic function. Horm Behav. 2010;58:122–37.
129. Wooten GF, Currie LJ, Bovbjerg VE, Lee JK, Patrie J. Are men at greater risk for Parkinson's disease than women? J Neurol Neurosurg Psychiatry. 2004;75:637–9.
130. Chisholm NC, Kim T, Juraska JM. Males, but not females, lose tyrosine hydroxylase fibers in the medial prefrontal cortex and are impaired on a delayed alternation task during aging. Behav Brain Res. 2013;243:239–46.
131. Dreher JC, Schmidt PJ, Kohn P, Furman D, Rubinow D, Berman KF. Menstrual cycle phase modulates reward-related neural function in women. Proc Natl Acad Sci U S A. 2007;104(7):2465–70.
132. Anker JJ, Carroll ME. The role of progestins in the behavioral effects of cocaine and other drugs of abuse: human and animal research. Neurosci Biobehav Rev. 2010;35(2):315–33.
133. Vallée M1, Vitiello S, Bellocchio L, Hébert-Chatelain E, Monlezun S, Martin-Garcia E, Kasanetz F, Baillie GL, Panin F, Cathala A, Roullot-Lacarrière V, Fabre S, Hurst DP, Lynch DL, Shore DM, Deroche-Gamonet V, Spampinato U, Revest JM, Maldonado R, Reggio PH, Ross RA, Marsicano G, Piazza PV. Pregnenolone can protect the brain from cannabis intoxication. Science. 2014;343(6166):94–8.
134. Terner JM, de Wit H. Menstrual cycle phase and responses to drugs of abuse in humans. Drug Alcohol Depend. 2006;84(1):1–13.
135. Justice AJ, de Wit H. Acute effects of d-amphetamine during the follicular and luteal phases of the menstrual cycle in women. Psychopharmacology (Berl). 1999;145(1):67–75.
136. Carroll ME, Anker JJ. Sex differences and ovarian hormones in animal models of drug dependence. Horm Behav. 2010;58(1):44–56.
137. Quinones-Jenab V, Jenab S. Progesterone attenuates cocaine-induced responses. Horm Behav. 2010;58(1):22–32.
138. Mello NK. Hormones, nicotine, and cocaine: clinical studies. Horm Behav. 2010;58(1):57–71.
139. Almirez RG, Smith CG, Asch RH. The effects of marijuana extract and delta9-tetrahydrocannabinol on luteal function in the rhesus monkey. Fertil Steril. 1983;39(2):212–7.
140. Huang GZ, Woolley CS. Estradiol acutely suppresses inhibition in the hippocampus through a sex-specific endocannabinoid and mGluR-dependent mechanism. Neuron. 2012;74(5):801–8.
141. Hill MN, Karacabeyli ES, Gorzalka BB. Estrogen recruits the endocannabinoid system to modulate emotionality. Psychoneuroendocrinology 2007;32(4):350–7.
142. Winsauer PJ, Daniel JM, Filipeanu CM, Leonard ST, Hulst JL, Rodgers SP, Lassen-Greene CL, Sutton JL. Long-term behavioral and pharmacodynamic effects of delta-9-tetrahydrocannabinol in female rats depend on ovarian hormone status. Addict Biol. 2011;16(1):64–81.
143. Fattore L, Spano MS, Altea S, Angius F, Fadda P, Fratta W. Cannabinoid self-administration in rats: sex differences and the influence of ovarian function. Br J Pharmacol. 2007;152(5):795–804.
144. Fattore L, Spano MS, Altea S, Fadda P, Fratta W. Drug- and cue induced reinstatement of cannabinoid-seeking behaviour in male and female rats: influence of ovarian hormones. Br J Pharmacol. 2010;160(3):724–35.
145. El-Talatini MR, Taylor AH, Konje JC. The relationship between plasma levels of the endocannabinoid, anandamide, sex steroids, and gonadotrophins during the menstrual cycle. Fertil Steril. 2010;93(6):1989–96.
146. Martel MM. Sexual selection and sex differences in the prevalence of childhood externalizing and adolescent internalizing disorders. Psychol Bull. 2013;139:1221–59.
147. Bethlehem RAI, van Honk J, Auyeung B, Baron-Cohen S. Oxytocin, brain physiology, and functional connectivity: a review of intranasal oxytocin fMRI studies. Psychoneuroendocrinol. 2013;38:962–74.

148. Marsh AA, Yu HH, Pine DS, Gorodetsky EK, Goldman D, Blair RJR. The influence of oxytocin administration on responses to infant faces and potential modertation by OXTR gene. Psychopharmacol. 2012;224:469–76.
149 Carson DS, Guastella AJ, Taylor ER, McGregor IS. A brief history of oxytocin and its role in modulating psychostimulant effects. J Psychopharmacol. 2013;27:231–47.
150. Lischke A, Gamer M, Berger C, Grossmann A, Hauenstein K, Heinrichs M, et al. Oxytocin increases amygdala reactivity to threatening scenes in females. Psychoneuroendocrinol. 2012;37:1431–8.
151. Baracz SJ, Rouke PI, Pardey MC, Hunt GE, McGregor IS, Cornish JL. Oxytocin directly administered into the nucleus accumbens core of subthalamic nucleus attenuates methamphetamine-induced place preference. Behav Brain Res. 2012;228:185–93.
152. McGregor IS, Callaghan PD, Hunt GE. From ultrasocial to antisocial: a role for oxytocin in the acute reinforcing effects and long-term adverse consequences of drug use? Br J Pharmacol. 2008;154(2):358–68.
153. Butovsky E, Juknat A, Elbaz J, Shabat-Simon M, Eilam R, Zangen A Altstein M, Vogel Z. Chronic exposure to Delta9-tetrahydrocannabinol downregulates oxytocin and oxytocin-associated neurophysin in specific brain areas. Mol Cell Neurosci. 2006;31(4):795–804.
154. Cui SS, Bowen RC, Gu GB, Hannesson DK, Yu PH, Zhang X. Prevention of cannabinoid withdrawal syndrome by lithium: involvement of oxytocinergic neuronal activation. J Neurosci. 2001;21(24):9867–76.
155. Verty AN, McFarlane JR, McGregor IS, Mallet PE. Evidence for an interaction between CB1 cannabinoid and oxytocin receptors in food and water intake. Neuropharmacology. 2004;47(4):593–603.
156. Schneiderman I, Zagoory-Sharon O, Leckman JF, Feldman R. Oxytocin during the initial stages of romantic attachment: relations to couples' interactive reciprocity. Psychoneuroendocrinol. 2012;37:1277–85.
157. Burkett JP, Young LJ. The behavioral, anatomical and pharmacological parallels between social attachment, love and addiction. Psychopharmacology. 2012;224:1–26.
158. Arnold AP, Breedlove SM. Organizational and activational effects of sex steroids on brain and behavior—a reanalysis. Horm Behav. 1985;19:469–98.
159. Lancaster FE. Gender differences in the brain: implications for the study of human alcoholism. Alcohol Clin Exp Res. 1994;18(3):740–6.
160. Becker JB, Molenda H, Hummer DL. Gender differences in the behavioral responses to cocaine and amphetamine. Implications for mechanisms mediating gender differences in drug abuse. Ann N Y Acad Sci. 2001;937:172–87.
161. Hu M, Crombag HS, Robinson TE, Becker JB. Biological basis of sex differences in the propensity to self-administer cocaine. Neuropsychopharmacology. 2004;29(1):81–5.
162. Craft RM, Leitl MD. Gonadal hormone modulation of the behavioral effects of Delta9-tetrahydrocannabinol in male and female rats. Eur J Pharmacol. 2008;578(1):37–42.
163. Castelli MP, Fadda P, Casu A, Spano MS, Casti A, Fratta W, Fattore L. Male and female rats differ in brain cannabinoid CB1 receptor density and function and in behavioural traits predisposing to drug addiction: effect of ovarian hormones. Curr Pharm Des. 2013. [Epub ahead of print]
164. Kritzer MF. Selective colocalization of immunoreactivity for intracellular gonadal hormone receptors and tyrosine hydroxylase in the ventral tegmental area, substantia nigra, and retrorubral fields in the rat. J Comp Neurol. 1997;379(2):247–60.
165. Creutz LM, Kritzer MF. Estrogen receptor-beta immunoreactivity in the midbrain of adult rats: regional, subregional, and cellular localization in the A10, A9, and A8 dopamine cell groups. J Comp Neurol. 2002;446(3):288–300.
166. Creutz LM, Kritzer MF. Mesostriatal and mesolimbic projections of midbrain neurons immunoreactive for estrogen receptor beta or androgen receptors in rats. J Comp Neurol. 2004;476(4):348–62.
167. Mani SK. Signaling mechanisms in progesterone-neurotransmitter interactions. Neuroscience. 2006;138(3):773–81.

168. Woolley CS. Acute effects of estrogen on neuronal physiology. Annu Rev Pharmacol Toxicol. 2007;47:657–80.
169. Fink G, Sumner BE, Rosie R, Grace O, Quinn JP. Estrogen control of central neurotransmission: effect on mood, mental state, and memory. Cell Mol Neurobiol. 1996;16(3):325–44.
170. Díaz-Véliz G, Butrón S, Benavides MS, Dussaubat N, Mora S. Gender, estrous cycle, ovariectomy, and ovarian hormones influence the effects of diazepam on avoidance conditioning in rats. Pharmacol Biochem Behav. 2000;66(4):887–92.
171. Kim J, Alger BE. Inhibition of cyclooxygenase-2 potentiates retrograde endocannabinoid effects in hippocampus. Nat Neurosci. 2004;7(7):697–8.
172. Busch L, Sterin-Borda L, Borda E. Effects of castration on cannabinoid CB1 receptor expression and on the biological actions of cannabinoid in the parotid gland. Clin Exp Pharmacol Physiol. 2006;33(3):258–63.
173. González S, Bisogno T, Wenger T, Manzanares J, Milone A, Berrendero F, Di Marzo V, Ramos JA, Fernández-Ruiz JJ. Sex steroid influence on cannabinoid CB(1) receptor mRNA and endocannabinoid levels in the anterior pituitary gland. Biochem Biophys Res Commun. 2000;270(1):260–6.
174. Mize AL, Alper RH. Acute and long-term effects of 17beta-estradiol on G(i/o) coupled neurotransmitter receptor function in the female rat brain as assessed by agonist-stimulated [35S]GTPgammaS binding. Brain Res. 2000;859(2):326–33.
175. Bradshaw HB, Rimmerman N, Krey JF, Walker JM. Sex and hormonal cycle differences in rat brain levels of pain-related cannabimimetic lipid mediators. Am J Physiol Regul Integr Comp Physiol. 2006;291(2):349–58.
176. Rodríguez de Fonseca F Cebeira M Ramos JA Martín M Fernández-Ruiz JJ. Cannabinoid receptors in rat brain areas: sexual differences, fluctuations during estrous cycle and changes after gonadectomy and sex steroid replacement. Life Sci 1994;54(3):159–70.
177. Schoenbaum G1, Chiba AA, Gallagher M. Orbitofrontal cortex and basolateral amygdala encode expected outcomes during learning. Nat Neurosci. 1998;1(2):155–9.
178. Bechara A, Damasio H, Damasio AR, Lee GP. Different contributions of the human amygdala and ventromedial prefrontal cortex to decision-making. J Neurosci. 1999;19(13):5473–81.
179. Diaz S, Farhang B, Hoien J, Stahlman M, Adatia N, Cox JM, Wagner EJ. Sex differences in the cannabinoid modulation of appetite, body temperature and neurotransmission at POMC synapses. Neuroendocrinology. 2009;89(4):424–40.
180. Romero J, Lastres-Becker I, de Miguel R, Berrendero F, Ramos JA, Fernández-Ruiz J. The endogenous cannabinoid system and the basal ganglia biochemical, pharmacological, and therapeutic aspects. Pharmacol Ther. 2002;95(2):137–52.
181. Wiley JL, O'connell MM, Tokarz ME, Wright MJ Jr. Pharmacological effects of acute and repeated administration of Delta(9)-tetrahydrocannabinol in adolescent and adult rats. J Pharmacol Exp Ther. 2007;320(3):1097–105.
182. Lynch WJ, Roth ME, Carroll ME. Biological basis of sex differences in drug abuse: preclinical and clinical studies. Psychopharmacology (Berl). 2002;164(2):121–37.
183. Das A, Chaudhuri SK. Effects of sex steroids on the concentrations of some brain neurotransmitters in male and female rats: some new observations. Indian J Physiol Pharmacol. 1995;39(3):223–30.
184. Volkow ND, Fowler JS, Wang GJ. Role of dopamine in drug reinforcement and addiction in humans: results from imaging studies. Behav Pharmacol. 2002;13(5–6):355–66.
185. Volkow ND, Fowler JS, Wang GJ, Swanson JM. Dopamine in drug abuse and addiction: results from imaging studies and treatment implications. Mol Psychiatry. 2004;9(6):557–69.
186. Mozley LH, Gur RC, Mozley PD, Gur RE. Striatal dopamine transporters and cognitive functioning in healthy men and women. Am J Psychiatry. 2001;158(9):1492–9.
187. Staley JK, Krishnan-Sarin S, Zoghbi S, Tamagnan G, Fujita M, Seibyl JP, Maciejewski PK, O'Malley S, Innis RB. Sex differences in [123I]beta-CIT SPECT measures of dopamine and serotonin transporter availability in healthy smokers and nonsmokers. Synapse. 2001;41(4):275–84.

188. Melis M, De Felice M, Lecca S, Fattore L, Pistis M. Sex-specific tonic 2-arachidonoylglycerol signaling at inhibitory inputs onto dopamine neurons of lister hooded rats. Front Integr Neurosci. 2013;7:93.
189. Galvan A, Rahdar A. The neurobiological effects of stress on adolescent decision making. Neurosci. 2013;249:223–1.
190. DeVito EE, Meda SA, Jiantonio R, Potenza MN, Krystal JH, Pearlson GD. Neural correlates of impulsivity in healthy males and females with family histories of alcoholism. Neuropsychopharmacol. 2013;38:1854–63.
191. Seo D, Jia Z, Lacadie CM, Tsou KA, Bergquist K, Sinha R. Sex differences in neural responses to stress and alcohol context cues. Hum Brain Map. 2011;32:1998–2013.
192. Sinha R, Garcia M, Paliwel P, Kreek MJ, Rounsaville BJ. Stress-induced cocaine craving and hypothalamic-pituitary-adrenal responses are predictive of cocaine relapse outcome. Arch Gen Psychiatry. 2006;63:324–31.
193. Galvan A, McGlennen KM. Daily stress increases risky decision-making in adolesents: a preliminary study. Develop Psychobiol. 2012;54:433–440.
194. Young EA. Sex differences and the HPA axis: implications for psychiatric disease. J Gend Specif Med. 1998;1:21–7.
195. Bourke CH, Neigh GN. Behavioral effects of chronic adolescent stress are sustained and sexually dimorphic. Horm Behav. 2011;60:112–20.
196. Sinha R, Fox H, Hong K-I, Sofuoglu M, Morgan PT, Bergquist KT. Sex steroid hormones, stress response, and drug craving in cocaine-dependent woman: implications for relapse susceptibility. Exper Clin Psychopharmacol. 2007;15:445–52.
197. Zitman FMP, Richter-Levin G. Age and sex-dependent differences in activity, plasticity and response to stress in the dentate gyrus. Neurosci. 2013;249:21–30.
198. van den Bos R Harteveld M Stoop H. Stress and decision-making in humans: performance is related to cortisol reactivity, albeit differently in man and women. Psychoneuroendocrinol. 2009;34:1449–58.
199. Cota D. The role of the endocannabinoid system in the regulation of hypothalamic-pituitary-adrenal axis activity. J Neuroendocrinol. 2008;20(Suppl 1):35–8.
200. Gorzalka BB, Hill MN, Hillard CJ. Regulation of endocannabinoid signaling by stress: implications for stress-related affective disorders. Neurosci Biobehav Rev. 2008;32(6):1152–60.
201. Dlugos A, Childs E, Stuhr KL, Hillard CJ, de Wit H. Acute stress increases circulating anandamide and other N-acylethanolamines in healthy humans. Neuropsychopharmacology 2012;37(11):2416–27.
202. Hill MN, Miller GE, Carrier EJ, Gorzalka BB, Hillard CJ. Circulating endocannabinoids and N-acyl ethanolamines are differentially regulated in major depression and following exposure to social stress. Psychoneuroendocrinology. 2009;34(8):1257–62.
203. Hill MN, Miller GE, Ho WS, Gorzalka BB, Hillard CJ. Serum endocannabinoid content is altered in females with depressive disorders: a preliminary report. Pharmacopsychiatry. 2008;41(2):48–53.
204. Weafer J, de Wit H. Sex differences in impulsive action and impulsive choice. Addict Behav. 2013. pii:S0306-4603(13)00367–5. [Epub ahead of print]
205. Nederkoorn C, Baltus M, Guerrieri R, Wiers RW. Heavy drinking is associated with deficient response inhibition in women but not in men. Pharmacol Biochem Behav. 2009;93(3):331–6.
206. Anker JJ, Gliddon LA, Carroll ME. Impulsivity on a go/no-go task for intravenous cocaine or food in male and female rats selectively bred for high and low saccharin intake. Behav Pharmacol. 2008;19(5–6):615–29.
207. Perry JL, Nelson SE, Carroll ME. Impulsive choice as a predictor of acquisition of IV cocaine self- administration and reinstatement of cocaine-seeking behavior in male and female rats. Exp Clin Psychopharmacol. 2008;16(2):165–77.
208. Bayless DW, Darling JS, Daniel JM. Mechanisms by which neonatal testosterone exposure mediates sex differences in impulsivity in prepubertal rats. Horm Behav. 2013;64(5):764–9.
209. Kollins SH. Delay discounting is associated with substance use in college students. Addict Behav. 2003;28(6):1167–73.

210. Vangsness L, Bry BH, LaBouvie EW. Impulsivity, negative expectancies, and marijuana use: a test of the acquired preparedness model. Addict Behav. 2005;30(5):1071–6.
211. Simons JS, Carey KB. Risk and vulnerability for marijuana use problems: the role of affect dysregulation. Psychol Addict Behav. 2002;16(1):72–5.
212. Day AM, Metrik J, Spillane NS, Kahler CW. Working memory and impulsivity predict marijuana-related problems among frequent users. Drug Alcohol Depend. 2013;131(1–2):171–4.
213. McDonald J, Schleifer L, Richards JB, de Wit H. Effects of THC on behavioral measures of impulsivity in humans. Neuropsychopharmacology. 2003;28(7):1356–65.
214. Kampov-Polevoy AB, Garbutt JC, Davis CE, Janowsky DS. Preference for higher sugar concentrations and Tridimensional Personality Questionnaire scores in alcoholic and nonalcoholic men. Alcohol Clin Exp Res. 1998;22(3):610–4.
215. Kampov-Polevoy AB, Ziedonis D, Steinberg ML, Pinsky I, Krejci J, Eick C, Boland G, Khalitov E, Crews FT. Association between sweet preference and paternal history of alcoholism in psychiatric and substance abuse patients. Alcohol Clin Exp Res. 2003;27(12):1929–36.
216. Nolan LJ, Scagnelli LM. Preference for sweet foods and higher body mass index in patients being treated in long-term methadone maintenance. Subst Use Misuse. 2007;42(10):1555–66.
217. Maone TR, Mattes RD, Beauchamp GK. Cocaine-exposed newborns show an exaggerated sucking response to sucrose. Physiol Behav. 1992;51(3):487–91.
218. Lieblich I, Yirmiya R, Liebeskind JC. Intake of and preference for sweet solutions are attenuated in morphine-withdrawn rats. Behav Neurosci. 1991;105(6):965–70.
219. Copersino ML, Boyd SJ, Tashkin DP, Huestis MA, Heishman SJ, Dermand JC, Simmons MS, Gorelick DA. Sociodemographic characteristics of cannabis smokers and the experience of cannabis withdrawal. Am J Drug Alcohol Abuse. 2010;36(6):311–9.
220. Agrawal A, Pergadia ML, Lynskey MT. Is there evidence for symptoms of cannabis withdrawal in the national epidemiologic survey of alcohol and related conditions? Am J Addict. 2008;17(3):199–208.
221. Robbins SJ, Ehrman RN, Childress AR, O'Brien CP. Comparing levels of cocaine cue reactivity in male and female outpatients. Drug Alcohol Depend. 1999;53(3):223–30.
222. Yu J, Zhang S, Epstein DH, Fang Y, Shi J, Qin H, Yao S, Le Foll B, Lu L. Gender and stimulus difference in cue-induced responses in abstinent heroin users. Pharmacol Biochem Behav. 2007;86(3):485–92.
223. Perkins KA, Gerlach D, Vender J, Grobe J, Meeker J, Hutchison S. Sex differences in the subjective and reinforcing effects of visual and olfactory cigarette smoke stimuli. Nicotine Tob Res. 2001;3(2):141–50.
224. Field M, Duka T. Cue reactivity in smokers: the effects of perceived cigarette availability and gender. Pharmacol Biochem Behav. 2004;78(3):647–52.
225. Saladin ME, Gray KM, Carpenter MJ, LaRowe SD, DeSantis SM, Upadhyaya HP. Gender differences in craving and cue reactivity to smoking and negative affect/stress cues. Am J Addict. 2012;21(3):210–20.
226. Willner P, Field M, Pitts K, Reeve G. Mood, cue and gender influences on motivation, craving and liking for alcohol in recreational drinkers. Behav Pharmacol. 1998;9(7):631–42.
227. Lundahl LH, Johanson CE. Cue-induced craving for marijuana in cannabis-dependent adults. Exp Clin Psychopharmacol. 2011;19(3):224–30.
228. Kreek MJ, Nielsen DA, Butelman ER, LaForge KS. Genetic influences on impulsivity, risk taking, stress responsivity and vulnerability to drug abuse and addiction. Nat Neurosci. 2005;8(11):1450–7.
229. Solinas M, Thiriet N, Chauvet C, Jaber M. Prevention and treatment of drug addiction by environmental enrichment. Prog Neurobiol. 2010;92(4):572–92.
230. El Rawas R Thiriet N Nader J Lardeux V Jaber M Solinas M. Early exposure to environmental enrichment alters the expression of genes of the endocannabinoid system. Brain Res. 2011;1390:80–9.

231. Skwara AJ, Karwoski TE, Czambel RK, Rubin RT, Rhodes ME. Influence of environmental enrichment on hypothalamic-pituitary-adrenal (HPA) responses to single-dose nicotine, continuous nicotine by osmotic mini-pumps, and nicotine withdrawal by mecamylamine in male and female rats. Behav Brain Res. 2012;234(1):1–10
232. Altman J, Everitt BJ, Glautier S, Markou A, Nutt D, Oretti R, Phillips GD, Robbins TW. The biological, social and clinical bases of drug addiction: commentary and debate. Psychopharmacology (Berl). 1996;125(4):285–345.
233. Becker JB, Hu M Sex differences in drug abuse. Front Neuroendocrinol. 2008;29(1):36–47.
234. Anker JJ, Carroll ME. Females are more vulnerable to drug abuse than males: evidence from preclinical studies and the role of ovarian hormones. Curr Top Behav Neurosci. 2011;8:73–96.
235. Heppner CC, Kemble ED, Cox MW. Effects of food deprivation on caffeine consumption in male and female rats. Pharmacol Biochem Behav. 1986;24(6):1555–9.
236. Morse AC, Erwin VG, Jones, BC. Strain and housing affect cocaine self-selection and open-field locomotor activity in mice. Pharmacol Biochem Behav. 1993;45(4):905–12.
237. Matthews KM, Robbins TW, Everitt BJ, Caine SB. Repeated maternal separation alters intravenous cocaine self-administration in rats. Psychopharmacology (Berl). 1999;141(2):123–34.
238. Carroll ME, Campbell UC, Heideman P. Ketoconazole inhibits food restriction induced increases in i.v. heroin self-administration in rats: sex differences. Exp Clin Psychopharm. 2001;9(3):307–16.
239. Hill SY, Powell BJ. Cocaine and morphine self-administration: effects of differential nose poke. Pharmacol Biochem Behav. 1976;5(6):701–4.
240. Alexander BK, Coambs RB, Hadaway PF. The effect of housing and gender on morphine self-administration in rats. Psychopharmacology 1978;58(2):175–9.
241. Klein LC, Popke JE, Grunberg NE. Sex differences in effects of predictable and unpredictable footshock on fentanyl self-administration in rats. Exp Clin Psychopharmacol. 1997;5(2):99–106.
242 Silverman JL, Koenig JI. Evidence for the involvement of ERbeta and RGS9-2 in 17-beta estradiol enhancement of amphetamine-induced place preference behavior. Horm Behav 2007;52(2):146–55.
243. Segarra AC, Agosto-Rivera JL, Febo M, Lugo-Escobar N, Menendez-Delmestre R, Puig-Ramos A, Torres-Diaz YM. Estradiol: A key biological substrate mediating the response to cocaine in female rats. Horm Behav. 2010;58(1):33–43.
244 Eriksson K. Genetic selection for voluntary alcohol consumption in albino rats. Science. ;159(3816):739–41.
245. Eriksson K, Pikkarainen PH. Differences between the sexes in voluntary alcohol consumption and liver ADH-activity in inbred strains of mice. Metabolism. ;17(11):1037–42.
246. Juarez J, Guzman-Flores C, Ervin FR, Palmour RM. Voluntary alcohol consumption in vervet monkeys: individual, sex and age differences. Pharmacol Biochem Behav. 1993;46(4):985–8.
247. Andreano JM, Cahill L. Sex influences on the neurobiology of learning and memory. Learn Mem. 2009;16(4):248–66.
248. McQueeny T, Padula CB, Price J, Medina KL, Logan P, Tapert SF. Gender effects on amygdala morphometry in adolescent marijuana users. Behav Brain Res. 2011;224(1):128–34.
249. Medina KL, McQueeny T, Nagel BJ, Hanson KL, Yang TT, Tapert SF. Prefrontal cortex morphometry in abstinent adolescent marijuana users: subtle gender effects. Addict Biol. 2009;14(4):457–68.
250. Filbey FM, Schacht JP, Myers US, Chavez RS, Hutchison KE. Marijuana craving in the brain. Proc Natl Acad Sci U S A. 2009;106(31):13016–21.
251. Spiga S, Lintas A, Migliore M, Diana M. Altered architecture and functional consequences of the mesolimbic dopamine system in cannabis dependence. Addict Biol. 2010;15(3):266–76.

252.. Hungund BL, Vinod KY, Kassir SA, Basavarajappa BS, Yalamanchili R, Cooper TB, Mann JJ, Arango V. Upregulation of CB1 receptors and agonist-stimulated [35S]GTP-gammaS binding in the prefrontal cortex of depressed suicide victims. Mol Psychiatry. 2004;9(2):184–90.
253. Vinod KY, Arango V, Xie S, Kassir SA, Mann JJ, Cooper TB, Hungund BL. Elevated levels of endocannabinoids and CB1 receptor-mediated G-protein signaling in the prefrontal cortex of alcoholic suicide victims. Biol Psychiatry. 2005;57(5):480–6.
254. Koethe D, Llenos IC, Dulay JR, Hoyer C, Torrey EF, Leweke FM, Weis S. Expression of CB1 cannabinoid receptor in the anterior cingulate cortex in schizophrenia, bipolar disorder, and major depression. J Neural Transm. 2007;114(8):1055–63.
255. Castelli MP, Piras AP, D'Agostino A, Pibiri F, Perra S, Gessa GL, Maccarrone M, Pistis M. Dysregulation of the endogenous cannabinoid system in adult rats prenatally treated with the cannabinoid agonist WIN 55,212-2. Eur J Pharmacol. 2007;573(1–3):11–9.
256. Wiley JL, Kendler SH, Burston JJ, Howard DR, Selley DE, Sim-Selley LJ. Antipsychotic-induced alterations in CB1 receptor-mediated G-protein signaling and in vivo pharmacology in rats. Neuropharmacology. 2008;55(7):1183–90.
257. Mateos B, Borcel E, Loriga R, Luesu W, Bini V, Llorente R, Castelli MP, Viveros MP. Adolescent exposure to nicotine and/or the cannabinoid agonist CP 55,940 induces gender-dependent long-lasting memory impairments and changes in brain nicotinic and CB(1) cannabinoid receptors. J Psychopharmacol. 2011;25(12):1676–90.
258. Burston JJ, Wiley JL, Craig AA, Selley DE, Sim-Selley LJ. Regional enhancement of cannabinoid CB1 receptor desensitization in female adolescent rats following repeated Delta-tetrahydrocannabinol exposure. Br J Pharmacol. 2010;161(1):103–12.
259. Van Laere K Goffin K Casteels C Dupont P Mortelmans L de Hoon J Bormans G. Gender-dependent increases with healthy aging of the human cerebral cannabinoid-type 1 receptor binding using [(18)F]MK-9470 PET. Neuroimage. 2008;39(4):1533–41.
260. Riebe CJ, Hill MN, Lee TT, Hillard CJ, Gorzalka BB. Estrogenic regulation of limbic cannabinoid receptor binding. Psychoneuroendocrinology. 2010;35(8):1265–9.
261. Perry JL, Joseph JE, Jiang Y, Zimmerman RS, Kelly TH, Darna M, Huettl P, Dwoskin LP, Bardo MT. Prefrontal cortex and drug abuse vulnerability: translation to prevention and treatment interventions. Brain Res Rev. 2011;65(2):124–49.
262. Davis M, Rainnie D, Cassell M. Neurotransmission in the rat amygdala related to fear and anxiety. Trends Neurosci. 1994;17(5):208–14.
263. Graeff FG, Guimarães FS, De Andrade TG, Deakin JF. Role of 5-HT in stress, anxiety, and depression. Pharmacol Biochem Behav. 1996;54(1):129–41.
264. Cahill L, McGaugh JL. Mechanisms of emotional arousal and lasting declarative memory. Trends Neurosci. 1998;21(7):294–9.
265. Plappert CF, Rodenbucher AM, Pilz PK. Effects of sex and estrous cycle on modulation of the acoustic startle response in mice. Physiol Behav. 2005;84(4):585–94.
266. La Greca AM Lopez N. Social anxiety among adolescents: linkages with peer relations and friendships. J Abnorm Child Psychol. 1998;26(2):83–94.
267. Buckner JD, Mallott MA, Schmidt NB, Taylor J. Peer influence and gender differences in problematic cannabis use among individuals with social anxiety. J Anxiety Disord. 2006;20(8):1087–102.
268. Buckner JD, Schmidt NB. Social anxiety disorder and marijuana use problems: the mediating role of marijuana effect expectancies. Depress Anxiety. 2009;26(9):864–70.
269. Redolat R, Pérez-Martínez A, Carrasco MC, Mesa P. Individual differences in novelty-seeking and behavioral responses to nicotine: a review of animal studies. Curr Drug Abuse Rev. 2009;2(3):230–42.
270. Quinones-Jenab V. Why are women from Venus and men from Mars when they abuse cocaine? Brain Res. 2006;1126(1):200–3.
271. Milesi-Hallé A, McMillan DE, Laurenzana EM, Byrnes-Blake KA, Owens SM. Sex differences in (+)-amphetamine- and (+)-methamphetamine-induced behavioral response in male and female Sprague-Dawley rats. Pharmacol Biochem Behav. 2007;86(1):140–9.

272. Gorzalka BB, Hill MN, Chang SC. Male-female differences in the effects of cannabinoids on sexual behavior and gonadal hormone function. Horm Behav. 2010;58(1):91–9.
273. Farhang B, Diaz S, Tang SL, Wagner EJ. Sex differences in the cannabinoid regulation of energy homeostasis. Psychoneuroendocrinology. 2009;34(Suppl 1):237–46.
274. Biscaia M, Marín S, Fernández B, Marco EM, Rubio M, Guaza C, Ambrosio E, Viveros MP. Chronic treatment with CP 55,940 during the peri-adolescent period differentially affects the behavioural responses of male and female rats in adulthood. Psychopharmacology (Berl). 2003;170(3):301–8.
275. Corchero J, García-Gil L, Manzanares J, Fernández-Ruiz JJ, Fuentes JA, Ramos JA. Perinatal delta9-tetrahydrocannabinol exposure reduces proenkephalin gene expression in the caudate-putamen of adult female rats. Life Sci. 1998;63(10):843–50.
276. Vela G, Martín S, García-Gil L, Crespo JA, Ruiz-Gayo M, Fernández-Ruiz JJ, García-Lecumberri C, Pélaprat D, Fuentes JA, Ramos JA, Ambrosio E. Maternal exposure to delta9-tetrahydrocannabinol facilitates morphine self-administration behavior and changes regional binding to central *mu* opioid receptors in adult offspring female rats. Brain Res. 1998;807(1–2):101–9.
277. Rubino T, Vigano' D, Realini N, Guidali C, Braida D, Capurro V, Castiglioni C, Cherubino F, Romualdi P, Candeletti S, Sala M, Parolaro D. Chronic delta9-tetrahydrocannabinol during adolescence provokes sex dependent changes in the emotional profile in adult rats: behavioral and biochemical correlates. Neuropsychopharmacology. 2008;33(11):2760–71.
278. Fattore L, Viganò D, Fadda P, Rubino T, Fratta W, Parolaro D. Bidirectional regulation of mu-opioid and CB1-cannabinoid receptor in rats self-administering heroin or WIN 55,212-2. Eur J Neurosci. 2007b;25(7):2191–200.
279. Lynch WJ. Sex differences in vulnerability to drug self-administration. Exp Clin Psychopharmacol. 2006;14(1):34–41.
280. Freund TF, Katona I, Piomelli D. Role of endogenous cannabinoids in synaptic signaling. Physiol Rev. 2003;83(3):1017–66.
281. Weiser MJ, Foradori CD, Handa RJ. Estrogen receptor beta in the brain: from form to function. Brain Res Rev. 2008;57(2):309–20.
282. Bonnin A, Fernandez-Ruiz JJ, Martin M, Rodriguez de Fonseca F, Hernandez ML, Ramos JA. Delta9-Tetrahydrocannabinol affects mesolimbic dopaminergic activity in the female rat brain: interactions with estrogens. J Neural Transm Gen Sect. 1993;92(2–3):81–95.
283. Quinn HR, Matsumoto I, Callaghan PD, Long LE, Arnold JC, Gunasekaran N, Thompson MR, Dawson B, Mallet PE, Kashem MA, Matsuda-Matsumoto H, Iwazaki T, McGregor IS. Adolescent rats find repeated delta(9)-THC less aversive than adult rats but display greater residual cognitive deficits and changes in hippocampal protein expression following exposure. Neuropsychopharmacology. 2008;33(5):1113–26.
284. Tanda G, Munzar P, Goldberg SR. Self-administration behavior is maintained by the psychoactive ingredient of marijuana in squirrel monkeys. Nat Neurosci. 2000;3(11):1073–4.
285. Fattore L, Cossu G, Martellotta CM, Fratta W. Intravenous self-administration of the cannabinoid CB1 receptor agonist WIN 55,212–2 in rats. Psychopharmacology (Berl). 2001;156(4):410–6.
286. Lepore M, Vorel SR, Lowinson J, Gardner EL. Conditioned place preference induced by delta9-tetrahydrocannabinol: comparison with cocaine, morphine, and food reward. Life Sci. 1995;56(23–24):2073–80.
287. Gardner EL. Addictive potential of cannabinoids: the underlying neurobiology. Chem Phys Lipids. 2002;121(1–2):267–90.
288. Arnold JC, Topple AN, Mallet PE, Hunt GE, McGregor IS. The distribution of cannabinoid-induced Fos expression in rat brain: differences between the Lewis and Wistar strain. Brain Res. 2001;921(1–2):240–55.
289. Deiana S, Fattore L, Spano MS, Cossu G, Porcu E, Fadda P, Fratta W. Strain and schedule-dependent differences in the acquisition, maintenance and extinction of intravenous cannabinoid self-administration in rats. Neuropharmacology. 2007;52(2):646–54.

290. Fattore L, Altea S, Fratta W. Sex differences in drug addiction: a review of animal and human studies. Womens Health. 2008;4:51.
291. Mello NK, Mendelson JH. Operant acquisition of marihuana by women. J Pharmacol Exp Ther. 1985;235(1):162–71.
292. Griffin ML, Mendelson JH, Mello NK, Lex BW. Marihuana use across the menstrual cycle. Drug Alcohol Depend. 1986;18(2):213–24.
293. Belin D, Belin-Rauscent A, Murray JE, Everitt BJ. Addiction: failure of control over maladaptive incentive habits. Curr Opin Neurobiol. 2013;23(4):564–72.
294. Tronson NC, Taylor JR. Addiction: a drug-induced disorder of memory reconsolidation. Curr Opin Neurobiol. 2013;23(4):573–80.
295. Cahill L, van Stegeren A. Sex-related impairment of memory for emotional events with beta-adrenergic blockade. Neurobiol Learn Mem. 2003;79(1):81–8.
296. Flint RW Jr, Valentine S, Papandrea D Jr. Reconsolidation of a long-term spatial memory is impaired by cycloheximide when reactivated with a contextual latent learning trial in male and female rats. Neuroscience. 2007;148(4):833–44.
297. Lonergan MH, Olivera-Figueroa LA, Pitman RK, Brunet A. Propranolol's effects on the consolidation and reconsolidation of long-term emotional memory in healthy participants: a meta-analysis. J Psychiatry Neurosci. 2013;38(4):222–31.
298. Cahill L, Haier RJ, White NS, Fallon J, Kilpatrick L, Lawrence C, Potkin SG, Alkire MT. Sex-related difference in amygdala activity during emotionally influenced memory storage. Neurobiol Learn Mem. 2001;75(1):1–9.
299. Drevets WC, Price JL, Bardgett ME, Reich T, Todd RD, Raichle ME. Glucose metabolism in the amygdala in depression: relationship to diagnostic subtype and plasma cortisol levels. Pharmacol Biochem Behav. 2002;71(3):431–47.
300. McNally GP. Extinction of drug seeking: neural circuits and approaches to augmentation. Neuropharmacology. 2014;76(Pt B):528–32.
301. Gil-Rivas V, Fiorentine R, Anglin MD. Sexual abuse, physical abuse, and posttraumatic stress disorder among women participating in outpatient drug abuse treatment. J Psychoactive Drugs. 1996;28(1):95–102.

Part III
Cannabinoid Interactions in Modulating Emotions and Reward

Chapter 13
Cannabinoid-Nicotine Interactions

Alessia Auber, Zuzana Justinova, Maria Scherma, Steven R. Goldberg and Leigh V. Panlilio

Abstract Although nicotine and delta-9-tetrahydrocannabinol (THC), the main psychoactive ingredients in tobacco and cannabis, bind to different receptors in the brain, they have several features in common. Each of these drugs is typically self-administered by smoking, and this behavior is maintained by rewarding effects that are mediated by brain circuitry that at least partially overlaps between the drugs. Each tends to be used chronically, leading to dependence and addiction in many users. Perhaps most importantly, the nicotinic and cannabinergic systems appear to interact in such a way that modulating one system can enhance or counteract effects of the other system. Therefore, studying cannabinoid-nicotine interactions could potentially lead to new treatments for addiction. This chapter considers such interactions, focusing on recent preclinical work that suggests manipulating the cannabinoid system can counteract the addictive effects of nicotine that manipulating the nicotinic system can counteract the addictive effects of cannabis, and that prior or concurrent use of cannabis has the detrimental effect of increasing the addictive effects of nicotine. Interactive effects of these systems on anxiety and memory are also considered, although these have been studied less than addiction-related effects.

Keywords Nicotine · THC · Tobacco · Marijuana · Rimonabant · Fatty acid amide hydrolase (FAAH) · Peroxisome proliferator-activated receptor (PPAR) · Drug self-administration · Nicotine dependence · Addiction

L. V. Panlilio (✉)
Preclinical Pharmacology Section, Behavioral Neuroscience Research Branch,
Intramural Research Program, Department of Health and Human Services,
National Institute on Drug Abuse, IRP, National Institutes of Health, Baltimore, MD, USA
e-mail: LPANLILI@INTRA.NIDA.NIH.GOV

A. Auber · Z. Justinova · S. R. Goldberg
Preclinical Pharmacology Section, Behavioral Neuroscience Research Branch,
Department of Health and Human Services, National Institute on Drug Abuse,
IRP, Baltimore, MD, USA

M. Scherma
Department of Biomedical Science, Section of Neuroscience and Clinical Pharmacology,
University of Cagliari, Cagliari, Italy

P. Campolongo, L. Fattore (eds.), *Cannabinoid Modulation of Emotion, Memory, and Motivation,* DOI 10.1007/978-1-4939-2294-9_13

Neurobiology of Nicotine and Nicotine Addiction

Tobacco smoking is the leading cause of preventable, premature death in the developed world [1]. Smoking-related diseases including cancer, lung disease, and cardiovascular disease are a consequence of prolonged exposure to toxins in tobacco smoke [2]. Although these adverse effects are well known, many smokers who attempt to quit are unable to do so, or have some success but eventually relapse. Accordingly, tobacco addiction is considered a behavioral disorder involving chronic exposure to the psychoactive substance, nicotine [3, 4], and tobacco use disorder is defined in The Diagnostic and Statistical Manual of Mental Disorders (DSM)-5 by difficulty in quitting, continued use despite adverse effects, and withdrawal symptoms, essentially the same criteria used with other addictive substances such as opioids and psychomotor stimulants.

After a first experience of smoking, many individuals elect to repeat the experience as a result of both pharmacological and non-pharmacological factors [5]. Like other addictive drugs, nicotine has a positive reinforcing effect, increasing the likelihood of the behavior that leads to its ingestion [6–9]. Nicotine can also enhance the effects of environmental stimuli that have weak reinforcing effects of their own, presumably including the sensory aspects of smoking [10]. Once tobacco smoking is established, prolonged exposure to nicotine induces neuroadaptations in the brain that increase its reinforcing effects [11]. When nicotine levels in the central nervous system decrease abruptly following smoking cessation, it produces temporary imbalances in neurological systems before compensatory mechanisms are triggered to restore homeostasis [12]. These imbalances are associated with unpleasant withdrawal effects such as irritability, headache, nausea, constipation or diarrhea, falling heart rate and blood pressure, fatigue, drowsiness or insomnia, depression, increased hunger and energy, lack of concentration, anxiety, and cravings for cigarettes [13], which are powerful incentives to start smoking again [14–16]. Thus, the basis of tobacco addiction is a combination of positive reinforcement (rewarding effects) and negative reinforcement (avoidance of and escape from withdrawal symptoms).

Smoked nicotine enters the brain in 7–10 s, where it binds to nicotinic cholinergic receptors (nAChRs) [17]. The nAChR is a ligand-gated ion channel that normally binds the neurotransmitter acetylcholine [18]. nAChRs consists of five peptidic subunits arranged around a central pore. The mammalian brain expresses twelve different types of subunit (nine α and three β). NAChRs containing $\alpha4$ and $\beta2$ subunits (i.e., $\alpha4\beta2$ nAChRs) have been shown to play a pivotal role in the addictive effects of nicotine [19]. Ligand binding occurs via the α subunit, with an agonist producing a conformation change that opens the cationic channel and allows sodium and calcium ion influx. After a few milliseconds the channel closes and becomes desensitized. In the absence of further agonist binding, the receptor returns to a standby stage where it is closed but activatable [20]. Chronic nicotine exposure increases nicotine or acetylcholine (ACh) binding in the brain, a phenomenon known as up-regulation [21]. Other types of nAChRs (e.g., containing $\alpha3$, $\alpha5$, or $\alpha6$ subunits) are also believed to modulate the addictive effects of nicotine [22–31].

Smokers typically maintain near-complete saturation of α4β2 receptors during their waking hours, which desensitizes the receptors. When chronic nicotine use is discontinued, up-regulated nicotinic receptors return from a desensitized state to a standby state, leading to hyperexcitability of the cholinergic system that is associated with withdrawal effects. For example, symptoms of craving and withdrawal begin in smokers when the up-regulated, desensitized α4β2 receptors become responsive during a period of abstinence as short as one night [32, 33]. Since nicotine binding to these receptors during smoking alleviates craving and withdrawal [33], returning to a desensitized state by smoking may be negatively reinforced through escape from these aversive symptoms.

Like all addictive drugs, nicotine is believed to produce rewarding effects by increasing dopamine signaling in the mesocorticolimbic system. Specifically, nicotine increases signaling of dopamine neurons that are located in the ventral tegmental area (VTA) and project to the striatum, amygdala, prefrontal cortex and the shell of nucleus accumbens. Nicotine binding to α4β2 nAChRs localized on these VTA neurons has been shown to stimulate the release of dopamine in the nucleus accumbens shell [34–36].

NAChRs are localized mainly at the presynaptic level on a number of different types of neurons, including not only acetylcholine and dopamine, but also glutamate, norepinephrine and gamma aminobutyric acid (GABA) neurons in the VTA, substantia nigra, and striatum. Thus, nicotine augments both glutamate and GABA release [37–50]. Since dopamine release is facilitated by glutamate and inhibited by GABA, the activity of VTA dopaminergic neurons is determined by the functional balance between these excitatory and inhibitory inputs to the dopamine cell, in addition to the direct effect of nicotine on the dopamine cell.

Chronic exposure to nicotine induces desensitization of some types of nAChR, but not all. GABA-mediated inhibition diminishes with extended nicotine exposure, while glutamate-mediated excitation persists, leading to an increase in firing of dopaminergic neurons and an enhancement in responsiveness to nicotine [43, 51, 52]. Chronic nicotine exposure also produces a selective decrease in the concentration of serotonin in the hippocampus [53], possibly underlying negative affective symptoms of nicotine withdrawal such as depressed mood and irritability [54]. Nicotine also affects the release of endogenous opioid peptides involved in mood regulation, decreased response to stress, conservation of energy, and relaxation [55].

Cholinergic and cannabinergic systems appear to interact bi-directionally. Cannabinoid CB_1 and nicotinic ACh receptors are both expressed in brain areas relevant to addiction and emotional effects, including forebrain, amygdala, striatum and hippocampus [56–62]. Glutamatergic and GABAergic neurons express both nicotinic and CB_1 receptors in several brain areas, and activation of these receptors has opposite effects on these neurons. Nicotinic α7 receptors and CB_1 receptors are both expressed pre-synaptically on glutamatergic neurons [52, 59, 63, 64]. Activation of α7 receptors enhances excitatory glutamate release [51], whereas activation of CB_1 receptors has been suggested to decrease excitatory glutamate release [65]. GABAergic neurons can also express both CB_1 receptors [64, 66] and nAChRs, mostly α4β2 [51], α6β2 [67] or α7 receptors [68]. The activation of nicotinic receptors potentiates [51], whereas the activation of CB_1 receptors inhibits GABAergic neurotransmission [69]. Recent studies have also indicated that CB_2 receptors are located in the central nervous system [70, 71], but their function is not yet known. Interestingly, Navarrete et al. [72], showed that in mice CB_2 and α3- or α4-nACh receptors are localized on the same neurons in the nucleus accumbens shell and VTA.

Although, the mechanisms have only been partially elucidated, findings such as these clearly indicate that the cannabinoid and nicotinic systems interact. In the sections below we evaluate evidence that behavior is affected by two kinds of cannabinoid-nicotine interactions, respectively: the endocannabinoid system modulating the effects of exogenous nicotine, and the cholinergic system modulating the effects of cannabinoid ligands.

Cannabinoid Modulation of Nicotine's Addictive Effects

Accumulating evidence from pre-clinical and clinical research indicates that the cannabinoid system plays a critical role in nicotine reward and relapse. This section of the chapter first describes how the addictive effects of nicotine can be influenced by direct modulation of cannabinoid receptors or by enhancement of the actions of endogenously-released cannabinoids and other members of the extended endocannabinoid family (i.e., ligands for peroxisome proliferator-activated receptors, PPAR). We then consider recent evidence indicating that prior exposure to exogenous cannabinoids increases the likelihood of becoming dependent on tobacco. Finally, we consider recent research on interactions between cannabinoid and nicotinic systems in modulating learning, memory, and anxiety.

Role of CB_1 Receptors in Nicotine Addiction

One of the first studies investigating interactions between nicotine and cannabinoids in modulating reward came from Valjent et al. [73], who used a conditioned place preference (CPP) procedure (Fig. 13.1). In this procedure, the effects of a drug are repeatedly paired with a distinctive environmental context and saline injections are paired with a different context; during a subsequent test in which no injections are given, the subjects (usually rodents) are allowed to move between the two contexts, and the relative amount of time spent in each context is used as an indirect measure of the reinforcing effects of the drug. Valjent et al. showed that sub-threshold doses of nicotine and Δ^9-tetrahydrocannabinol (THC), which had no effect on place preference when given separately, did induce a place preference in mice when given together. In agreement, Scherma et al. [74] found that THC potentiated the effects of a sub-threshold dose of nicotine; that is, this dose of nicotine failed to induce a significant CPP when tested alone, but induced a significant CPP after pretreatment with THC during the acquisition phase. This finding is consistent with the fact that humans often combine nicotine and THC (in the form of cannabis) to obtain an enhanced rewarding effect [75, 76].

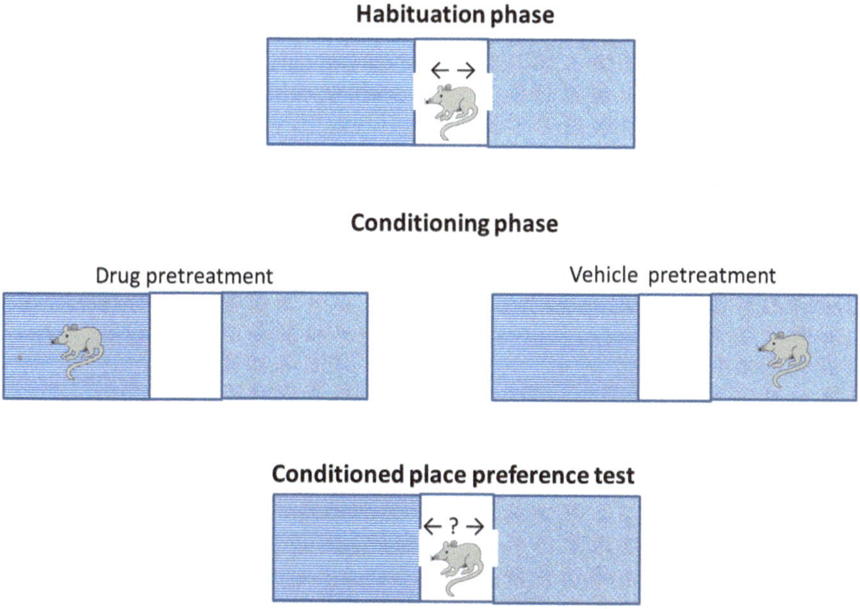

Fig. 13.1 Conditioned Place Preference. The place preference apparatus consists of 3 different compartments. Animals are initially placed in the middle compartment and allowed to explore both the right and left compartments (habituation phase). In the subsequent conditioning phase, animals are treated with a drug (e.g. nicotine, the unconditioned stimulus) or vehicle. They are always placed in either the *left* or the *right* compartment without access to the other compartments, e.g. in the *right* compartment after nicotine administration or in the *left* compartment after vehicle administration. This conditioning phase generally consists of 8–10 consecutive sessions during which drug sessions or vehicle sessions are randomly conducted. Thus, during the conditioning phase one compartment becomes associated with the drug's effects and the other associated with vehicle. After the conditioning phase, a place preference test is performed, during which animals are placed in the middle compartment and allowed access to both the drug-associated and the vehicle-associated compartments. The relative amount of time spent in the drug-associated compartment is considered a measure of the drug's reinforcing effects

Further evidence for interaction between the endocannabinoid system and pharmacological effects of nicotine has been provided by Gonzales et al. [77], who showed that chronic nicotine exposure increases levels of the endogenous CB_1 receptor agonists 2-arachidonoyl glycerol and anandamide in the limbic forebrain and brainstem, but not in the hippocampus, striatum or cerebral cortex. Given that the limbic forebrain is a key area mediating reward processes, these findings suggest that the endocannabinoid system is implicated in the modulation of nicotine reward. In the same study, Gonzales and colleagues found that prolonged exposure to nicotine did not change expression levels or binding capacity of CB_1 receptors. Like nicotine, chronic THC exposure can also enhance anandamide content in the limbic forebrain [78]. Moreover, these studies showed that chronic exposure to nicotine or THC leads to a decrease in 2-arachidonoyl glycerol content in the striatum, which may play a role in the control of motor behavior.

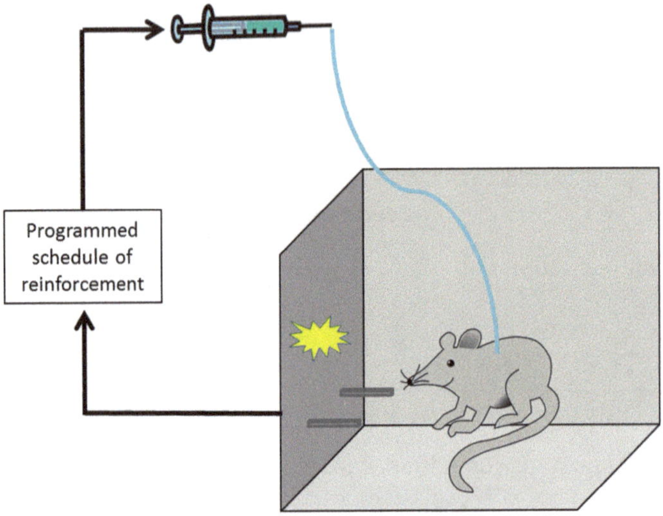

Fig. 13.2 Self administration. Rats are surgically prepared with chronic venous catheters (typically in a jugular vein). After a period of recovery, animals are placed in an operant cage (Skinner box), equipped with two levers (one active and one inactive) or two holes (one active and one inactive) centered on the front panel. Lever presses or nose-pokes into the holes are registered by an automatic control system (e.g. Med PC). Initially, animals press the lever or make a nose poke into the holes by chance, as a result of exploratory activity. Pressing the active lever, or nose-poking into the active hole, can activate an infusion pump with a syringe containing the drug under study (e.g. nicotine unconditioned stimulus). The syringe is connected to the animal's venous catheter by tubing. Active lever presses or nose pokes, according to a programmed schedule of reinforcement (e.g. Fixed Ratio, Fixed Interval etc.…), result in an infusion of a determined volume of drug. Each drug infusion is associated with illumination of a cue light and/or a tone (conditioned stimuli). If the drug has reinforcing effects, the animals will become motivated to repeatedly press the active lever press or nose-poke the active hole to self-administer the drug (conditioned response). The inactive lever or the inactive hole serves as a control for the motivated behavior

CB_1 Receptors and Nicotine Reward

The role of the CB_1 receptor in mediating nicotine reward has been studied with nicotine-induced CPP and with intravenous nicotine self-administration procedures (see Table 13.1, and Fig. 13.2). Whereas CPP involves conditioned reinforcement, in which initially-neutral features of a context come to have reinforcing effects through association with drug effects, drug self-administration procedures [79] provide a more direct measure of the reinforcing effects of a drug by allowing the subject to press a lever that produces both an intravenous drug injection and an associated stimulus (e.g. a light or tone). The basic self-administration procedure can be modified to focus on drug seeking or drug taking, and treatment compounds can be tested to determine whether they alter the direct rewarding effects of the drug, the conditioned reinforcing effects of drug-paired stimuli, or the motivation to receive the drug.

Table 13.1 Role of CB1 receptors in nicotine addiction: summary of the behavioral data

<table>
<tr><th></th><th>CB1 Receptor agonist</th><th colspan="2">CB1 Receptor antagonist</th><th colspan="2">CB1 Genetic deletion</th></tr>
<tr><td>Conditioned place preference</td><td>↑
Mice (73)</td><td colspan="2">↓
Rats (81, 82)</td><td colspan="2">↓
Mice (80)</td></tr>
<tr><td>Nicotine self-administration</td><td>↑
Rats (90)</td><td colspan="2">↓
Rats (87,88,91)</td><td colspan="2">= Mice (86)</td></tr>
<tr><td>Reinstatement</td><td>↑
Rats (90)</td><td colspan="2">↓
Rats (101, 102)</td><td colspan="2">N.A.</td></tr>
<tr><td>Nicotine discrimination</td><td>CB1/CB2 agonist substitutes for nicotine
Rats (105)</td><td>=
Rats (87, 81, 104)</td><td>↓
Rats (105)</td><td colspan="2">N.A.</td></tr>
<tr><td>Relapse in smokers</td><td>N.A.</td><td colspan="2">↓
(106)</td><td colspan="2">N.A.</td></tr>
<tr><td>Withdrawal</td><td>↓
Mice (143)</td><td colspan="2">=
Mice (125)</td><td>=
Mice (80)</td><td>↓
Mice (141)</td></tr>
</table>

CPP Studies

Two ways to evaluate the role of CB_1 receptors in behavioral effects of nicotine are to use CB_1 receptor knockout mice, in which the receptor is genetically deleted, or to pharmacologically block CB_1 receptors by administering a selective antagonist. Castañé et al. [80] found that a nicotine dose (0.5 mg/kg) that was able to produce a reinforcing effect in wild-type mice, as measured by CPP, was unable to produce the same effect in CB_1 receptor knockout mice. Interestingly, no differences between the two genotypes were observed in the severity of nicotine withdrawal symptoms precipitated by the nAChR antagonist mecamylamine after chronic nicotine exposure in mice [80].

Evidence that blockade of the CB_1 receptor reduces the reinforcing effect of nicotine in CCP procedures comes from a study by Le Foll and Goldberg [81] in which acute administration of a CB_1 receptor antagonist/inverse agonist, rimonabant, blocked the expression of nicotine-induced CPP. Similar results have been obtained by Forget et al. [82], who showed that rimonabant, administered during the training of nicotine-induced CPP (i.e., before each nicotine injection) prevented the acquisition of nicotine-induced CPP. However, when rats were tested 3–12 weeks after conditioning, a single injection of rimonabant before the CPP test no longer antagonized the expression of nicotine-CPP [82, 83]. Rimonabant has also been shown to disrupt the reconsolidation of nicotine memory after re-exposure to the drug-paired context and significantly blocked the reinstatement of nicotine-induced CPP [84, 85]. Taken together, these data indicate that the endocannabinoid system plays a pivotal role in the initial conditioning value of nicotine-associated cues. On the other hand, once it has been established, the long-term expression of nicotine-CPP seems to be independent of CB_1 receptors.

Self-Administration and Relapse Studies

Although the absence of CB_1 receptors did not disrupt the acquisition of nicotine self-administration in knock-out mice [86], several studies have demonstrated that antagonism of CB_1 receptors in normal rats can decrease nicotine self-administration. Consistent with the finding that pharmacological blockade of CB_1 receptors abolished the expression of nicotine CPP [87], Cohen et al. [87] showed that rimonabant reduced nicotine self-administration in rats. Similarly, Shoaib [88] and Wing and Shoaib [89] found that AM251 (a selective CB_1 receptor antagonist/inverse agonist structurally related to rimonabant) dose-dependently suppressed nicotine self-administration. Moreover, AM251 dose-dependently reduced relapse-like effects when rats were passively re-exposed to nicotine after a period of forced abstinence in a reinstatement model of relapse [88].

Consistent with the finding that sub-threshold doses of nicotine and THC can support CPP when given in combination [73], there is evidence that a cannabinoid receptor agonist can enhance the rewarding effects of nicotine in self-administration procedures. Gamaleddin et al. [90] showed that WIN 55,212-2 (a full agonist at both CB_1 and CB_2 receptors) increased nicotine self-administration in a progressive ratio schedule, indicating an increase in the reinforcing efficacy of nicotine. [In this type of schedule the number of operant responses (e.g., lever presses or nose-poke entries) required to obtain nicotine increases with each delivery of the drug until the rat fails to emit the required number of responses; the highest response requirement that is met, termed the breaking point, is taken as a measure of the reinforcing efficacy of the drug.] Moreover, Gamaleddin et al. [90] showed that treatment with WIN 55,212-2 by itself actually induced reinstatement of nicotine seeking (Fig. 13.3); this effect was blocked by a CB_1 receptor antagonist but not a CB_2 receptor antagonist. Treatment with WIN 55,212-2 also enhanced the relapse-like behavior induced by re-exposure to nicotine-associated cues (i.e., cue-induced reinstatement). Taken together, these data suggest that stimulating CB_1 receptors by smoking cannabis could facilitate nicotine addiction by not only enhancing the rewarding effects of nicotine, but also by enhancing the relapse-inducing effects of nicotine-related environmental cues, and even by directly triggering relapse to nicotine-seeking behavior. But, more encouragingly, the findings with rimonabant and AM251 suggest that medications that block CB_1 receptors could have the opposite effects, decreasing the addictive effects of nicotine and preventing relapse; as described below, this possibility has been borne out in clinical trials with rimonabant.

Mechanisms

The neuronal mechanisms through which CB_1 receptors regulate nicotine reward have been partially elucidated. Cohen et al. [87] showed that a dose of rimonabant

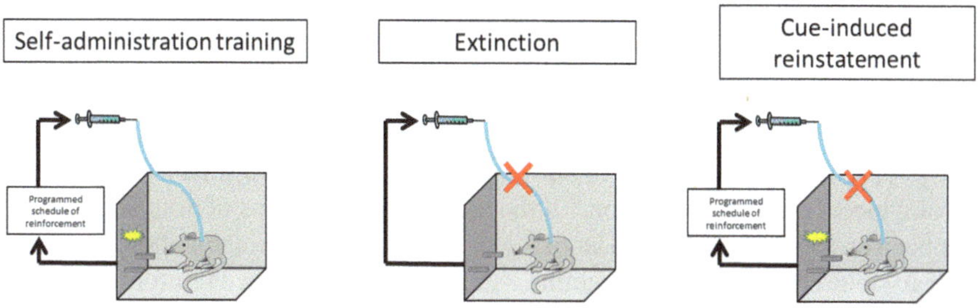

Fig. 13.3 Relapse to drug seeking behavior. *Cue-induced reinstatement*: Animals are trained to intravenously self-administer a training drug (e.g., nicotine) as described in Fig. 13.2. When they reach stable levels of self-administration responding, they enter the Extinction phase. In this phase, lever presses or nose pokes have no programmed consequences. The lack of reinforcement leads to a cessation of lever press or nose poke responding. After responding reaches stable low levels, a test session is conducted in which animals are exposed to the conditioned stimuli previously associated with administration of the training drug (e.g., light and/or tone) which may trigger reinstatement of drug-seeking behavior (lever presses or nose pokes = conditioned response). *Priming-induced reinstatement:* Self-administration and extinction phases are performed as described for the cue-induced reinstatement. Before the test session, rats receive a non-contingent priming injection of a drug and this priming may induce reinstatement of drug-seeking behavior. *Context-induced reinstatement*: Self-administration is performed as previously described. Extinction is performed in a context other than the self-administration context (e.g. a modified operant chamber with a different texture on the wall, different floor grid, different odor, etc.). After Extinction, animals are placed back in the self-administration context. The re-exposure to the drug-associated context may trigger reinstatement of drug-seeking behavior

that can reduce nicotine reward blocks nicotine-induced dopamine release in the shell of nucleus accumbens, as well as in the bed nucleus of the stria terminalis, an area that is interconnected with the nucleus accumbens and VTA and seems to play a role in the acquisition and maintenance of drug addiction [34]. Simonnet et al. [91] showed that a local infusion of the CB_1 receptor antagonist/inverse agonist AM251 into the VTA, but not into the nucleus accumbens, dose-dependently reduced nicotine self-administration in rats. Together, these data suggest that VTA CB_1 receptors mediate the reinforcing effects of nicotine. Moreover, given evidence that nicotine reward depends on dopaminergic terminal modulation [92] and that dopaminergic terminals in the nucleus accumbens lack CB_1 receptors [93, 94], it seems unlikely that the ability of cannabinoid antagonists to prevent nicotine from inducing dopamine release in the nucleus accumbens is mediated by CB_1 receptors localized within the accumbens. On the other hand, electrophysiological recording of the activity of VTA dopaminergic neurons in anaesthetized rats showed that rimonabant does not modify the excitatory response of dopaminergic neurons to nicotine [95]. Therefore the role of VTA CB_1 receptors in mediating the reinforcing effects of nicotine is still unclear.

CB_1 Receptors and Nicotine Conditioned Stimuli

Conditioned stimuli (CS) are important for the maintenance of nicotine dependence and can trigger relapse even after prolonged nicotine abstinence. Although nicotine's reinforcing effects seem to be weak compared to drugs such as cocaine and heroin, tobacco dependence is one of the most persistent forms of drug addiction. It has been proposed that tobacco smoking is maintained by a combination of: (1) the primary reinforcing effects of nicotine itself; (2) the conditioned reinforcing effects of nicotine-associated stimuli; and (3) nicotine's ability to enhance the primary reinforcing effects of sensory stimuli that accompany the act of smoking [10, 96–98].

Cohen et al. [99] have shown that rimonabant, besides altering nicotine-taking behavior [87], reduces the conditioned-reinforcing effects of stimuli that have been associated with nicotine. Rats were trained to self-administer nicotine by pressing a lever, with each nicotine injection accompanied by an audiovisual stimulus (discrete CS). Then, nicotine infusions were discontinued, and lever presses were reinforced only by CS presentations. Under this condition, responding persisted for 3 months. But, rimonabant, given systemically one month following nicotine withdrawal, decreased this responding. In a subsequent study, the same authors investigated the brain circuitry involved in the effect of rimonabant. They found that rimonabant, injected bilaterally into either the nucleus accumbens shell, basolateral amygdala, or pre-limbic cortex decreased the CS-induced nicotine seeking behavior [100]. Consistent with these findings, rimonabant dose-dependently blocked both discrete CS-induced and context-induced reinstatement in rats previously trained to self-administer nicotine [101, 102]. Taken together these data suggest that the CB_1 receptor plays a role in responding for both discrete and contextual nicotine conditioned stimuli.

Wing and Shoaib [103] evaluated the reinforcement-enhancing effect of nicotine in rats and the involvement of the endocannabinoid system in mediating this effect. They used a second-order schedule of food reinforcement, in which responding is ultimately maintained by food delivery, but the responding occurs at a higher rate than it otherwise would because it also produces conditioned reinforcers. Specifically, responding produces brief stimulus presentation under one schedule, and these stimulus presentations are accompanied by food according to another schedule. They found that a nicotine injection, given just before the second-order session, dose-dependently enhanced responding maintained by the CS that had previously been paired with food under the second-order schedule. Nicotine also enhanced responding maintained by food when no stimuli were presented. Nicotine's ability to enhance responding was attenuated when cannabinoid receptors were blocked with AM251, indicating that CB_1 receptors are involved in the reinforcement-enhancing effects of nicotine. In another study using a similar second-order schedule in which responding produced intravenous nicotine instead of food, AM251 decreased behavior maintained by the cues that had previously been paired with nicotine [103]. Taken together, these findings support the hypothesis that nicotine enhances the effects of both primary and conditioned reinforcers, and that this enhancement is mediated by CB_1 receptors.

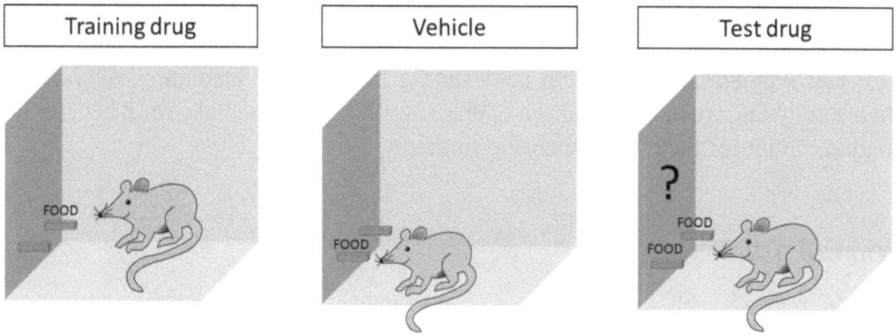

Fig. 13.4 Drug Discrimination. Before each training session, animals are treated with either a drug under study (e.g., THC) or a vehicle according to a daily schedule. Treatment with drug produces distinct interoceptive stimulus effects that the animals learn to discriminate. When the animals are treated with the training drug before the session, depression of one of two levers (e.g.: *right* lever=drug-associated lever) results in the delivery of a food pellet, whereas depression of the other lever does not produce food. When the animals are treated with vehicle before the training session, depression of the drug-associated lever does not produce food, whereas depression of the other lever (e.g. *left* lever=vehicle-associated lever) results in the delivery of food. Once the animal has learned to discriminate between delivery of the training drug and delivery of vehicle by pressing the correct lever with high accuracy, a different drug or different doses of the training drug can be administered before test sessions. During the test session, depression of either lever results in delivery of food. If the test drug produces interoceptive effects similar to the training drug, animals will predominantly press the drug-associated lever during the test. Then it can be stated that the test drug produced discriminative-stimulus effects similar to those of the training drug (e.g., THC-like discriminative-stimulus effects)

CB_1 Receptors and Nicotine's Subjective Effects

The effects of rimonabant and other cannabinoid ligands have been evaluated in animal models of the subjective effects of nicotine (Fig. 13.4). In the drug-discrimination procedure, the interoceptive effects of a drug serve as a discriminative stimulus, cueing the performance of a behavioral task. Typically, there are two levers available, and a rat learns that pressing one of the levers only produces sugar pellets in sessions when the rat has been given nicotine, and pressing the other lever only produces sugar pellets in sessions when the rat has not been given nicotine. Several studies have shown that, when given to rats trained to detect the effects of nicotine, cannabinoid ligands do not have nicotine-like discriminative effects, nor do they alter the discriminative effects of nicotine when administered prior to nicotine [81, 87, 104]. However, opposite results have been obtained by Murray et al. [105], who investigated endocannabinoid system involvement in the interoceptive effects of nicotine by using a Pavlovian discrimination procedure that differs from the drug-discrimination procedure described above. In this procedure the intermittent delivery of water occurred only during sessions when rats received nicotine; no operant response was required to produce the water, but the number of times the rat placed its head into the water dipper was measured. In this procedure, nicotine increased the number of dipper entries, but pretreatment with rimonabant blocked this conditioned response. In addition, the CB_1 and CB_2 receptor agonist CP 55,940 produced a nicotine-like discriminative effect when given alone. The authors suggest that a potential dissociation between the mechanisms mediating behavior in their Pavlovian procedures and the operant studies mentioned above [81, 87, 104] might account for the discrepancies in findings.

Human Studies

Based on findings that CB_1 receptor antagonists can alter the reinforcing effects of nicotine in rodents, rimonabant was tested as an aid to smoking cessation and relapse prevention. In addition, since the endocannabinoid system appears to play a crucial role in controlling food consumption, and since rimonabant was initially developed as a possible treatment for obesity, rimonabant was also tested to determine whether it could counteract the weight-gain that is often associated with smoking cessation. Two main clinical trials were conducted to assess smoking cessation: STRATUS-EU (Studies with Rimonabant and Tobacco Use-Europe) and STRATUS-US (Studies with Rimonabant and Tobacco Use-USA). Another clinical trial STRATUS-WW (World Wilde) focused on the efficacy of rimonabant as an aid for relapse prevention. A Cochrane review, conducted in 2007 and updated in 2011 [106], revealed that the 20 mg dose of rimonabant increased by 1.5 fold the chance of not smoking for 1 year, whereas the 5 mg dose had no effect. Moreover, weight gain was significantly lower in the 20 mg condition than in 5 mg or placebo conditions. However, despite showing efficacy as a smoking cessation-relapse prevention aid, rimonabant was found to produce gastrointestinal and psychiatric side effects, including depression and suicidal ideation. Due to these side effects, the US Food and Drug Administration (FDA) USA did not approve rimonabant for commercial use, and the European Medicine Agency (EMEA) in 2006 required the manufacturers to withdraw the drug.

It is possible that neutral CB_1 receptor antagonists that do not share the inverse-agonist effects of rimonabant and AM251 could provide a safe but effective approach to smoking cessation. In fact, the neutral cannabinoid antagonist AM4113 has been found to produce milder side effects than rimonabant [107, 108]. To our knowledge there have been no published studies evaluating the effectiveness of a neutral cannabinoid antagonists in counteracting the addiction-related effects of nicotine in humans. However, there is a study that investigated effects of cannabidiol on nicotine addiction in tobacco smokers that wished to quit smoking [109]. Cannabidiol is a non-psychoactive component of cannabis with action at several receptors that also acts as an antagonist at CB_1 and CB_2 receptors. In the Morgan et al. [109] study, cannabidiol treatment reduced the number of cigarettes smoked over a 7 day period. There are several possible explanations for these effects of cannabidiol. Apart from its antagonism at CB_1/CB_2 receptors, cannabidiol also acts as an inhibitor of fatty acid amide hydrolase (FAAH—enzyme that degrades anandamide) [77, 110], but also as an antagonist at $\alpha 7$ nAChRs [111].

Table 13.2 Role of CB2 receptors in nicotine addiction: summary of the behavioral data

	CB1 Receptor agonist	CB1 Receptor antagonist		CB1 Genetic Deletion	
Conditioned place preference	↑ Mice (73)	↓ Rats (81, 82)		↓ Mice (80)	
Nicotine Self-administration	↑ Rats (90)	Rats (87,88,91)		= Mice (86)	
Reinstatement	↑ Rats (90)	↓ Rats (101, 102)		N.A.	
Nicotine discrimination	CB1/CB2 agonist substitutes for nicotine Rats (105)	= Rats (87, 81, 104)	↓ RATS (105)	N.A.	
Relapse in smokers	N.A.	↓ (106)		N.A.	
Withdrawal	↓ Mice (143)	= Mice (125)		= Mice (80)	↓ Mice (141)

Role of CB_2 Receptors in Nicotine Addiction

It was long thought that CB_2 receptors had little effect on behavior because they are expressed mainly in immune tissues, outside the central nervous system. However, in recent years, potential involvement of the CB_2 receptor in drug addiction has been revealed [112–117]. Three studies have evaluated the role of CB_2 receptor in mediating nicotine reward.

Gamaleddin et al. [118] explored the effects of a CB_2 receptor antagonist (AM630) and a CB_2 receptor agonist (AM1241) on nicotine taking behavior in rats by using the nicotine self-administration paradigm under two schedules of reinforcement (fixed ratio and progressive ratio) (see Table 13.2). Administration of the CB_2 receptor ligands did not affect nicotine self-administration under either of these schedules. Moreover, neither AM630 nor AM1241 affected nicotine-induced reinstatement or cue-induced reinstatement of nicotine seeking.

In contrast, Navarrete et al. [72] and Ignatowska-Jankowska et al. [119] did find evidence that CB_2 receptors can mediate the reinforcing effects of nicotine. They found that nicotine did not produce conditioned place preference in CB_2 receptor-deficient mice, and that CB_2 receptor antagonists (such as AM630 or SR144528) blocked nicotine-induced CPP in wild-type mice. Moreover, CB_2 receptor knockout mice self-administered less nicotine compared to their littermates, and AM630 reduced nicotine self-administration in wild-type mice, under both fixed-ratio and progressive-ratio schedules [72].

It should be noted that gene expression of α3 and α4-nAChRs was significantly reduced in the VTA of the CB_2 receptor-deficient mice studied by Navarrete et al. [72], but not the mice studied by Ignatowska-Jankowska et al. [119]. It is generally accepted that nicotine exerts its reinforcing effects by binding to nAChRs in the VTA, leading to a release of dopamine in the nucleus accumbens, and some studies have suggested that α3 and α4-nAChRs play an important role in the reinforcing effects of nicotine [28, 120, 121]. Therefore it cannot be excluded that the lack of nicotine CPP and the low rate of nicotine self-administration in the CB_2 receptor knock-out mice in the study by Navarrete et al. [72] were due to a lack of the α3 and α4 receptor subtypes. It is also possible that the negative findings of Gamaleddin et al. [118] in rats differ from the those of the Navarrete et al. in mice [72] and Ignatowska-Jankowska et al. [119] because of species differences. Further studies are clearly needed to better understand the role of the CB_2 receptor in nicotine addiction.

Alteration of Nicotine's Addictive Effects by Modulation of the Endocannabinoid System

CB_1 receptor agonists have potentially beneficial effects such as antiemesis, and medicinal use of cannabis is becoming increasingly prevalent. However, CB_1 receptor agonists have side effects that limit their usefulness as medications [122]. A potentially safer and more effective approach to manipulating this system is to enhance the levels of endogenously-released cannabinoids. This approach offers the advantage of enhancing the actions of endocannabinoids only when and where they are endogenously released. Two possible ways to enhance the effect of endogenously-released cannabinoid ligands are to prolong their effect by inhibiting their degradation (Sect. 5.2) or by inhibiting the transport mechanism that removes the endocannabinoid ligands from the synapse (Sect. 5.4). Although the research described above (summarized in Table 13.1) indicates that exogenous CB_1 receptor agonists such as THC enhance the addictive effects of nicotine, the evidence evaluated below surprisingly suggests that enhancing the actions of endogenous cannabinoids and non-CB_1 members of the extended cannabinoid family can counteract nicotine's addictive effects.

Inhibiting the Degradation of Endocannabinoids by Fatty Acid Amide Hydrolase (FAAH)

The main degradation mechanism for anandamide is enzymatic breakdown by fatty acid amide hydrolase (FAAH). FAAH is abundantly expressed throughout the brain and is responsible for the degradation of anandamide, oleamide, N-acylamines oleoylethanolamide (OEA) and palmitoylethanolamide (PEA). Among these, only anandamide and oleamide act on CB_1 receptors, but all are considered members of the extended endocannabinoid family [123, 124]. There are several highly-selective FAAH inhibitors available, such as URB694, OL-135, PF3845 and AM3506, but

Table 13.3 Modulation of the endocannabinoid system and nicotine addiction: summary of the behavioral data

	Faah inhibitor		PPAR receptor agonist	Anandamide uptake-inhibitor	
Conditioned place preference	↑ Mice (125)	↓ Rats (127)	N.A.	↓ Rats (74)	
Nicotine self-administration	↓ Rats (127)		↓ Rats and Squirrel monkeys (131,132)	= Rats (137)	↓ Rats (74)
Reinstatement	↓ Rats (127,128)		↓ Rats and Squirrel monkeys (131,132)	↓ Rats (137, 138)	
Withdrawal	↑ Mice (125)	↓ Rats (142)	N.A.	N.A.	

the effects of FAAH inhibition on addiction-related effects of nicotine have mostly been evaluated with URB597.

The fact that FAAH inhibition substantially increases levels of the CB_1 receptor agonist anandamide, coupled with findings that exogenous CB_1 receptor agonists can enhance the rewarding effects of nicotine [73, 90], suggests that FAAH inhibition might increase the rewarding effects of nicotine. Indeed, Merritt et al. [125] showed that the genetic deletion of FAAH or the pharmacological blockade of FAAH by URB597 enhances the expression of nicotine CPP in mice (see Table 13.3). However, this might be species specific [126] since a series of experiments has shown that FAAH inhibition can decrease addiction-related effects of nicotine in rats.

Scherma et al. [127] found that URB597 prevented the development of nicotine induced CPP and the acquisition of nicotine self-administration in rats. Moreover, URB597 reduced reinstatement induced by nicotine and cues in both CPP-based and self-administration-based models of relapse in rats [127, 128]. Taken together these data suggest that FAAH inhibition may counteract the reinforcing effects of nicotine. However, Forget et al. [128] found that URB597 did not reduce the breaking point for nicotine self-administration in a progressive ratio schedule in rats. It is notable that the progressive-ratio experiment was conducted after nicotine self-administration had already been established, so it cannot be excluded that FAAH inhibition decreases the acquisition of nicotine self-administration in rats, but does not block nicotine's effects once self-administration has been established.

In agreement with their data showing that URB597 reversed nicotine's behavioral effects in rats, Scherma et al. [127] used microdialysis to show that that URB597 decreased nicotine's reward-related neurochemical effect of elevating extracellular dopamine levels in the nucleus accumbens shell. Consistent with this finding, URB597 blocked nicotine-induced increases in the rate of firing and the incidence of burst firing of VTA dopamine neurons in rats [95], and it also reduced the effects of nicotine on GABAergic medium spiny neurons in the shell of the nucleus accumbens [129].

Importantly, Melis et al. [95] found that the effect of URB597 on nicotine-induced excitation of VTA dopamine neurons does not appear to be entirely mediated by the activation of CB_1 receptors since both rimonabant and AM251 are not able to reverse it. Also, rimonabant by itself did not suppress nicotine-induced activation of VTA DA neurons. However, they found that alpha-type peroxisome proliferator-activated receptors (PPAR-α) contribute to URB597's reversal of nicotine's effects on dopamine cell firing, both in vivo and in vitro; that is, the PPAR-α antagonist MK886 reversed URB597's blockade of nicotine induced bursting in dopamine neurons. The effect of URB597 was mimicked by the administration the PPAR-α agonists oleoylethanolamine (OEA) and palmitoylethanolamide (PEA) but not methanandamide, a hydrolysis resistant analog of anandamide. Both OEA and PEA actions were blocked by MK886 confirming the involvement of PPAR-α receptors. Consistent with these findings, Luchicchi et al. [129] demonstrated that URB597 blocks nicotine's effect on firing of medium spiny neurons of the nucleus accumbens shell via both CB_1 and PPAR-α receptor-dependent mechanisms. These finding converge to suggest that URB597's ability to modulate the rewarding effects of nicotine are due largely, but perhaps not solely, to enhancements of the endogenously released PPAR-α ligands OEA and PEA.

Modulation of Nicotine's Effects by Peroxisome Proliferator-Activated Receptors (PPARs)

PPARs are ligand-activated nuclear receptors that function as a transcription factor and regulate gene expression. PPARs play essential roles in the regulation of cellular differentiation, development, metabolism (carbohydrate, lipid, protein), and tumorigenesis. In addition, PPARs have recently been recognized to have non-genomic extracellular effects, such as affecting protein kinase. PPARs are expressed in tissues throughout the body, including many areas in the brain [130].

Extending the electrophysiology findings that endogenous PPAR ligands such as OEA and PEA might have a role in mediating the addictive effects of nicotine [95], Mascia et al. [131] and Panlilio et al. [132] investigated the effects of administering exogenous PPAR-α agonists (WY14643, clofibrate, and methyl-oleoylethanolamide, a long-lasting form of OEA) on the addiction-related effects of nicotine in rats and squirrel monkeys (see Table 13.3). Each of these PPAR-α agonists substantially reduced nicotine self-administration in both rats and monkeys. These effects were reversed by the administration of the PPAR-α antagonist MK886 and were specific for nicotine reward, since the agonists had no effect on cocaine or food self-administration. Moreover, PPAR-agonists prevented the reinstatement induced by nicotine or nicotine-associated stimuli. In agreement with the behavioral data, electrophysiology and microdialysis experiments showed that PPAR-α agonists dose-dependently decreased nicotine-induced excitation of dopamine neurons in the VTA [95] and nicotine-induced elevation of dopamine levels in the nucleus accumbens of rats [131].

Melis at al. [95] have provided insight into the mechanism of PPAR-nicotine interactions. They found that the activation of PPARs modulates dopamine activity through hydrogen peroxide production, which in turn activates tyrosine kinase and increases phosphorylation of nicotinic receptors containing the β2 subunit [133]. The phosphorylation of the α4 β2 nicotinic receptors on dopamine cells diminishes ionic conductance [134] reducing responses to nicotine.

Taken together these data suggest that PPAR-α might provide a valuable target for the development of novel smoking cessation medications. Notably, clofibrate is a representative of the fibrate class of medications, several of which are already approved for human use and in fact have been used for decades in the treatment of lipid disorders such as hypercholesterolemia [135]. As smoking is associated with increased hypercholesterolemia [136], fibrate medications might counteract this smoking effect and simultaneously reduce the reinforcing effects of nicotine leading to smoking cessation.

Anandamide Uptake Inhibitors

As described above, FAAH inhibitors provide a means of increasing endogenous levels of agonists for CB_1 receptors (i.e., anandamide) and PPAR-α (i.e., OEA and PEA). It is also possible to assess the effects of specifically increasing endogenous levels of anandamide, using anandamide uptake inhibitors such as AM404 or VDM11, which do not affect OEA or PEA. Several studies have used these pharmacological tools to investigate the specific role of anandamide in interactions between the endocannabinoid and nicotinic systems.

In reinstatement models of relapse in rats, anandamide uptake inhibitors dose-dependently reduced nicotine seeking induced by nicotine-conditioned stimuli or by nicotine priming, [137, 138] (see Table 13.3). Moreover, AM404 prevented the development of nicotine-induced CPP and impeded nicotine-induced reinstatement of extinguished CPP [74]. Consistent with the behavioral data, microdialysis experiments revealed that AM404 reduced nicotine's ability to elevate dopamine levels in the nucleus accumbens shell [74]. These findings suggest that anandamide might counteract the addictive effect of nicotine. However, inhibition of anandamide uptake failed to reduce nicotine taking behavior in rats trained to self-administer nicotine under fixed-ratio or progressive-ratio schedules [137, 138].

Comparing the CPP and self-administration studies on methodological grounds, it is notable that with CPP the anandamide transport inhibitors were administered during the acquisition phase, whereas in the self-administration study AM404 or VDM11 were only given after nicotine self-administration had already been established. This situation parallels the situation described above, in which Scherma et al. [127] and Forget et al. [128] tested the effects of the FAAH inhibitor URB597 on acquisition of nicotine-induced CPP or maintenance of nicotine taking behavior in a self-administration paradigm, respectively. It can be argued that anandamide decreases the reinforcing effects of nicotine that drive the acquisition of CPP and the reinstatement of the extinguished nicotine seeking behavior, but it does not decrease the maintenance of established nicotine-taking behavior in rodents. There is evidence that different neuronal mechanisms underlie initial drug use and subsequent drug seeking and taking, which are more habitual and compulsive [139, 140]. Given that the CB_1 receptor agonist WIN 55,212–2 increased nicotine self-administration in rats [90], it is somewhat puzzling that inhibiting uptake of the endogenous CB_1 receptor agonist anandamide did not have a similar effect. This might be explained by exogenous ligands having global effects on CB_1 receptors throughout the brain, in contrast to endogenous anandamide—even when enhanced by a reuptake inhibitor or FAAH inhibitor—having more local effects, limited to areas where it is naturally being released. In any case, given the apparent species differences in the effects of FAAH manipulations between rats and mice, it will be important to test the effects of FAAH inhibitors and anandamide uptake inhibitors in nonhuman primates, and eventually in humans.

Cannabinoid Modulation of Nicotine Withdrawal

There have been several studies investigating the role of the cannabinoid system in nicotine withdrawal symptoms, with some showing involvement and others a lack of involvement (see Tables 13.1 13.2). A first study by Castañé et al. [80] found no differences between CB_1 receptor knockout and wild type mice in the severity of withdrawal precipitated by the nAChR antagonist mecamylamine after chronic treatment with nicotine. Similar results were found in a model of spontaneous nicotine withdrawal [125]. Bura et al. [141] investigated the role of CB_1 receptors in mediating nicotine withdrawal-associated anxiety using the light-dark box test, an animal model of anxiety. This test is based on the conflict between rodents' natural aversion to open, illuminated areas and their propensity to explore new environments. In wild-type mice, nicotine withdrawal was associated with decreased exploration of a lighted area, suggesting an increase in anxiety, but this effect was not observed in CB_1 receptor knock-out mice. It should be noted that the CB_1 receptor knock-out mice had a lower baseline level of exploration (possibly suggesting higher anxiety levels), so it cannot be excluded that the lack of an anxiogenic-like effect of nicotine withdrawal was due to a ceiling effect.

Some studies suggest that inhibiting endocannabinoid function might offer therapeutic advantages for the treatment of nicotine withdrawal, but that enhancing endocannabinoid function might exacerbate it. Merritt et al. [125] found that acute administration of rimonabant ameliorated somatic symptoms of nicotine withdrawal in wild-type mice, but that genetic deletion or pharmacological blockade of FAAH (which increases levels of anandamide) had the opposite effect, increasing the severity of somatic signs of withdrawal. Navarrete et al. [72] found that genetic deletion or pharmacological blockade of CB_2 receptors significantly decreased somatic signs of mecamylamine-induced nicotine withdrawal in mice.

Other studies suggest that activation of the cannabinoid system alleviates rather than exacerbates nicotine withdrawal. Cippitelli et al. [142] found that the FAAH inhibitor UR597 decreased anxiety-like behavior associated with protracted nicotine withdrawal, but did not modify the expression of somatic withdrawal symptoms in rats. Furthermore, Balerio et al. [143] found that an acute injection of THC attenuated somatic signs of nicotine withdrawal induced by mecamylamine or naltrexone in mice.

In summary, the findings on the role of the cannabinoid system in nicotine withdrawal are inconsistent. This might be due to complex or indirect relationships between the underlying mechanisms, and it might also be due to differences in the experimental procedures used in the various studies, including mice vs. rats, spontaneous vs. precipitated withdrawal, direct-acting cannabinoid ligands vs. FAAH manipulations, and somatic vs. behavioral measures. However, there is some consistency in the finding that FAAH inhibition has little effect on somatic symptoms of nicotine withdrawal, but may reduce withdrawal-induced anxiety.

THC and Nicotine Pre-Exposure

It has long been noted that there is a typical progression of drug use from tobacco and alcohol to cannabis, and then to drugs such as cocaine and heroin [144, 145]. However, epidemiological evidence also indicates that a substantial percentage of people use cannabis regularly before becoming regular users of tobacco [146]. This kind of progression could be due to a number of factors, but perhaps the most testable hypothesis is that lasting effects of cannabinoid exposure on the brain—such as changes in cannabinoid and opioid receptors [147–149]—enhance the rewarding effects of other drugs in animal models of drug abuse. Extending a series of experiments designed to assess the effects of prior cannabis exposure on the subsequent taking of other drugs, Panlilio et al. [150] exposed rats to THC for three days, with the final THC exposure given one week before the rats were allowed to self-administer nicotine. Surprisingly, given earlier findings that the same regimen of THC exposure did not enhance the acquisition of heroin or cocaine self-administration [151, 152], a history of THC exposure increased the likelihood that rats acquired nicotine self-administration behavior. Furthermore, consistent with an increased level of addiction, rats previously exposed to THC would pay a higher "price" for nicotine when the response requirement was increased in a behavioral economics procedure.

Nicotine exposure can also produce lasting changes in the brain and alter the effects of other drugs taken at a later time. For example, prior exposure to nicotine increased rats' locomotor sensitization to cocaine and also increased reward-related synaptic plasticity induced by cocaine [153]. With regards to cannabinoid-nicotine interactions, Le Foll et al. [154] found that rats that had previously been exposed to nicotine actually showed reduced rewarding effects of THC in a conditioned place-preference procedure, and appeared to be sensitized to aversive effects of THC; however, rats that had been exposed to nicotine also did not develop tolerance or behavioral sensitization to locomotor effects of THC. Continued application of the general approach used by these preliminary studies of cannabinoid-nicotine interaction involving previous exposure to one type of drug [150, 154] and also the approach used in the studies involving concurrent administration of nicotine and cannabinoid agonists [73, 90] should provide further insights into “gateway” effects and the prevalent co-abuse of cannabis and nicotine.

Role of Cholinergic Receptors in Cannabinoid Addiction

The rewarding effects of cannabis, like those of nicotine and other addictive drugs, are believed to involve facilitation of dopamine signaling in the nucleus accumbens [155, 156]. Although the mechanisms by which THC and other cannabinoids have this effect are still uncertain, it is likely that they involve cannabinoid receptors on dopaminergic and glutamatergic neurons that project to the accumbens from the VTA and prefrontal cortex. CB_1 receptors are present in the VTA, the nucleus accumbens and several areas projecting to these two structures, including the prefrontal cortex, central amygdala and hippocampus, where they appear to play an important role in brain reinforcement and reward processes [157]. These receptors are mainly located at the presynaptic level, and an important functional consequence of their activation is the inhibition of neurotransmitter release [158]. Activation of CB_1 receptors on axon terminals of GABAergic neurons in the VTA and on glutamatergic neurons in both the VTA and Nucleus accumbens should inhibit GABAergic and glutamatergic neurotransmission [65, 93]. The final effect on VTA dopaminergic activity will depend upon the relative level of activation of these inputs under distinct behavioral circumstances [93]. Several studies have demonstrated that cannabinoid agonists can increase the activity of VTA DA neurons [159–162]. This cannabinoid-induced increase in DA neuron activity was apparently responsible for increased extracellular levels of DA observed in the Nucleus accumbens [155, 156, 163]. Even though THC shares dopaminergic mechanisms with other drugs of abuse, it is not intravenously self-administered by rodents. Reliable self-administration behavior has now been demonstrated in laboratory animals for almost all drugs abused by humans, including psychostimulants, opiates, ethanol, nicotine [8, 164–167]. However, THC has only been showed to be persistently self-administered by squirrel monkeys [168–173]. However, a rodent model of intravenous cannabinoid self-administration can be achieved by training rats to self-administer synthetic CB_1/CB_2 receptor agonist WIN 55,212-2 [173–176].

Symmetrically to the finding that cannabinoids do not produce nicotine-like interoceptive effects [81, 87, 104], nicotine and other cholinergic drugs have not been found to produce THC-like effects in drug discrimination procedures.

Solinas et al. [177] found that AChR agonists did not produce THC-like effects in rats trained to discriminate THC. However, nicotine, and also the muscarinic AChR agonist pilocarpine, potentiated the discriminative effects of THC, producing a leftward shift of the THC dose-response curve. Interestingly, rimonabant reversed the potentiation induced by nicotine, but not by pilocarpine, suggesting that nicotinic and muscarinic AChRs influence the THC discriminative stimulus though different mechanisms. Accordingly, the FAAH inhibitor URB597, given in conjunction with nicotine, but not when given alone or in combination with pilocarpine, produced significant THC-like discriminative effects, presumably due to nicotine potentiating the effects of endogenous anandamide.

In another study, the same authors [178] investigated the specific role of the α7 nAChR in mediating addiction-related effects of THC. In rats, the selective α7nAChR antagonist methyllycaconitine, but not the non α7 nAChR antagonist dihydrobetaerythroidine, blocked the THC discriminative stimulus. The same dose of methyllycaconitine had no THC like discriminative effect by itself, but reduced self-administration of the CB_1/CB_2 receptor agonist WIN 55,212-2 and also attenuated THC-induced dopamine elevations in the nucleus accumbens shell.

Findings that manipulating α7 nAChRs can counteract the rewarding effects of CB_1 receptor agonists suggest that medications targeting these receptors could provide a means of treating cannabis abuse disorder. Unfortunately, directly antagonizing α7 nAChRs with drugs such as methyllycaconitine produces side effects that preclude their use in humans. However, a safer way to manipulate this system might be through allosteric modulation instead of direct antagonism. It was recently discovered that enhancing brain levels of kynurenic acid, an endogenous allosteric modulator of α7nAChR, is highly effective at blocking the addiction-related effects of cannabinoid agonists [173]. Enhancing levels of kynurenic acid: (1) attenuated the ability of THC and WIN 55,212-2 to increase extracellular dopamine levels in reward related areas in rats; (2) decreased WIN 55,212-2 self-administration in rats; (3) decreased THC self-administration in squirrel monkeys; (4) prevented drug-induced reinstatement of cannabinoid-seeking behavior in rats and monkeys; and (5) blocked reinstatement induced by re-exposure to THC-associated cues in monkeys, showing that it decreased not only the direct effects of THC, but also the conditioned effects of THC-associated stimuli. In contrast, enhancing kynurenic acid levels had no effect on cocaine or food self-administration, indicating that the behavioral effects were specific to cannabinoids and not due to a general suppression of operant behavior. These findings suggest that α7 nAChR play an important role in mediating the reinforcing effects of THC and that allosteric modulation of these receptors might provide an effective treatment strategy.

Cannabinoid-Nicotine Interaction and Anxiety

Nicotine has been shown to have effects on anxiety in humans and in animal models, and these effects can be modulated by cannabinoid-receptor ligands. Nicotine can have an anxiolytic effect, and it has been suggested that smokers might regulate their nicotine intake in order to control their anxiety level [179, 180]. However, the effects of nicotine on anxiety–related behavior have been inconsistent in laboratory studies, with some studies showing anxiolytic effects but others showing anxiogenic effects. The direction of effects of nicotine on anxiety in both humans and animals is probably dependent on the regimen of administration (acute, chronic, withdrawal), the route of administration (i.p., s.c., i.v., smoked) and the behavioral state of the experimental subjects (relaxed, stressed, in nicotine withdrawal) [181].

Balerio et al. [182] showed that a low dose of nicotine (0.05 mg/kg subcutaneously-SC) induces an anxiolytic-like effect, whereas at a higher dose (0.8 mg/kg SC) is anxiogenic. This anxiolytic-like effect was decreased by rimonabant in a dose-dependent manner, but was not affected by THC. Moreover an ineffective dose of nicotine, given in conjunction with THC, induced an anxiolytic-like effect. On the other hand, the anxiogenic effect of nicotine was increased by rimonabant and attenuated by THC [182]. This suggests that activation of the cannabinoid system may enhance nicotine's anxiolytic effects and reduce its anxiogenic effects. Hayase [183] found that repeated nicotine injections (0.8 mg/kg s.c., 4 days) produced anxiogenic effects in the elevated plus maze, which is designed to take advantage of the conflict between rodents' natural aversion to open spaces and their propensity to explore. This effect of nicotine was attenuated by pretreatment with either the nonspecific cannabinoid agonist CP 55,940 or the CB_1 receptor partial agonist/antagonist virodhamine. Intermittent nicotine exposure during the peripubertal-juvenile period in rats resulted in an increase in social anxiety when nicotine treatment was discontinued (nicotine-withdrawal); administration of the CB_1 receptor antagonist AM251 during the late phase of nicotine abstinence increased social anxiety [184]. In summary, it seems that the endocannabinoid system plays a role in nicotine's effects on anxiety, but more research is need to better understand this interaction. Given that nicotine can have either anxiolytic- or anxiogenic-like effects, it is crucial to test the cannabinoid ligands with procedures sensitive to both kinds of effect.

Cannabinoid Nicotine Interactions on Memory

Nicotine can have profound effects on cognition, but, as already described for anxiety, the direction of this effect varies between conditions. Methodological issues such as different species (rats, mice or humans), different doses and different nicotine-exposure state (chronic or acute nicotine administration, withdrawal etc.) may account for the discrepancies in the results. The effects of nicotine on learning and memory processes have been reviewed by Heishman and colleagues [185, 186].

There have been a few studies investigating the role of the endocannabinoid system in the memory-related effects of nicotine.

Biala and Kruk [187] found that an acute dose of nicotine improved memory in an elevated plus-maze test, which is usually used to study anxiety; by exposing mice just once, the procedure was adapted to study learning. Mice were placed into an open arm of the maze, and the time elapsed in moving to the enclosed arm was recorded. The same test was repeated 24 h later, and a decrease in the time spent to reach the enclosed arm was taken as a measure of memory. Nicotine significantly decreased the time of transfer latency, possibly by enhancing memory, and surprisingly this effect was reversed by pre-treatment with either the CB_1 receptor antagonist AM251 or the CB_1/CB_2 receptor agonist WIN 55,212-2.

On the other hand, a study by Hayase [183] found that repeated administration of nicotine (0.8 mg/kg/day for 4 days, a higher dose than the 0.5 mg/kg dose used in the study by Biala and Kruk [187]) impaired working memory in mice in a Y-maze test. In this test, which is based on spontaneous exploratory behavior, the number of successive entries into each of three arms, without any repeated entries, was taken as a measure of working-memory-dependent performance. Pretreatment with either the CB_1 receptor antagonist AM251 or the CB_1 partial receptor agonist/antagonist virodhamine, at a dose that had no effect by itself, significantly attenuated the nicotine-induced impairments in working memory. Hayase [183] noted that there were differences in which cannabinoid ligands blocked the memory-impairing effects of nicotine in this test versus the ligands that blocked the anxiogenic effects of nicotine in the elevated-plus maze test of anxiety (described above in Sect. 9, 'Cannabinoid-nicotine interaction and anxiety'). Interestingly, nicotine had effects in both the anxiety and memory procedures that were similar to those of immobilization stress.

Summary and Conclusions

Interactions between the endocannabinoid and nicotinic systems are clearly important with respect to cannabis and tobacco use disorders. These substances are often co-abused, and this probably results from cannabis enhancing the addictive effects of nicotine, and vice versa. The endocannabinoid system can modulate the rewarding effects of nicotine, affecting ongoing tobacco smoking, and it can also alter the effects of nicotine-associated environmental cues, affecting relapse in tobacco smokers who are trying to achieve or maintain abstinence. The evidence generally indicates that CB_1 receptor agonists facilitate nicotine use and CB_1 receptor antagonist/inverse agonists diminish it, although CB_1 receptor agonists may alleviate the symptoms of nicotine withdrawal. Unfortunately, the CB_1 receptor inverse-agonist/agonist rimonabant, which was found to have efficacy for the treatment of tobacco dependence in humans, was hampered by adverse side effects. It remains to be seen whether CB_1 receptor antagonists without inverse-agonist activity will be safe and effective for this purpose. Another potential way to decrease tobacco use through manipulation of the endocannabinoid system is with FAAH inhibitors, which enhance levels of the endogenous CB_1 receptor agonist anandamide as well as non-CB_1 members of the extended endocannabinoid family (specifically, PPARα agonists). FAAH inhibitors show promise in preclinical research in models of ongoing nicotine use and relapse. It should be noted that there appear to be species differences with respect to FAAH-nicotine interactions, with rats and monkeys but not mice showing potentially beneficial effects of FAAH inhibitions, so it will be interesting to determine the effects of not only FAAH inhibitors but PPARα agonists as smoking-cessation treatments in humans. There is some evidence that CB_2 receptor antagonists might be useful for reducing the rewarding effects of nicotine and alleviating nicotine withdrawal. Finally, interactive effects of nicotine and cannabinoid ligands on anxiety and memory have begun to be studied, and it appears that the prominent effects of cannabinoids on these processes indeed interact with the effects of nicotine.

Acknowledgments The research activity of the authors is supported by the Intramural Research Program, National Institute on Drug Abuse, National Institutes of Health, Department of Health and Human Services.

References

1. World Health Organization. Global health risks: mortality and burden of disease attributable to selected major risks. Geneva:World Health Organization; 2004.
2. U.S. Department of Health and Human Services. How Tobacco Smoke Causes Disease: The Biology and Behavioral Basis for Smoking-Attributable Disease: A Report of the Surgeon General. Atlanta (GA): U.S. Department of Health and Human Services, Centers for Disease Control and Prevention, National Center for Chronic Disease Prevention and Health Promotion, Office on Smoking and Health; 2010.
3. Abrams DB. Nicotine addiction: paradigms for research in the 21st century. Nicotine Tob Res. 1999;1(Suppl 2):S211–S5.
4. Stolerman IP, Jarvis MJ. The scientific case that nicotine is addictive. Psychopharmacology (Berl). 1995;117(1):2–10.
5. Rose JE. Nicotine and nonnicotine factors in cigarette addiction. Psychopharmacology (Berl). 2006;184(3–4):274–85.
6. DeNoble VJ, Mele PC. Intravenous nicotine self-administration in rats: effects of mecamylamine, hexamethonium and naloxone. Psychopharmacology (Berl). 2006;184(3–4):266–72.
7. Goldberg SR, Henningfield JE. Reinforcing effects of nicotine in humans and experimental animals responding under intermittent schedules of i.v. drug injection. Pharmacol Biochem Behav. 1988;30(1):227–34.
8. Goldberg SR, Spealman RD, Goldberg DM. Persistent behavior at high rates maintained by intravenous self-administration of nicotine. Science. 1981;214(4520):573–5.
9. Le Foll B, Goldberg SR. Effects of nicotine in experimental animals and humans: an update on addictive properties. Handb Exp Pharmacol. 2009;(192):335–67. doi:10.1007/978-3-540-69248-5_12.
10. Chaudhri N, Caggiula AR, Donny EC, Palmatier MI, Liu X, Sved AF. Complex interactions between nicotine and nonpharmacological stimuli reveal multiple roles for nicotine in reinforcement. Psychopharmacology (Berl). 2006;184(3–4):353–66.

11. Soria R, Stapleton JM, Gilson SF, Sampson-Cone A, Henningfield JE, London ED. Subjective and cardiovascular effects of intravenous nicotine in smokers and non-smokers. Psychopharmacology (Berl). 1996;128(3):221–6.
12. De Biasi M, Salas R. Influence of neuronal nicotinic receptors over nicotine addiction and withdrawal. Exp Biol Med (Maywood). 2008;233(8):917–29.
13. Benowitz NL. Drug therapy. Pharmacologic aspects of cigarette smoking and nicotine addiction. N Engl J Med. 1988;319(20):1318–30.
14. Hughes JR, Hatsukami DK, Pickens RW, Krahn D, Malin S, Luknic A. Effect of nicotine on the tobacco withdrawal syndrome. Psychopharmacology (Berl). 1984;83(1):82–7.
15. Hughes JR. Tobacco withdrawal in self-quitters. J Consult Clin Psychol. 1992;60(5):689–97.
16. Hughes JR, Gust SW, Skoog K, Keenan RM, Fenwick JW. Symptoms of tobacco withdrawal. A replication and extension. Arch Gen Psychiatry. 1991;48(1):52–9.
17. Hukkanen J, Jacob P III, Benowitz NL. Metabolism and disposition kinetics of nicotine. Pharmacol Rev. 2005;57(1):79–115.
18. Albuquerque EX, Pereira EF, Alkondon M, Rogers SW. Mammalian nicotinic acetylcholine receptors: from structure to function. Physiol Rev. 2009;89(1):73–120.
19. Picciotto MR, Zoli M, Rimondini R, Lena C, Marubio LM, Pich EM, et al. Acetylcholine receptors containing the beta2 subunit are involved in the reinforcing properties of nicotine. Nature. 1998;391(6663):173–7.
20. Dani JA, Bertrand D. Nicotinic acetylcholine receptors and nicotinic cholinergic mechanisms of the central nervous system. Annu Rev Pharmacol Toxicol. 2007;47:699–729.
21. Buisson B, Bertrand D. Chronic exposure to nicotine upregulates the human (alpha)4(beta)2 nicotinic acetylcholine receptor function. J Neurosci. 2001;21(6):1819–29.
22. Bousman CA, Rivard C, Haese JD, Ambrosone C, Hyland A. Alpha-5 and -3 nicotinic receptor gene variants predict nicotine dependence but not cessation: findings from the COMMIT cohort. Am J Med Genet B Neuropsychiatr Genet. 2012;159B(2):227–35.
23. Exley R, Maubourguet N, David V, Eddine R, Evrard A, Pons S, et al. Distinct contributions of nicotinic acetylcholine receptor subunit alpha4 and subunit alpha6 to the reinforcing effects of nicotine. Proc Natl Acad Sci U S A. 2011;108(18):7577–82.
24. Fowler CD, Tuesta L, Kenny PJ. Role of alpha5* nicotinic acetylcholine receptors in the effects of acute and chronic nicotine treatment on brain reward function in mice. Psychopharmacology (Berl). 2013;229(3):503–13.
25. Gold AB, Lerman C. Pharmacogenetics of smoking cessation: role of nicotine target and metabolism genes. Hum Genet. 2012;131(6):857–76.
26. Gotti C, Guiducci S, Tedesco V, Corbioli S, Zanetti L, Moretti M, et al. Nicotinic acetylcholine receptors in the mesolimbic pathway: primary role of ventral tegmental area alpha6beta2* receptors in mediating systemic nicotine effects on dopamine release, locomotion, and reinforcement. J Neurosci. 2010;30(15):5311–25.
27. Jackson KJ, McIntosh JM, Brunzell DH, Sanjakdar SS, Damaj MI. The role of alpha6-containing nicotinic acetylcholine receptors in nicotine reward and withdrawal. J Pharmacol Exp Ther. 2009;331(2):547–54.
28. Liu L, Zhao-Shea R, McIntosh JM, Gardner PD, Tapper AR. Nicotine persistently activates ventral tegmental area dopaminergic neurons via nicotinic acetylcholine receptors containing alpha4 and alpha6 subunits. Mol Pharmacol. 2012;81(4):541–8.
29. Morel C, Fattore L, Pons S, Hay YA, Marti F, Lambolez B, et al. Nicotine consumption is regulated by a human polymorphism in dopamine neurons. Mol Psychiatry. 2013. doi:10.1038/mp.2013.158.
30. Picciotto MR, Kenny PJ. Molecular mechanisms underlying behaviors related to nicotine addiction. Cold Spring Harb Perspect Med. 2013;3(1):a012112.
31. Yang K, Buhlman L, Khan GM, Nichols RA, Jin G, McIntosh JM, et al. Functional nicotinic acetylcholine receptors containing alpha6 subunits are on GABAergic neuronal boutons adherent to ventral tegmental area dopamine neurons. J Neurosci. 2011;31(7):2537–48.

32. Balfour DJ. Neural mechanisms underlying nicotine dependence. Addiction. 1994;89(11):1419–23.
33. Dani JA, Heinemann S. Molecular and cellular aspects of nicotine abuse. Neuron. 1996;16(5):905–8.
34. Di Chiara G, Imperato A. Drugs abused by humans preferentially increase synaptic dopamine concentrations in the mesolimbic system of freely moving rats. Proc Natl Acad Sci U S A. 1988;85(14):5274–8.
35. Mereu G, Yoon KW, Boi V, Gessa GL, Naes L, Westfall TC. Preferential stimulation of ventral tegmental area dopaminergic neurons by nicotine. Eur J Pharmacol. 1987;141(3):395–9.
36. Pidoplichko VI, DeBiasi M, Williams JT, Dani JA. Nicotine activates and desensitizes midbrain dopamine neurons. Nature. 1997;390(6658):401–4.
37. Albuquerque EX, Pereira EF, Alkondon M, Schrattenholz A, Maelicke A. Nicotinic acetylcholine receptors on hippocampal neurons: distribution on the neuronal surface and modulation of receptor activity. J Recept Signal Transduct Res. 1997;17(1–3):243–66.
38. Alkondon M, Pereira EF, Barbosa CT, Albuquerque EX. Neuronal nicotinic acetylcholine receptor activation modulates gamma-aminobutyric acid release from CA1 neurons of rat hippocampal slices. J Pharmacol Exp Ther. 1997;283(3):1396–411.
39. Gray R, Rajan AS, Radcliffe KA, Yakehiro M, Dani JA. Hippocampal synaptic transmission enhanced by low concentrations of nicotine. Nature. 1996;383(6602):713–6.
40. Guo JZ, Tredway TL, Chiappinelli VA. Glutamate and GABA release are enhanced by different subtypes of presynaptic nicotinic receptors in the lateral geniculate nucleus. J Neurosci. 1998;18(6):1963–9.
41. Jones IW, Wonnacott S. Precise localization of alpha7 nicotinic acetylcholine receptors on glutamatergic axon terminals in the rat ventral tegmental area. J Neurosci. 2004;24(50):11244–52.
42. Li X, Rainnie DG, McCarley RW, Greene RW. Presynaptic nicotinic receptors facilitate monoaminergic transmission. J Neurosci. 1998;18(5):1904–12.
43. Mansvelder HD, McGehee DS. Long-term potentiation of excitatory inputs to brain reward areas by nicotine. Neuron. 2000;27(2):349–57.
44. McGehee DS, Heath MJ, Gelber S, Devay P, Role LW. Nicotine enhancement of fast excitatory synaptic transmission in CNS by presynaptic receptors. Science. 1995;269(5231):1692–6.
45. McGehee DS, Role LW. Physiological diversity of nicotinic acetylcholine receptors expressed by vertebrate neurons. Annu Rev Physiol. 1995;57:521–46.
46. Radcliffe KA, Dani JA. Nicotinic stimulation produces multiple forms of increased glutamatergic synaptic transmission. J Neurosci. 1998;18(18):7075–83.
47. Radcliffe KA, Fisher JL, Gray R, Dani JA. Nicotinic modulation of glutamate and GABA synaptic transmission of hippocampal neurons. Ann N Y Acad Sci. 1999;868:591–610.
48. Role LW, Berg DK. Nicotinic receptors in the development and modulation of CNS synapses. Neuron. 1996;16(6):1077–85.
49. Wonnacott S. Presynaptic nicotinic ACh receptors. Trends Neurosci. 1997;20(2):92–8.
50. Yin R, French ED. A comparison of the effects of nicotine on dopamine and non-dopamine neurons in the rat ventral tegmental area: an in vitro electrophysiological study. Brain Res Bull. 2000;51(6):507–14.
51. Mansvelder HD, Keath JR, McGehee DS. Synaptic mechanisms underlie nicotine-induced excitability of brain reward areas. Neuron. 2002;33(6):905–19.
52. Mansvelder HD, McGehee DS. Cellular and synaptic mechanisms of nicotine addiction. J Neurobiol. 2002;53(4):606–17.
53. Benwell ME, Balfour DJ. Effects of nicotine administration and its withdrawal on plasma corticosterone and brain 5-hydroxyindoles. Psychopharmacology (Berl). 1979;63(1):7–11.
54. Schwartz RD, Lehmann J, Kellar KJ. Presynaptic nicotinic cholinergic receptors labeled by [3H]acetylcholine on catecholamine and serotonin axons in brain. J Neurochem. 1984;42(5):1495–8.
55. Cesselin F. Opioid and anti-opioid peptides. Fundam Clin Pharmacol. 1995;9(5):409–33.

56. Kesner RP, Hopkins RO. Mnemonic functions of the hippocampus: a comparison between animals and humans. Biol Psychol. 2006;73(1):3–18.
57. Martin BR, Aceto MD. Nicotine binding sites and their localization in the central nervous system. Neurosci Biobehav Rev. 1981;5(4):473–8.
58. Herkenham M, Lynn AB, Johnson MR, Melvin LS, de Costa BR, Rice KC. Characterization and localization of cannabinoid receptors in rat brain: a quantitative in vitro autoradiographic study. J Neurosci. 1991;11(2):563–83.
59. Gotti C, Zoli M, Clementi F. Brain nicotinic acetylcholine receptors: native subtypes and their relevance. Trends Pharmacol Sci. 2006;27(9):482–91.
60. Moldrich G, Wenger T. Localization of the CB1 cannabinoid receptor in the rat brain. An immunohistochemical study. Peptides. 2000;21(11):1735–42.
61. Levey AI, Kitt CA, Simonds WF, Price DL, Brann MR. Identification and localization of muscarinic acetylcholine receptor proteins in brain with subtype-specific antibodies. J Neurosci. 1991;11(10):3218–26.
62. Degroot A, Kofalvi A, Wade MR, Davis RJ, Rodrigues RJ, Rebola N, et al. CB1 receptor antagonism increases hippocampal acetylcholine release: site and mechanism of action. Mol Pharmacol. 2006;70(4):1236–45.
63. Gerdeman G, Lovinger DM. CB1 cannabinoid receptor inhibits synaptic release of glutamate in rat dorsolateral striatum. J Neurophysiol. 2001;85(1):468–71.
64. Kano M, Ohno-Shosaku T, Hashimotodani Y, Uchigashima M, Watanabe M. Endocannabinoid-mediated control of synaptic transmission. Physiol Rev. 2009;89(1):309–80.
65. Melis M, Pistis M, Perra S, Muntoni AL, Pillolla G, Gessa GL. Endocannabinoids mediate presynaptic inhibition of glutamatergic transmission in rat ventral tegmental area dopamine neurons through activation of CB1 receptors. J Neurosci. 2004;24(1):53–62.
66. Marsicano G, Lutz B. Expression of the cannabinoid receptor CB1 in distinct neuronal subpopulations in the adult mouse forebrain. Eur J Neurosci. 1999;11(12):4213–25.
67. Endo T, Yanagawa Y, Obata K, Isa T. Nicotinic acetylcholine receptor subtypes involved in facilitation of GABAergic inhibition in mouse superficial superior colliculus. J Neurophysiol. 2005;94(6):3893–902.
68. Pidoplichko VI, Prager EM, Aroniadou-Anderjaska V, Braga MF. alpha7-Containing nicotinic acetylcholine receptors on interneurons of the basolateral amygdala and their role in the regulation of the network excitability. J Neurophysiol. 2013;110(10):2358–69.
69. Szabo B, Siemes S, Wallmichrath I. Inhibition of GABAergic neurotransmission in the ventral tegmental area by cannabinoids. Eur J Neurosci. 2002;15(12):2057–61.
70. Gong JP, Onaivi ES, Ishiguro H, Liu QR, Tagliaferro PA, Brusco A, et al. Cannabinoid CB2 receptors: immunohistochemical localization in rat brain. Brain Res. 2006;1071(1):10–23.
71. Onaivi ES, Ishiguro H, Gong JP, Patel S, Perchuk A, Meozzi PA, et al. Discovery of the presence and functional expression of cannabinoid CB2 receptors in brain. Ann N Y Acad Sci. 2006;1074:514–36.
72. Navarrete F, Rodriguez-Arias M, Martin-Garcia E, Navarro D, Garcia-Gutierrez MS, Aguilar MA, et al. Role of CB2 cannabinoid receptors in the rewarding, reinforcing, and physical effects of nicotine. Neuropsychopharmacology. 2013;38(12):2515–24.
73. Valjent E, Mitchell JM, Besson MJ, Caboche J, Maldonado R. Behavioural and biochemical evidence for interactions between Delta9-tetrahydrocannabinol and nicotine. Br J Pharmacol. 2002;135(2):564–78.
74. Scherma M, Justinova Z, Zanettini C, Panlilio LV, Mascia P, Fadda P, et al. The anandamide transport inhibitor AM404 reduces the rewarding effects of nicotine and nicotine-induced dopamine elevations in the nucleus accumbens shell in rats. Br J Pharmacol. 2012;165(8):2539–48.
75. Amos A, Wiltshire S, Bostock Y, Haw S, McNeill A. 'You can't go without a fag… you need it for your hash'-a qualitative exploration of smoking, cannabis and young people. Addiction. 2004;99(1):77–81.
76. Lee JP, Battle RS, Lipton R, Soller B. 'Smoking': use of cigarettes, cigars and blunts among Southeast Asian American youth and young adults. Health Educ Res. 2010;25(1):83–96.

77. Gonzalez S, Cascio MG, Fernandez-Ruiz J, Fezza F, Di Marzo V, Ramos JA. Changes in endocannabinoid contents in the brain of rats chronically exposed to nicotine, ethanol or cocaine. Brain Res. 2002;954(1):73–81.
78. Di Marzo V, Berrendero F, Bisogno T, Gonzalez S, Cavaliere P, Romero J, et al. Enhancement of anandamide formation in the limbic forebrain and reduction of endocannabinoid contents in the striatum of delta9-tetrahydrocannabinol-tolerant rats. J Neurochem. 2000;74(4):1627–35.
79. Panlilio LV, Goldberg SR. Self-administration of drugs in animals and humans as a model and an investigative tool. Addiction. 2007;102(12):1863–70.
80. Castane A, Valjent E, Ledent C, Parmentier M, Maldonado R, Valverde O. Lack of CB1 cannabinoid receptors modifies nicotine behavioural responses, but not nicotine abstinence. Neuropharmacology. 2002;43(5):857–67.
81. Le Foll B, Goldberg SR. Rimonabant, a CB1 antagonist, blocks nicotine-conditioned place preferences. Neuroreport. 2004;15(13):2139–43.
82. Forget B, Hamon M, Thiebot MH. Cannabinoid CB1 receptors are involved in motivational effects of nicotine in rats. Psychopharmacology (Berl). 2005;181(4):722–34.
83. Forget B, Barthelemy S, Saurini F, Hamon M, Thiebot MH. Differential involvement of the endocannabinoid system in short- and long-term expression of incentive learning supported by nicotine in rats. Psychopharmacology (Berl). 2006;189(1):59–69.
84. Biala G, Budzynska B, Staniak N. Effects of rimonabant on the reinstatement of nicotine-conditioned place preference by drug priming in rats. Behav Brain Res. 2009;202(2):260–5.
85. Fang Q, Li FQ, Li YQ, Xue YX, He YY, Liu JF, et al. Cannabinoid CB1 receptor antagonist rimonabant disrupts nicotine reward-associated memory in rats. Pharmacol Biochem Behav. 2011;99(4):738–42.
86. Cossu G, Ledent C, Fattore L, Imperato A, Bohme GA, Parmentier M, et al. Cannabinoid CB1 receptor knockout mice fail to self-administer morphine but not other drugs of abuse. Behav Brain Res. 2001;118(1):61–5.
87. Cohen C, Perrault G, Voltz C, Steinberg R, Soubrie P. SR141716, a central cannabinoid (CB(1)) receptor antagonist, blocks the motivational and dopamine-releasing effects of nicotine in rats. Behav Pharmacol. 2002;13(5–6):451–63.
88. Shoaib M. The cannabinoid antagonist AM251 attenuates nicotine self-administration and nicotine-seeking behaviour in rats. Neuropharmacology. 2008;54(2):438–44.
89. Wing VC, Shoaib M. Second-order schedules of nicotine reinforcement in rats: effect of AM251. Addict Biol. 2010;15(4):393–402.
90. Gamaleddin I, Wertheim C, Zhu AZ, Coen KM, Vemuri K, Makryannis A, et al. Cannabinoid receptor stimulation increases motivation for nicotine and nicotine seeking. Addict Biol. 2012;17(1):47–61.
91. Simonnet A, Cador M, Caille S. Nicotine reinforcement is reduced by cannabinoid CB1 receptor blockade in the ventral tegmental area. Addict Biol. 2013;18(6):930–6.
92. Exley R, Clements MA, Hartung H, McIntosh JM, Cragg SJ. Alpha6-containing nicotinic acetylcholine receptors dominate the nicotine control of dopamine neurotransmission in nucleus accumbens. Neuropsychopharmacology. 2008;33(9):2158–66.
93. Lupica CR, Riegel AC. Endocannabinoid release from midbrain dopamine neurons: a potential substrate for cannabinoid receptor antagonist treatment of addiction. Neuropharmacology. 2005;48(8):1105–16.
94. Riegel AC, Lupica CR. Independent presynaptic and postsynaptic mechanisms regulate endocannabinoid signaling at multiple synapses in the ventral tegmental area. J Neurosci. 2004;24(49):11070–8.
95. Melis M, Pillolla G, Luchicchi A, Muntoni AL, Yasar S, Goldberg SR, et al. Endogenous fatty acid ethanolamides suppress nicotine-induced activation of mesolimbic dopamine neurons through nuclear receptors. J Neurosci. 2008;28(51):13985–94.
96. Caggiula AR, Donny EC, White AR, Chaudhri N, Booth S, Gharib MA, et al. Environmental stimuli promote the acquisition of nicotine self-administration in rats. Psychopharmacology (Berl). 2002;163(2):230–7.

97. Caggiula AR, Donny EC, Chaudhri N, Perkins KA, Evans-Martin FF, Sved AF. Importance of nonpharmacological factors in nicotine self-administration. Physiol Behav. 2002;77(4–5):683–7.
98. Donny EC, Chaudhri N, Caggiula AR, Evans-Martin FF, Booth S, Gharib MA, et al. Operant responding for a visual reinforcer in rats is enhanced by noncontingent nicotine: implications for nicotine self-administration and reinforcement. Psychopharmacology (Berl). 2003;169(1):68–76.
99. Cohen C, Perrault G, Griebel G, Soubrie P. Nicotine-associated cues maintain nicotine-seeking behavior in rats several weeks after nicotine withdrawal: reversal by the cannabinoid (CB1) receptor antagonist, rimonabant (SR141716). Neuropsychopharmacology. 2005;30(1):145–55.
100. Kodas E, Cohen C, Louis C, Griebel G. Cortico-limbic circuitry for conditioned nicotine-seeking behavior in rats involves endocannabinoid signaling. Psychopharmacology (Berl). 2007;194(2):161–71.
101. De Vries TJ, Schoffelmeer AN. Cannabinoid CB1 receptors control conditioned drug seeking. Trends Pharmacol Sci. 2005;26(8):420–6.
102. Diergaarde L, de Vries W, Raaso H, Schoffelmeer AN, De Vries TJ. Contextual renewal of nicotine seeking in rats and its suppression by the cannabinoid-1 receptor antagonist Rimonabant (SR141716A). Neuropharmacology. 2008;55(5):712–6.
103. Wing VC, Shoaib M. A second-order schedule of food reinforcement in rats to examine the role of CB1 receptors in the reinforcement-enhancing effects of nicotine. Addict Biol. 2010;15(4):380–92.
104. Zaniewska M, McCreary AC, Przegalinski E, Filip M. Evaluation of the role of nicotinic acetylcholine receptor subtypes and cannabinoid system in the discriminative stimulus effects of nicotine in rats. Eur J Pharmacol. 2006;540(1–3):96–106.
105. Murray JE, Wells NR, Lyford GD, Bevins RA. Investigation of endocannabinoid modulation of conditioned responding evoked by a nicotine CS and the Pavlovian stimulus effects of CP 55,940 in adult male rats. Psychopharmacology (Berl). 2009;205(4):655–65.
106. Cahill K, Ussher M. Cannabinoid type 1 receptor antagonists (rimonabant) for smoking cessation. Cochrane Database Syst Rev. 2007(3):CD005353.
107. Salamone JD, McLaughlin PJ, Sink K, Makriyannis A, Parker LA. Cannabinoid CB1 receptor inverse agonists and neutral antagonists: effects on food intake, food-reinforced behavior and food aversions. Physiol Behav. 2007;91(4):383–8.
108. Sink KS, McLaughlin PJ, Wood JA, Brown C, Fan P, Vemuri VK, et al. The novel cannabinoid CB1 receptor neutral antagonist AM4113 suppresses food intake and food-reinforced behavior but does not induce signs of nausea in rats. Neuropsychopharmacology. 2008;33(4):946–55.
109. Morgan CJ, Das RK, Joye A, Curran HV, Kamboj SK. Cannabidiol reduces cigarette consumption in tobacco smokers: preliminary findings. Addict Behav. 2013;38(9):2433–6.
110. Leweke FM, Piomelli D, Pahlisch F, Muhl D, Gerth CW, Hoyer C, et al. Cannabidiol enhances anandamide signaling and alleviates psychotic symptoms of schizophrenia. Transl Psychiatry. 2012;2:e94.
111. Mahgoub M, Keun-Hang SY, Sydorenko V, Ashoor A, Kabbani N, Al KL, et al. Effects of cannabidiol on the function of alpha7-nicotinic acetylcholine receptors. Eur J Pharmacol. 2013;720(1–3):310–9.
112. Ishiguro H, Iwasaki S, Teasenfitz L, Higuchi S, Horiuchi Y, Saito T, et al. Involvement of cannabinoid CB2 receptor in alcohol preference in mice and alcoholism in humans. Pharmacogenomics J. 2007;7(6):380–5.
113. Ishiguro H, Carpio O, Horiuchi Y, Shu A, Higuchi S, Schanz N, et al. A nonsynonymous polymorphism in cannabinoid CB2 receptor gene is associated with eating disorders in humans and food intake is modified in mice by its ligands. Synapse. 2010;64(1):92–6.
114. Onaivi ES, Ishiguro H, Gong JP, Patel S, Meozzi PA, Myers L, et al. Brain neuronal CB2 cannabinoid receptors in drug abuse and depression: from mice to human subjects. PLoS One. 2008;3(2):e1640.

115. Xi ZX, Peng XQ, Li X, Song R, Zhang HY, Liu QR, et al. Brain cannabinoid CB(2) receptors modulate cocaine's actions in mice. Nat Neurosci. 2011;14(9):1160–6.
116. Adamczyk P, Miszkiel J, McCreary AC, Filip M, Papp M, Przegalinski E. The effects of cannabinoid CB1, CB2 and vanilloid TRPV1 receptor antagonists on cocaine addictive behavior in rats. Brain Res. 2012;1444:45–54.
117. Aracil-Fernandez A, Trigo JM, Garcia-Gutierrez MS, Ortega-Alvaro A, Ternianov A, Navarro D, et al. Decreased cocaine motor sensitization and self-administration in mice overexpressing cannabinoid CB(2) receptors. Neuropsychopharmacology. 2012;37(7):1749–63.
118. Gamaleddin I, Zvonok A, Makriyannis A, Goldberg SR, Le Foll B. Effects of a selective cannabinoid CB2 agonist and antagonist on intravenous nicotine self administration and reinstatement of nicotine seeking. PLoS One. 2012;7(1):e29900.
119. Ignatowska-Jankowska BM, Muldoon PP, Lichtman AH, Damaj MI. The cannabinoid CB2 receptor is necessary for nicotine-conditioned place preference, but not other behavioral effects of nicotine in mice. Psychopharmacology (Berl). 2013;229(4):591–601.
120. Pons S, Fattore L, Cossu G, Tolu S, Porcu E, McIntosh JM, et al. Crucial role of alpha4 and alpha6 nicotinic acetylcholine receptor subunits from ventral tegmental area in systemic nicotine self-administration. J Neurosci. 2008;28(47):12318–27.
121. Glick SD, Sell EM, McCallum SE, Maisonneuve IM. Brain regions mediating alpha3beta4 nicotinic antagonist effects of 18-MC on nicotine self-administration. Eur J Pharmacol. 2011;669(1–3):71–5.
122. Hall W, Degenhardt L. Adverse health effects of non-medical cannabis use. Lancet. 2009;374(9698):1383–91.
123. O'Sullivan SE. Cannabinoids go nuclear: evidence for activation of peroxisome proliferator-activated receptors. Br J Pharmacol. 2007;152(5):576–82.
124. Pistis M, Melis M. From surface to nuclear receptors: the endocannabinoid family extends its assets. Curr Med Chem. 2010;17(14):1450–67.
125. Merritt LL, Martin BR, Walters C, Lichtman AH, Damaj MI. The endogenous cannabinoid system modulates nicotine reward and dependence. J Pharmacol Exp Ther. 2008;326(2):483–92.
126. Muldoon PP, Lichtman AH, Parsons LH, Damaj MI. The role of fatty acid amide hydrolase inhibition in nicotine reward and dependence. Life Sci. 2013;92(8–9):458–62.
127. Scherma M, Panlilio LV, Fadda P, Fattore L, Gamaleddin I, Le Foll B, et al. Inhibition of anandamide hydrolysis by cyclohexyl carbamic acid 3'-carbamoyl-3-yl ester (URB597) reverses abuse-related behavioral and neurochemical effects of nicotine in rats. J Pharmacol Exp Ther. 2008;327(2):482–90.
128. Forget B, Coen KM, Le Foll B. Inhibition of fatty acid amide hydrolase reduces reinstatement of nicotine seeking but not break point for nicotine self-administration-comparison with CB(1) receptor blockade. Psychopharmacology (Berl). 2009;205(4):613–24.
129. Luchicchi A, Lecca S, Carta S, Pillolla G, Muntoni AL, Yasar S, et al. Effects of fatty acid amide hydrolase inhibition on neuronal responses to nicotine, cocaine and morphine in the nucleus accumbens shell and ventral tegmental area: involvement of PPAR-alpha nuclear receptors. Addict Biol. 2010;15(3):277–88.
130. Moreno S, Farioli-Vecchioli S, Ceru MP. Immunolocalization of peroxisome proliferator-activated receptors and retinoid X receptors in the adult rat CNS. Neuroscience. 2004;123(1):131–45.
131. Mascia P, Pistis M, Justinova Z, Panlilio LV, Luchicchi A, Lecca S, et al. Blockade of nicotine reward and reinstatement by activation of alpha-type peroxisome proliferator-activated receptors. Biol Psychiatry. 2011;69(7):633–41.
132. Panlilio LV, Justinova Z, Mascia P, Pistis M, Luchicchi A, Lecca S, et al. Novel use of a lipid-lowering fibrate medication to prevent nicotine reward and relapse: preclinical findings. Neuropsychopharmacology. 2012;37(8):1838–47.
133. Melis M, Carta S, Fattore L, Tolu S, Yasar S, Goldberg SR, et al. Peroxisome proliferator-activated receptors-alpha modulate dopamine cell activity through nicotinic receptors. Biol Psychiatry. 2010;68(3):256–64.

134. Charpantier E, Wiesner A, Huh KH, Ogier R, Hoda JC, Allaman G, et al. Alpha7 neuronal nicotinic acetylcholine receptors are negatively regulated by tyrosine phosphorylation and Src-family kinases. J Neurosci. 2005;25(43):9836–49.
135. Jackevicius CA, Tu JV, Ross JS, Ko DT, Carreon D, Krumholz HM. Use of fibrates in the United States and Canada. JAMA. 2011;305(12):1217–24.
136. Hata Y, Nakajima K. Life-style and serum lipids and lipoproteins. J Atheroscler Thromb. 2000;7(4):177–97.
137. Gamaleddin I, Guranda M, Scherma M, Fratta W, Makriyannis A, Vadivel SK, et al. AM404 attenuates reinstatement of nicotine seeking induced by nicotine-associated cues and nicotine priming but does not affect nicotine- and food-taking. J Psychopharmacol. 2013;27(6):564–71.
138. Gamaleddin I, Guranda M, Goldberg SR, Le Foll B. The selective anandamide transport inhibitor VDM11 attenuates reinstatement of nicotine seeking behaviour, but does not affect nicotine intake. Br J Pharmacol. 2011;164(6):1652–60.
139. Everitt BJ, Robbins TW. From the ventral to the dorsal striatum: devolving views of their roles in drug addiction. Neurosci Biobehav Rev. 2013;37(9 Pt A):1946–54.
140. Everitt BJ, Robbins TW. Neural systems of reinforcement for drug addiction: from actions to habits to compulsion. Nat Neurosci. 2005;8(11):1481–9.
141. Bura SA, Burokas A, Martin-Garcia E, Maldonado R. Effects of chronic nicotine on food intake and anxiety-like behaviour in CB(1) knockout mice. Eur Neuropsychopharmacol. 2010;20(6):369–78.
142. Cippitelli A, Astarita G, Duranti A, Caprioli G, Ubaldi M, Stopponi S, et al. Endocannabinoid regulation of acute and protracted nicotine withdrawal: effect of FAAH inhibition. PLoS One. 2011;6(11):e28142.
143. Balerio GN, Aso E, Berrendero F, Murtra P, Maldonado R. Delta9-tetrahydrocannabinol decreases somatic and motivational manifestations of nicotine withdrawal in mice. Eur J Neurosci. 2004;20(10):2737–48.
144. Kandel D. Stages in adolescent involvement in drug use. Science. 1975;190(4217):912–4.
145. Kandel DB, Yamaguchi K, Klein LC. Testing the Gateway Hypothesis. Addiction. 2006;101(4):470–2.
146. Vaughn M, Wallace J, Perron B, Copeland V, Howard M. Does marijuana use serve as a gateway to cigarette use for high-risk African-American youth? Am J Drug Alcohol Abuse. 2008;34(6):782–91.
147. Ellgren M, Artmann A, Tkalych O, Gupta A, Hansen HS, Hansen SH, et al. Dynamic changes of the endogenous cannabinoid and opioid mesocorticolimbic systems during adolescence: THC effects. Eur Neuropsychopharmacol. 2008;18(11):826–34.
148. Ellgren M, Spano SM, Hurd YL. Adolescent cannabis exposure alters opiate intake and opioid limbic neuronal populations in adult rats. Neuropsychopharmacology. 2007;32(3):607–15.
149. Maldonado R. Study of cannabinoid dependence in animals. Pharmacol Ther. 2002;95(2):153–64.
150. Panlilio LV, Zanettini C, Barnes C, Solinas M, Goldberg SR. Prior exposure to THC increases the addictive effects of nicotine in rats. Neuropsychopharmacology. 2013;38(7):1198–208.
151. Solinas M, Panlilio LV, Goldberg SR. Exposure to delta-9-tetrahydrocannabinol (THC) increases subsequent heroin taking but not heroin's reinforcing efficacy: a self-administration study in rats. Neuropsychopharmacology. 2004;29(7):1301–11.
152. Panlilio LV, Solinas M, Matthews SA, Goldberg SR. Previous exposure to THC alters the reinforcing efficacy and anxiety-related effects of cocaine in rats. Neuropsychopharmacology. 2007;32(3):646–57.
153. Levine A, Huang Y, Drisaldi B, Griffin EA Jr, Pollak DD, Xu S, et al. Molecular mechanism for a gateway drug: epigenetic changes initiated by nicotine prime gene expression by cocaine. Sci Transl Med. 2011;3(107):107ra109.

154. Le Foll B, Wiggins M, Goldberg SR. Nicotine pre-exposure does not potentiate the locomotor or rewarding effects of Delta-9-tetrahydrocannabinol in rats. Behav Pharmacol. 2006;17(2):195–9.
155. Chen JP, Paredes W, Li J, Smith D, Lowinson J, Gardner EL. Delta9-tetrahydrocannabinol produces naloxone-blockable enhancement of presynaptic basal dopamine efflux in nucleus accumbens of conscious, freely-moving rats as measured by intracerebral microdialysis. Psychopharmacology (Berl). 1990;102(2):156–62.
156. Tanda G, Pontieri FE, Di Chiara G. Cannabinoid and heroin activation of mesolimbic dopamine transmission by a common mu1 opioid receptor mechanism. Science. 1997;276(5321):2048–50.
157. Gardner EL. Endocannabinoid signaling system and brain reward: emphasis on dopamine. Pharmacol Biochem Behav. 2005;81(2):263–84.
158. Schlicker E, Kathmann M. Modulation of transmitter release via presynaptic cannabinoid receptors. Trends Pharmacol Sci. 2001;22(11):565–72.
159. French ED. delta9-Tetrahydrocannabinol excites rat VTA dopamine neurons through activation of cannabinoid CB1 but not opioid receptors. Neurosci Lett. 1997;226(3):159–62.
160. French ED, Dillon K, Wu X. Cannabinoids excite dopamine neurons in the ventral tegmentum and substantia nigra. Neuroreport. 1997;8(3):649–52.
161. Gessa GL, Melis M, Muntoni AL, Diana M. Cannabinoids activate mesolimbic dopamine neurons by an action on cannabinoid CB1 receptors. Eur J Pharmacol. 1998;341(1):39–44.
162. Wu X, French ED. Effects of chronic delta9-tetrahydrocannabinol on rat midbrain dopamine neurons: an electrophysiological assessment. Neuropharmacology. 2000;39(3):391–8.
163. Ng Cheong Ton JM, Gerhardt GA, Friedemann M, Etgen AM, Rose GM, Sharpless NS, et al. The effects of delta9-tetrahydrocannabinol on potassium-evoked release of dopamine in the rat caudate nucleus: an in vivo electrochemical and in vivo microdialysis study. Brain Res. 1988;451(1–2):59–68.
164. Collins RJ, Weeks JR, Cooper MM, Good PI, Russell RR. Prediction of abuse liability of drugs using IV self-administration by rats. Psychopharmacology (Berl). 1984;82(1–2):6–13.
165. Koob GF, Weiss F. Pharmacology of drug self-administration. Alcohol. 1990;7(3):193–7.
166. Yokel RA. Intravenous self-administration: response rates, the effects of pharmacological challenges, and drug preference. In: Bozarth MA, editor. Methods of assessing the reinforcing properties of abused drugs. New York: Springer-Verlag; 1987. pp. 1–33.
167. Young AM, Herling S. Drugs as reinforcers: Studies in laboratory animals. In: Goldberg SR, Stolerman IP, editors. Behavioral analysis of drug dependence. Orlando: Academic; 1986. pp. 9–67.
168. Justinova Z, Tanda G, Redhi GH, Goldberg SR. Self-administration of Delta(9)-tetrahydrocannabinol (THC) by drug naive squirrel monkeys. Psychopharmacology (Berl). 2003;169(2):135–40.
169. Justinova Z, Goldberg SR, Heishman SJ, Tanda G. Self-administration of cannabinoids by experimental animals and human marijuana smokers. Pharmacol Biochem Behav. 2005;81(2):285–99.
170. Justinova Z, Mangieri RA, Bortolato M, Chefer SI, Mukhin AG, Clapper JR, et al. Fatty acid amide hydrolase inhibition heightens anandamide signaling without producing reinforcing effects in primates. Biol Psychiatry. 2008;64(11):930–7.
171. Justinova Z, Munzar P, Panlilio LV, Yasar S, Redhi GH, Tanda G, et al. Blockade of THC-seeking behavior and relapse in monkeys by the cannabinoid CB(1)-receptor antagonist rimonabant. Neuropsychopharmacology. 2008;33(12):2870–7.
172. Justinova Z, Ferre S, Redhi GH, Mascia P, Stroik J, Quarta D, et al. Reinforcing and neurochemical effects of cannabinoid CB1 receptor agonists, but not cocaine, are altered by an adenosine A2A receptor antagonist. Addict Biol. 2011;16(3):405–15.
173. Justinova Z, Mascia P, Wu HQ, Secci ME, Redhi GH, Panlilio LV, et al. Reducing cannabinoid abuse and preventing relapse by enhancing endogenous brain levels of kynurenic acid. Nat Neurosci. 2013;16(11):1652–61.

174. Fattore L, Cossu G, Martellotta CM, Fratta W. Intravenous self-administration of the cannabinoid CB1 receptor agonist WIN 55,212-2 in rats. Psychopharmacology (Berl). 2001;156(4):410–6.
175. Martellotta MC, Cossu G, Fattore L, Gessa GL, Fratta W. Self-administration of the cannabinoid receptor agonist WIN 55,212-2 in drug-naive mice. Neuroscience. 1998;85(2):327–30.
176. Spano MS, Fattore L, Cossu G, Deiana S, Fadda P, Fratta W. CB1 receptor agonist and heroin, but not cocaine, reinstate cannabinoid-seeking behaviour in the rat. Br J Pharmacol. 2004;143(3):343–50.
177. Solinas M, Scherma M, Tanda G, Wertheim CE, Fratta W, Goldberg SR. Nicotinic facilitation of delta9-tetrahydrocannabinol discrimination involves endogenous anandamide. J Pharmacol Exp Ther. 2007;321(3):1127–34.
178. Solinas M, Scherma M, Fattore L, Stroik J, Wertheim C, Tanda G, et al. Nicotinic alpha 7 receptors as a new target for treatment of cannabis abuse. J Neurosci. 2007;27(21):5615–20.
179. Pomerleau OF. Nicotine as a psychoactive drug: anxiety and pain reduction. Psychopharmacol Bull. 1986;22(3):865–9.
180. Gilbert DG, Robinson JH, Chamberlin CL, Spielberger CD. Effects of smoking/nicotine on anxiety, heart rate, and lateralization of EEG during a stressful movie. Psychophysiology 1989;26(3):311–20.
181. Picciotto MR, Brunzell DH, Caldarone BJ. Effect of nicotine and nicotinic receptors on anxiety and depression. Neuroreport. 2002;13(9):1097–106.
182. Balerio GN, Aso E, Maldonado R. Role of the cannabinoid system in the effects induced by nicotine on anxiety-like behaviour in mice. Psychopharmacology (Berl). 2006;184(3–4):504–13.
183. Hayase T. Working memory- and anxiety-related behavioral effects of repeated nicotine as a stressor: the role of cannabinoid receptors. BMC Neurosci. 2013;14:20.
184. Aydin C, Oztan O, Isgor C. Nicotine-induced anxiety-like behavior in a rat model of the novelty-seeking phenotype is associated with long-lasting neuropeptidergic and neuroplastic adaptations in the amygdala: effects of the cannabinoid receptor 1 antagonist AM251. Neuropharmacology. 2012;63(8):1335–45.
185. Heishman SJ. Behavioral and cognitive effects of smoking: relationship to nicotine addiction. Nicotine Tob Res. 1999;1(Suppl 2):S143–S7.
186. Heishman SJ, Kleykamp BA, Singleton EG. Meta-analysis of the acute effects of nicotine and smoking on human performance. Psychopharmacology (Berl). 2010;210(4):453–69.
187. Biala G, Kruk M. Cannabinoid receptor ligands suppress memory-related effects of nicotine in the elevated plus maze test in mice. Behav Brain Res. 2008;192(2):198–202.

Chapter 14
Cannabinoid-Alcohol Interactions

Luis A. Natividad, Paola Maccioni, Loren H. Parsons and Giancarlo Colombo

Abstract Marijuana and alcohol are two of the most commonly abused substances worldwide. Although the psychotropic effects of these drugs are mediated by distinct pharmacodynamic mechanisms, the similar physiological and neuropsychological effects of these drugs have long been recognized. This chapter focuses on interactive effects between cannabinoids and alcohol with a particular emphasis on the influence of endogenous cannabinoid (eCB) signaling on the motivation for alcohol consumption and the etiology of alcohol use disorders. Topics include the modulation of alcohol consumption by exogenous cannabinoid receptor agonists and antagonists, the influence of alcohol exposure on brain eCB formation and alcohol-induced disruptions in eCB-mediated synaptic plasticity. Associations between genetic variants in eCB signaling with alcohol use disorders are discussed along with a possible role of dysregulated eCB signaling in symptoms of alcohol dependence and protracted withdrawal. The chapter concludes with a brief consideration of the eCB system as a viable treatment target for alcohol dependence.

Keywords Endocannabinoid · CB1 · CNR1 · Anandamide · 2-Arachidonoylglycerol (2-AG) · Alcohol · Alcoholism · Reward · Dependence · Addiction

The original version of this chapter was revised. An erratum to this chapter can be found at DOI 10.1007/978-1-4939-2294-9_18

L. H. Parsons (✉)
Committee on the Neurobiology of Addictive Disorders,
The Scripps Research Institute, La Jolla, CA, USA
e-mail: lparsons@scripps.edu

L. A. Natividad
Research Associate, Committee on the Neurobiology of Addictive Disorders,
The Scripps Research Institute, La Jolla, CA, USA

P. Maccioni
Institute of Neuroscience-Cagliari, National Research Council (CNR), Monserrato, Italy

G. Colombo
Institute of Neuroscience-Cagliari, National Research Council (CNR), Monserrato, Italy

P. Campolongo, L. Fattore (eds.), *Cannabinoid Modulation of Emotion, Memory, and Motivation,* DOI 10.1007/978-1-4939-2294-9_14

Introduction

Alcohol is one of the most widely used psychoactive drugs worldwide. Marijuana is the most commonly used illicit drug [1] and is also the most commonly used illicit drug by alcohol dependent individuals. Recent studies indicate that 25–55 % of alcohol dependent subjects use both alcohol and marijuana, frequently on the same day and often simultaneously (2–6). The prevalence of this co-abuse is more than 2-fold higher than the concurrent use of alcohol with opioids, cocaine or other substances.

Evidence linking the neuropsychological effects of alcohol and cannabis has existed for more than 40 years. Early studies in humans revealed evidence of cross-tolerance to alcohol among cannabis users [7, 8] and subsequent studies in rodents definitively demonstrated a symmetrical cross-tolerance between these abused drugs [9–16].

Although the subjective effects of alcohol and cannabis are distinct, these drugs induce a number of similar physiological and behavioral effects. Both substances produce depressant effects including hypolocomotion, hypothermia and ataxia [17–20], and both substances impair memory, attention and psychomotor performance [21–23] and do so in a synergistic manner [8, 24, 25]. Alcohol and cannabis produce comparable disruptions of sleep patterns [26–28] and both drugs decrease subjective time estimation [[7, 29, 30] but see [23]].

This chapter will describe evidence demonstrating that exogenous cannabinoid receptor agonists enhance the motivational properties of alcohol while cannabinoid receptor antagonists diminish alcohol reward. This relationship has sparked the hypothesis that alcohol consumption engages the brain endogenous cannabinoid (eCB) system in a manner that contributes to the motivational effects of acute and chronic alcohol consumption. Accordingly, we will review existing evidence that clinically relevant levels of alcohol exposure disrupt brain eCB formation and eCB-mediated synaptic plasticity along with evidence that manipulations of eCB processing influence the motivation for alcohol consumption. Evidence implicating genetic disruptions in cannabinoid signaling in the etiology of alcohol use disorders is discussed along with a review of clinical studies on the efficacy of therapeutics targeting the cannabinoid system for problem alcohol use.

Effect of Cannabinoid Agonists on Alcohol Consumption

Data collected to date invariably suggest that cannabinoid receptor agonists stimulate alcohol drinking and alcohol self-administration in laboratory rodents. Acute or repeated intraperitoneal (i.p.) or subcutaneous (s.c.) administration of the synthetic cannabinoid type-1 (CB_1) receptor agonists CP 55,940 (3–50 μg/kg, i.p.) and WIN 55,212–2 (0.5–2 mg/kg, i.p. or s.c.) has been reported to substantially stimulate alcohol drinking in selectively bred Sardinian alcohol-preferring (sP) rats

(31) and mice [32, 33] exposed to the standard, homecage 2-bottle "alcohol *vs* water" choice regimen. Under this experimental procedure, rodents are exposed to the choice between a relatively low concentrated alcohol solution (usually 10 % v/v) and water and freely allowed to consume alcohol. This procedure—largely used because of its relative simplicity—provides information of the mere consumption of alcohol. It possesses considerable predictive validity, especially when applied to rats selected for high alcohol intake. More recently, acute treatment with 0.5 mg/kg WIN 55,212–2 (i.p.) has also been found to increase alcohol intake in an interesting mouse model of binge-like drinking [34].

Several studies have also tested the capacity of cannabinoid agonists to modulate operant responding for orally available alcohol self-administration in rats. Under these procedures, alcohol is made available to rats *via* completion of a behavioral response (usually, responding on a lever for a given number of times), to allow the reinforcing and motivational properties of alcohol to be assessed (in addition to its mere consumption). The most common form of operant alcohol self-administration employs a fixed ratio (FR) schedule of reinforcement, in which the response requirement (RR; i.e., the "cost" of each alcohol presentation in terms of number of responses on the lever) is predetermined and kept fixed throughout the session (providing a measure of both alcohol consumption and reinforcing properties of alcohol). Under an FR schedule of reinforcement pre-treatment with synthetic CB_1 receptor selective agonists such as CP 55,940 (1–50 μg/kg, i.p.) or WIN 55,212–2 (0.5–2 mg/kg, i.p.) increased operant responding for alcohol in rat lines selectively bred for high alcohol preference and consumption including the alcohol-preferring Alko Alcohol (AA) line [35] and Indiana P line [36]. Similar CB_1 receptor agonist-induced enhancement of FR responding for alcohol has been observed in unselected Wistar rats [37]. Moreover, CB_1 receptor agonists increase responding for alcohol under progressive ratio (PR) schedules of reinforcement. Under the PR schedule the RR is progressively increased after the delivery of each reinforcer and the lowest ratio not completed (termed the "breakpoint" of responding) is taken as a measure of the animals' motivation to obtain the alcohol reinforcer. Pre-treatment with either CP 55,940 or WIN 55,212–2 has been shown to increase the breakpoint of responding for alcohol by alcohol-preferring AA and P rats as well as non-selected Wistar rats [35–37]. Additional studies have investigated the effect of CB_1 receptor agonists in a rat model of alcohol relapse named "alcohol deprivation effect" (ADE). ADE is defined as the temporary increase in voluntary alcohol intake or operant self-administration—often doubling baseline—occurring after a period of forced abstinence, or deprivation, from alcohol [38, 39]. Treatment with WIN 55,212–2 (0.4–10 mg/kg, s.c.) increased the ADE in rats tested with operant procedures of alcohol self-administration [40–42]. Collectively, these findings demonstrate that potent, synthetic CB_1 receptor agonists enhance the motivation for alcohol and increase overall alcohol consumption.

Similar data have also been collected with Δ9-Tetrahydrocannabinol (Δ9-THC; the primary psychoactive constituent of *Cannabis sativa*. Acute Δ9-THC administration (0.3–3 mg/kg, i.p.) dose-dependently stimulated alcohol intake (up to 40–45 %) in alcohol-preferring sP rats exposed to the homecage 2-bottle "alcohol

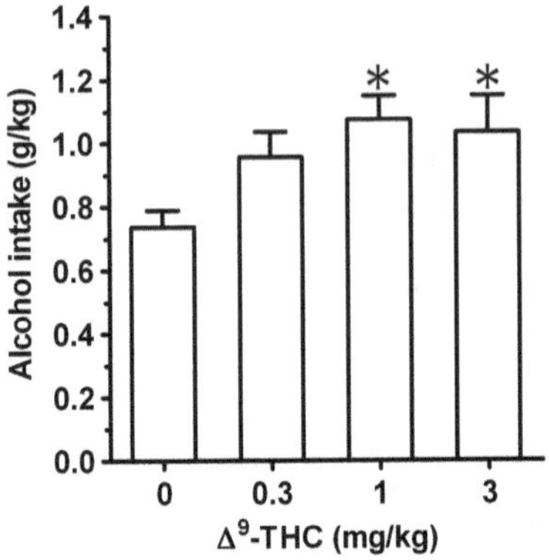

Fig. 14.1 Stimulating effect of the cannabinoid receptor agonist, Δ9-Tetrahydrocannabinol (Δ9-THC), on alcohol intake in selectively bred, Sardinian alcohol-preferring (sP) rats. Alcohol (10% v/v) was offered under the standard, homecage 2-bottle "alcohol *vs* water" choice regimen with unlimited access for 24 hours/day. The test day with Δ9-THC was conducted after approximately 4 weeks of exposure to the 2-bottle choice regimen ("alcohol-experienced" rats, experimental model of the active drinking phase of human alcoholism). On the test day, alcohol and water intake was recorded 60 min after lights off. Δ9-THC (0.3–3 mg/kg, i.p.) was administered acutely 20 min before lights off. Δ9-THC was suspended in 3 ml/kg in a 1:1:18 mixture made up of ethanol, Tween 80, and saline. Each bar is the mean ± SEM of $n = 12$ rats. ANOVA results: (a) alcohol intake: [$F(3;44) = 3.38$, $P < 0.05$]; (b) water intake: [$F(3;44) = 0.69$, $P > 0.05$].*: $P < 0.05$ with respect to vehicle-treated rats (Newman-Keuls test)

vs water" choice regimen (Fig. 14.1); the effect of Δ9-THC tended to be selective for alcohol intake, as water intake was not affected by treatment with Δ9-THC. A previous study [13] reported that acute intragastric (i.g.) administration of relatively high doses of Δ9-THC (3–30 mg/kg) resulted in a non-specific decrease in alcohol self-administration in unselected rats. It is likely that sedative effects produced by these doses of Δ9-THC may have affected the normal rates of drinking, contributing to the observed reduction in alcohol intake. Accordingly, when Δ9-THC was repeatedly administered for 3 consecutive weeks—and tolerance to some effects of Δ9-THC, including sedation, likely developed—alcohol intake increased above control levels [13]. Notably, acute treatment with 1 mg/kg Δ9-THC reinstated unreinforced alcohol-seeking behavior in rats [43].

Increased alcohol consumption evident following CB_1 receptor agonist pretreatment has been shown to be associated with a proportional increase in blood alcohol concentrations [31]. This discounts the possibility that rodents alter their alcohol intake in compensation for diminished alcohol absorption secondary to cannabinoid-induced disruptions in gastric emptying and intestinal peristalsis, and suggests that cannabinoid agonists likely induce an upward shift in the set-point controlling alcohol drinking behavior. Enhancement of alcohol consumption by CP 55,940, WIN 55,212–2 and Δ9-THC are abolished by pre-treatment with *per se* ineffective doses of the CB_1 receptor antagonist Rimonabant (SR141716A) [31, 32,

37], suggesting that the effects of these cannabinoid agonists are indeed secondary to activation of the cannabinoid CB_1 receptor. Additionally, the enhancement of alcohol self-administration by CP 55,940-, WIN 55,212–2-, and Δ9-THC is prevented by pre-treatment with (a) the opiate receptor antagonist naloxone [31, 37], (b) the $GABA_B$ receptor agonist baclofen [44], (c) the serotonin 5-HT_3 receptor antagonist tropisetron [45], and (d) the serotonin 5-HT_{1A} receptor agonist, 8-OH-DPAT [32]. This suggests a complex interplay of neural mechanisms contributing to the effects of cannabinoids on alcohol-related behaviors.

Pertaining to the mechanism by which WIN 55,212–2, CP 55,940, and Δ9-THC exert their stimulating effect on alcohol intake and alcohol self-administration, it can be hypothesized that cannabinoids produced a "priming" effect on alcohol seeking and taking. This hypothesis is based on the notion that cannabinoids and alcohol share several central effects and interactions [46], including activation of the mesolimbic dopamine neurons of the ventral tegmental area and stimulation of dopamine release in the nucleus accumbens [47–50], i.e. events likely associated to the "rewarding" properties of drugs of abuse. Thus, it can be proposed that cannabinoid-induced stimulation of the mesolimbic dopamine system may function as a "primer" on alcohol intake, mimicking the effect of alcohol, and triggering alcohol intake to additionally stimulate the system.

Effect of Cannabinoid Antagonists on Alcohol Consumption

The majority of studies characterizing the effects of cannabinoid receptor antagonism on alcohol-related effects have employed the prototypic cannabinoid CB_1 receptor antagonist/inverse agonist, Rimonabant (also known as SR141716 or Acomplia®). However, as discussed below some results obtained with Rimonabant have been reproduced using distinct CB_1 receptor antagonists/inverse agonists, including Surinabant, LH-21, AM251, and SLV330.

Since the very first studies in the late 1990's [51, 52], numerous studies have investigated the effect of acutely or repeatedly administered rimonabant on alcohol drinking behavior in rats and mice exposed to the standard, homecage 2-bottle "alcohol *vs* water" choice regimen. When repeatedly administered during the initial period of alcohol exposure, Rimonabant (0.3–10 mg/kg, i.p.) has consistently been reported to block the acquisition of alcohol drinking behavior in (a) alcohol-preferring (sP) [53, 54] and P [36] rats, (b) non-selected rats made alcohol-dependent by prolonged exposure to alcohol vapors [56], and (c) C57BL/6 mice known for high levels of alcohol consumption [51, 57]. These data suggest that Rimonabant diminishes the psychopharmacological effects that sustain alcohol drinking in rats and mice and implicate CB_1 receptors in the development of high levels of alcohol preference and consumption. Indeed, CB_1 receptor knockout mice exhibit diminished acquisition of alcohol drinking behavior and lower levels of daily alcohol preference and consumption in comparison to wild type mice ([33, 57–61] but see also [62]). Thus, diminished CB_1 receptor function—either through pharmacological or genetic approaches—diminishes the acquisition of alcohol drinking behaviors in rats and mice[34].

Several studies have also evaluated the effects of Rimonabant on alcohol consumption by rodents with already established patterns of alcohol consumption resulting in pharmacologically-relevant blood alcohol levels (e.g. models of the active, or on-going, drinking phase of human alcoholism). Acute or repeated administration of Rimonabant (0.15–10 mg/kg, i.p.) reduced alcohol intake—with relatively good selectivity with respect to water and food intake—in (a) alcohol-preferring sP [45, 52, 54], P [55], and Warsaw High Preferring [63] rats, (b) unselected Wistar rats [64], (c) and C57BL/6 mice [32, 33, 59, 65]. These results demonstrate the capacity of Rimonabant to decrease alcohol intake and preference in rats and mice displaying consolidated high alcohol consumption and preference.

Additional studies have investigated the effects of Rimonabant on the motivation for alcohol consumption using operant alcohol self-administration procedures. Virtually all studies conducted to date have reported that acute and repeated, systemic treatment with Rimonabant (0.3–10 mg/kg, i.p.) suppressed alcohol self-administration under an FR schedule of reinforcement (varying from FR1 to FR5) by (a) selectively bred alcohol-preferring AA [66], P [36], and sP [67] rats as well as (b) unselected Wistar rats [68–70]. Rimonabant pre-treatment (0.3–3 mg/kg, i.p.) also significantly reduces the breakpoint of responding for alcohol under a PR schedule of reinforcement by unselected Wistar rats [68] and suppresses non-reinforced extinction responding (ER) for alcohol by alcohol-preferring sP rats—a different measure of the motivation for alcohol reward [71]. Rimonabant exhibits relatively modest selectivity for reducing alcohol self-administration vs. the self-administration of non-drug reinforcers such as food, sucrose or saccharin [66, 68]. This is not wholly unexpected, however, as Rimonabant is known to suppress the motivation for even highly palatable foods [72].

Several studies have sought to characterize the neural substrates through which Rimonabant diminishes alcohol self-administration. In alcohol-preferring AA rats, alcohol self-administration was reduced following localized Rimonabant infusions into the nucleus accumbens [35], ventral tegmental area (VTA) [35] and prefrontal cortex (PFC) [66]. Similarly, alcohol self-administration by non-selected Wistar rats is reduced following Rimonabant infusions into the nucleus accumbens [73, 74] and posterior portion of the VTA [74], though no disruptions in alcohol self-administration were evident following infusions into the anterior VTA or medial PFC [74]. Together, these data suggest that CB_1 receptors in key areas of the mesolimbic dopamine system (nucleus accumbens, VTA) [the brain "reward" system [75]], as well as the PFC (in some rat lines) contribute to the motivation for alcohol self-administration.

Several studies have also characterized the effects of CB_1 receptor antagonism on alcohol-seeking behaviors in animal models of relapse. In the ADE model of alcohol-seeking, Rimonabant (0.3–3 mg/kg, i.p.) has been shown to completely suppress excessive alcohol intake by alcohol-preferring sP [54, 76] and P [55] rats in the 2-bottle choice paradigm, and to reduce the ADE in P rats in the operant paradigm [36]. In another model of relapse-like behavior, lever-pressing is extinguished by the removal of the alcohol reinforcer and associated conditioned cues, following which lever-pressing can be reinstated by the presentation of alcohol-related conditioned cues predictive of alcohol availability [77–80]. Using this model, Rimonabant has been shown to block cue-induced reinstatement of alcohol-seeking in both alcohol-preferring P rats [36] and non-selected Wistar rats [68]. Collectively, these findings demonstrate that Rimonabant suppresses alcohol-seeking behavior in a manner consistent with the "*anti*-alcohol" profile of this CB_1 receptor antagonist/inverse agonist.

Several of the observations made using Rimonabant have been reproduced using distinct CB_1 receptor antagonists/inverse agonists. For example, Surinabant (SR147778; structurally analogous to Rimonabant), LH-21, AM251, and SLV330 have each been reported to suppress alcohol drinking in the 2-bottle choice paradigm (both acquisition and maintenance phases), operant alcohol self-administration, the ADE and the reinstatement of alcohol-seeking behavior in rats [69, 82–86]. These observations underscore the likely involvement of CB_1 receptors in modulating alcohol-related behaviors, and discount the likelihood of off-target influences of Rimonabant in the many studies described above.

It is worth noting that a synergistic effect of Rimonabant and opioid receptor antagonists has been observed in the modulation of alcohol-related behaviors. For example, combined pre-treatment with *per se* ineffective doses of Rimonabant and either naloxone or naltrexone reduced 2-bottle choice alcohol consumption by alcohol-preferring sP rats [87, 88] and non-selected Wistar rats [64], reduced operant alcohol self-administration by Wistar rats [64] and blocked the ADE in Wistar rats [64]. These results are consistent with a series of experimental data suggesting the existence of functional links between the actions of opioids and cannabinoids and the hypothesis that this interaction extends to the control of different alcohol-related behaviors [89].

Role of the Endogenous Cannabinoid System in Alcohol-Related Behavior and Physiology

As reviewed above, a substantial literature indicates that CB_1 receptors exert a facilitory influence on alcohol preference and consumption. The consistent reduction in alcohol reward produced by CB_1 receptor antagonism has led to the hypothesis that alcohol consumption engages the brain eCB system in a manner that contributes to the motivational effects of acute and chronic alcohol consumption. This section addresses this hypothesis by reviewing evidence that alcohol exposure alters brain eCB formation and eCB-mediated mechanisms of synaptic plasticity, and that the behavioral effects of alcohol are influenced by modulation of eCB processing mechanisms.

Introduction to the Endocannabinoid System

The primary endogenous ligands of cannabinoid receptors (i.e. CB_1 and CB_2) are N-arachidonyl-ethanolamine (anandamide or AEA [90], and 2-archidonoyl-glycerol or 2-AG [91, 92]). Although a host of other endogenous lipid molecules also exert cannabinoid-like effects [93–99] there is presently little known regarding the influence of these substances on alcohol-related behaviors and the sections that follow will focus primarily on the effects produced by AEA and 2-AG.

Due to their lipophilic nature, eCBs cannot be stored by vesicles and are synthesized "on demand" via cleavage from membrane lipid precursors and immediate extrusion from neurons through distinct calcium-dependent mechanisms. AEA derives from the phospholipid precursor N-arachidonoyl-phosphatidylethanolamine [100, 101]] though several different mechanisms have been proposed for the conversion of NAPE to AEA [102–105] and the precise biosynthetic pathways contributing to AEA formation remain controversial. 2-AG is derived from the hydrolytic metabolism of 1,2-diacylglycerol (DAG) by two sn-1-selective DAG lipases, DAGL-α and DAGL-β [93, 101, 106], although alternate pathways have been described [92].

Inactivation of eCB signaling is mediated via cellular reuptake into both neurons and glial cells [107, 108] followed by intracellular hydrolysis. AEA and other ethanolamides are degraded by fatty acid amide hydrolase [109, 110]], and 2-AG is mainly degraded by monoacylglycerol lipase [MAGL [111]], although other enzymes such as ABHD6 and ABHD12 also participate in 2-AG degradation [111, 112]. Interestingly, while FAAH is present in postsynaptic cells (e.g. near the origin of eCB synthesis), both MAGL and CB_1 receptors are localized in presynaptic terminals [113–115]. This organization suggests that AEA may be more involved in tonic activation of CB_1 receptors, while 2-AG participates in more rapid, short-lived responses that occur in response to discrete stimuli [116].

The physiological and behavioral effects produced by eCBs result from interactions with cannabinoid CB_1 and CB_2 receptors as well as non-CB receptors. In general, CB_1 receptors are found on presynaptic terminals of neurons in the brain and peripheral tissues [117–123], while CB_2 receptors are mostly found on immune cells in peripheral tissues [124, 125]. However, recent evidence suggests that CB_2 receptors are also expressed in the CNS on microglia [126–128] and neurons [129–133] where these receptors can be markedly induced by a variety of insults. CB_1 and CB_2 receptors are coupled to signal transduction systems associated with G_i/G_o proteins, and their activation reduces adenylate cyclase activity, reduces Ca^{2+} influx, stimulates inwardly rectifying potassium channels and activates the mitogen-activated protein kinase pathway [134–136].

AEA binds with slightly higher affinity to CB_1 than CB_2 receptors, and is a partial agonist at both CB receptors with lesser efficacy at CB_2 than CB_1 receptors [137]. 2-AG binds with more or less equal affinity at both receptor types, and has greater potency and efficacy than AEA at these receptors [137]. AEA and 2-AG also function as agonists at orphan G-protein coupled receptors that may be members of the cannabinoid receptor family such as G-protein coupled receptor 55 (GPR55) and 119 (GPR119) [137–141]. AEA also potently activates non-cannabinoid receptors such as transient receptor potential vanilloid type-1 (TRPV1) receptors [142]. The functional relevance of AEA activation of TRPV1 receptors is the subject of ongoing studies, and remains to be characterized.

Evidence of Alcohol-Induced Alterations in Brain eCB Formation

Several observations have led to the hypothesis that alcohol consumption increases brain eCB formation and that long-term alcohol exposure can thereby lead to disruptions in eCB signaling. Chronic alcohol exposure down-regulates CB_1 receptor expression and function in rodent brain [54, 143–148]. Several *post-mortem* studies of brain tissue from alcohol dependent patients have demonstrated disrupted CB_1 receptor expression in the PFC and ventral striatum [149, 150]. *In vivo* imaging studies have reported decreased CB_1 receptor availability in heavy drinking alcoholics that persist for at least 1 month of abstinence [151, 152] but see [153]]. Further, CB_2 and GPR55 expression are significantly higher in human monocyte-derived dendritic cells from heavy drinkers vs. controls (these cells are pivotal antigen presenting cells of the immune system) [154]. Although these findings in humans may be related to variants in the genes encoding cannabinoid receptors (see section 5 below), a common interpretation is that these CB_1 receptor adaptations result in part from prolonged alcohol-induced increases in brain eCB levels, similar to CB_1 down-regulation that occurs following long-term exposure to synthetic CB_1 receptor agonists [155–157]. This interpretation is further supported by evidence of a transient recovery (and perhaps up-regulation) of CB_1 receptor function during protracted alcohol abstinence [145, 147]. Chronic alcohol exposure also disrupts gene expression for the primary eCB clearance enzymes, FAAH and MAGL, in a manner that is sensitive to the intermittent nature of alcohol exposure and post-alcohol abstinence period [158]. Collectively there is substantial biochemical evidence that long-term alcohol exposure disrupts eCB processing and signaling mechanisms.

Several studies have sought to directly evaluate alcohol-induced alterations in brain eCB formation. Early studies by Basalingappa Hungund demonstrated that chronic alcohol exposure increases both AEA and 2-AG formation in human neuroblastoma cells and primary cultures of rodent neurons [144, 159, 160]. Subsequent studies evaluated alterations in eCB levels extracted from *post-mortem* brain tissue following acute and chronic alcohol exposure and while the results clearly demonstrate alcohol-induced alterations in brain eCBs, substantial inconsistencies among studies make it difficult to draw clear conclusions on the direction of change and regional nature of the effects. For example, while some studies report increased AEA content in the limbic forebrain and nucleus accumbens (NAc) of alcohol-exposed rats [145, 161–163] consistent with a reduction in FAAH activity following chronic alcohol [145], other studies report significant decreases in AEA in components of the limbic forebrain such as the amygdala, hippocampus, and PFC [163, 164].

Moreover, alcohol consumption is reported to both increase and decrease AEA content in the caudate putamen [163, 164]. With regard to 2-AG, alcohol is reported to both increase and decrease 2-AG content in striatal tissue [54, 161, 162], and to decrease 2-AG content in PFC tissue [163, 164]. Recently, one study reported that 2-AG striatal content was increased during the acquisition phase of alcohol drinking in sP rats, whereas the maintenance phase was associated with increases in both AEA and 2-AG [54]. Also, withdrawal from chronic intermittent alcohol exposure produced an increase in levels of both eCBs in the rat hippocampus [147]. These distinct findings may result from differences in rat strain, gender, amount of daily alcohol consumption, duration of alcohol exposure, and post-exposure time that was evaluated. Moreover, quantification of tissue eCB content is highly sensitive to the procedures employed for tissue preparation, eCB extraction and analysis [165], and these and other methodological differences likely underlie the many discrepant literature reports on alcohol-induced alterations in brain eCB levels.

The effects of acute and chronic alcohol have also been studied using *in vivo* microdialysis procedures. Voluntary alcohol self-administration is reported to robustly increase nucleus accumbens dialysate 2-AG levels without altering AEA levels, with the rise and fall of dialysate 2-AG content aligning with the pattern of blood alcohol concentrations typically observed following oral consumption [73]. Interestingly, involuntary administration of moderate alcohol doses produces a mild increase in dialysate 2-AG levels, while reducing AEA levels in the NAc of alcohol-naïve rats [166, 167] though higher doses induce a mild increase in dialysate AEA levels [148]. As compared with voluntary alcohol self-administration these findings suggest that the volitional nature of alcohol exposure (e.g. voluntary vs. forced administration) may differentially impact eCB responses. Available evidence also demonstrates that alcohol induces region-specific alterations in brain eCBs as significant alcohol-induced disruptions in eCB levels are consistently observed in striatal regions [73, 148, 166, 167] while alcohol-induced disruptions in frontal cortical eCB levels are not evident [74].

Evidence of Alcohol-Induced Disruptions in eCB-Mediated Synaptic Plasticity

It is well established that eCBs serve as retrograde messengers at neuronal synapses in the CNS [168–171] and contribute to different forms of short and long-term synaptic plasticity [171–177]. Synaptic plasticity is considered one of the primary mechanisms underlying learning and memory processes, and is thought to be vital for experience-dependent modifications in neural function that underlies behavioral flexibility. In this context, several conceptualizations of addiction theorize that alcohol and drug exposure disrupt plasticity mechanisms resulting in pathology of learning and memory mechanisms in brain reward circuits [178–182].

Acute and repeated alcohol exposure disrupts eCB-mediated synaptic plasticity. Short periods of low-frequency stimulation produce a CB_1 receptor-dependent long-lasting disinhibition (DLL) of striatal output neurons as a result of reduced synaptic strength at inhibitory synapses [183]. Acute exposure of striatal slices to moderate alcohol doses substantially reduces eCB-mediated DLL in the dorsolateral striatum [184]. DLL is also significantly reduced in the dorsolateral striatum of rats following long-term voluntary alcohol consumption [185]. Endocannabinoid-mediated long-term depression (LTD) at inhibitory striatal synapses is also reduced by acute alcohol, though no significant alcohol effects are evident on LTD of excitatory synapses [184]. Because the dorsal striatum is involved in reward-guided learning and habitual behavior [186, 187] it is possible that alcohol-induced interference in eCB-LTD contributes to maladaptive habitual behavior associated with addiction. Indeed, recent work from the Holmes lab demonstrates that CB_1 receptor-dependent long-term depression is absent in the dorsolateral striatum of mice following chronic intermittent alcohol exposure [188]. This disruption in eCB-mediated plasticity was associated with enhanced neuronal encoding of striatal-based behaviors, which may drive aberrant reward learning and modulation of rewarded behavior in a manner that contributes to the progression of alcoholism.

Effects of Altered eCB Clearance on Alcohol Consumption, Alcohol Reward and Alcohol-Seeking Behavior

As reviewed earlier, substantial evidence demonstrates that CB_1 receptor agonists increase the rewarding effects of alcohol and increase alcohol consumption. Conversely, reductions in CB_1 receptor function (either pharmacologically or genetically) decrease alcohol preference and consumption, and prevent alcohol-induced activation of the mesolimbic dopamine system. Although these findings demonstrate a facilitory CB_1 receptor influence on alcohol consumption, they do not necessarily implicate alcohol-induced alterations in eCB signaling in the motivation for alcohol consumption. To more directly address this issue several studies have characterized the effects of eCB clearance inhibition on alcohol consumption.

Acute administration of the FAAH inhibitor URB597 increases alcohol preference and consumption in wildtype mice, though this compound does not produce these effects in either CB_1 receptor or FAAH knockout mice [189–191]. This indicates that the effects of URB597 are mediated through its actions on FAAH, and that the resultant effect on alcohol consumption relies on CB_1 receptor signaling. FAAH knockout mice also exhibit enhanced alcohol preference and consumption relative to wildtype mice, and the lack of genotypic differences in the consumption of non-alcoholic tastants such as saccharin or quinine suggests an alteration in the pharmacological effects of alcohol [189–191]. However, FAAH deficient mice are less sensitive to alcohol-induced motor incoordination and intoxication [190] that may allow these animals to consume more alcohol without behavioral impairment. As such, it is not clear whether FAAH deletion/inhibition confers increased alcohol reward or simply a greater capacity for alcohol intake. In contrast to these observations in mice, systemic URB597 administration was not found to alter alcohol consumption under 2-bottle choice or operant procedures or to affect reinstatement of alcohol-seeking behavior in either Marchigian Sardinian alcohol-preferring (msP) rats or non-selected Wistar rats [66, 70]. However, more regionally specific impairments in FAAH activity may exert greater influences on alcohol consumption by rats based on evidence that alcohol-preferring AA rats exhibit reduced FAAH expression and activity in the PFC, and that intra-PFC URB597 administration increases alcohol consumption by non-selected Wistar rats [66].

Other than inhibitors of FAAH activity very few pharmacological tools have been available for selectively manipulating eCB clearance. One study evaluated the effects of the putative eCB reuptake inhibitor AM404 and observed a reduction in operant alcohol self-administration by Wistar rats at AM404 doses that did not alter saccharin self-administration [192]. However, this effect was not blocked by CB_1, CB_2 or TRPV1 receptor antagonists suggesting that the effects of this compound were not mediated through enhanced eCB signaling. Despite evidence of alcohol-induced increases in brain 2-AG formation [73, 74, 167] there are to date no published experiments characterizing the effects MAGL inhibition on alcohol consumption and related alcohol-induced effects. This is anticipated to be alleviated in the near future however in light of the recent availability of selective and efficacious MAGL inhibitors [193, 194].

As previously reviewed, exogenous CB_1 receptor agonists increase alcohol-seeking behavior following periods of abstinence [40–43] and CB_1 receptor antagonists attenuate the reinstatement of alcohol-seeking [36, 40, 41, 54, 55, 68, 76, 82, 84–86, 195]. At the present time the effects of eCB clearance inhibition on alcohol-seeking behavior have not been extensively evaluated. Cippitelli and colleagues have reported that neither the FAAH inhibitor URB597 nor the putative eCB reuptake inhibitor AM404 alter cue- or stress-induced reinstatement of alcohol-seeking behavior [70, 192]. There are presently no reports characterizing the effects of MAGL inhibition on the reinstatement of alcohol-seeking behavior.

Genetic Variants in Endocannabinoid Signaling and Alcohol use Disorders

Several genetic polymorphisms of eCB-related genes have been linked to alcohol abuse. Most studies have focused on the CNR1 gene that encodes the cannabinoid receptor CB_1. Among the first polymorphisms reported in this system was a triplet repeat of varying number (AAT)n in the 3′-flanking region of the CNR1 (CB_1) genetic locus [196]. A significant relationship between this triplet repeat marker and decreased P300 event-related potential amplitudes has been reported in subjects with alcohol and substance abuse disorders [197]. Further, a 6-repeat allele of the triplet repeat polymorphism (AATn/A6) is significantly associated with impulsivity in a Native American population demonstrating a high lifetime prevalence of substance dependence [198]. A similar polymorphism was observed in male Spanish alcoholics with a history of childhood ADHD [199]. Other studies have linked this triplet repeat marker with psychostimulant and cannabis dependence in non-Hispanic Caucasians and an African-Caribbean population [200, 201]. However, several studies have failed to find a significant linkage between the (AAT)n polymorphism and drug dependence [202–205].

Another CNR1 gene polymorphism is a silent mutation that results in the substitution of G to A at nucleotide position 1359 in codon 435 (Thr) [206, 207]. A modest association between the A/A genotype and alcohol withdrawal delirium has been reported in German patients who are A/A homozygous [208], though further studies in a similar population did not replicate this finding or correlate this polymorphism with alcohol dependence [209, 210].

CNR1-associated SNPs, particularly rs6454674, rs806368, rs1049353 and rs1535355 are also associated with alcohol dependence in European Americans, African-Americans and Europeans [210–213]. The C allele of rs2023239 appears to be associated with greater CB_1 receptor binding in the PFC, greater alcohol cue elicited brain activation in the midbrain and PFC and greater subjective reward when consuming alcohol [214, 215]. Interestingly, this allele is also associated with more positive outcomes in individuals receiving Olanzapine, a medication that targets mesocorticolimbic circuitry [214]. Four SNPs in or near the CNR1 gene (rs1535225, rs2023239, rs1049353, rs806368) are associated with increased impulsivity in a Native American population characterized by a high prevalence of alcoholism and substance abuse [198].

Several recent reports suggest that genetic alterations in FAAH expression or function confer susceptibility to problem alcohol and drug use. For example, FAAH expression and function are reduced in the PFC of rats selectively bred for high alcohol preference and consumption [66]. Consistently, genetic deletion of FAAH results in increased alcohol preference and consumption in mice [189–191]. Animals lacking FAAH are also less sensitive to alcohol-induced motor incoordination and exhibit reduced signs of alcohol withdrawal, leading to the theory that reductions in the aversive effects of alcohol intake confer increased motivation for alcohol consumption. FAAH knockout mice also display enhanced sensitivity to both the rewarding effects of nicotine and the aversive effects of nicotine withdrawal [216], though Δ9-THC dependence and precipitated withdrawal appear to be unaltered by FAAH deletion [217]. Collectively these findings in rodents suggest that genetic disruption of FAAH function results in altered sensitivity to some abused drugs such as alcohol and nicotine.

In humans, a missense SNP in the FAAH DNA sequence has been identified (C385A, rs324420) that leads to a conserved proline to threonine (P129T) conversion in the FAAH amino acid sequence that results in reduced FAAH activity and a presumed increase in levels of AEA and other FAAH substrates (218). Initial reports described a significant association between the C385A polymorphism and problem drug use [219, 220]. Further characterization has shown this FAAH mutation to be strongly associated with risk for frequent sedative use, though individuals with this SNP do not appear to be at greater risk for alcohol, nicotine or methamphetamine use or dependence [221, 222]. However, carriers of the FAAH C385A SNP display increased ventral striatal reactivity associated with delay discounting, a behavioral index of impulsivity and reward sensitivity [223]. Moreover, C385A carriers exhibit a markedly decreased relationship between threat-related amygdala reactivity and trait anxiety, similar to patterns observed in individuals with high familial risk for alcoholism [224]. These findings suggest that dysregulation of FAAH function through the C385A polymorphism confers increased impulsivity and decreased threat perception that result in increased risk-taking behavior associated with addiction. Moreover, genotype comparisons of the FAAH C385A polymorphism showed that CC carriers were over-represented among risky drinkers relative to A allele carriers [225].

Clinical Trials of Rimonabant for Treatment of Alcohol Dependence

As reviewed throughout this chapter a great deal of evidence implicates an involvement of the eCB system in the neurobiology of alcohol dependence. In particular, substantial preclinical evidence indicates that CB_1 receptor antagonism reduces alcohol preference, consumption, the motivation to obtain alcohol, the vigor of cue-induced alcohol-seeking behavior and the magnitude of the ADE upon return to alcohol consumption following periods of abstinence. Based on this evidence two clinical trials have been conducted in human alcoholics to evaluate the efficacy of Rimonabant as a therapeutic for alcohol dependence.

The first trial was a 12-week Phase IIa (proof of concept), double-blind, placebo-controlled study of the efficacy of Rimonabant to prevent alcohol relapse in recently detoxified alcoholics [226]. To this end, 131 patients received Rimonabant (20 mg/day) while 127 patients received placebo. Compliance tended to be higher in Rimonabant than placebo group (71.8 % *vs* 62.2 %), the safety and tolerability of Rimonabant was good and the rate of adverse events was comparable between treatment groups. A statistically non-significant tendency toward a protective effect of Rimonabant was observed: at the end of the 12-week study, (a) 41.5 % and 47.7 % patients in the Rimonabant and placebo groups, respectively, had relapsed to drinking, and (b) 27.7 % and 35.6 % patients in the Rimonabant and placebo groups, respectively, had relapsed to heavy drinking. A similar non-significant tendency toward beneficial effects in the Rimonabant patients was observed in the secondary outcomes (cumulative abstinence duration, percentage of drinking days, and percentage of heavy drinking days). The authors concluded that the unusually high positive response rate in the placebo group (e.g. number of control subjects maintaining abstinence) confounded the evidence of a beneficial effect of Rimonabant, and suggested that longer duration trials may be informative.

The second trial was a Phase I/II, double-blind, placebo-controlled study designed to assess the effect of Rimonabant on alcohol consumption in 49 non-treatment seeking heavy alcohol drinkers [227]. The initial phase entailed a 2-week outpatient evaluation during which 18 patients received 20 mg/day Rimonabant and 21 patients received placebo. Rimonabant was not found to be efficacious for reducing daily alcohol consumption in this phase of the trial. The second phase of the trial entailed an inpatient study evaluating the efficacy of 20 mg Rimonabant for reducing alcohol consumption in a laboratory "self-administration" paradigm. After consuming a priming dose of alcohol, patients had the option of consuming up to 8 alcoholic drinks or receiving a small amount of money for each drink that was declined. Rimonabant failed to significantly reduce the number of drinks consumed.

During the time in which Rimonabant was approved for clinical use in Europe (primarily for treating obesity), significant adverse psychiatric events were observed including depression, anxiety and suicidal ideation [228] and based on this profile, the approved use of this drug was revoked [229].

Conclusions and Future Directions

As described above, substantial and long-standing evidence demonstrates strong interactive effects between cannabinoid signaling and the behavioral effects produced by alcohol. A number of genetic variants in eCB signaling are associated with both alcohol dependence and psychiatric disturbances that are risk factors for the development of problematic alcohol use. In particular, CB_1 receptors play a facilitory role with regard to the motivation for alcohol and the regulation of alcohol consumption and substantial preclinical evidence demonstrates that CB_1 receptor antagonism diminishes the motivation for alcohol, reduces alcohol consumption and attenuates alcohol-seeking behavior in animal models of relapse. Despite these promising observations, initial clinical trials failed to observe significant therapeutic efficacy of the CB_1 antagonist Rimonabant for reducing alcohol consumption or prolonging abstinence in alcohol dependent patients. Although improved experimental design may reveal beneficial effects of CB_1 receptor antagonism for treating alcohol dependence, these pharmacological agents are no longer approved for clinical use due to presentation of adverse psychiatric side-effects.

Despite some inconsistencies in the literature, a preponderance of evidence suggests that alcohol exposure increases eCB formation in brain [54, 73, 145, 148, 161–163, 166, 167]. Moreover, chronic alcohol exposure disrupts eCB clearance mechanisms [158], impairs eCB-mediated forms of synaptic plasticity [184, 185, 188, 230] and down-regulates cannabinoid receptor function and expression in a manner that persists well into protracted alcohol abstinence [54, 143–148, 150–152, 154]. Accordingly, alcohol dependence and protracted withdrawal may be characterized by disruption in eCB signaling. This may be of relevance given that alcohol dependence and protracted withdrawal are characterized by increased prevalence of anxiety, depression and sensitivity to stressors [231–234] and eCB signaling plays a prominent role in the maintenance of affective state and the constraint of stress responsivity [235–243]. These symptoms of protracted withdrawal play a prominent role in the relapse to heavy alcohol consumption [232, 234, 244] and if these symptoms derive in part from impaired eCB signaling, then therapeutic approaches aimed at restoring or bolstering cannabinoid signaling may have clinical benefit for treating alcohol dependence. The use of exogenous CB_1 receptor agonists for this purpose would likely be problematic as these compounds can promote alcohol-seeking behavior, increase the magnitude of excessive alcohol consumption evident following periods of abstinence (e.g. the ADE) and can exacerbate withdrawal-related affective disorders when administered at high doses. However, enhancement of eCB tone through inhibition of eCB clearance may be a viable therapeutic approach in light of evidence that these manipulations constrain stress responses and produce anxiolytic- and antidepressive-like effects. Indeed, Cippitelli and colleagues have reported that acute FAAH inhibition ameliorates alcohol withdrawal-related anxiety-like behavior without promoting alcohol-seeking behavior [70].

Importantly, because eCBs are generally produced in response to specific stimuli, the behavioral effects of moderate eCB clearance inhibition may be preferentially evident in limited circumstances such as exposure to stress or drug-associated conditioned cues. eCB clearance inhibition may also facilitate region-specific increases in brain eCB signaling as a result of regionally distinct stimulus-induced eCB production, thereby producing fewer unwanted behavioral effects than those produced by exogenous cannabinoid agonists that induce widespread cannabinoid receptor activation. Although initial studies have evaluated the effects of acute FAAH inhibition on alcohol dependence-related behaviors [70, 192], the effects of long-term FAAH inhibition have not been characterized. Moreover, there is sparse knowledge regarding the effects of MAGL inhibition on alcohol dependence-related behaviors. These mechanisms reflect an important area of study involving the eCB system as a viable therapeutic target for alcohol use disorders and alcoholism.

References

1. SAMHSA. Results from the 2012 National survey on drug use and health: summary of national findings. Rockville: Office of Applied Studies, DHHS; 2013.
2. White HR, Morgan TJ, Pugh LA, Celinska K, Labouvie EW, Pandina RJ. Evaluating two brief substance-use interventions for mandated college students. J Stud Alcohol. 2006;67(2):309–17.
3. Welfare AloHa. 2007 National Drug Strategy Household Survey: First results. Canberra: Australian Institute of Health and Welfare; 2008.
4. Hedden SL, Martins SS, Malcolm RJ, Floyd L, Cavanaugh CE, Latimer WW. Patterns of illegal drug use among an adult alcohol dependent population: results from the national survey on drug use and health. Drug Alcohol Depend. 2010;106(2–3):119–25.
5. Magill M, Barnett NP, Apodaca TR, Rohsenow DJ, Monti PM. The role of marijuana use in brief motivational intervention with young adult drinkers treated in an emergency department. J Stud Alcohol Drugs. 2009;70(3):409–13.
6. Peters EN, Leeman RF, Fucito LM, Toll BA, Corbin WR, O'Malley SS. Co-occurring marijuana use is associated with medication nonadherence and nonplanning impulsivity in young adult heavy drinkers. Addict Behav. 2012;37(4):420–6.
7. Jones RT, Stone GC. Psychological studies of marijuana and alcohol in man. Psychopharmacologia. 1970;18(1):108–17.
8. Macavoy MG, Marks DF. Divided attention performance of cannabis users and non-users following cannabis and alcohol. Psychopharmacologia. 1975;44(2):147–52.

9. Newman LM, Lutz MP, Gould MH, Domino EF. 9-Tetrahydrocannabinol and ethyl alcohol: evidence for cross-tolerance in the rat. Science. 1972;175(4025):1022–3.
10. Newman LM, Lutz MP, Domino EF. Delta9-tetrahydrocannabinol and some CNS depressants: evidence for cross-tolerance in the rat. Arch Int Pharmacodyn Ther. 1974;207(2):254–9.
11. Sprague GL, Craigmill AL. Ethanol and delta-9-tetrahydrocannabinol: mechanism for cross-tolerance in mice. Pharmacol Biochem Behav. 1976;5(4):409–15.
12. Siemens AJ, Doyle OL. Cross-tolerance between delta9-tetrahydrocannabinol and ethanol: the role of drug disposition. Pharmacol Biochem Behav. 1979;10(1):49–55.
13. McMillan DE, Snodgrass SH. Effects of acute and chronic administration of delta9-tetrahy-drocannabinol or cocaine on ethanol intake in a rat model. Drug Alcohol Depend. 1991;27(3):263–74.
14. da Silva GE, Morato GS, Takahashi RN. Rapid tolerance to Delta(9)-tetrahydrocannabinol and cross-tolerance between ethanol and Delta(9)-tetrahydrocannabinol in mice. Eur J Pharmacol. 2001;431(2):201–7.
15. Lemos JI, Takahashi RN, Morato GS. Effects of SR141716 and WIN 55,212-2 on tolerance to ethanol in rats using the acute and rapid procedures. Psychopharmacology (Berl). 2007;194(2):139–49.
16. Pava MJ, Blake EM, Green ST, Mizroch BJ, Mulholland PJ, Woodward JJ. Tolerance to cannabinoid-induced behaviors in mice treated chronically with ethanol. Psychopharmacology (Berl). 2012;219(1):137–47.
17. Hollister LE, Gillespie HK. Marihuana, ethanol, and dextroamphetamine. Mood and mental function alterations. Arch Gen Psychiatry. 1970;23(3):199–203.
18. Kalant H, LeBlanc AE. Effect of acute and chronic pre-treatment with delta1-tetrahydrocannabinol on motor impairment by ethanol in the rat. Can J Physiol Pharmacol. 1974;52(2):291–7.
19. Martin BR, Compton DR, Thomas BF, Prescott WR, Little PJ, Razdan RK, Johnson MR, Melvin LS, Mechoulam R, Ward SJ. Behavioral, biochemical, and molecular modeling evaluations of cannabinoid analogs. Pharmacol Biochem Behav. 1991;40(3):471–8.
20. Crabbe JC, Metten P, Cameron AJ, Wahlsten D. An analysis of the genetics of alcohol intoxication in inbred mice. Neurosci Biobehav Rev. 2005;28(8):785–802.
21. Heishman SJ, Stitzer ML, Bigelow GE. Alcohol and marijuana: comparative dose effect profiles in humans. Pharmacol Biochem Behav. 1988;31(3):649–55.
22. Chait LD, Perry JL. Acute and residual effects of alcohol and marijuana, alone and in combination, on mood and performance. Psychopharmacology (Berl). 1994;115(3):340–9.
23. Heishman SJ, Arasteh K, Stitzer ML. Comparative effects of alcohol and marijuana on mood, memory, and performance. Pharmacol Biochem Behav. 1997;58(1):93–101.
24. Manno JE, Kiplinger GF, Scholz N, Forney RB. The influence of alcohol and marihuana on motor and mental performance. Clin Pharmacol Ther. 1971;12(2):202–11.
25. Marks DF, MacAvoy MG. Divided attention performance in cannabis users and non-users following alcohol and cannabis separately and in combination. Psychopharmacology (Berl). 1989;99(3):397–401.
26. Phillips RN, Brown DJ, Forney RB. Enhancement of depressant properties of alcohol or barbiturate in combination with aqueous suspended delta9-tetrahydrocannabinol in rats. J Forensic Sci. 1971;16(2):152–61.
27. Zarcone VP, Jr. Marijuana and ethanol: effects on sleep. Int J Psychiatry Med. 1973;4(2):201–12.
28. Friedman E, Gershon S. Effect of delta8-THC on alcohol-induced sleeping time in the rat. Psychopharmacologia. 1974;39(3):193–8.
29. Bech P, Rafaelsen L, Rafaelsen OJ. Cannabis and alcohol: effects on estimation of time and distance. Psychopharmacologia. 1973;32(4):373–81.
30. Tinklenberg JR, Roth WT, Kopell BS. Marijuana and ethanol: differential effects on time perception, heart rate, and subjective response. Psychopharmacology (Berl). 1976;49(3):275–9.
31. Colombo G, Serra S, Brunetti G, Gomez R, Melis S, Vacca G, Carai MAM, Gessa GL. Stimulation of voluntary ethanol intake by cannabinoid receptor agonists in ethanol-preferring sP rats. Psychopharmacology (Berl). 2002;159(2):181–7.

32. Kelai S, Hanoun N, Aufrere G, Beauge F, Hamon M, Lanfumey L. Cannabinoid-serotonin interactions in alcohol-preferring vs. alcohol-avoiding mice. J Neurochem. 2006;99(1):308–20.
33. Vinod KY, Yalamanchili R, Thanos PK, Vadasz C, Cooper TB, Volkow ND, Hungund BL. Genetic and pharmacological manipulations of the CB(1) receptor alter ethanol preference and dependence in ethanol preferring and nonpreferring mice. Synapse. 2008;62(8):574–81.
34. Linsenbardt DN, Boehm SL. 2nd. Agonism of the endocannabinoid system modulates binge-like alcohol intake in male C57BL/6J mice: involvement of the posterior ventral tegmental area. Neuroscience. 2009;164(2):424–34.
35. Malinen H, Hyytiä P. Ethanol self-administration is regulated by CB1 receptors in the nucleus accumbens and ventral tegmental area in alcohol-preferring AA rats. Alcohol Clin Exp Res. 2008;32(11):1976–83.
36. Getachew B, Hauser SR, Dhaher R, Katner SN, Bell RL, Oster SM, McBride WJ, Rodd ZA. CB1 receptors regulate alcohol-seeking behavior and alcohol self-administration of alcohol-preferring (P) rats. Pharmacol Biochem Behav. 2011;97(4):669–75.
37. Gallate JE, Saharov T, Mallet PE, McGregor IS. Increased motivation for beer in rats following administration of a cannabinoid CB1 receptor agonist. Eur J Pharmacol. 1999;370(3):233–40.
38. Spanagel R. How to measure relapse in animals. In: Spanagel R, Mann KF, editors. Drugs for relapse prevention of alcoholism. Basel: Birkhäuser; 2005. p. 13–21.
39. Martin-Fardon R, Weiss F. Modeling relapse in animals. Curr Top Behav Neurosci. 2013;13:403–32.
40. Lopez-Moreno JA, Gonzalez-Cuevas G, Rodríguez de Fonseca F, Navarro M. Long-lasting increase of alcohol relapse by the cannabinoid receptor agonist WIN 55,212-2 during alcohol deprivation. J Neurosci. 2004;24(38):8245–52.
41. Alén F, Moreno-Sanz G, Isabel de Tena A, Brooks RD, López-Jimenez A, Navarro M, López-Moreno JA. Pharmacological activation of CB1 and D2 receptors in rats: predominant role of CB1 in the increase of alcohol relapse. Eur J Neurosci. 2008;27(12):3292–8.
42. Alén F, Santos A, Moreno-Sanz G, González-Cuevas G, Giné E, Franco-Ruiz L, Navarro M, López-Moreno JA. Cannabinoid-induced increase in relapse-like drinking is prevented by the blockade of the glycine-binding site of N-methyl-D-aspartate receptors. Neuroscience. 2009;158(2):465–73.
43. McGregor IS, Dam KD, Mallet PE, Gallate JE. Delta9-THC reinstates beer- and sucrose-seeking behaviour in abstinent rats: comparison with midazolam, food deprivation and predator odour. Alcohol Alcohol. 2005;40(1):35–45.
44. Colombo G, Serra S, Vacca G, Gessa GL, Carai MAM. Suppression by baclofen of the stimulation of alcohol intake induced by morphine and WIN 55,212-2 in alcohol-preferring rats. Eur J Pharmacol. 2004;492(2–3):189–93.
45. Colombo G, Orrù A, Lai P, Cabras C, Maccioni P, Rubio M, Gessa GL, Carai MAM. The cannabinoid CB1 receptor antagonist, rimonabant, as a promising pharmacotherapy for alcohol dependence: preclinical evidence. Mol Neurobiol. 2007;36(1):102–12.
46. Mechoulam R, Parker L. Cannabis and alcohol-a close friendship. Trends Pharmacol Sci. 2003;24(6):266–8.
47. Gessa GL, Muntoni F, Collu M, Vargiu L, Mereu G. Low doses of ethanol activate dopaminergic neurons in the ventral tegmental area. Brain Res. 1985;348(1):201–3.
48. Imperato A, Di Chiara G. Preferential stimulation of dopamine release in the nucleus accumbens of freely moving rats by ethanol. J Pharmacol Exp Ther. 1986;239(1):219–28.
49. Tanda G, Pontieri FE, Di Chiara G. Cannabinoid and heroin activation of mesolimbic dopamine transmission by a common mu1 opioid receptor mechanism. Science. 1997;276(5321):2048–50.
50. Gessa GL, Melis M, Muntoni AL, Diana M. Cannabinoids activate mesolimbic dopamine neurons by an action on cannabinoid CB1 receptors. Eur J Pharmacol. 1998;341(1):39–44.
51. Arnone M, Maruani J, Chaperon F, Thiébot MH, Poncelet M, Soubrié P, Le Fur G. Selective inhibition of sucrose and ethanol intake by SR 141716, an antagonist of central cannabinoid (CB1) receptors. Psychopharmacology (Berl). 1997;132(1):104–6.

52. Colombo G, Agabio R, Fà M, Guano L, Lobina C, Loche A, Reali R, Gessa GL. Reduction of voluntary ethanol intake in ethanol-preferring sP rats by the cannabinoid antagonist SR-141716. Alcohol Alcohol. 1998;33(2):126–30.
53. Serra S, Carai MAM, Brunetti G, Gomez R, Melis S, Vacca G, Colombo G, Gessa GL. The cannabinoid receptor antagonist SR 141716 prevents acquisition of drinking behavior in alcohol-preferring rats. Eur J Pharmacol. 2001;430(2–3):369–71.
54. Vinod KY, Maccioni P, Garcia-Gutierrez MS, Femenia T, Xie S, Carai MAM, Manzanares J, Cooper TB, Hungund BL, Colombo G. Innate difference in the endocannabinoid signaling and its modulation by alcohol consumption in alcohol-preferring sP rats. Addict Biol. 2012;17(1):62–75.
55. Bell RL, Rodd ZA, Sable HJK, McQueen VK, Davids MR, Murphy JM, Lumeng L, McBride WJ. Cannabinoid CB1 antagonist reduces alcohol intake during acquisition, maintenance, and relapse in inbred alcohol preferring (IP) rats. Soc Neurosci. 2004. Program No. 489.6.
56. Lallemand F, Soubrie PH, De Witte PH. Effects of CB1 cannabinoid receptor blockade on ethanol preference after chronic ethanol administration. Alcohol Clin Exp Res. 2001;25(9):1317–23.
57. Poncelet M, Maruani J, Calassi R, Soubrie P. Overeating, alcohol and sucrose consumption decrease in CB1 receptor deleted mice. Neurosci Lett. 2003;343(3):216–8.
58. Hungund BL, Szakall I, Adam A, Basavarajappa BS, Vadasz C. Cannabinoid CB1 receptor knockout mice exhibit markedly reduced voluntary alcohol consumption and lack alcohol-induced dopamine release in the nucleus accumbens. J Neurochem. 2003;84(4):698–704.
59. Wang L, Liu J, Harvey-White J, Zimmer A, Kunos G. Endocannabinoid signaling via cannabinoid receptor 1 is involved in ethanol preference and its age-dependent decline in mice. Proc Natl Acad Sci USA. 2003;100(3):1393–8.
60. Naassila M, Pierrefiche O, Ledent C, Daoust M. Decreased alcohol self-administration and increased alcohol sensitivity and withdrawal in CB1 receptor knockout mice. Neuropharmacology. 2004;46(2):243–53.
61. Lallemand F, de Witte P. Ethanol induces higher BEC in CB1 cannabinoid receptor knockout mice while decreasing ethanol preference. Alcohol Alcohol. 2005;40(1):54–62.
62. Racz I, Bilkei-Gorzo A, Toth ZE, Michel K, Palkovits M, Zimmer A. A critical role for the cannabinoid CB1 receptors in alcohol dependence and stress-stimulated ethanol drinking. J Neurosci. 2003;23(6):2453–8.
63. Dyr W, Ligieza J, Kostowski W. The effect of cannabinoid CB(1) receptor antagonist rimonabant (SR-141716) on ethanol drinking in high-preferring rats. Alcohol. 2008;42(6):509–12.
64. Gallate JE, Mallet PE, McGregor IS. Combined low dose treatment with opioid and cannabinoid receptor antagonists synergistically reduces the motivation to consume alcohol in rats. Psychopharmacology (Berl). 2004;173(1–2):210–6.
65. Thanos PK, Dimitrakakis ES, Rice O, Gifford A, Volkow ND. Ethanol self-administration and ethanol conditioned place preference are reduced in mice lacking cannabinoid CB1 receptors. Behav Brain Res. 2005;164(2):206–13.
66. Hansson AC, Bermúdez-Silva FJ, Malinen H, Hyytiä P, Sanchez-Vera I, Rimondini R, Rodríguez de Fonseca F, Kunos G, Sommer WH, Heilig M. Genetic impairment of frontocortical endocannabinoid degradation and high alcohol preference. Neuropsychopharmacology. 2007;32(1):117–26.
67. Maccioni P, Fantini N, Carai MAM, Gessa GL, Colombo G. Suppressing effect of the cannabinoid CB1 receptor antagonist, rimonabant, on alcohol self-administration in alcohol-preferring rats. Open Neuropsychopharmacol. 2009;2:40–4.
68. Economidou D, Mattioli L, Cifani C, Perfumi M, Massi M, Cuomo V, Trabace L, Ciccocioppo R. Effect of the cannabinoid CB1 receptor antagonist SR-141716A on ethanol self-administration and ethanol-seeking behaviour in rats. Psychopharmacology (Berl). 2006;183(4):394–403.
69. Pavon FJ, Bilbao A, Hernández-Folgado L, Cippitelli A, Jagerovic N, Abellán G, Rodríguez-Franco MA, Serrano A, Macias M, Gómez R, Navarro M, Goya P, Rodríguez de Fonseca F. Antiobesity effects of the novel in vivo neutral cannabinoid receptor antagonist

5-(4-chlorophenyl)-1-(2,4-dichlorophenyl)-3-hexyl-1H-1,2,4-triazole-LH21. Neuropharmacology. 2006;51(2):358–66.
70. Cippitelli A, Cannella N, Braconi S, Duranti A, Tontini A, Bilbao A, Rodríguez de Fonseca F, Piomelli D, Ciccocioppo R. Increase of brain endocannabinoid anandamide levels by FAAH inhibition and alcohol abuse behaviours in the rat. Psychopharmacology (Berl). 2008;198(4):449–60.
71. Colombo G, Vacca G, Serra S, Carai MAM, Gessa GL. Suppressing effect of the cannabinoid CB1 receptor antagonist, SR 141716, on alcohol's motivational properties in alcohol-preferring rats. Eur J Pharmacol. 2004;498(1–3):119–23.
72. Maccioni P, Pes D, Carai MAM, Gessa GL, Colombo G. Suppression by the cannabinoid CB1 receptor antagonist, rimonabant, of the reinforcing and motivational properties of a chocolate-flavoured beverage in rats. Behav Pharmacol. 2008;19(3):197–209.
73. Caille S, Alvarez-Jaimes L, Polis I, Stouffer DG, Parsons LH. Specific alterations of extracellular endocannabinoid levels in the nucleus accumbens by ethanol, heroin, and cocaine self-administration. J Neurosci. 2007;27(14):3695–702.
74. Alvarez-Jaimes L, Polis I, Parsons LH. Regional influence of cannabinoid CB1 receptors in the regulation of ethanol self-administration by Wistar rats. Open Neuropsychopharmacol. 2009;2:77–85.
75. Koob GF, Le Moal M. Neurobiology of Addiction. London:Elsevier; 2006.
76. Serra S, Brunetti G, Pani M, Vacca G, Carai MAM, Gessa GL, Colombo G. Blockade by the cannabinoid CB(1) receptor antagonist, SR 141716, of alcohol deprivation effect in alcohol-preferring rats. Eur J Pharmacol. 2002;443(1–3):95–7.
77. Shaham Y, Erb S, Stewart J. Stress-induced relapse to heroin and cocaine seeking in rats: a review. Brain Res Rev. 2000;33(1):13–33.
78. Stewart J. Pathways to relapse: the neurobiology of drug- and stress-induced relapse to drug-taking. J Psychiatr Neurosci. 2000;25(2):125–36.
79. See RE. Neural substrates of cocaine-cue associations that trigger relapse. Eur J Pharmacol. 2005;526(1–3):140–6.
80. Weiss F. Neurobiology of craving, conditioned reward and relapse. Curr Opin Pharmacol. 2005;5(1):9–19.
81. Ren K, Bristow LJ, Fong T, Lorrain DS, Morse AC. The cannabinoid CB1 inverse agonist AM251 attenuates alcohol self-administration and the alcohol deprivation effect. Soc Neurosci. 2004. Program No. 117.11.
82. Gessa GL, Serra S, Vacca G, Carai MAM, Colombo G. Suppressing effect of the cannabinoid CB1 receptor antagonist, SR147778, on alcohol intake and motivational properties of alcohol in alcohol-preferring sP rats. Alcohol Alcohol. 2005;40(1):46–53.
83. Lallemand F, De Witte P. SR147778, a CB1 cannabinoid receptor antagonist, suppresses ethanol preference in chronically alcoholized Wistar rats. Alcohol. 2006;39(3):125–34.
84. Femenia T, Garcia-Gutierrez MS, Manzanares J. CB1 receptor blockade decreases ethanol intake and associated neurochemical changes in fawn-hooded rats. Alcohol Clin Exp Res. 2010;34(1):131–41.
85. de Bruin NM, Lange JH, Kruse CG, Herremans AH, Schoffelmeer AN, van Drimmelen M, De Vries TJ. SLV330, a cannabinoid CB(1) receptor antagonist, attenuates ethanol and nicotine seeking and improves inhibitory response control in rats. Behav Brain Res. 2011;217(2):408–15.
86. Dean RL, Eyerman D, Todtenkopf MS, Turncliff RZ, Bidlack JM, Deaver DR. Effects of oral loperamide on efficacy of naltrexone, baclofen and AM-251 in blocking ethanol self-administration in rats. Pharmacol Biochem Behav. 2012;100(3):530–7.
87. Carai MAM, Lobina C, Gessa GL, Colombo G. Cannabinoid receptor antagonists: a perspective. In: Spanagel R, Mann KF, editors. Drugs for relapse prevention of alcoholism. Basel: Birkhauser; 2005. pp. 181–7.
88. Colombo G, Serra S, Vacca G, Carai MAM, Gessa GL. Effect of the combination of naltrexone and baclofen, on acquisition of alcohol drinking behavior in alcohol-preferring rats. Drug Alcohol Depend. 2005;77(1):87–91.

89. Manzanares J, Ortiz S, Oliva JM, Perez-Rial S, Palomo T. Interactions between cannabinoid and opioid receptor systems in the mediation of ethanol effects. Alcohol Alcohol. 2005;40(1):25–34.
90. Devane WA, Hanus L, Breuer A, Pertwee RG, Stevenson LA, Griffin G, Gibson D, Mandelbaum A, Etinger A, Mechoulam R. Isolation and structure of a brain constituent that binds to the cannabinoid receptor. Science. 1992;258(5090):1946–9.
91. Mechoulam R, Ben-Shabat S, Hanus L, Ligumsky M, Kaminski NE, Schatz AR, GopherA, Almog S, Martin BR, Compton DR. Identification of an endogenous 2-monoglyceride, present in canine gut, that binds to cannabinoid receptors. Biochem Pharmacol. 1995;50(1):83–90.
92. Sugiura T, Kondo S, Sukagawa A, Nakane S, Shinoda A, Itoh K, Yamashita A. WakuK. 2-Arachidonoylglycerol: a possible endogenous cannabinoid receptor ligand in brain. Biochem Biophys Res Commun. 1995;215(1):89–97.
93. Bisogno T, Melck D, Bobrov M, Gretskaya NM, Bezuglov VV, De Petrocellis L, Di Marzo V. N-acyl-dopamines: novel synthetic CB(1) cannabinoid-receptor ligands and inhibitors of anandamide inactivation with cannabimimetic activity in vitro and in vivo. Biochem J. 2000;351(3):817–24.
94. Hanus L, Gopher A, Almog S, Mechoulam R. Two new unsaturated fatty acid ethanolamides in brain that bind to the cannabinoid receptor. J Med Chem. 1993;36(20):3032–4.
95. Hanus L, Abu-Lafi S, Fride E, Breuer A, Vogel Z, Shalev DE, Kustanovich I. Mechoulam R. 2-arachidonyl glyceryl ether. an endogenous agonist of the cannabinoid CB1 receptor. Proc Natl Acad Sci USA. 2001;98(7):3662–5.
96. Pertwee R, Griffin G, Hanus L, Mechoulam R. Effects of two endogenous fatty acid ethanolamides on mouse vasa deferentia. Eur J Pharmacol. 1994;259(2):115–20.
97. Porter AC, Sauer JM, Knierman MD, Becker GW, Berna MJ, Bao J, Nomikos GG, Carter P, Bymaster FP, Leese AB, Felder CC. Characterization of a novel endocannabinoid, virodhamine, with antagonist activity at the CB1 receptor. J Pharmacol Exp Ther. 2002;301(3):1020–4.
98. Re G, Barbero R, Miolo A, Di Marzo V. Palmitoylethanolamide, endocannabinoids and related cannabimimetic compounds in protection against tissue inflammation and pain: potential use in companion animals. Vet J. 2007;173(1):21–30.
99. Rodríguez de Fonseca F, Navarro M, Gomez R, Escuredo L, Nava F, Fu J, Murillo -RodríguezE, Giuffrida A, LoVerme J, Gaetani S, Kathuria S, Gall C, Piomelli D. An anorexic lipid mediator regulated by feeding. Nature. 2001;414(6860):209–12.
100. Cadas H, Gaillet S, Beltramo M, Venance L, Piomelli D. Biosynthesis of an endogenous cannabinoid precursor in neurons and its control by calcium and cAMP. J Neurosci. 1996;16(12):3934–42.
101. Piomelli D. The molecular logic of endocannabinoid signalling. Nat Rev Neurosci. 2003;4(11):873–84.
102. Okamoto Y, Morishita J, Tsuboi K, Tonai T, Ueda N. Molecular characterization of a phospholipase D generating anandamide and its congeners. J Biol Chem. 2004;279(7):5298–305.
103. Liu J, Wang L, Harvey-White J, Huang BX, Kim HY, Luquet S, Palmiter RD, Krystal G, Rai R, Mahadevan A, Razdan RK, Kunos G. Multiple pathways involved in the biosynthesis of anandamide. Neuropharmacology. 2008;54(1):1–7.
104. Simon GM, Cravatt BF. Endocannabinoid biosynthesis proceeding through glycerophospho-N-acyl ethanolamine and a role for alpha/beta-hydrolase 4 in this pathway. J Biol Chem. 2006;281(36):26465–72.
105. Simon GM, Cravatt BF. Anandamide biosynthesis catalyzed by the phosphodiesterase GDE1 and detection of glycerophospho-N-acyl ethanolamine precursors in mouse brain. J Biol Chem. 2008;283(14):9341–9.
106. Stella N, Schweitzer P, Piomelli D. A second endogenous cannabinoid that modulates long-term potentiation. Nature. 1997;388(6644):773–8.

107. Beltramo M, Stella N, Calignano A, Lin SY, Makriyannis A, Piomelli D. Functional role of high-affinity anandamide transport, as revealed by selective inhibition. Science. 1997;277(5329):1094–7.
108. Hillard CJ, Jarrahian A. The movement of N-arachidonoylethanolamine (anandamide) across cellular membranes. Chem Phys Lipids. 2000;108(1–2):123–34.
109. McKinney MK, Cravatt BF. Structure and function of fatty acid amide hydrolase. Ann Rev Biochem. 2005;74:411–32.
110. Fezza F, De Simone C, Amadio D, Maccarrone M. Fatty acid amide hydrolase: a gate-keeper of the endocannabinoid system. Subcell Biochem. 2008;49:101–32.
111. Blankman JL, Simon GM, Cravatt BF. A comprehensive profile of brain enzymes that hydrolyze the endocannabinoid 2-arachidonoylglycerol. Chem Biol. 2007;14(12):1347–56.
112. Marrs WR, Blankman JL, Horne EA, Thomazeau A, Lin YH, Coy J, Bodor AL, Muccioli GG, Hu SS, Woodruff G, Fung S, Lafourcade M, Alexander JP, Long JZ, Li W, Xu C, Möller T, Mackie K, Manzoni OJ, Cravatt BF, Stella N. The serine hydrolase ABHD6 controls the accumulation and efficacy of 2-AG at cannabinoid receptors. Nat Neurosci. 2010;13(8):951–7.
113. Gulyas AI, Cravatt BF, Bracey MH, Dinh TP, Piomelli D, Boscia F, Freund TF. Segregation of two endocannabinoid-hydrolyzing enzymes into pre- and postsynaptic compartments in the rat hippocampus, cerebellum and amygdala. Eur J Neurosci. 2004;20(2):441–58.
114. Ludanyi A, Hu SS, Yamazaki M, Tanimura A, Piomelli D, Watanabe M, Kano M, Sakimura K, Maglóczky Z, Mackie K, Freund TF, Katona I. Complementary synaptic distribution of enzymes responsible for synthesis and inactivation of the endocannabinoid 2-arachidonoylglycerol in the human hippocampus. Neuroscience. 2011;174:50–63.
115. Straiker A, Hu SS, Long JZ, Arnold A, Wager-Miller J, Cravatt BF, Mackie K. Monoacylglycerol lipase limits the duration of endocannabinoid-mediated depolarization-induced suppression of excitation in autaptic hippocampal neurons. Mol Pharmacol. 2009;76(6):1220–7.
116. Kim J, Alger BE. Reduction in endocannabinoid tone is a homeostatic mechanism for specific inhibitory synapses. Nat Neurosci. 2010;13(5):592–600.
117. Batkai S, Jarai Z, Wagner JA, Goparaju SK, Varga K, Liu J, Wang L, Mirshahi F, Khanolkar AD, Makriyannis A, Urbaschek R, Garcia N Jr, Sanyal AJ, Kunos G. Endocannabinoids acting at vascular CB1 receptors mediate the vasodilated state in advanced liver cirrhosis. Nat Med. 2001;7(7):827–32.
118. Herkenham M. Characterization and localization of cannabinoid receptors in brain: an in vitro technique using slide-mounted tissue sections. NIDA Res Monogr. 1991;112:129–45.
119. Herkenham M, Groen BG, Lynn AB, De Costa BR, Richfield EK. Neuronal localization of cannabinoid receptors and second messengers in mutant mouse cerebellum. Brain Res. 1991;552(2):301–10.
120. Herkenham M, Lynn AB, de Costa BR, Richfield EK. Neuronal localization of cannabinoid receptors in the basal ganglia of the rat. Brain Res. 1991;547(2):267–74.
121. Herkenham M, Lynn AB, Johnson MR, Melvin LS, de Costa BR, Rice KC. Characterization and localization of cannabinoid receptors in rat brain: a quantitative in vitro autoradiographic study. J Neurosci. 1991;11(2):563–83.
122. Herkenham M, Lynn AB, Little MD, Johnson MR, Melvin LS, de Costa BR, Rice KC. Cannabinoid receptor localization in brain. Proc Natl Acad Sci USA. 1990;87(5):1932–6.
123. Howlett AC, Bidaut-Russell M, Devane WA, Melvin LS, Johnson MR, Herkenham M. The cannabinoid receptor: biochemical, anatomical and behavioral characterization. Trends Neurosci. 1990;13(10):420–3.
124. Galiègue S, Mary S, Marchand J, Dussossoy D, Carrière D, Carayon P, Bouaboula M, Shire D, Le Fur G, Casellas P. Expression of central and peripheral cannabinoid receptors in human immune tissues and leukocyte subpopulations. Eur J Biochem. 1995;232(1):54–61.
125. Munro S, Thomas KL, Abu-Shaar M. Molecular characterization of a peripheral receptor for cannabinoids. Nature. 1993;365(6441):61–5.
126. Cabral GA, Raborn ES, Griffin L, Dennis J, Marciano-Cabral F. CB2 receptors in the brain: role in central immune function. Br J Pharmacol. 2008;153(2):240–51.

127. Maresz K, Pryce G, Ponomarev ED, Marsicano G, Croxford JL, Shriver LP, Ledent C, Cheng XD, Carrier EJ, Mann MK, Giovannoni G, Pertwee RG, Yamamura T, Buckley NE, Hillard CJ, Lutz B, Baker D, Dittel BN. Direct suppression of CNS autoimmune inflammation via the cannabinoid receptor CB1 on neurons and CB2 on autoreactive T cells. Nat Med. 2007;13(4):492–7.
128. Stella N. Cannabinoid signaling in glial cells. Glia. 2004;48(4):267–77.
129. Ashton JC, Glass M. The cannabinoid CB2 receptor as a target for inflammation-dependent neurodegeneration. Curr Neuropharmacol. 2007;5(2):73–80.
130. Onaivi ES, Ishiguro H, Gong JP, Patel S, Meozzi PA, Myers L, Perchuk A, MoraZ, Tagliaferro PA, Gardner E, Brusco A, Akinshola BE, Hope B, Lujilde J, Inada T, Iwasaki S, Macharia D, Teasenfitz L, Arinami T, Uhl GR. Brain neuronal CB2 cannabinoid receptors in drug abuse and depression: from mice to human subjects. PloS One. 2008;3(2):e1640.
131. Onaivi ES, Ishiguro H, Gong JP, Patel S, Perchuk A, Meozzi PA, Myers L, MoraZ, Tagliaferro P, Gardner E, Brusco A, Akinshola BE, Liu QR, Hope B, Iwasaki S, Arinami T, Teasenfitz L, Uhl GR. Discovery of the presence and functional expression of cannabinoid CB2 receptors in brain. Ann New York Acad Sci. 2006;1074:514–36.
132. Van Sickle MD, Duncan M, Kingsley PJ, Mouihate A, Urbani P, Mackie K, StellaN, Makriyannis A, Piomelli D, Davison JS, Marnett LJ, Di Marzo V, Pittman QJ, Patel KD, Sharkey KA. Identification and functional characterization of brainstem cannabinoid CB2 receptors. Science. 2005;310(5746):329–32.
133. Aracil-Fernandez A, Trigo JM, Garcia-Gutierrez MS, Ortega-Alvaro A, Ternianov A, Navarro D, Robledo P, Berbel P, Maldonado R, Manzanares J. Decreased cocaine motor sensitization and self-administration in mice overexpressing cannabinoid CB(2) receptors. Neuropsychopharmacology. 2012;37(7):1749–63.
134. Mackie K, Hille B. Cannabinoids inhibit N-type calcium channels in neuroblastoma-glioma cells. Proc Natl Acad Sci USA. 1992;89(9):3825–9.
135. Mackie K, Lai Y, Westenbroek R, Mitchell R. Cannabinoids activate an inwardly rectifying potassium conductance and inhibit Q-type calcium currents in AtT20 cells transfected with rat brain cannabinoid receptor. J Neurosci. 1995;15(10):6552–61.
136. Twitchell W, Brown S, Mackie K. Cannabinoids inhibit N- and P/Q-type calcium channels in cultured rat hippocampal neurons. J Neurophysiol. 1997;78(1):43–50.
137. Pertwee RG. Receptors and channels targeted by synthetic cannabinoid receptor agonists and antagonists. Curr Med Chem. 2010;17(14):1360–81.
138. Lauckner JE, Jensen JB, Chen HY, Lu HC, Hille B, Mackie K. GPR55 is a cannabinoid receptor that increases intracellular calcium and inhibits M current. Proc Natl Acad Sci USA. 2008;105(7):2699–704.
139. Overton HA, Babbs AJ, Doel SM, Fyfe MC, Gardner LS, Griffin G, Jackson HC, Procter MJ, Rasamison CM, Tang-Christensen M, Widdowson PS, Williams GM, Reynet C. Deorphanization of a G protein-coupled receptor for oleoylethanolamide and its use in the discovery of small-molecule hypophagic agents. Cell Metab. 2006;3(3):167–75.
140. Ryberg E, Larsson N, Sjogren S, Hjorth S, Hermansson NO, Leonova J, ElebringT, Nilsson K, Drmota T, Greasley PJ. The orphan receptor GPR55 is a novel cannabinoid receptor. Br J Pharmacol. 2007;152(7):1092–101.
141. Sharir H, Abood ME. Pharmacological characterization of GPR55, a putative cannabinoid receptor. Pharmacol Ther. 2010;126(3):301–13.
142. Zygmunt PM, Petersson J, Andersson DA, Chuang H, Sorgard M, Di Marzo V, JuliusD, Högestätt ED. Vanilloid receptors on sensory nerves mediate the vasodilator action of anandamide. Nature. 1999;400(6743):452–7.
143. Basavarajappa BS, Cooper TB, Hungund BL. Chronic ethanol administration downregulates cannabinoid receptors in mouse brain synaptic plasma membrane. Brain Res. 1998;793(1–2):212–8.
144. Basavarajappa BS, Hungund BL. Chronic ethanol increases the cannabinoid receptor agonist anandamide and its precursor N-arachidonoylphosphatidylethanolamine in SK-N-SH cells. J Neurochem. 1999;72(2):522–8.

145. Vinod KY, Yalamanchili R, Xie S, Cooper TB, Hungund BL. Effect of chronic ethanol exposure and its withdrawal on the endocannabinoid system. Neurochem Int. 2006;49(6):619–25.
146. Moranta D, Esteban S, Garcia-Sevilla JA. Ethanol desensitizes cannabinoid CB1 receptors modulating monoamine synthesis in the rat brain in vivo. Neurosci Lett. 2006;392(1–2):58–61.
147. Mitrirattanakul S, Lopez-Valdes HE, Liang J, Matsuka Y, Mackie K, Faull KF, Spigelman I. Bidirectional alterations of hippocampal cannabinoid 1 receptors and their endogenous ligands in a rat model of alcohol withdrawal and dependence. Alcohol Clin Exp Res. 2007;31(5):855–67.
148. Ceccarini J, Casteels C, Koole M, Bormans G, Van Laere K. Transient changes in the endocannabinoid system after acute and chronic ethanol exposure and abstinence in the rat: a combined PET and microdialysis study. Eur J Nucl Med Mol Imaging. 2013;40(10):1582–94.
149. Vinod KY, Arango V, Xie S, Kassir SA, Mann JJ, Cooper TB, Hungund BL. Elevated levels of endocannabinoids and CB1 receptor-mediated G-protein signaling in the prefrontal cortex of alcoholic suicide victims. Biol Psychiatry. 2005;57(5):480–6.
150. Vinod KY, Kassir SA, Hungund BL, Cooper TB, Mann JJ, Arango V. Selective alterations of the CB1 receptors and the fatty acid amide hydrolase in the ventral striatum of alcoholics and suicides. J Psychiatr Res. 2010;44(9):591–7.
151. Hirvonen J, Zanotti-Fregonara P, Umhau JC, George DT, Rallis-Frutos D, Lyoo CH, Li CT, Hines CS, Sun H, Terry GE, Morse C, Zoghbi SS, Pike VW, Innis RB, Heilig M. Reduced cannabinoid CB1 receptor binding in alcohol dependence measured with positron emission tomography. Mol Psychiatry. 2013;18(8):916–21.
152. Ceccarini J, Hompes T, Verhaeghen A, Casteels C, Peuskens H, Bormans G, Claes S, Van Laere K. Changes in cerebral CB1 receptor availability after acute and chronic alcohol abuse and monitored abstinence. J Neurosci. 2014;34(8):2822–31.
153. Neumeister A, Normandin MD, Murrough JW, Henry S, Bailey CR, Luckenbaugh DA, Tuit K, Zheng MQ, Galatzer-Levy IR, Sinha R, Carson RE, Potenza MN, Huang Y. Positron emission tomography shows elevated cannabinoid CB1 receptor binding in men with alcohol dependence. Alcohol Clin Exp Res. 2012;36(12):2104–9.
154. Agudelo M, Yndart A, Morrison M, Figueroa G, Munoz K, Samikkannu T, Nair MP. Differential expression and functional role of cannabinoid genes in alcohol users. Drug Alcohol Depend. 2013;133(2):789–93.
155. Oviedo A, Glowa J, Herkenham M. Chronic cannabinoid administration alters cannabinoid receptor binding in rat brain: a quantitative autoradiographic study. Brain Res. 1993;616(1–2):293–302.
156. Romero J, Berrendero F, Garcia-Gil L, Ramos JA, Fernandez-Ruiz JJ. Cannabinoid receptor and WIN-55,212-2-stimulated [35S]GTP gamma S binding and cannabinoid receptor mRNA levels in the basal ganglia and the cerebellum of adult male rats chronically exposed to delta9-tetrahydrocannabinol. J Mol Neurosci. 1998;11(2):109–19.
157. Romero J, Garcia L, Fernandez-Ruiz JJ, Cebeira M, Ramos JA. Changes in rat brain cannabinoid binding sites after acute or chronic exposure to their endogenous agonist, anandamide, or to delta9-tetrahydrocannabinol. Pharmacol Biochem Behav. 1995;51(4):731–7.
158. Serrano A, Rivera P, Pavon FJ, Decara J, Suarez J, Rodríguez de Fonseca F, Parsons LH. Differential effects of single versus repeated alcohol withdrawal on the expression of endocannabinoid system-related genes in the rat amygdala. Alcohol Clin Exp Res. 2012;36(6):984–94.
159. Basavarajappa BS, Saito M, Cooper TB, Hungund BL. Stimulation of cannabinoid receptor agonist 2-arachidonylglycerol by chronic ethanol and its modulation by specific neuromodulators in cerebellar granule neurons. Biochim Biophys Acta. 2000;1535(1):78–86.
160. Basavarajappa BS, Saito M, Cooper TB, Hungund BL. Chronic ethanol inhibits the anandamide transport and increases extracellular anandamide levels in cerebellar granule neurons. Eur J Pharmacol. 2003;466(1–2):73–83.

161. Gonzalez S, Cascio MG, Fernandez-Ruiz J, Fezza F, Di Marzo V, Ramos JA. Changes in endocannabinoid contents in the brain of rats chronically exposed to nicotine, ethanol or cocaine. Brain Res. 2002;954(1):73–81.
162. Gonzalez S, Valenti M, de Miguel R, Fezza F, Fernandez-Ruiz J, Di Marzo V, Ramos JA. Changes in endocannabinoid contents in reward-related brain regions of alcohol-exposed rats, and their possible relevance to alcohol relapse. Br J Pharmacol. 2004;143(4):455–64.
163. Malinen H, Lehtonen M, Hyytiä P. Modulation of brain endocannabinoid levels by voluntary alcohol consumption in alcohol-preferring AA rats. Alcohol Clin Exp Res. 2009;33(10):1711–20.
164. Rubio M, McHugh D, Fernandez-Ruiz J, Bradshaw H, Walker JM. Short-term exposure to alcohol in rats affects brain levels of anandamide, other N-acylethanolamines and 2-arachidonoyl-glycerol. Neurosci Lett. 2007;421(3):270–4.
165. Buczynski MW, Parsons LH. Quantification of brain endocannabinoid levels: methods, interpretations and pitfalls. Br J Pharmacol. 2010;160(3):423–42.
166. Ferrer B, Bermudez-Silva FJ, Bilbao A, Alvarez-Jaimes L, Sanchez-Vera I, Giuffrida A, Serrano A, Baixeras E, Khaturia S, Navarro M, Parsons LH, Piomelli D, Rodríguez de Fonseca F. Regulation of brain anandamide by acute administration of ethanol. Biochem J. 2007;404(1):97–104.
167. Alvarez-Jaimes L, Stouffer DG, Parsons LH. Chronic ethanol treatment potentiates ethanol-induced increases in interstitial nucleus accumbens endocannabinoid levels in rats. J Neurochem. 2009;111(1):37–48.
168. Kreitzer AC, Regehr WG. Retrograde inhibition of presynaptic calcium influx by endogenous cannabinoids at excitatory synapses onto Purkinje cells. Neuron. 2001;29(3):717–27.
169. Maejima T, Hashimoto K, Yoshida T, Aiba A, Kano M. Presynaptic inhibition caused by retrograde signal from metabotropic glutamate to cannabinoid receptors. Neuron. 2001;31(3):463–75.
170. Ohno-Shosaku T, Maejima T, Kano M. Endogenous cannabinoids mediate retrograde signals from depolarized postsynaptic neurons to presynaptic terminals. Neuron. 2001;29(3):729–38.
171. Wilson RI, Nicoll RA. Endocannabinoid signaling in the brain. Science. 2002;296(5568):678–82.
172. Alger BE. Retrograde signaling in the regulation of synaptic transmission: focus on endocannabinoids. Prog Neurobiol. 2002;68(4):247–86.
173. Diana MA, Marty A. Endocannabinoid-mediated short-term synaptic plasticity: depolarization-induced suppression of inhibition (DSI) and depolarization-induced suppression of excitation (DSE). Br J Pharmacol. 2004;142(1):9–19.
174. Chevaleyre V, Takahashi KA, Castillo PE. Endocannabinoid-mediated synaptic plasticity in the CNS. Ann Rev Neurosci. 2006;29:37–76.
175. Marsicano G, Lutz B. Neuromodulatory functions of the endocannabinoid system. J Endocrinol Invest. 2006;29(3 Suppl):27–46.
176. Lovinger DM. Presynaptic modulation by endocannabinoids. Handb Exp Pharmacol. 2008;184:435–77.
177. Kano M, Ohno-Shosaku T, Hashimotodani Y, Uchigashima M, Watanabe M. Endocannabinoid-mediated control of synaptic transmission. Physiol Rev. 2009;89(1):309–80.
178. Everitt BJ, Dickinson A, Robbins TW. The neuropsychological basis of addictive behaviour. Brain Res Rev. 2001;36(2–3):129–38.
179. Hyman SE, Malenka RC, Nestler EJ. Neural mechanisms of addiction: the role of reward-related learning and memory. Ann Rev Neurosci. 2006;29:565–98.
180. Kauer JA, Malenka RC. Synaptic plasticity and addiction. Nat Rev Neurosci. 2007;8(11):844–58.
181. Kalivas PW, O'Brien C. Drug addiction as a pathology of staged neuroplasticity. Neuropsychopharmacology. 2008;33(1):166–80.
182. Torregrossa MM, Corlett PR, Taylor JR. Aberrant learning and memory in addiction. Neurobiol Learn Mem. 2011;96(4):609–23.

183. Adermark L, Lovinger DM. Frequency-dependent inversion of net striatal output by endocannabinoid-dependent plasticity at different synaptic inputs. J Neurosci. 2009 Feb 4;29(5):1375–80.
184. Clarke RB, Adermark L. Acute ethanol treatment prevents endocannabinoid-mediated long-lasting disinhibition of striatal output. Neuropharmacology. 2010;58(4–5):799–805.
185. Adermark L, Jonsson S, Ericson M, Söderpalm B. Intermittent ethanol consumption depresses endocannabinoid-signaling in the dorsolateral striatum of rat. Neuropharmacology. 2011;61(7):1160–5.
186. Volkow ND, Fowler JS, Wang GJ, Swanson JM, Telang F. Dopamine in drug abuse and addiction: results of imaging studies and treatment implications. Arch Neurol. 2007;64(11):1575–9.
187. Yin HH, Ostlund SB, Balleine BW. Reward-guided learning beyond dopamine in the nucleus accumbens: the integrative functions of cortico-basal ganglia networks. Eur J Neurosci. 2008;28(8):1437–48.
188. DePoy L, Daut R, Brigman JL, MacPherson K, Crowley N, Gunduz-Cinar O, Pickens CL, Cinar R, Saksida LM, Kunos G, Lovinger DM, Bussey TJ, Camp MC, Holmes A. Chronic alcohol produces neuroadaptations to prime dorsal striatal learning. Proc Natl Acad Sci USA. 2013;110(36):14783–8.
189. Basavarajappa BS, Yalamanchili R, Cravatt BF, Cooper TB, Hungund BL. Increased ethanol consumption and preference and decreased ethanol sensitivity in female FAAH knockout mice. Neuropharmacology. 2006;50(7):834–44.
190. Blednov YA, Cravatt BF, Boehm SL, 2nd, Walker D, Harris RA. Role of endocannabinoids in alcohol consumption and intoxication: studies of mice lacking fatty acid amide hydrolase. Neuropsychopharmacology. 2007;32(7):1570–82.
191. Vinod KY, Sanguino E, Yalamanchili R, Manzanares J, Hungund BL. Manipulation of fatty acid amide hydrolase functional activity alters sensitivity and dependence to ethanol. J Neurochem. 2008;104(1):233–43.
192. Cippitelli A, Bilbao A, Gorriti MA, Navarro M, Massi M, Piomelli D, Ciccocioppo R, Rodríguez de Fonseca F. The anandamide transport inhibitor AM404 reduces ethanol self-administration. Eur J Neurosci. 2007;26(2):476–86.
193. Niphakis MJ, Cognetta AB 3rd, Chang JW, Buczynski MW, Parsons LH, Byrne F, Burston JJ, Chapman V, Cravatt BF. Evaluation of NHS carbamates as a potent and selective class of endocannabinoid hydrolase inhibitors. ACS Chem Neurosci. 2013;4(9):1322–32.
194. Chang JW, Niphakis MJ, Lum KM, Cognetta AB 3rd, Wang C, Matthews ML, NiessenS, Buczynski MW, Parsons LH, Cravatt BF. Highly selective inhibitors of monoacylglycerol lipase bearing a reactive group that is bioisosteric with endocannabinoid substrates. Chem Biol. 2012;19(5):579–88.
195. Cippitelli A, Bilbao A, Hansson AC, del Arco I, Sommer W, Heilig M, Massi M, Bermúdez-Silva FJ, Navarro M, Ciccocioppo R, Rodríguez de Fonseca F. Cannabinoid CB1 receptor antagonism reduces conditioned reinstatement of ethanol-seeking behavior in rats. Eur J Neurosci. 2005;21(8):2243–51.
196. Dawson E, Gill M, Curtis D, Castle D, Hunt N, Murray R, Powell J. Genetic association between alleles of pancreatic phospholipase A2 gene and bipolar affective disorder. Psychiatr Genet. 1995;5(4):177–80.
197. Johnson JP, Muhleman D, MacMurray J, Gade R, Verde R, Ask M, Kelley J, ComingsDE. Association between the cannabinoid receptor gene (CNR1) and the P300 event-related potential. Mol Psychiatry. 1997;2(2):169–71.
198. Ehlers CL, Slutske WS, Lind PA, Wilhelmsen KC. Association between single nucleotide polymorphisms in the cannabinoid receptor gene (CNR1) and impulsivity in southwest California Indians. Twin Res Hum Genet. 2007;10(6):805–11.
199. Ponce G, Hoenicka J, Rubio G, Ampuero I, Jiménez-Arriero MA, Rodríguez-JiménezR, Palomo T, Ramos JA. Association between cannabinoid receptor gene (CNR1) and childhood attention deficit/hyperactivity disorder in Spanish male alcoholic patients. Mol Psychiatry. 2003;8(5):466–7.

200. Comings DE, Muhleman D, Gade R, Johnson P, Verde R, Saucier G, MacMurray J. Cannabinoid receptor gene (CNR1): association with i.v. drug use. Mol Psychiatry. 1997;2(2):161–8.
201. Ballon N, Leroy S, Roy C, Bourdel MC, Charles-Nicolas A, Krebs MO, Poirier MF. (AAT)n repeat in the cannabinoid receptor gene (CNR1): association with cocaine addiction in an African-Caribbean population. Pharmacogenomics J. 2006;6(2):126–30.
202. Li T, Liu X, Zhu ZH, Zhao J, Hu X, Ball DM, Sham PC, Collier DA. No association between (AAT)n repeats in the cannabinoid receptor gene (CNR1) and heroin abuse in a Chinese population. Mol Psychiatry. 2000;5(2):128–30.
203. Covault J, Gelernter J, Kranzler H. Association study of cannabinoid receptor gene (CNR1) alleles and drug dependence. Mol Psychiatry. 2001;6(5):501–2.
204. Heller D, Schneider U, Seifert J, Cimander KF, Stuhrmann M. The cannabinoid receptor gene (CNR1) is not affected in German i.v. drug users. Addict Biol. 2001;6(2):183–7.
205. Zhang PW, Ishiguro H, Ohtsuki T, Hess J, Carillo F, Walther D, Onaivi ES, Arinami T, Uhl GR. Human cannabinoid receptor 1: 5' exons, candidate regulatory regions, polymorphisms, haplotypes and association with polysubstance abuse. Mol Psychiatry. 2004;9(10):916–31.
206. Hoehe MR, Caenazzo L, Martinez MM, Hsieh WT, Modi WS, Gershon ES, Bonner TI. Genetic and physical mapping of the human cannabinoid receptor gene to chromosome 6q14-q15. New Biol. 1991;3(9):880–5.
207. Gadzicki D, Muller-Vahl K, Stuhrmann M. A frequent polymorphism in the coding exon of the human cannabinoid receptor (CNR1) gene. Mol Cell Probes. 1999;13(4):321–3.
208. Schmidt LG, Samochowiec J, Finckh U, Fiszer-Piosik E, Horodnicki J, Wendel B, Rommelspacher H, Hoehe MR. Association of a CB1 cannabinoid receptor gene (CNR1) polymorphism with severe alcohol dependence. Drug Alcohol Depend. 2002;65(3):221–4.
209. Preuss UW, Koller G, Zill P, Bondy B, Soyka M. Alcoholism-related phenotypes and genetic variants of the CB1 receptor. Eur Arch Psychiatry Clin Neurosci. 2003;253(6):275–80.
210. Zuo L, Kranzler HR, Luo X, Covault J, Gelernter J. CNR1 variation modulates risk for drug and alcohol dependence. Biol Psychiatry. 2007;62(6):616–26.
211. Zuo L, Kranzler HR, Luo X, Yang BZ, Weiss R, Brady K, Poling J, Farrer L, Gelernter J. Interaction between two independent CNR1 variants increases risk for cocaine dependence in European Americans: a replication study in family-based sample and population-based sample. Neuropsychopharmacology. 2009;34(6):1504–13.
212. Herman AI, Kranzler HR, Cubells JF, Gelernter J, Covault J. Association study of the CNR1 gene exon 3 alternative promoter region polymorphisms and substance dependence. Am J Med Genet B Neuropsychiatr Genet. 2006;141B(5):499–503.
213. Marcos M, Pastor I, de la CC, Barrio-Real L, Laso FJ, Gonzalez-Sarmiento R. Cannabinoid receptor 1 gene is associated with alcohol dependence. Alcohol Clin Exp Res. 2012;36(2):267–71.
214. Hutchison KE, Haughey H, Niculescu M, Schacht J, Kaiser A, Stitzel J, HortonWJ, Filbey F. The incentive salience of alcohol: translating the effects of genetic variant in CNR1. Arch Gen Psychiatry. 2008;65(7):841–50.
215. van den Wildenberg E, Janssen RG, Hutchison KE, van Breukelen GJ, Wiers RW. Polymorphisms of the dopamine D4 receptor gene (DRD4 VNTR) and cannabinoid CB1 receptor gene (CNR1) are not strongly related to cue-reactivity after alcohol exposure. Addict Biol. 2007;12(2):210–20.
216. Merritt LL, Martin BR, Walters C, Lichtman AH, Damaj MI. The endogenous cannabinoid system modulates nicotine reward and dependence. J Pharmacol Exp Ther. 2008;326(2):483–92.
217. Schlosburg JE, Carlson BL, Ramesh D, Abdullah RA, Long JZ, Cravatt BF, Lichtman AH. Inhibitors of endocannabinoid-metabolizing enzymes reduce precipitated withdrawal responses in THC-dependent mice. AAPS J. 2009;11(2):342–52.
218. Chiang KP, Gerber AL, Sipe JC, Cravatt BF. Reduced cellular expression and activity of the P129T mutant of human fatty acid amide hydrolase: evidence for a link between defects in the endocannabinoid system and problem drug use. Hum Mol Genet. 2004;13(18):2113–9.

219. Sipe JC, Chiang K, Gerber AL, Beutler E, Cravatt BF. A missense mutation in human fatty acid amide hydrolase associated with problem drug use. Proc Natl Acad Sci USA. 2002;99(12):8394–9.
220. Flanagan JM, Gerber AL, Cadet JL, Beutler E, Sipe JC. The fatty acid amide hydrolase 385 A/A (P129T) variant: haplotype analysis of an ancient missense mutation and validation of risk for drug addiction. Hum Genet. 2006;120(4):581–8.
221. Iwasaki S, Ishiguro H, Higuchi S, Onaivi ES, Arinami T. Association study between alcoholism and endocannabinoid metabolic enzyme genes encoding fatty acid amide hydrolase and monoglyceride lipase in a Japanese population. Psychiatr Genet. 2007;17(4):215–20.
222. Tyndale RF, Payne JI, Gerber AL, Sipe JC. The fatty acid amide hydrolase C385A (P129T) missense variant in cannabis users: studies of drug use and dependence in Caucasians. Am J Med Genet B Neuropsychiatr Genet. 2007;144B(5):660–6.
223. Hariri AR, Gorka A, Hyde LW, Kimak M, Halder I, Ducci F, Ferrell RE, GoldmanD, Manuck SB. Divergent effects of genetic variation in endocannabinoid signaling on human threat- and reward-related brain function. Biol Psychiatry. 2009;66(1):9–16.
224. Glahn DC, Lovallo WR, Fox PT. Reduced amygdala activation in young adults at high risk of alcoholism: studies from the Oklahoma family health patterns project. Biol Psychiatry. 2007;61(11):1306–9.
225. Buhler KM, Huertas E, Echeverry-Alzate V, Gine E, Moltó E, Montoliu L, López-Moreno JA. Risky alcohol consumption in young people is associated with the fatty acid amide hydrolase gene polymorphism C385A and affective rating of drug pictures. Mol Genet Genomics. 2014;289(3):279–89.
226. Soyka M, Koller G, Schmidt P, Lesch OM, Leweke M, Fehr C, Gann H, Mann KF. Cannabinoid receptor 1 blocker rimonabant (SR 141716) for treatment of alcohol dependence: results from a placebo-controlled, double-blind trial. J Clin Psychopharmacol. 2008;28(3):317–24.
227. George DT, Herion DW, Jones CL, Phillips MJ, Hersh J, Hill D, Heilig M, Ramchandani VA, Geyer C, Spero DE, Singley ED, O'Malley SS, Bishai R, RawlingsRR, Kunos G. Rimonabant (SR141716) has no effect on alcohol self-administration or endocrine measures in nontreatment-seeking heavy alcohol drinkers. Psychopharmacology (Berl). 2010;208(1):37–44.
228. Christensen R, Kristensen PK, Bartels EM, Bliddal H, Astrup A. Efficacy and safety of the weight-loss drug rimonabant: a meta-analysis of randomised trials. The Lancet. 2007;370(9600):1706–13.
229. European Medicines Agency. Public statement on Zimulti (Rimonabant) withdrawal of the marketing authorisation in the European Union. European Medicines Agency EMEA/39560/2009. 2009. http://www.ema.europa.eu/docs/en_GB/document_library/Public_statement/2009/11/WC500012181.pdf.
230. Sidhpura N, Parsons LH. Endocannabinoid-mediated synaptic plasticity and addiction-related behavior. Neuropharmacology. 2011;61(7):1070–87.
231. McKinley RA, Moorhead HH. Alcoholism. Prog Neurol Psychiatry. 1967;22:459–68.
232. Hershon HI. Alcohol withdrawal symptoms and drinking behavior. J Stud Alcohol. 1977;38(5):953–71.
233. De Soto CB, O'Donnell WE, Allred LJ, Lopes CE. Symptomatology in alcoholics at various stages of abstinence. Alcohol Clin Exp Res. 1985;9(6):505–12.
234. Koob GF. Alcoholism: allostasis and beyond. Alcohol Clin Exp Res. 2003;27(2):232–43.
235. Navarro M, Hernandez E, Munoz RM, del Arco I, Villanua MA, Carrera MR, Rodríguez de Fonseca F. Acute administration of the CB1 cannabinoid receptor antagonist SR 141716A induces anxiety-like responses in the rat. Neuroreport. 1997;8(2):491–6.
236. Arevalo C, de Miguel R, Hernandez-Tristan R. Cannabinoid effects on anxiety-related behaviours and hypothalamic neurotransmitters. Pharmacol Biochem Behav. 2001;70(1):123–31.
237. Martin M, Ledent C, Parmentier M, Maldonado R, Valverde O. Involvement of CB1 cannabinoid receptors in emotional behaviour. Psychopharmacology (Berl). 2002;159(4):379–87.

238. Haller J, Bakos N, Szirmay M, Ledent C, Freund TF. The effects of genetic and pharmacological blockade of the CB1 cannabinoid receptor on anxiety. Eur J Neurosci. 2002;16(7):1395–8.
239. Kathuria S, Gaetani S, Fegley D, Valino F, Duranti A, Tontini A, Mor M, TarziaG, La Rana G, Calignano A, Giustino A, Tattoli M, Palmery M, Cuomo V, Piomelli D. Modulation of anxiety through blockade of anandamide hydrolysis. Nat Med. 2003;9(1):76–81.
240. Uriguen L, Perez-Rial S, Ledent C, Palomo T, Manzanares J. Impaired action of anxiolytic drugs in mice deficient in cannabinoid CB1 receptors. Neuropharmacology. 2004;46(7):966–73.
241. Bortolato M, Campolongo P, Mangieri RA, Scattoni ML, Frau R, Trezza V, La RanaG, Russo R, Calignano A, Gessa GL, Cuomo V, Piomelli D. Anxiolytic-like properties of the anandamide transport inhibitor AM404. Neuropsychopharmacology. 2006;31(12):2652–9.
242. Rutkowska M, Jamontt J, Gliniak H. Effects of cannabinoids on the anxiety-like response in mice. Pharmacol Rep. 2006;58(2):200–6.
243. Moreira FA, Kaiser N, Monory K, Lutz B. Reduced anxiety-like behaviour induced by genetic and pharmacological inhibition of the endocannabinoid-degrading enzyme fatty acid amide hydrolase (FAAH) is mediated by CB1 receptors. Neuropharmacology. 2008;54(1):141–50.
244. Annis HM, Sklar SM, Moser AE. Gender in relation to relapse crisis situations, coping, and outcome among treated alcoholics. Addict Behav. 1998;23(1):127–31.

Chapter 15
Cannabinoid-Opioid Interactions

Michael L. Miller, Benjamin Chadwick, Claudia V. Morris, Michael Michaelides and Yasmin L. Hurd

Abstract The growing use of cannabis in western societies is of significant concern especially considering that its consumption has increased in teens who are most at risk for developing addictive disorders and other psychiatric illnesses linked to early cannabis exposure. Unfortunately recent years have also seen a surge in the abuse of opioid drugs, a phenomenon often predated by early cannabis use. Moreover, many abusers find greater reward by combining use of cannabis with opioids such as heroin. These patterns suggest possible synergistic interactions between cannabinoid and opioid systems underlying addiction and related psychiatric disorders. This chapter reviews neurobiological systems relevant to reward, motivation and emotional regulation in which cannabinoids and opioids interact.

Keywords CB1 receptor · THC · Endocannabinoids · Heroin · Cannabis dependence · Co-abuse · Endorphin · Enkephalin · Dynorphin · PENK · PDYN

Introduction

It is estimated that over 7% of adults and 6.5% of teens smoke marijuana in the USA [1, 2], numbers that are unfortunately also now being mirrored in other western countries [3]. European and USA population surveys show that cannabis use in teenagers ranges between 10–40% and is on an increased trajectory [1]. Epidemiological studies consistently report that adolescent cannabis exposure precedes the use of heavy drugs of abuse such as heroin [4–6]. Indeed, over a quarter of individuals who progressed to illicit drug use had previous experience with marijuana, whereas only 2–3% of legal drug users (*e.g.*, alcohol) without previous marijuana

Y. L. Hurd (✉)
Icahn School of Medicine at Mount Sinai, Hess Center for Science and Medicine, New York, NY, USA
e-mail: Yasmin.hurd@mssm.edu

M. L. Miller · B. Chadwick · C. V. Morris · M. Michaelides · Y. L. Hurd
Departments of Neuroscience and Psychiatry, Friedman Brain Institute, Icahn School of Medicine at Mount Sinai, New York, NY, USA

P. Campolongo, L. Fattore (eds.), *Cannabinoid Modulation of Emotion, Memory, and Motivation,* DOI 10.1007/978-1-4939-2294-9_15

experience progressed to illicit drug use. Evident limitations of human studies make it difficult to imply direct 'gateway' associations between cannabis and future use of other drugs of abuse given the many inherent confounds (*e.g.*, culture, family history and peer-pressure, accessibility to drugs) that are not possible to control in humans. However, animal experiments have provided direct evidence for a causal relationship between prior cannabiniod exposure and subsequent opioid use [reviewed in 7–9]. In addition to the suggested sequential sensitization of prior cannabis exposure for subsequent opiate use, many substance abusers tout the potentiation of drug reward by the co-use of cannabinoids and opiates: "I barely enjoy my opiate high until I smoke weed:) It is my favorite drug combo." [10] and "… opiates and ganja are my personal favorite drug-combo as well.. It almost seems impossible how well they go together…" [10].

Neurobiological studies have emphasized the necessity of the cannabinoid system for the rewarding effects of heroin. For example, administration of antagonists at cannabinoid receptors block opioid reward as measured by drug self-administration and conditioned place preference (CPP) models [11]. In addition, animals lacking the cannabinoid receptor 1 gene neither self-administer heroin [12, 13] nor develop morphine-induced CPP [14]. Other converging lines of research from animal studies have documented strong neurobiological interactions between the cannabinoid and opioid systems, in particular with respect to drug reward sensitivity. For example, animal studies have confirmed that prior exposure to Δ9-Tetrahydrocannabinol (THC), marijuana's main psychoactive component, increases heroin self-administration behavior [15–17]. Moreover, exposure to synthetic cannabinoid agonists increases heroin-induced CPP [18, 19]. The following sections provide an overview of the neurobiological systems highly implicated in drug abuse in which functional interactions between cannabinoid and opioid systems have been implicated.

Endocannabinoid System

The direct pharmacological effects of THC are mediated by the endocannabinoid (eCB) system which consists of at least two $G_{i/o}$ protein-coupled receptors, CB_1Rs and CB_2Rs that are targeted by exogenous and endogenous cannabinoids. The CB_1R is one of the most abundant G protein-coupled receptors in the brain with widespread expression in multiple regions [20, 21]. Once thought to be localized primarily to the periphery, recent studies also show a significant contribution of CB_2R to behavior with its expression primarily in glia and lower abundance in certain neuronal populations [22–24]. Given the limited information currently known about the neuronal contribution of the CB_2R subtype particularly in relation to cannabinoid-opioid interactions, this review focuses on the CB_1R.

The endocannabinoid ligands include anandamide (arachidonoyl ethanolamide) (AEA), 2-arachidonoyl glycerol (2-AG), and 2-arachidonyl glyceryl ether [25], which are liberated by the enzymes *N*-acyl-phosphatidyl-ethanolamine-selective

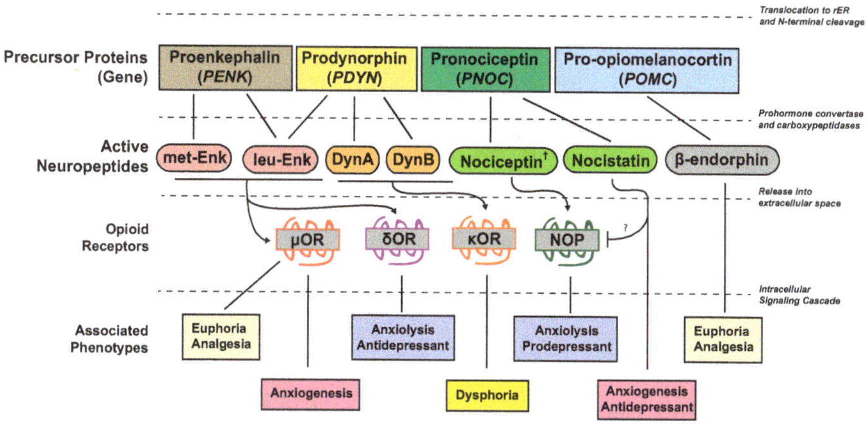

Fig. 15.1 Overview of the endogenous opioid system and functional associations. Precursor proteins, derived from four known genes, give rise to numerous active peptides; only the major peptides are depicted. The active ligands act on at least four Gi-coupled opioid receptors to regulate cognition, mood and behavior. †, also known as orphanin FQ

phospholipase D (NAPE-PLD) [26] and *sn*-1-diacylglycerol lipase α and β [27], respectively. They are subsequently degraded to biologically inactive intermediates by the catabolic enzymes fatty-acid amide hydrolase (FAAH) [28] and monoacylglycerol lipase (MAGL) [29–31]. The consumption of cannabis which produces supraphysiological effects at eCB-targeted receptors usurps the tightly regulated biosynthetic and degradative pathways that normally ensure proper signaling of the eCB system in reward, motivation, cognitive and motor function. Indeed, even enhancing endocannabinoid signaling by inhibition of AEA metabolism has been shown to promote CB_1R-dependent enhancement of morphine reward as measured with CPP [32].

Endogenous Opioid System

The endogenous opioid system is centrally involved in nociception, stress and reward, and mediates these processes through an evolutionarily conserved set of neuropeptides and G-protein coupled receptors (GPCRs). The ligands of this system are neuropeptides derived from precursor proteins encoded by at least four genes, namely proenkephalin (PENK), prodynoprhin (PDYN), pro-opiomelanocortin (POMC) and pronociceptin (PNOC) [33, 34]. These neuropeptides act on opioid receptors, which, similar to CB_1Rs, signal primarily through heterotrimic $G_{i/o}$ proteins. Several ligand-receptor relationships have been characterized, but given the complex post-translation processing of pro-neuropeptides, and the potential for receptor-receptor heterodimers, these relationships are complex (as summarized in Fig. 15.1). However, in large part, enkephalins and β-endorphin exhibit high affinity for mu opioid receptors (μOR) and delta opioid receptors (δOR) and are associated

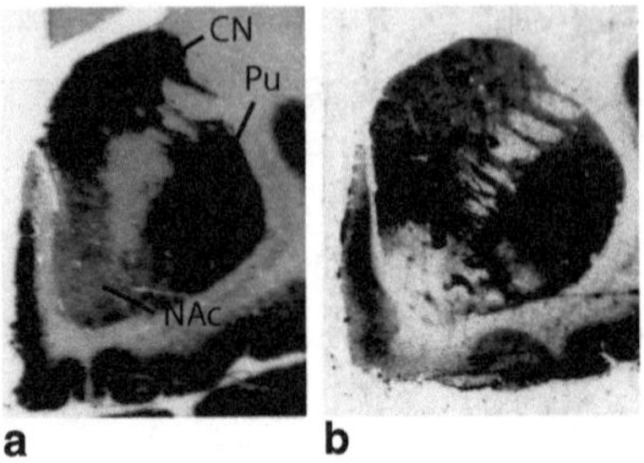

Fig. 15.2 Expression pattern of CNR1 mRNA, which encodes CB1R (**a**), and μOR binding (**b**) in the human striatum. *CN* caudate nucleus, *Pu* putamen, *NAc* nucleus accumbens

with reward [35, 36], whereas dynorphins preferentially bind to kappa opioid receptors (κOR) and are linked to dysphoria and negative mood states [35, 37].

Cannabinoid-Opioid Interactions in the Striatum

Anatomical Interactions

The neurobiological evidence is overwhelmingly in support of a strong interaction between the endocannabinoid and opioid systems, especially in relation to reward, cognition, emotional regulation and addictive behaviors [12, 38, 39]. Most of the data accumulated to date are based on studies of the dorsal and ventral striatum, given the critical role of this brain region in integrating cognitive, sensory and motor function. Additionally, the striatum is characterized by a high abundance of CB_1Rs and μORs (Fig. 15.2). The ventral striatum (nucleus accumbens; NAc) is a key neuroanatomical substrate for reward [40, 41], central for the acquisition of appetitive responses via reinforcement learning and also contributes to motivational control, whereas the dorsal striatum is critical for the formation of stimulus-response contingencies which underlie habits and compulsive behaviors [42]. Both the CB_1R and μOR show a very complementary dorsoventral gradient with their expression being more prominent in the dorsal as compared to ventral striatum (Fig. 15.2). [Note: There are species differences in opioid peptide and receptor expression in the brain. For example, primates have low expression of the μOR in the NAc, whereas rodents have more abundant receptor levels.]

The predominant striatal cells are the medium spiny neurons, which in their vast majority contain the inhibitory amino acid γ-aminobutyric acid (GABA), and are subdivided into two distinct striatal output pathways [43–45] that differentially express PDYN and PENK neuropeptides. Neurons expressing PDYN project predominantly to the substantia nigra pars reticulata and internal globus pallidus and constitute the striatonigral pathway [46, 47], while PENK-expressing neurons project predominantly to the external globus pallidus and constitute the striatopal-

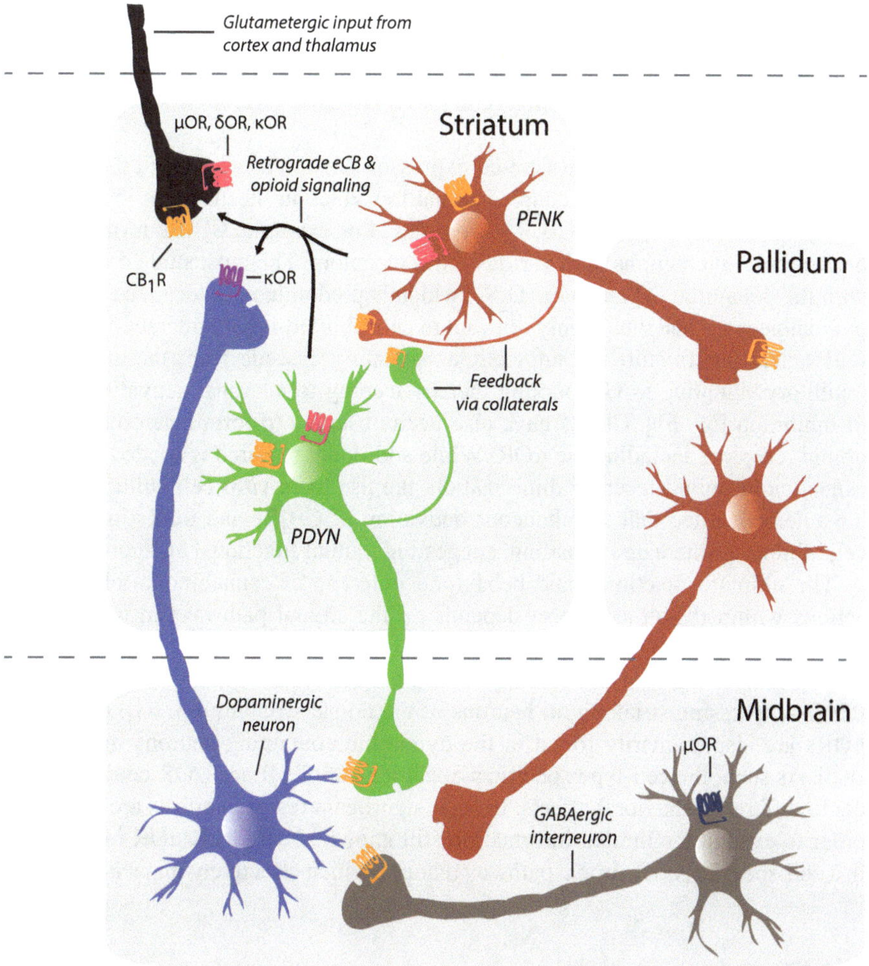

Fig. 15.3 Schematic diagram illustrating CB1R and opioid receptors localization within striatal circuits. The figure highlights the two major striatal output pathways—striatonigral and striatopallidal—along with dopaminergic (*blue neuron*) and glutamatergic (*black neuron*) inputs

lidal pathway [48] (Fig. 15.3). A similar organization exists in the NAc where the ventral striatonigral pathway innervates the ventral tegmental area (VTA) [49]. The striato-nigral/-tegmental circuit constitutes the "Go" (positive choice) pathway that facilitates behavioral responses and reward sensitivity, whereas the striatopallidal circuit constitutes the "No-Go" (inhibitory control) pathway involved in suppressing inappropriate responses [50–54].

CB_1Rs are located both on presynaptic axons and on striatal dendrites. CB_1Rs and μORs are localized on similar medium spiny neurons in the striatum [55, 56]. Ultrastructural examination of the dorsal striatum has revealed that CB_1Rs and μORs colocalize to specific patches of post-synaptic neurons with approximately 50% of all CB_1R-positive dendrites containing the μOR [55]. The NAc shell, which is a central component of the reward neural circuit, appears to have a greater prevalence of CB_1R/μOR coexpression, compared with the NAc core subcomparment [56].

Although CB_1Rs and μORs are co-expressed on similar neurons in the striatum and share similar G-protein coupled signal transduction mechanisms [57, 58], the functional relevance of the CB_1R is complex. For example, CB_1Rs form physical heterodimer interactions with various other receptors. The most studied of these is with the dopamine D2 receptor (D_2R), which is predominantly localized to striatopallidal neurons and thus highly relevant to cannabinoid-opioid interactions. CB_1R/D_2R heterodimers shift the intracellular signaling cascade from the normal $G_{i/o}$ inhibitory signaling to G_s coupling thereby leading to neuronal activation instead of inhibition [59–61]. CB_1Rs have also been observed to form heterodimers with opioid receptors including the μOR. While additional research is needed regarding cannabinoid-opioid receptor dimerization, the use of *in vitro* cell culture systems have demonstrated that simultaneous activation of CB_1Rs and μORs in the same cell inhibits G_i-mediated signaling, suggesting mutual functional antagonism [62].

The ultimate functional and behavioral outcome of cannabinoid-opioid interactions within the striatum also depends on the striatal pathways in which these interactions occur. While μOR is expressed within both striatal projection neurons, there is greater regional and cellular overlap between this receptor's expression and PDYN-expressing striatonigral neurons of the dorsal striatum [63, 64]. Moreover, μORs are also primarily found in the dynorphin-containing neurons in the NAc shell. As such, the cell-type-specific associations of CB_1R and μOR could contribute to different behavioral effects. In sum, significant research efforts are needed in order to expand the limited information still known about CB_1R/μOR interactions in a cell-specific and neuronal pathway manner, which effectively impacts behavior.

Functional Interactions

Functionally, administration of THC can modulate the endogenous opioid system by directly altering the release of opioid peptides. For example, acute administration of a CB_1R agonist, THC, elevates β-endorphin and enkephalin peptide levels in the NAc [39]. The acute exposure of other cannabinoid agonists also increases extracellular levels of endogenous opioids [65–68]. Moreover, chronic cannabinoid exposure increases the levels of endogenous opioid peptide precursors [69–71]. Such cannabinoid-induced regulation of opioid levels would be expected to alter the sensitivity of the cells to the subsequent exposure to opiate drugs.

Indeed, many studies have now demonstrated significant sensitivity to opiate drugs as a consequence of prior exposure to cannabinoids. The most studied functional interaction involves the investigation of early developmental exposure to cannabinoids due to the question regarding potential 'gateway' neurobiological effects of cannabis. Cannabis or THC exposure during prenatal and adolescent brain development have been demonstrated to have signficant effects particularly on PENK striatopallidal cells. For example, in human fetal subjects with *in utero* cannabis exposure, striatal PENK mRNA levels are decreased in the striatum, whereas PDYN levels were not significantly related to maternal cannabis use [72]. This finding in the human fetal brain was replicated in rodent models with prenatal THC exposure at a similar midgestation ontogenic period of neurodevelopment as that studied in the human [73]. Disturbances of PENK mRNA expression were also evident with adolescent exposure to THC [9, 15, 16]. An important observation of both the prenatal and adolescent animal studies was the protracted effects in adulthood emphasizing the long-term alteration of the opioid system due to the early cannabinoid effects on the brain. The fact that most of the data to date shows a greater effect on the PENK system might relate to dose and frequency of the THC administered. Given that both CB_1Rs and μORs are present on striatopallidal and striatonigral pathways it is expected that exposure to higher amounts of THC would impact other neuronal pathways.

The μOR and D_2R enriched-striatopallidal circuit has been implicated in developmental cannabinoid exposure in animal studies [15, 39, 73–77]. These findings are particularly intriuging given that in striatopallidal neurons μOR colocalize with the D_2R, a dopamine receptor subtype highly implicated in the vulnerability to addictive disorders [78–82]. Furthermore, genetic polymorphism of the PENK gene is not only associated with cannabis dependence [83], but also heroin abuse [84]. Thus, there may also be interactions at the genetic level that link the neurobiological organization and function to enhance vulnerability for synergistic interactions at the behavioral level.

Cannabinoid-Opioid Interactions in the Amygdala

The amygdaloid complex is a neurochemically heterogeneous subcortical region critical for the regulation of affect, emotion and stress. The lateral nucleus is the main input nucleus and receives sensory information from the cortex and thalamus, while the basal and central nuclei are primarily output nuclei, mediating instrumental learning and sympathetic arousal, respectively. Most studies to date involving the amygdala have focused on its role in mediating negative affect, such as fear and anxiety, and these emotional responses are fine-tuned by eCB-mediated neuroplasticity [85]. Within this framework, the cannabinoid and opioid systems interact within the amygdala to mediate adaptive stress-responses [86]. Disruption of the interaction between these two systems could therefore contribute to affective disorders such as depression, anxiety disorders and post-traumatic stress disorder.

Anatomical Interactions

CB_1Rs are expressed in several subregions of amygdala, but its highest expression is in the deeper basolateral nucleus [87]. Within the basolateral nucleus of the amygdala, CB_1R expression is primarily reserved to presynaptic termini of cholecystokinin-expressing GABAergic interneurons, a cellular distribution consistent with this receptor's neocortical and hippocampal expression [87–89].

Opioid peptides and receptors are also expressed in several regions of the amygdala. While the μOR is most abundant in the intercalated nuclei and posteromedial cortical amygdala [90], it is moderately expressed in the central amygdala (CeA), a region containing GABAergic neurons enriched with enkephalin peptides [91, 92].

Unlike the striatum, in the amygdala there is a clear spatial separation between components of the cannabinoid and opioid systems. Whereas CB_1R is enriched in the interneurons of the basolateral nucleus, components of the opioid system reside in the more medial and superficial nuclei of the amygdala, such as the CeA. Therefore the interaction between these two systems is likely explained via indirect intercellular connections between adjacent amygdala territories, which give rise to a complex network that integrates sensory inputs and emotional valence.

Functional Interactions

To date, there is a paucity of studies explicitly characterizing the cannabinoid-opioid interaction within the amygdala, perhaps because of this region's complex nature. In the amygdala, cannabinoids have been primarily studied in the context of the fear-conditioning and the hypothalamic-pituitary-adrenal (HPA) axis. Interestingly, our study investigating the prenatal effects of THC revealed a marked up-regulation of PENK mRNA in the CeA of adult rats that also had enhanced heroin-seeking behavior upon exposure to mild stress [73]. The direct causal link of apparent CB_1R-induced dysregulation of the opioid system in the amygdala to enhance heroin intake awaits to be established. However, the fact that the amygdala plays a central role in anxiety and drug-seeking behavior [93] suggests that the developmental cannabinoid exposure leads to impaired amygdala opioid function.

It is important to emphasize that it is enkephalinergic neurons within the CeA that have been well documented to be critically involved in anxiety and stress responsivity [94]. The abundant expression of PENK in the CeA and its role in anxiety is intriguing in light of the finding that polymorphisms of the PENK gene is strongly associated with expression of PENK mRNA levels in the human CeA nucleus and that there is synergism between anxiety behavioral trait and genetic polymorphism of PENK that increases the odds for cannabis dependence [95]. Thus, the CeA might be a potential central target for opioid-cannabinoid interactions relevant to negative affects traits predictive of addiction vulnerability.

Cannabinoid-Opioid Interactions in the Midbrain

The midbrain plays an important role in reward and nociception via neuromodulatory projections to the forebrain and spinal cord and opioid-cannabinoid interactions within the mesencephalon contributes to these processes. We focus this section on midbrain processes most relevant to the forebrain structures discussed above since mesencephalic inputs to these structures have been well studied for their significant contributions to the regulation of addiction, motivation and emotions.

Anatomical Interactions

Within the substantia nigra and VTA, CB_1Rs are predominantly expressed on GABAergic forebrain terminals and local interneurons [96] that directly regulate dopaminergic neurons innervating the dorsal and ventral striatum, respectively. Midbrain dopaminergic neurons also target, and are in turn innervated by amygdala nuclei including the CeA [97, 98]. Mesencephalic GABAergic interneurons also express μORs which suggests the possibility for CB_1R-μOR intracellular interactions in cells that directly modulate dopamine neurons, the origin of the mesocorticolimbic circuit that mediates reward, mood and cognition. Very limited research, however, exists on the cannabinoid-opioid functional intracellular signaling mechanisms in the midbrain.

Functional Interactions

Different lines of evidence underscore the significant functional midbrain interactions between the cannabinoid and opioid systems. An intriguing relationship is the finding that CB_1R binding in the substantia nigra is positively correlated with heroin intake behavior emphasizing that the tone of the endogenous cannabinoid system directly modulates opioid reward [15]. In addition, the use of cannabinoids has been shown to alter the midbrain endogenous opioid system. For example, μOR coupling is significantly increased in the VTA and substantia nigra in adult animals as a consequence of repeated adolescent THC exposure [15]. Such alteration in the μOR coupling would be expected to potentiate dopamine levels in the striatum since stimulation of midbrain μORs would disinhibit midbrain dopaminergic activity [99]. The acute administration of THC also alters endogenous opioids as evidenced by elevation of β-endorphin peptide levels in the VTA [100]. Additionally, direct infusion of WIN55,212-2, a CB_1R agonist, into the VTA enhances morphine CPP [101], while antagonism of CB_1R reduces morphine-induced c-Fos in the VTA [102]. Overall, these studies emphasize the significant contribution of the midbrain eCB signaling in opiate-reward.

It is important to emphasize that the midbrain cannabinoid-opioid relationship is bidirectional as evident by the fact that opioid regulation also impacts the endocannabinoid system. For instance, rats self-administering heroin exhibit elevated CB_1R density in the VTA [103]. Furthermore, THC's impact on c-Fos activity in the VTA is prevented by μOR antagonists [104] while blocking VTA μORs prevents THC-induced increases in extracellular NAc dopamine [105]. These studies suggest a strong interaction within midbrain cannabinoid and opioid systems to regulate key pathways central to reward and emotional regulation.

Summary

There are significant cannabinoid-opioid interactions in key brain regions that mediate reward, motivation, emotional regulation and compulsive habitual behavior relevant to addiction risk. Significant research is still needed about the specific cells and neuronal pathways in which CB_1Rs and μORs interact to impact behavior. Moreover, gene × gene interactions to ascertain genetic contributions within these neuronal systems are still unknown and could have important implications regarding individual vulnerability for the synergistic cannabis and opiate abuse.

References

1. Johnston LD, O'Malley PM, Bachman JG, Schulenberg JE. Monitoring the Future national results on drug use: 2012 Overview, Key Findings on Adolescent Drug Use. Institute for Social Research, The University of Michigan. Ann Arbor. 2013.
2. SAMHSA. Results from the 2012 National Survey on Drug Use and Health: summary of National Findings. Substance Abuse and Mental Health Services Administration, Rockville, MD. 2013.
3. EMCDDA. Annual report on the state of the drugs problem in Europe. Publications Office of the European Union, Luxembourg; 2012.
4. Lynskey MT, Heath AC, Bucholz KK, Slutske WS, Madden PA, Nelson EC, Statham DJ, Martin NG. Escalation of drug use in early-onset cannabis users vs co-twin controls. JAMA. 2003;289(4):427–33.
5. Fergusson DM, Horwood LJ. Does cannabis use encourage other forms of illicit drug use? Addiction. 2000;95(4):505–20.
6. Agrawal A, Neale MC, Prescott CA, Kendler KS. A twin study of early cannabis use and subsequent use and abuse/dependence of other illicit drugs. Psychol Med. 2004;34(7):1227–37.
7. Spano MS, Fadda P, Fratta W, Fattore L. Cannabinoid-opioid interactions in drug discrimination and self-administration: effect of maternal, postnatal, adolescent and adult exposure to the drugs. Curr Drug Targets. 2010;11(4):450–61.
8. Morris CV, DiNieri JA, Szutorisz H, Hurd YL. Molecular mechanisms of maternal cannabis and cigarette use on human neurodevelopment. Eur J Neurosci. 2011;34(10):1574–83.
9. Chadwick B, Miller ML, Hurd YL. Cannabis use during adolescent development: susceptibility to psychiatric illness. Front Psychiatry. 2013;4:129.
10. Bluelight.org. Forum on "Cannabis action on opioid receptors", Cannabis Discussion Forum. http://www.bluelight.org/vb/archive/index.php/t-455925.html (2009). Accessed Jan 2014.

11. Navarro M, Carrera MRA, Fratta W, Valverde O, Cossu G, Fattore L, Chowen JA, Gomez R, del Arco I, Villanua MA, Maldonado R, Koob GF, de Fonseca FR. Functional interaction between opioid and cannabinoid receptors in drug self-administration. J Neurosci. 2001;21(14):5344–50.
12. Cossu G, Ledent C, Fattore L, Imperato A, Bohme GA, Parmentier M, Fratta W. Cannabinoid CB_1 receptor knockout mice fail to self-administer morphine but not other drugs of abuse. Behav Brain Res. 2001;118(1):61–5.
13. Ledent C, Valverde O, Cossu G, Petitet F, Aubert JF, Beslot F, Bohme GA, Imperato A, Pedrazzini T, Roques BP, Vassart G, Fratta W, Parmentier M. Unresponsiveness to cannabinoids and reduced addictive effects of opiates in CB_1 receptor knockout mice. Science. 1999;283(5400):401–4.
14. Martin M, Ledent C, Parmentier M, Maldonado R, Valverde O. Cocaine, but not morphine, induces conditioned place preference and sensitization to locomotor responses in CB1 knockout mice. Eur J Neurosci. 2000;12(11):4038–46.
15. Ellgren M, Spano SM, Hurd YL. Adolescent cannabis exposure alters opiate intake and opioid limbic neuronal populations in adult rats. Neuropsychopharmacology. 2007;32(3):607–15.
16. Tomasiewicz HC, Jacobs MM, Wilkinson MB, Wilson SP, Nestler EJ, Hurd YL. Proenkephalin mediates the enduring effects of adolescent cannabis exposure associated with adult opiate vulnerability. Biol Psychiatry. 2012;72(10):803–10.
17. Solinas M, Panlilio LV, Goldberg SR. Exposure to Δ-9-tetrahydrocannabinol (THC) increases subsequent heroin taking but not heroin's reinforcing efficacy: a self-administration study in rats. Neuropsychopharmacology. 2004;29(7):1301–11.
18. Singh ME, McGregor IS, Mallet PE. Perinatal exposure to Δ^9-tetrahydrocannabinol alters heroin-induced place conditioning and fos-immunoreactivity. Neuropsychopharmacology. 2006;31(1):58–69.
19. Biscaia M, Fernandez B, Higuera-Matas A, Miguens M, Viveros MP, Garcia-Lecumberri C, Ambrosio E. Sex-dependent effects of periadolescent exposure to the cannabinoid agonist CP-55,940 on morphine self-administration behaviour and the endogenous opioid system. Neuropharmacology. 2008;54(5):863–73.
20. Herkenham M, Lynn AB, Little MD, Johnson MR, Melvin LS, de Costa BR, Rice KC. Cannabinoid receptor localization in brain. Proc Natl Acad Sci U S A. 1990;87(5):1932–6.
21. Westlake TM, Howlett AC, Bonner TI, Matsuda LA, Herkenham M. Cannabinoid receptor binding and messenger RNA expression in human brain: an in vitro receptor autoradiography and in situ hybridization histochemistry study of normal aged and Alzheimer's brains. Neuroscience. 1994;63(3):637–52.
22. den Boon FS, Chameau P, Schaafsma-Zhao Q, van Aken W, Bari M, Oddi S, Kruse CG, Maccarrone M, Wadman WJ, Werkman TR. Excitability of prefrontal cortical pyramidal neurons is modulated by activation of intracellular type-2 cannabinoid receptors. Proc Natl Acad Sci U S A. 2012;109(9):3534–9.
23. Cabral GA, Raborn ES, Griffin L, Dennis J, Marciano-Cabral F. CB_2 receptors in the brain: role in central immune function. Br J Pharmacol. 2008;153(2):240–51.
24. Brusco A, Tagliaferro PA, Saez T, Onaivi ES. Ultrastructural localization of neuronal brain CB2 cannabinoid receptors. Ann NY Acad Sci. 2008;1139:450–7.
25. Freund TF, Katona I, Piomelli D. Role of endogenous cannabinoids in synaptic signaling. Physiol Rev. 2003;83(3):1017–66.
26. Hillard CJ, Campbell WB. Biochemistry and pharmacology of arachidonylethanolamide, a putative endogenous cannabinoid. J Lipid Res. 1997;38(12):2383–98.
27. Bisogno T, Howell F, Williams G, Minassi A, Cascio MG, Ligresti A, Matias I, Schiano-Moriello A, Paul P, Williams EJ, Gangadharan U, Hobbs C, Di Marzo V, Doherty P. Cloning of the first sn1-DAG lipases points to the spatial and temporal regulation of endocannabinoid signaling in the brain. J Cell Biol. 2003;163(3):463–8.
28. Cravatt BF, Giang DK, Mayfield SP, Boger DL, Lerner RA, Gilula NB. Molecular characterization of an enzyme that degrades neuromodulatory fatty-acid amides. Nature. 1996;384(6604):83–7.
29. Dinh TP, Freund TF, Piomelli D. A role for monoglyceride lipase in 2-arachidonoylglycerol inactivation. Chem Phys Lipids. 2002;121(1–2):149–58.

30. Dinh TP, Carpenter D, Leslie FM, Freund TF, Katona I, Sensi SL, Kathuria S, Piomelli D. Brain monoglyceride lipase participating in endocannabinoid inactivation. Proc Natl Acad Sci U S A. 2002;99(16):10819–24.
31. Gulyas AI, Cravatt BF, Bracey MH, Dinh TP, Piomelli D, Boscia F, Freund TF. Segregation of two endocannabinoid-hydrolyzing enzymes into pre- and postsynaptic compartments in the rat hippocampus, cerebellum and amygdala. Eur J Neurosci. 2004;20(2):441–58.
32. De Carvalho CR, Pamplona FA, Cruz JS, Takahashi RN. Endocannabinoids underlie reconsolidation of hedonic memories in Wistar rats. Psychopharmacology (Berl). 2013;231(7):1417–25.
33. Akil H, Watson SJ, Young E, Lewis ME, Khachaturian H, Walker JM. Endogenous opioids: biology and function. Annu Rev Neurosci. 1984;7:223–55.
34. Calò G, Lambert DG, Guerrini R. Chapter 215—Nociceptin/Orphanin FQ. In: Kastin AJ, editor. Handbook of biologically active peptides. 2nd edn. Boston: Academic; 2013. pp. 1577–1585.
35. Bals-Kubik R, Ableitner A, Herz A, Shippenberg TS. Neuroanatomical sites mediating the motivational effects of opioids as mapped by the conditioned place preference paradigm in rats. J Pharmacol Exp Ther. 1993;264(1):489–95.
36. Shippenberg TS, Bals-Kubik R, Herz A. Motivational properties of opioids: evidence that an activation of δ-receptors mediates reinforcement processes. Brain Res. 1987;436(2):234–9.
37. Pfeiffer A, Brandt V, Herz A. Psychotomimesis mediated by κ opiate receptors. Science. 1986;233:774–6.
38. Ghozland S, Matthes HW, Simonin F, Filliol D, Kieffer BL, Maldonado R. Motivational effects of cannabinoids are mediated by μ-opioid and κ-opioid receptors. J Neurosci. 2002;22(3):1146–54.
39. Valverde O, Noble F, Beslot F, Dauge V, Fournie-Zaluski MC, Roques BP. Δ^9-tetrahydrocannabinol releases and facilitates the effects of endogenous enkephalins: reduction in morphine withdrawal syndrome without change in rewarding effect. Eur J Neurosci. 2001;13(9):1816–24.
40. Koob GF. Neural mechanisms of drug reinforcement. Ann NY Acad Sci. 1992;654:171–91.
41. Everitt BJ, Wolf ME. Psychomotor stimulant addiction: a neural systems perspective. J Neurosci. 2002;22(9):3312–20.
42. Haber SN, Fudge JL. The primate substantia nigra and VTA: integrative circuitry and function. Crit Rev Neurobiol. 1997;11(4):323–42.
43. Le Moine C, Normand E, Bloch B. Phenotypical characterization of the rat striatal neurons expressing the D1 dopamine receptor gene. Proc Natl Acad Sci U S A. 1991;88:4205–9.
44. Le Moine C, Normand E, Guitteny AF, Fouque B, Teoule R, Bloch B. Dopamine receptor gene expression by enkephalin neurons in rat forebrain. Proc Natl Acad Sci U S A. 1990;87(1):230–4.
45. Gerfen CR, Enber TM, Susel Z, Chase TN, Monsma FJ, Mahan LC, Sibley DR. D1 and D2 dopamine receptor regulated gene expression of striatonigral and striatopallidal neurons. Science. 1990;250:1429–32.
46. Brownstein MJ, Mroz EA, Tappaz ML, Leeman SE. On the origin of substance P and glutamic acid decarboxylase (GAD) in the substantia nigra. Brain Res. 1977;135(2):315–23.
47. Vincent SR, Hokfelt T, Christensson I, Terenius L. Dynorphin-immunoreactive neurons in the central nervous system of the rat. Neurosci Lett. 1982;33(2):185–90.
48. Cuello AC, Paxinos G. Evidence for a long Leu-enkephalin striopallidal pathway in rat brain. Nature. 1978;271(5641):178–80.
49. Zhou L, Furuta T, Kaneko T. Chemical organization of projection neurons in the rat accumbens nucleus and olfactory tubercle. Neuroscience. 2003;120(3):783–98.
50. Durieux PF, Bearzatto B, Guiducci S, Buch T, Waisman A, Zoli M, Schiffmann SN, de Kerchove d'Exaerde A. D_2R striatopallidal neurons inhibit both locomotor and drug reward processes. Nat Neurosci. 2009;12(4):393–5.
51. Frank MJ, Moustafa AA, Haughey HM, Curran T, Hutchison KE. Genetic triple dissociation reveals multiple roles for dopamine in reinforcement learning. Proc Natl Acad Sci U S A. 2007;104(41):16311–16.

52. Klein TA, Neumann J, Reuter M, Hennig J, von Cramon DY, Ullsperger M. Genetically determined differences in learning from errors. Science. 2007;318(5856):1642–5.
53. Sano H, Yasoshima Y, Matsushita N, Kaneko T, Kohno K, Pastan I, Kobayashi K. Conditional ablation of striatal neuronal types containing dopamine D2 receptor disturbs coordination of basal ganglia function. J Neurosci. 2003;23(27):9078–88.
54. Lobo MK, Covington HE 3rd, Chaudhury D, Friedman AK, Sun H, Damez-Werno D, Dietz DM, Zaman S, Koo JW, Kennedy PJ, Mouzon E, Mogri M, Neve RL, Deisseroth K, Han MH, Nestler EJ. Cell type-specific loss of BDNF signaling mimics optogenetic control of cocaine reward. Science. 2010;330(6002):385–90.
55. Rodriguez JJ, Mackie K, Pickel VM. Ultrastructural localization of the CB_1 cannabinoid receptor in μ-opioid receptor patches of the rat Caudate Putamen nucleus. J Neurosci. 2001;21:823–33.
56. Pickel VM, Chan J, Kash TL, Rodriguez JJ, MacKie K. Compartment-specific localization of cannabinoid 1 (CB1) and μ-opioid receptors in rat nucleus accumbens. Neuroscience. 2004;127(1):101–12.
57. Canals M, Milligan G. Constitutive activity of the cannabinoid CB1 receptor regulates the function of co-expressed Mu opioid receptors. J Biol Chem. 2008;283(17):11424–34.
58. Blume LC, Bass CE, Childers SR, Dalton GD, Roberts DC, Richardson JM, Xiao R, Selley DE, Howlett AC. Striatal CB_1 and D_2 receptors regulate expression of each other, CRIP1A and delta opioid systems. J Neurochem. 2013;124(6):808–20.
59. Glass M, Felder CC. Concurrent stimulation of cannabinoid CB_1 and dopamine D_2 receptors augments cAMP accumulation in striatal neurons: evidence for a Gs linkage to the CB1 receptor. J Neurosci. 1997;17(14):5327–33.
60. Kearn CS, Blake-Palmer K, Daniel E, Mackie K, Glass M. Concurrent stimulation of cannabinoid CB1 and dopamine D2 receptors enhances heterodimer formation: a mechanism for receptor cross-talk? Mol Pharmacol. 2005;67(5):1697–704.
61. Jarrahian A, Watts VJ, Barker EL. D_2 dopamine receptors modulate G_α-subunit coupling of the CB_1 cannabinoid receptor. J Pharmacol Exp Ther. 2004;308(3):880–6.
62. Rios C, Gomes I, Devi LA. μ opioid and CB1 cannabinoid receptor interactions: reciprocal inhibition of receptor signaling and neuritogenesis. Br J Pharmacol. 2006;148(4):387–95.
63. Hurd YL, Herkenham M. The human neostriatum shows compartmentalization of neuropeptide gene expression in dorsal and ventral regions: an in situ hybridization histochemical analysis. Neuroscience. 1995;64(3):571–86.
64. Guttenberg ND, Klop H, Minami M, Satoh M, Voorn P. Co-localization of μ opioid receptor is greater with dynorphin than enkephalin in rat striatum. Neuroreport. 1996;7(13):2119–24.
65. Pugh G Jr, Abood ME, Welch SP. Antisense oligodeoxynucleotides to the κ-1 receptor block the antinociceptive effects of Δ^9-THC in the spinal cord. Brain Res. 1995;689(1):157–8.
66. Pugh G Jr, Smith PB, Dombrowski DS, Welch SP. The role of endogenous opioids in enhancing the antinociception produced by the combination of Δ^9-tetrahydrocannabinol and morphine in the spinal cord. J Pharmacol Exp Ther. 1996;279(2):608–16.
67. Pugh G Jr, Mason DJ Jr, Combs V, Welch SP. Involvement of dynorphin B in the antinociceptive effects of the cannabinoid CP55,940 in the spinal cord. J Pharmacol Exp Ther. 1997;281(2):730–7.
68. Houser SJ, Eads M, Embrey JP, Welch SP. Dynorphin B and spinal analgesia: induction of antinociception by the cannabinoids CP55,940, Δ^9-THC and anandamide. Brain Res. 2000;857(1–2):337–42.
69. Corchero J, Avila MA, Fuentes JA, Manzanares J. Δ-9-Tetrahydrocannabinol increases prodynorphin and proenkephalin gene expression in the spinal cord of the rat. Life Sci. 1997;61(4):PL 39–43.
70. Manzanares J, Corchero J, Romero J, Fernandez-Ruiz JJ, Ramos JA, Fuentes JA. Chronic administration of cannabinoids regulates proenkephalin mRNA levels in selected regions of the rat brain. Brain Res Mol Brain Res. 1998;55(1):126–32.

71. Corchero J, Fuentes JA, Manzanares J. Δ^9-Tetrahydrocannabinol increases proopiomelanocortin gene expression in the arcuate nucleus of the rat hypothalamus. Eur J Pharmacol. 1997;323(2–3):193–5.
72. Wang X, Dow-Edwards D, Anderson V, Minkoff H, Hurd YL. Discrete opioid gene expression impairment in the human fetal brain associated with maternal marijuana use. Pharmacogenomics J. 2006;6(4):255–64.
73. Spano MS, Ellgren M, Wang X, Hurd YL. Prenatal cannabis exposure increases heroin seeking with allostatic changes in limbic enkephalin systems in adulthood. Biol Psychiatry. 2007;61(4):554–63.
74. Perez-Rosado A, Manzanares J, Fernandez-Ruiz J, Ramos JA. Prenatal Δ^9-tetrahydrocannabinol exposure modifies proenkephalin gene expression in the fetal rat brain: sex-dependent differences. Brain Res Dev Brain Res. 2000;120(1):77–81.
75. Morel LJ, Giros B, Dauge V. Adolescent exposure to chronic delta-9-tetrahydrocannabinol blocks opiate dependence in maternally deprived rats. Neuropsychopharmacology. 2009;34(11):2469–76.
76. Corchero J, Garcia-Gil L, Manzanares J, Fernandez-Ruiz JJ, Fuentes JA, Ramos JA. Perinatal delta9-tetrahydrocannabinol exposure reduces proenkephalin gene expression in the caudate-putamen of adult female rats. Life Sci. 1998;63(10):843–50.
77. Corchero J, Romero J, Berrendero F, Fernandez-Ruiz J, Ramos JA, Fuentes JA, Manzanares J. Time-dependent differences of repeated administration with Δ^9-tetrahydrocannabinol in proenkephalin and cannabinoid receptor gene expression and G-protein activation by μ-opioid and CB_1-cannabinoid receptors in the caudate-putamen. Brain Res Mol Brain Res. 1999;67(1):148–57.
78. Heinz A, Siessmeier T, Wrase J, Hermann D, Klein S, Grusser SM, Flor H, Braus DF, Buchholz HG, Grunder G, Schreckenberger M, Smolka MN, Rosch F, Mann K, Bartenstein P. Correlation between dopamine D_2 receptors in the ventral striatum and central processing of alcohol cues and craving. Am J Psychiatry. 2004;161(101927):1783–9.
79. Volkow ND, Chang L, Wang GJ, Fowler JS, Ding YS, Sedler M, Logan J, Franceschi D, Gatley J, Hitzemann R, Gifford A, Wong C, Pappas N. Low level of brain dopamine D_2 receptors in methamphetamine abusers: association with metabolism in the orbitofrontal cortex. Am J Psychiatry. 2001;158(12):2015–21.
80. Volkow ND, Wang GJ, Fowler JS, Logan J, Gatley SJ, Gifford A, Hitzemann R, Ding YS, Pappas N. Prediction of reinforcing responses to psychostimulants in humans by brain dopamine D_2 receptor levels. Am J Psychiatry. 1999;156(9):1440–3.
81. Wang GJ, Volkow ND, Fowler JS, Logan J, Abumrad NN, Hitzemann RJ, Pappas NS, Pascani K. Dopamine D_2 receptor availability in opiate-dependent subjects before and after naloxone-precipitated withdrawal. Neuropsychopharmacology. 1997;16:174–82.
82. Volkow ND, Fowler JS, Wang GJ, Swanson JM. Dopamine in drug abuse and addiction: results from imaging studies and treatment implications. Mol Psychiatry. 2004;9(6):557–69.
83. Jutras-Aswad D, Jacobs MM, Yiannoulos G, Roussos P, Bitsios P, Nomura Y, Liu X, Hurd YL. Cannabis-dependence risk relates to synergism between neuroticism and proenkephalin snps associated with amygdala gene expression: case-control study. PLoS ONE. 2012;7(6):e39243.
84. Nikoshkov A, Drakenberg K, Wang X, Horvath MC, Keller E, Hurd YL. Opioid neuropeptide genotypes in relation to heroin abuse: dopamine tone contributes to reversed mesolimbic proenkephalin expression. Proc Natl Acad Sci U S A. 2008;105(2):786–91.
85. Gunduz-Cinar O, Hill MN, McEwen BS, Holmes A. Amygdala FAAH and anandamide: mediating protection and recovery from stress. Trends Pharmacol Sci. 2013;34(11):637–44.
86. Viveros MP, Marco EM, File SE. Endocannabinoid system and stress and anxiety responses. Pharmacol Biochem Behav. 2005;81(2):331–42.
87. Katona I, Rancz EA, Acsady L, Ledent C, Mackie K, Hajos N, Freund TF. Distribution of CB1 cannabinoid receptors in the amygdala and their role in the control of GABAergic transmission. J Neurosci. 2001;21(23):9506–18.

88. McDonald AJ, Mascagni F. Localization of the CB1 type cannabinoid receptor in the rat basolateral amygdala: high concentrations in a subpopulation of cholecystokinin-containing interneurons. Neuroscience. 2001;107(4):641–52.
89. Marsicano G, Lutz B. Expression of the cannabinoid receptor CB1 in distinct neuronal subpopulations in the adult mouse forebrain. Eur J Neurosci. 1999;11(12):4213–25.
90. Mansour A, Fox CA, Akil H, Watson SJ. Opioid-receptor mRNA expression in the rat CNS: anatomical and functional implications. Trends Neurosci. 1995;18(1):22–9.
91. Sar M, Stumpf WE, Miller RJ, Chang KJ, Cuatrecasas P. Immunohistochemical localization of enkephalin in rat brain and spinal cord. J Comp Neurol. 1978;182(1):17–37.
92. Veinante P, Stoeckel ME, Freund-Mercier MJ. GABA- and peptide-immunoreactivities co-localize in the rat central extended amygdala. Neuroreport. 1997;8(13):2985–9.
93. Erb S. Evaluation of the relationship between anxiety during withdrawal and stress-induced reinstatement of cocaine seeking. Prog Neuropsychopharmacol Biol Psychiatry. 2010;34(5):798–807.
94. Kung JC, Chen TC, Shyu BC, Hsiao S, Huang AC. Anxiety- and depressive-like responses and c-fos activity in preproenkephalin knockout mice: oversensitivity hypothesis of enkephalin deficit-induced posttraumatic stress disorder. J Biomed Sci. 2010;17:29.
95. Jutras-Aswad D, DiNieri JA, Harkany T, Hurd YL. Neurobiological consequences of maternal cannabis on human fetal development and its neuropsychiatric outcome. Eur Arch Psychiatry Clin Neurosci. 2009;259(7):395–412.
96. Fitzgerald ML, Shobin E, Pickel VM. Cannabinoid modulation of the dopaminergic circuitry: implications for limbic and striatal output. Prog Neuropsychopharmacol Biol Psychiatry. 2012;38(1):21–9.
97. Zahm DS, Jensen SL, Williams ES, Martin JR 3rd. Direct comparison of projections from the central amygdaloid region and nucleus accumbens shell. Eur J Neurosci. 1999;11(4):1119–26.
98. Fudge JL, Haber SN. The central nucleus of the amygdala projection to dopamine subpopulations in primates. Neuroscience. 2000;97(3):479–94.
99. Johnson SW, North RA. Opioids excite dopamine neurons by hyperpolarization of local interneurons. J Neurosci. 1992;12(2):483–8.
100. Solinas M, Zangen A, Thiriet N, Goldberg SR. β-endorphin elevations in the ventral tegmental area regulate the discriminative effects of Δ-9-tetrahydrocannabinol. Eur J Neurosci. 2004;19(12):3183–92.
101. Rashidy-Pour A, Pahlevani P, Vaziri A, Shaigani P, Zarepour L, Vafaei AA, Haghparast A. Involvement of CB1 receptors in the ventral tegmental area in the potentiation of morphine rewarding properties in acquisition but not expression in the conditioned place preference model. Behav Brain Res. 2013;247:259–67.
102. Singh ME, Verty AN, Price I, McGregor IS, Mallet PE. Modulation of morphine-induced Fos-immunoreactivity by the cannabinoid receptor antagonist SR 141716. Neuropharmacology. 2004;47(8):1157–69.
103. Fattore L, Vigano D, Fadda P, Rubino T, Fratta W, Parolaro D. Bidirectional regulation of mu-opioid and CB1-cannabinoid receptor in rats self-administering heroin or WIN 55,212-2. Eur J Neurosci. 2007;25(7):2191–200.
104. Allen K, McGregor IS, Hunt GE, Singh ME, Mallet PE. Regional differences in naloxone modulation of Δ^9-THC induced Fos expression in rat brain. Neuropharmacology. 2003;44(2):264–74.
105. Tanda G, Pontieri FE, Di Chiara G. Cannabinoid and heroin activation of mesolimbic dopamine transmission by a common μ_1 opioid receptor mechanism. Science. 1997;276(5321):2048–50.

Chapter 16
Interactions of Cannabis and Amphetamine-Type Stimulants

Simone Tambaro and Marco Bortolato

Abstract Amphetamine-type stimulants (ATSs) are a large family of substances of abuse, characterized by well-known mood- and performance-enhancing properties. This class encompasses several high-potency stimulants and entactogens, such as the precursor compound *d*-amphetamine (AMPH), its synthetic *N*-methylated derivatives methamphetamine (METH) and 3, 4-methylenedioxy-*N*-methylamphetamine (MDMA, or "ecstasy"), as well as novel designer drugs, based on substituted forms of the natural alkaloid cathinone. ATSs (and in particular METH) are among the most commonly abused substances worldwide, second only to *Cannabis sativa*; indeed, the rate of concurrent consumption of METH and cannabis has been increasing over the last decade, particularly among adolescents. Anecdotal evidence suggests that marijuana may offset some unpleasant subjective effects of ATSs, such as anxiety and paranoia. Both drugs have been shown to increase schizophrenia vulnerability in young vulnerable individuals, raising the possibility that their concurrent intake may have synergistic effects with respect to the development of psychotic manifestations. In addition, the combination of these two substances may affect their subjective effects and exacerbate their abuse liability. Although current evidence on the neurobiological interactions of cannabis and ATSs remains mostly elusive, initial studies in animal models suggest that the cannabinoid system may play a relevant role in the motivational and addictive properties of ATSs; furthermore, cannabinoids may modify the behavioral effects and even attenuate some untoward long-term consequences of ATSs. In this chapter we review the available evidence on these potential interactions and outline some key mechanisms that may account for the mutual modulatory influence of these substances.

Keywords Cannabis · Amphetamine-type stimulants · Methamphetamine · Dopamine · CB_1 receptors · Neuroprotection

---

M. Bortolato (✉) · S. Tambaro
Department of Pharmacology and Toxicology, School of Pharmacy, University of Kansas, Lawrence, Kansas, USA
e-mail: bortolato@ku.edu

P. Campolongo, L. Fattore (eds.), *Cannabinoid Modulation of Emotion, Memory, and Motivation*, DOI 10.1007/978-1-4939-2294-9_16

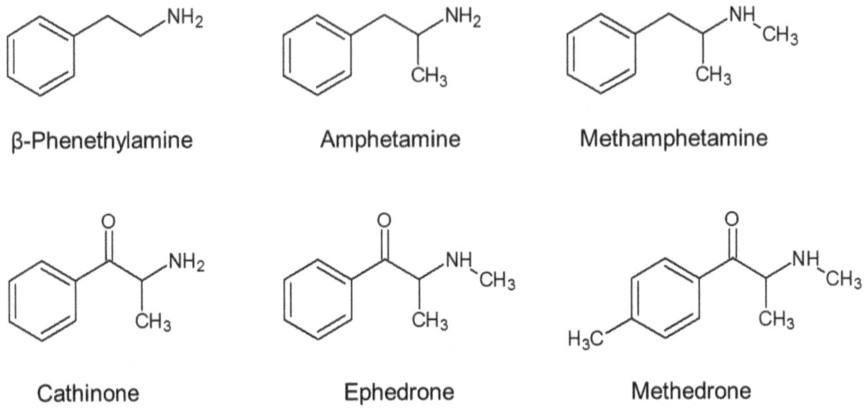

Fig. 16.1 Chemical structures of β-Phenethylamine and some of the major amphetamine-type stimulants (ATSs)

Introduction

Amphetamine-type stimulants (ATSs) are a large family of psychoactive drugs characterized by a common phenylethylamine core structure. The precursor of this class, d-amphetamine (AMPH; 1-phenylpropan-2-amine) and its N-methylated derivative methamphetamine (METH; N-methyl-1-phenylpropan-2-amine), were respectively synthesized in 1887 [1] and 1893, and marketed as decongestants under the commercial names of *Benzedrine* and *Methedrine* [2]. Following the discovery and characterization of the psychostimulant properties of these drugs, they were originally proposed and used for numerous illnesses, such as depression, migraine, alcoholism and obesity[1]. With the growing diffusion of AMPH and METH as therapeutic agents, it was recognized that high doses of these agents could lead to prominent euphoria and excitement, disinhibition, increased libido and arousal, sense of invincibility, fatigue resistance and sleeplessness; furthermore, it soon became apparent that both drugs had a high addiction liability, and that their abuse was associated with a higher risk for mania and psychosis. Nowadays, the class of ATSs is known to encompass a large variety of different synthetic compounds (the structures of the main ATSs are represented in Fig. 16.1, as well as the natural alkaloids ephedrine and cathinone, respectively obtained from the plants *Ephedra sinica* and *Catha edulis*. The behavioral properties of the ATSs vary depending on the chemical structure; for example, the effects of 3, 4-methylenedioxy-methylamphetamine (MDMA, also known as "ecstasy) and similar compounds induce effects typically

[1] Nowadays, the therapeutic applications of ATSs are mostly limited to low-potency compounds, which carry a very limited liability for dependence. Notably, low doses of the dextrorotatory enantiomers of AMPH and METH are still approved by the Food and Drug Administration for the treatment of narcolepsy and attention-deficit hyperactivity disorder (ADHD).

different from those elicited by AMPH and METH, typically described as an enhances sense of emotional closeness and empathy.

METH features greater potency and a significant longer half-life than cocaine or other common stimulants (ranging from 10–30 h) [3]; because of these properties, its misuse for recreational purposes (generally by smoking or snorting) gained momentum in the 1960s and has reached the proportions of a veritable epidemic in the past decades [4], particularly among adolescents of North America, East Asia and Oceania [5–9]. A recent report released by the United Nations Office on Drugs and Crime has ranked METH and other ATSs as the world's most widely abused type of illicit substance after cannabis [10].

Until the late 1980s, the concomitant abuse of cannabis products and ATSs was generally regarded as a relatively infrequent phenomenon [11], possibly due to the divergence in the sociocultural milieux traditionally associated with the consumption of either substance. In the early 1990s, however, this trend was rapidly reversed by the introduction of large amounts of high-purity METH by Mexican drug cartels in the illicit market of the Western and Midwestern regions of the United States. The increased availability of pure METH at lower prices led to its growing popularity among the local communities of cannabis users [12]. By 2002, it was estimated that cannabis was the most common secondary substance of abuse among METH-dependent individuals [13]. In striking contrast with the skyrocketing proportion of the comorbid abuse of cannabis and METH, research on the interactions of these two substances has considerably lagged behind. To the date of this writing (December 2013), only few systematic clinical studies on the combined effects of ATSs and cannabis have been published in peer-review publications.

In this chapter, we will outline the available evidence on the interactions of cannabis and ATSs, as well as their underlying neurobiological mechanisms. In particular, we will mainly focus on the interaction of AMPH and METH with the two most abundant ingredients of cannabis, namely its main psychoactive alkaloid Δ9-Tetrahydrocannabinol (THC), and cannabidiol (CBD). Nevertheless, it is worth noting that similar effects and mechanisms are predicted for the interaction of newly-developed synthetic cannabinoids ("Spice") and new-generation ATSs ("Bath salts"). The latter, which include mephedrone, methylone, methcathinone, amfepramone and pyrovalerone, are mainly synthetic cathinone derivatives [14]. Conversely, synthetic cannabinoids (including bicyclic compounds, benzopyrans and aminoalkylindole derivatives) [15] were originally developed as experimental drugs, but have recently reached the illicit market (see Chap. 10 of this book). Unfortunately, these compounds are often sold in combination; in particular, it has been recently reported that new drugs that combine the pharmacological properties of both categories may have already been developed [16]. This alarming scenario raises the urgency of a better understanding of the interactions of cannabinoids and ATSs, particularly with respect to their behavioral and toxic consequences.

Effects and Mechanisms of Action of AMPH and METH

Although the clinical and behavioral effects of AMPH and METH have been documented for longer than 50 years, their mechanism of action still remains partially elusive. Several molecular mechanisms of AMPH and METH are posited to mimic the actions of their endogenous analog β-phenylethylamine (β-PEA), a naturally occurring trace amine that acts as a neuromodulator of monoamine neurotransmitters, such as dopamine (DA), norepinephrine (NE) and serotonin (5-HT) [17]. β-PEA is mainly present in monoaminergic neurons, where it is synthesized by decarboxylation of the amino acid phenylalanine [17] and metabolized by monoamine oxidase (MAO) B [18]. Although the synthesis of β-PEA is thought to occur with a rate similar to that of DA and NE, its concentration are significantly lower than those of catecholamines, because of its significantly higher metabolism by MAO B [19]. Physiological concentrations of β-PEA play an important role in the modulation of DAergic neurotransmission, by inducing DA release, inhibiting its reuptake and limiting the responses of D_2 autoreceptors [20–23]. Most of these actions are ascribed to the activation of the main receptor of β-PEA, named trace amine associated receptor 1 (TAAR1) [24]. This G_s-protein-coupled receptor appears mainly located within intracellular membranes [25, 26] of monoaminergic neurons [24, 27].

Effects of AMPH and METH on DA Neurotransmission

In addition to their analogy to β-PEA, AMPH and METH bear a strong structural resemblance with DA, and compete with this neurotransmitter for their uptake into the presynaptic terminals of DAergic neurons by the DA transporter (DAT) [28, 29]. Indeed, the intracellular transport of AMPH and METH enhances DA concentrations in the extracellular space by reducing its uptake and facilitating its release through DAT-mediated antiport [30–32] (Fig. 16.2). Once AMPH and METH are carried in the cytosol, they activate TAAR1 [33], stimulating the protein kinases A (PKA) and protein kinases C delta (PKCΔ) [34–36]. The ensuing phosphorylation of DAT leads to its endocytosis, accumulation in endosomes and reduced recycling [37].

Although AMPH and METH are potent TAAR1 agonists, this receptor is not thought to play a primary role in the ability of these drugs to enhance the activity of DAergic neurons; accordingly, TAAR1 activation has been shown to reduce, rather than increase, the firing of DAergic neurons [38]. The mechanisms that likely support the psychostimulant properties of AMPH and METH are based on their ability to bind to the vesicular monoamine transporter 2 (VMAT2), which results in the inhibition of DA transport within the vesicles [39]. The inactivation of VMAT2 has been shown to counter some of the phenotypical effects of METH and AMPH, such as the enhancement of DA efflux [40–43], as well as the behavioral effects of these drugs [44, 45]. A third mechanism that contributes to the enhancement of extracellular DA levels by AMPH and METH is afforded by their inhibition of MAOs, which catalyze the metabolism of DA and other monoamines [46, 47].

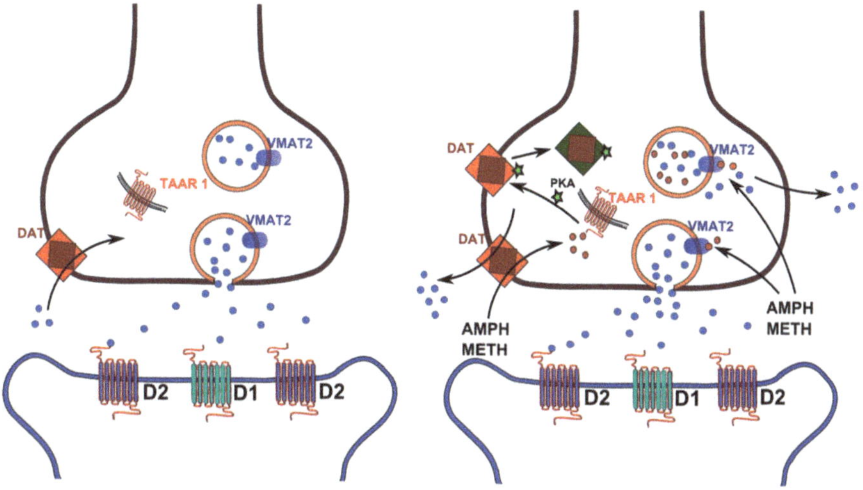

Fig. 16.2 Schematic model of the actions of amphetamine-type stimulants (ATSs) in the presynaptic terminal of the dopaminergic neuron. **a** In physiological condition DA is released in the synaptic cleft by a calcium-dependent system. The uptake is accomplished by a membrane carrier (1), which can transport DA into and out of the terminal depending on the existing concentration gradient. Cytoplasmic DA is transported into storage vesicle by vesicular monoamine transport 2 (VMAT2). **b** ATSs enter the presynaptic bouton across through DAT. AMPH and METH enhance DA concentrations in the extracellular space by reducing its uptake and facilitating its release through DAT-mediated antiport. Once AMPH and METH are carried in the cytosol, they activate TAAR1, stimulating the protein kinases PKA and PKC (not represented). The phosphorylation of DAT leads to its endocytosis. AMPH and METH interfere also with the vesicular carrier VMAT2, thereby reducing the intravesicular uptake of DA and facilitating its release in the cytosol

The combination of the mechanisms outlined above results in a general enhancement of DA neurotransmission in the nigrostriatal and mesocorticolimbic projections, the two main pathways of the DAergic system that regulate movement and reward-associated responses. These functions account for the marked stimulant effects of these substances, which lead to the enhancement of motoric and motivation-based activity.

It is worth noting that the effects of AMPH and METH are largely due to the increase of DA *volume transmission,* one of the two main modalities of DA neurotransmission. Volume transmission consists in the non-vesicular release of DA from non-junctional varicosities of its neurons, leading to the activation of DAergic receptors in the extrasynaptic and perisynaptic space [48–51]. In the striatum and nucleus accumbens, the fine regulation of the balance between volume and synaptic transmission of DA is considered to play a key role in encoding informational salience with respect to locomotor modulation or the execution of motivated behaviors.

Low doses of AMPH lead to a modest increase in volume transmission, which may be essential to enhance focused attention. Conversely, higher doses of ATSs are likely to lead to a more robust DA spillover, which may facilitate the development of psychotic responses through a generalized attribution of salience to irrelevant information and thoughts [52]. This impairment leads to a deficit of the signal-to-noise ratio, leading to deficits of sensorimotor gating and information filtering [53]. In confirmation of this theory, only high doses of AMPH have been found to result in the disruption of the prepulse inhibition of the startle reflex, the most common operational index for the measurement of sensorimotor gating [54].

The role of AMPH and METH in the modulation of attentional and cognitive processes may reflect the distribution of the two main classes of DA receptors, D_1 and D_2, in the nucleus accumbens and striatum. Low doses of AMPH are likely to lead to activation of D_2 receptors in the postsynaptic terminal or in direct contiguity with the presynaptic bouton. Accordingly, low doses of AMPH has been shown to affect D_2, but not D_1 binding [55]. The increase in DA release corresponding to these dosages may not be sufficient to stimulate D_1 receptors, which are generally localized in spines and dendrites of medium-spiny GABAergic neurons in proximity of glutamatergic synapses, further away from the synapse than D_2 receptors [56–60]. Conversely, higher doses of AMPH may lead to the joint activation of extrasynaptic D_1 and D_2 receptors. The hyperlocomotion and sensorimotor gating deficits of AMPH have been shown to be dependent on either class of receptors [61–65], even though their differential role may reflect specific differences in receptor distribution and sensitivity among different strains and species.

Higher doses of ATSs are thought to produce a marked increase of striatal extrasynaptic and perisynaptic DA levels [66]. The D_1 receptors are actually essential to induce a prolonged and robust excitatory action in the extrasynaptic terminals of the striatum [67]. Presynaptic D_2 receptors have been shown to mediate effects not only as DA autoreceptors, but also on γ-aminobutyric acid (GABA) [68–70] and glutamate [71] in the striatum. Thus, it is tempting to hypothesize that a robust DA spillover may have effects also on the release of GABA and glutamate through activation of these receptors. Future studies are needed to confirm this possibility and evaluate its functional significance.

Non-DAergic Mechanisms of AMPH and METH

The actions of AMPH and METH as analogs of β-PEA have repercussions also on the other monoamine neurotransmitters, namely 5-HT and NE. Both drugs have been found to reduce the uptake and increase the extracellular levels of these neurotransmitters, and these mechanisms have been shown to play an essential role in the behavioral effects of ATSs [72–74]. However, the role of TAAR1 in these processes has not been fully clarified yet. While it is assumed that the actions on 5-HT neurotransmission may be similar to those observed for DA, the mechanisms may differ with respect to NE. For example, the internalization of NE transporter has not been observed in response to AMPH. The effects of ATSs are likely not limited to the monoaminergic systems, but are likely to involve also other neurotransmitters, such as glutamate and GABA [75–77]. The details of these processes, however, await further clarification.

Neurotoxic Effects and Mechanisms of METH

One of the most problematic consequences of METH lies in the permanent damage of midbrain, striatal and cortical neurons, which leads to long-lasting depletions of striatal DA and 5-HT [78, 79]. The molecular mechanisms supporting the neurotoxic effects of METH are not fully understood, but they are generally thought to be related to excessive concentrations of DA (and, possibly, 5-HT) in the cytosol. In this compartment, DA undergoes non-enzymatic oxidation with the production of quinones and other oxyradicals [80, 81], which trigger oxidative stress, mitochondrial dysfunctions and the formation of oligomeric protein aggregates of DAT as well as α-synuclein [82–85], ultimately leading to the death of DAergic cells. One of the most common and dangerous effects of METH neurotoxicity is a life-threatening hyperthermia, which accompanies alterations of the blood-brain barrier and brain edema [86], and is a primary cause of lethality following METH overdose and toxicity [87, 88].

The neurotoxic mechanisms of METH, albeit still partially unclear, have been recently shown to be inversely related to the availability of VMAT2 [43, 89, 90] and involve the activation of D_2 receptors [91]. It should also be noted that METH-induced neurotoxicity has been shown to involve glutamatergic excitotoxicity in the striatum [92], likely due to the enhancement of glutamate level in the striatum, which may potentiate the oxidative stress induced by DA [93]. Other METH-induced neurotoxic mechanisms appear to involve the activation of caspase 3 and other apoptotic mechanisms [94–96].

Interactions of Cannabis and ATSs

The well-documented role of ATSs and cannabis in the pathogenesis of psychiatric disorders, ranging from anxiety-spectrum to psychotic and cognitive disorders [97, 98] raises serious concerns about the sequelae of their combined use. This issue may become even more problematic with the recent diffusion of synthetic designer drugs in both categories, which are often sold as mixtures.

As mentioned above, marijuana and other hemp products are the most common secondary substances of abuse among METH users [13], and, in particular, adolescents. This scenario is really concerning, in view of well-documented evidence linking both substances to psychotic manifestations in youth [99, 100]. Indeed, as widely reviewed in Chapt. 12 of this book, cannabis abuse in adolescence has been highlighted as a key risk factor for schizophrenia for genetically vulnerable individuals [101, 102]. Thus, it is likely that the combined consumption of cannabinoids and ATSs may be particularly dangerous and addictive.

Based on anecdotal reports, it has been suggested that cannabis may prolong and intensify the sensation of euphoria associated with consumption of ATSs [103] as well as other psychostimulants [104]. In addition, the calming and relaxing properties of cannabis may offset some of the psychological untoward consequences of ATS, such as anxiety and agitation. This possibility is supported by preclinical evidence in rodents subjected to both acute and chronic administration of METH [105]. However, cannabis has been shown to produce variable effects with respect to anxiety [106] and may even exacerbate some of the negative subjective sensations induced by ATSs, such as panic and paranoia.

Pending the expansion of research on the topic, the current state of the art generally relies on the assumption that cannabis may interact with ATSs in a fashion similar to those documented with cocaine, another psychostimulant whose mechanism of action is largely based on the enhancement of DAergic neurotransmission. Indeed, several behavioral and physiological effects of METH resemble those of cocaine. Nevertheless, this vista does not account for key mechanistic differences between cocaine and ATSs: while the latter blocks DA uptake, ATSs also increase the release and inhibit the metabolism of this neurotransmitter. In addition, as mentioned above, METH has a significantly slower clearance and wider brain distribution than cocaine [107]. Because of these differences, METH produces a much more prolonged sensation of "high" than cocaine. In addition, the higher amounts of extra-vesicular DA are likely to result in greater neurotoxicity levels, due to reactive oxygen species (ROSs) from non-enzymatic catabolism of DA.

In the next sections, we will summarize the available evidence on the interactions of cannabis and ATSs and elaborate on their putative mechanisms. This discussion will be preceded by a brief preamble on the endocannabinoid system and its relevance to the mechanisms of cannabis, specifically designed to facilitate the readers who may be unfamiliar with the key neurobiological mechanisms of THC and other cannabinoids. For a more thorough treatment of these topics, however, the interested reader is referred to the excellent reviews by De Petrocellis et al. [108] and Mechoulam and Parker [109].

A Brief Outline on the Endocannabinoid System

Most of the actions of THC are mediated by two major cannabinoid receptors, both coupled to $G_{i/o}$ proteins [110], respectively termed CB_1 [111] and CB_2 [112]. CB_1 receptors are abundantly expressed in the brain and implicated in the majority of the psychotropic actions of cannabis. These receptors are typically localized on the membrane of presynaptic terminals of GABAergic and glutamatergic neurons [113, 114], where they control the release of either neurotransmitter in response to retrograde activation from the postsynaptic terminal [115–118]; furthermore, these receptors are involved in other key plasticity mechanisms, such as short- and long-term synaptic depression [119]. CB_1 receptors also form heteromeric complexes with other G-protein complex receptors, such as dopamine D_2, μ-opioid and adenosine A_{2a} [120–122].

In contrast with CB_1 receptors, CB_2 receptors are abundantly expressed in most peripheral organs (and particularly in immune cells, where they regulate cytokine secretion and modulate cell trafficking) [123]. Although the presence of CB_2 receptors in neurons has been revealed [124, 125], their function remains poorly understood and awaits further characterization.

Cannabinoid receptors are bound by the two endogenous ligands (endocannabinoids) 2-arachidonoylglycerol (2-AG) [126, 127] and N-arachidonoylethanolamine (also termed anandamide) [128].

2-AG acts as a full agonist at both CB_1 and CB_2 receptors, and mediates the mechanisms of short-term control of glutamate and GABA release [116, 117]. Conversely, anandamide acts as a high-affinity partial agonist for both CB_1 and CB_2 receptors. In addition, this endocannabinoid interacts also with the vanilloid channel receptors 1 (TRPV1), which are abundantly distributed in DAergic neurons. Interestingly, TRPV1 receptors are also activated by the main non-psychotropic ingredient of cannabis, CBD, raising the interesting possibility that some of the therapeutic effects of this alkaloid may be mediated by these channels.

Anandamide is synthesized on demand by enzymatic hydrolysis of the membrane phospholipid N-arachidonoyl phosphatidylethanolamine (NAPE), a process catalyzed by several phospholipases [129, 130]. Following release and activation of CB receptors, anandamide is rapidly removed from the synaptic cleft by a carrier-mediated system [131–134] and subsequently hydrolyzed by the membrane enzyme fatty acid amide hydrolase (FAAH) [135–137]. Anandamide appears to be mainly involved in plastic mechanisms such as long-term depression (LTD) [138]. It should also be noted that, in the striatum, 2-AG and anandamide may serve different roles with respect to the release of glutamate and GABA. 2-AG acts preferentially on CB_1 receptors in GABAergic neurons; in fact, the simulation of its synthesis was found to reduce GABAergic, but not glutamatergic neurotransmission [139, 140]. Conversely, anandamide may inhibit the release of glutamate by activating CB_1 receptors in glutamatergic neurons [141].

Effects of METH and Cannabis on Schizophrenia

In comparison with other psychostimulants, such as cocaine, METH consumption is associated with a markedly high schizophrenia risk [100]. This aspect is particularly noteworthy in consideration of its potential interactions with cannabis, the only other substance of abuse that has been unequivocally linked to a significantly higher vulnerability for schizophrenia and other psychotic disorders [99, 100]. Based on this premise, several studies have been recently focused on the possibility of synergistic effects of cannabis and METH with respect to the pathogenesis of schizophrenia. Although the evidence on these interactions is still rudimentary, recent studies have shown that the combined abuse of cannabis and ATSs is indeed associated with an earlier age of schizophrenia onset, in comparison with consumption of either cannabis or ATSs alone [142].

To understand the nature of the interactions of cannabis and METH with respect to schizophrenia, it is useful to briefly review the evidence on the role of the endocannabinoid system in this disease (for an extensive overview of the topic, see [143–145]. Schizophrenia patients have been found to feature elevated anandamide levels in plasma [146] and cerebrospinal fluid (CSF) [144], as well as higher CB_1 receptor density in prefrontal and cingulate cortex [147–149]. Notably, the levels of CB_1 receptors in schizophrenia patients are down-regulated by antipsychotics [150].

The pathogenic mechanism whereby cannabis and METH may lie to schizophrenia has been posited to lie in DA-induced maladaptive interpretations of contextual cues, which may reflect developmental alterations in adolescence and may be further exacerbated by environmental and psychosocial adversity [151]. Whereas there is general consensus on the primary involvement of DA in the mechanisms supporting METH-induced acute psychotic states [152, 153], the role of this neurotransmitter in cannabis-induced psychosis is more controversial. It has been reported that, in schizophrenia patients, doses of THC that exacerbate psychotic symptoms are associated with a rapid reduction of D_2 receptor binding in the ventral striatum and precommissural dorsal putamen [154, 155]. However, similar phenomena were not observed in healthy volunteers [156]. Furthermore, the psychotomimetic effects of THC are not attenuated by the benchmark typical antipsychotic haloperidol, which acts as a D_2 receptor antagonist. Indeed, this drug was even found to exacerbate some of the cognitive deficits induced by THC, such as distractibility and reduced vigilance [157].

Irrespective of the mechanism, emerging evidence supports that METH's psychotomimetic properties may be modulated by cannabis through activation of CB_1 receptors. A recent genetic study [158] found that the latency to the onset of psychotic responses to METH consumption is associated with variants of a single-nucleotide polymorphism (Rs806379) of the gene CNR1 (encoding the CB_1 receptor). Notably, this gene has been associated with schizophrenia vulnerability in several studies [159, 160]. While preclinical results have suggested that antagonism of CB_1 may attenuate schizophrenia symptoms [161, 162], preliminary clinical trials have failed to support this possibility [163].

Effects of METH and Cannabis on Abuse and Dependence

Another important domain of investigation on the clinical interactions of ATSs and cannabis concerns the establishment of abuse and dependence. Cannabis has long been posited to serve as a "gateway" drug, which may facilitate the subsequent abuse and dependence of other substances [164–166]. This characteristic may be potentiated by METH abuse and dependence, which have been associated with a higher proclivity to engage in risky behaviors [167, 168]. Indeed, the concurrent abuse of cannabis and METH has been recently found to be associated with earlier initiation to ecstasy use [169].

A plausible interpretation for a combined effect of METH and cannabis on the initiation to the use of other drugs is likely to reflect abnormalities in DA striatal neurotransmission, resulting in abnormal decision-making processes related to motivational responses. In keeping with this interpretation, Churchwell and collaborators [170] found higher novelty-seeking and striatal volume in adolescents reporting comorbid abuse of METH and cannabis.

Several lines of evidence suggest that the endocannabinoid system may play a role in the subjective effects of ATSs, and therefore influence the risk for abuse and dependence [171]. Allelic variants of the *FAAH* gene have been associated with the subjective response to AMPH [172], as well as a higher risk for substance abuse and dependence [173, 174]. Similarly, genetic variants of the [AAT]n trinucleotide short-tandem repeat polymorphism of the *CNR1* gene (encoding CB_1 receptors) have been associated with an increased risk for intravenous use of AMPH [175] as well as other drugs [176]. The role of CB_1 receptors in the subjective properties of ATSs is also supported by the finding that its blockade in *Cebus* monkeys reduced the arousal induced by AMPH [177].

The facilitatory role of cannabis with respect to METH abuse and dependence is generally supported by several lines of evidence on rodent models. Indeed, while pre-exposure to THC does not appear to affect the self-administration of AMPH in rats [178], the blockade of CB_1 receptors attenuates METH self-administration [179, 180]. Notably, this effect may not reflect an intrinsic reduction of the rewarding properties of METH, but rather the acquisition and consolidation of the preference for this drug [181, 182], suggesting that the key effect of cannabis may modulate the plastic and adaptive response to repeated administrations of ATSs. In particular, these phenomena are likely to be mediated by CB_1 receptors in the nucleus accumbens, possibly in relation to the modulation of acetylcholine neurotransmission [183, 184]. In keeping with this hypothesized mechanism, CB_1 receptors may also affect the reinstatement of METH self-administration [185, 186]. Notably, CB_1 receptor antagonists have also been shown to reduce METH-induced deficits in operant responding [187] and inhibitory control [188]. Of note, THC has been shown to potentiate the extinction of AMPH-induced conditioned preference learning [189], but this mechanism was not reversed by CB_1 receptor blockade, supporting a possible role of CB_2 receptors.

Role of CB_1 Receptors in the Psychostimulant Properties of ATSs

As mentioned above, the clinical evidence on the role of CB_1 receptors on the psychostimulant effects of AMPH and METH is only limited to anecdotal evidence and indirect inferences based on genetic studies. Conversely, several studies on the topic have been performed in rodent models. In general, the bulk of evidence suggests that CB_1 receptors in the nucleus accumbens may contribute to some of the motor effects of ATSs, including AMPH and METH-induced hyperactivity and stereotyped behaviors [190–193]. These effects are posited to reflect a negative modulatory action on DAergic neurotransmission; in fact, CB_1 receptor blockade has been shown to potentiate, rather than attenuate, the stereotyped behavior induced by co-administration of D_1 and D_2 receptor antagonists [194]; furthermore, activation of CB_1 receptors has been shown to reduce the hyperactivity induced by D_2 receptor stimulation [195]. In apparent contrast with this evidence, several studies have indicated that the genetic inactivation of CB_1 receptors may attenuate the hyperactivity induced by AMPH [162]; these effects, however, have not been consistently replicated [196, 197], possibly in relation to different influences of genetic and environmental vulnerability factors. The alterations of ATS-induced effects in CB_1 knockout (KO) mice may reflect their lower levels of striatal DA [198], as well as abnormalities in D_2 (but not D_1) receptor binding in the striatum [196].

Finally, several studies have shown that CB_1 receptors play a role in the development of motor sensitization to AMPH. Chronic THC administration facilitates this phenomenon [199], while CB_1 receptor inactivation decreases the development of motor sensitization to AMPH [200–202].

Mechanisms of Interactions of Cannabinoids and ATSs

Cannabinoids and ATSs are posited to interact on multiple levels, through a complex array of intersecting mechanisms across several brain regions. In this synoptic overview, we will summarize the best-characterized lines of evidence on the collective implication of these substances on the regulation of DAergic neurotransmission and its behavioral and pathophysiological correlates. Nevertheless, we should point out that these mechanisms are part of a broader network that incorporates the actions of the other neurotransmitters affected by both ATSs and cannabinoids, such as NE and 5-HT. With respect to the role of DAergic pathways, the interplay of cannabis and ATSs is likely related to the endogenous modulatory mechanisms of this system, which involve both trace amines (such as β-PEA) and endocannabinoids as well as their attending receptors in DA neurotransmission and signaling [203–210].

As noted above, although CB_1 receptors are expressed in DAergic neurons [211–213], most of the effects of cannabinoids on DAergic neurotransmission are posited to be the result of indirect mechanisms, primarily mediated by the activation of CB_1 receptors on presynaptic terminals of neighboring GABAergic and glutamatergic neurons (Fig. 16.3). This modulation occurs both around the somata of DAergic neurons in the midbrain, as well as along their terminals, in the dorsal striatum, nucleus accumbens and prefrontal cortex [214]. The actions of cannabinoids mimic the molecular activity of 2-AG and anandamide on the activity of mesolimbic and mesocortical DAergic neurons.

In general, the effects of cannabinoids on the activity of these cells are highly variable, and may follow dose-dependent patterns, likely reflective of the progressive recruitment of different subpopulation of neurons subserving different modulatory roles in relation to DAergic activity. In line with this concept, the loss of GABAergic inhibition in CB_1-positive neurons has been recently shown to counter the DA-releasing properties of AMPH [215]. In addition, the variability of the effects of cannabis depends on a wide set of genetic and environmental vulnerability factors [216], including sex (see Chap. 12 of this book); some of these variable, such as stress, are known to affect the sensitivity to ATSs [217, 218]. In summary, the direction and verse of the modifications of DAergic activity ensuing the co-administration of ATSs and cannabinoids are heterogeneous, depending on specific individual characteristics as well as dose-dependent modalities of action on various circuitries associated with DAergic pathways. In spite of this high variability, pre-

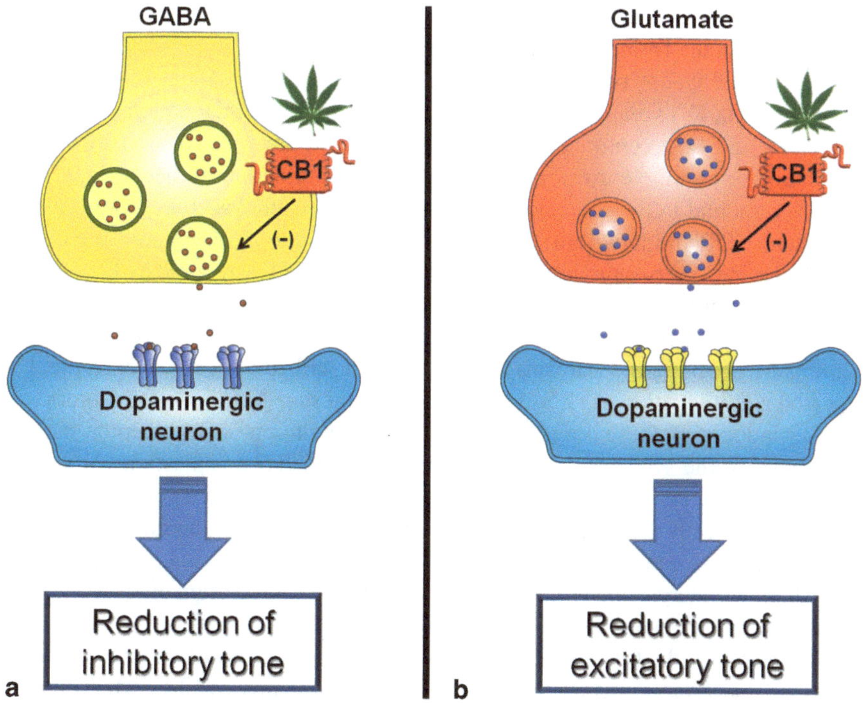

Fig. 16.3 Schematic representation of CB_1 receptor-mediated effects in GABAergic (**a**) and glutamatergic neurons (**b**). Activation of CB_1 receptors in these two neurons exert opposite effects with respect to the modulation of DAergic neurons. This mechanism, which plays a key role in the adaptive plasticity of the DAergic system, sets the stage for some of the most critical interactions between cannabinoids and ATSs (which act as DA releasers)

clinical studies in animal models have enabled a preliminary characterization of the key domains of mutual interaction between cannabinoids and ATSs.

The stimulation of D_2 receptors (one of the key direct molecular effects of ATS) triggers anandamide synthesis in the striatum [219]; it appears that such process is teleologically directed at the attenuation of DA release and the reduction of DAergic psychomotor activation [219]. Notably, this mechanism may be contributed by TRPV1 receptors [220], which are abundantly expressed in DAergic neurons. In contrast, higher doses of cannabinoids (which are posited to stimulate CB_1 and CB_2 receptors across multiple sites) generally increase the activity of mesolimbic DAergic system including neuronal firing, DA release and metabolism and expression of D_1 receptors [221]. Accordingly, CB_1 receptor stimulation leads to the exacerbation of DA release in the nucleus accumbens induced by METH and AMPH [187, 222].

The bulk of evidence indicates that the endocannabinoid system is one of the main orchestrators of the plasticity of DAergic neurons [223–227]; accordingly, DA deficiency leads to a pronounced up-regulation of CB_1 receptors [228–230]. Based on these premises, it is possible that the interactions of ATSs and cannabinoids may be supported by mechanisms aimed at shaping short-term and long-term adaptive plasticity of the DAergic system.

These processes are likely to be regulated also by trace amines, and particularly β-PEA. This trace amine is likely to exert a modulatory role on DAergic plasticity through modifications of DA efflux modality in response to salient environmental inputs. Indeed, DA volume transmission may lead to differential patterns of activation of D_1 and D_2 receptors across the spines of medium-spiny neurons, the main population of output neurons in the striatum. The stimulation of these targets is in turn instrumental to the enactment of plasticity phenomena, such as long-term potentiation (LTP) and long-term depression (LTD). D_1 and D_2 receptors have differential roles in these two processes within the striatum: LTP is favored by D_1 receptor stimulation, but inhibited by D_2 receptor activation [231–233].

It is highly likely that ATSs may influence the synaptic plasticity of DAergic neurons by adopting mechanisms akin to those described above. For example, the repeated administration of ATSs leads to behavioral sensitization to the motoric responses induced by these drugs [234–236]. The stimulation of DA receptor, in turn, contributes to the synthesis of endocannabinoids, which shape plasticity processes through their action on GABAergic and glutamatergic neurons in close proximity with DAergic cells (see Chapt. 19 of this book). For example, the modulatory role of CB_1 receptors on the firing and activity of DAergic neurons is largely mediated by both glutamatergic and GABAergic neurons in the ventral tegmental area (VTA) of the midbrain, where they are abundantly expressed [237, 238].

On one hand, the GABAergic neurons of the VTA are posited to exert a tonic inhibition of DAergic neurons; thus, the activation of presynaptic CB_1 receptors leads to a reduction of GABA release, thereby increasing the activity of DAergic neurons [239, 240]. Physiologically, these CB_1 receptors are activated by 2-AG synthesized by the somatodendritic compartments of the DAergic neurons in the VTA. This phenomenon appears to be instrumental for habit formation [241] and may be essential for the enactment of responses to chronic ATS administration, such as sensitization to AMPH.

On the other hand, the initiation of sensitization to AMPH-induced hyperactivity is related to changes in glutamatergic transmission within the VTA [242], which lead to alterations in plasticity of the DAergic neurons in this area [243–247]. Indeed, the sensitization to AMPH is contributed by enhancements in glutamate receptor expression and increased responsiveness to glutamate in the synapses of the VTA, with a resulting suppression of LTD mechanisms [248–252]. This process is likely shaped by endocannabinoids. Although the mechanisms of this involvement are not completely clear, it is known that the cell bodies of DAergic neurons in the VTA auto-regulate their firing and bursting activity through the synthesis of 2-AG in response to metabotropic glutamate receptor stimulation [223]; the newly-synthesized 2-AG activates presynaptic CB_1 receptors by retrograde action, leading to the reduction of glutamate release [223].

Cannabinoids may also interact with ATSs by affecting D_1 and D_2 receptor responses in medium-spiny neurons. These interactions are based on the role of endocannabinoids as key modulators of DAergic neurotransmission in the basal ganglia [253–255]. Notably, CB_1 receptors are abundantly expressed in striatal neurons [256–259] and interact with both D_1 and D_2 receptors [260, 261]. Preliminary evidence suggests that the combined activity of ATSs and cannabinoids may have differential effects on these two receptors. Accordingly, the transcripts of D_1 and D_2 receptors in striatum are respectively up-and down-regulated by the repeated treatment with METH and the anandamide analog methanandamide [262].

CB_1 receptor activation is posited to modulate the effects of striatal D_2 receptor signaling, as well as their effects on motor function [195, 219, 260, 263–265]. CB_1 receptors in the striatum are thought to condition the trafficking the D_2 receptors in response to activation [266]. The interaction of CB_1 and DAergic receptors is likely instrumental for the enactment of key plasticity processes, such as LTD and LTP. In the striatum, these mechanisms are actually influenced by both D_1 and D_2 receptors [235, 267–270]. The enactment of long-term plasticity at the striatal level is likely essential to shape the pattern of activation of this region in response to glutamatergic inputs from cortical neurons [271]. The effects of CB_1 receptors on D_1 and D_2 receptor signaling involve dopamine- and cAMP-regulated phosphoprotein of 32 kDa (DARPP-32) [265, 272], whose activation is also a fundamental requirement for LTD and LTP [273]. Furthermore, CB_1 and D_2 receptors are known to form heteromeric complexes, which, unlike the two individual receptors (which are coupled to G_I/G_o proteins), is coupled to a G_s protein [254, 274]. This suggests that the formation of these complexes can lead to significantly different phenotypical results than those produced by stimulation of each receptor [122].

One of the principal processes that may support the interactions of ATSs and cannabis with respect to the regulation of DAergic neurotransmission is LTD. This mechanism requires endocannabinoids in the striatum [224]. Several lines of evidence support the possibility that anandamide may be particularly implicated in this process. Indeed, this endocannabinoid has been shown to play an essential role in the developmental orchestration of LTD mechanisms [138]. Anandamide, but not 2-AG, is selectively produced by activation of D_2 receptors in striatum [219]. D_2 receptors have been shown to be necessary for LTD induction [232]. Notably, the implication of this anandamide in LTD is not limited to the striatum, but has also been attested in other brain regions, such as amygdala [275] and hippocampus [276]. In the latter region, it has been notably found that anandamide mediates LTD through activation of TRPV1, but not CB_1 receptors [276].

Indeed, it is interesting to note that some of the effects of cannabis may be mediated by TRPV1 receptors [277]. CBD and other ingredients of cannabis (such as cannabigerol, cannabigevarine and Δ9-Tetrahydrocannabinol) have been shown to activate these receptors [278–280]. Interestingly, Moreira and Guimarães [281] found that CBD countered the hyperlocomotive effects of AMPH, without inducing extrapyramidal-like effects. TRPV1 are activated by anandamide as well as N-arachidonoyl-dopamine (NADA), which is formed by DA linked to arachidonic acid by an amide bond, conferring properties of endocannabinoid and endovanilloid ligand [282]. This mechanism consists in the conjugation of arachidonic acid directly with DA [283]; while the role of this compound is not fully understood, recent evidence supports the possibility that it may be an antioxidant and exert neuroprotective properties [284]. Notably, anandamide has been shown to inhibit DAT through a mechanism not dependent on G-protein-coupled proteins [285], which may be related to the activation of TRPV1 receptors.

CB_2 receptors may be also implicated in the interactions of cannabis and ATSs. Accordingly, CB_2 receptors have been recently discovered in the brain and may play a role in certain mental disorders [124, 286]. The involvement of CB_2 receptors in the modulation of DAergic transmission is supported by the reduced expression of D_2 receptor in the prefrontal cortex of CB_2 knockout mice, as well as their enhanced responsiveness to cocaine [287]. Nevertheless, the involvement of CB_2 receptors in the outcomes of METH has been partially challenged by recent studies, finding the lack of implications of this receptor in the behavioral effects of METH [200].

Role of Cannabis in the Outcomes of METH Neurotoxicity

Another important theme of the potential combined role of cannabis and METH concerns the influence of cannabinoids on the neurotoxic sequelae induced by METH. In the striatum, METH neurotoxicity reflects deficits of DAergic, glutamatergic and GABAergic parvalbumin-positive neurons [288]; given the profound involvement of the endocannabinoid system in the regulation of GABA and glutamate signaling, it is expectable that METH neurotoxicity may be affected by cannabinoids and, in turn, alter the subjective responses to cannabis.

Although this important theme has been targeted by few clinical studies, a seminal contribution in this respect has been afforded by a study by Gonzalez and colleagues [289], who reported that heavy cannabis use did not exacerbate METH-induced cognitive impairments. On the contrary, users of both substances were found to display a milder severity of their neuropsychological deficits in comparison with users of METH alone, suggesting a protective role of cannabis against METH-induced abnormalities [289].

To verify the mutual interactions of cannabinoids and METH neurotoxicity, our group examined the effects of a "binge" schedule of METH (consisting in repeated administrations of high METH doses at short time intervals) on the expression and behavioral function of brain CB_1 receptors. METH neurotoxicity led to a significant increase of CB_1 receptor expression across key brain regions implicated in behavioral regulation (prefrontal cortex, striatum, amygdala and hippocampus) [216]. This up-regulation of CB_1 receptors following METH excitotoxicity is in line with previous evidence on similar phenomena consequent to neurotoxic insults [290, 291]. The bases of this phenomenon may be related to the well-characterized neuroprotective and anti-inflammatory actions of cannabinoids [292–296]. Thus, it is possible that the toxicity caused by ROSs may have stimulated CB_1 receptor upregulation as a countermeasure to curtail the deleterious impact of this drug. In addition, DA receptors may be involved in these phenomena. Interestingly, CB_1 receptor expression is increased by the lesion of DA terminals due to lesions [297]. Specifically, it is possible that the up-regulation of CB_1 receptors may limit glutamate efflux, which serves a key mediating role in METH-mediated neurotoxicity [117, 255].

The neuroprotective properties of cannabis may lie in the ability of THC to mimic the actions of 2-AG in inhibiting the release of glutamate by *depolarization-induced suppression of excitation* (DSE). Accordingly, CB_1 receptor agonists have been shown to reduce glutamate-mediated excitotoxicity in rodent brains [298–300]. Interestingly, both THC and CBD have been shown to have potent antioxidant properties [301, 302] and reduce the formation of ROSs. Cannabinoids have also been shown to reduce brain injury in ischemia models [292, 303–308], and may be therapeutically efficacious in the treatment of head trauma patients [309].

This background, together with the well-characterized neuroprotective and anti-inflammatory actions of CB_1 receptor agonists [292–296] highlights the possibility that CB_1 receptor synthesis may be stimulated by METH neurotoxicity in specific regions, as a countermeasure to curtail its deleterious impact. This may represent a compensatory mechanism to correct for the impaired GABA transmission.

Interestingly, the up-regulation of CB_1 receptors in METH-exposed rats were associated with an enhancement of anxiolytic properties of cannabinoids. This scenario suggests that METH neurotoxicity may result in altered responsiveness of CB_1 receptors, possibly due to selective damages of specific subpopulation of neurons and homeostatic imbalances of the endocannabinoid system in the brain areas that regulate the modality and intensity of environmental reactivity, such as amygdala, prefrontal cortex and hippocampus [106, 310].

Concluding Remarks

In this review, we have outlined the current knowledge and recent advances on the clinical and preclinical effects of cannabis and ATSs, a growing phenomenon that may have important negative repercussions particularly with respect to the development of psychotic disorders and addiction. We have also explored the mechanisms underlying these interactions, which represent the way for cannabinoids to interfere with the consequences of ATSs. As shown in the review, the interactions among these substances occur at multiple, highly integrated levels, reflecting a complex modulatory mechanism of endocannabinoids and dopamine, as well as other monoamines. Although the specific possibility of direct interactions between CB_1, TRPV1 and TAAR1 remains to be explored, it is likely that studies on these mechanisms may contribute to determine a number of pivotal discoveries in relation to the regulation of DA (also with respect to synaptic and extrasynaptic activation) and shed light on the neurobiological underpinnings of the psychiatric outcomes of the comorbid cannabis and ATS abuse.

Acknowledgments This chapter was partially supported by grants from the National Institute of Health (R21HD070611) and the Tourette Syndrome Association.

Glossary of Acronyms

2-AG	2-arachidonoylglycerol
5-HT	Serotonin
ADHD	Attention-deficit hyperactivity disorder
AMPH	d-amphetamine
ATSs	Amphetamine-type stimulants
CB1	Cannabinoid receptors type 1
CB2	Cannabinoid receptor type 2
CBD	Cannabidiol
CSF	Cerebrospinal fluid
DA	Dopamine
DARPP-32	Dopamine and cAMP-regulated phosphoprotein, 32 kDa
DAT	Dopamine transporter
DSE	Depolarization-induced suppression of excitation
FAAH	Fatty acid amide hydrolase
GABA	γ-aminobutyric acid
KO	Knockout
LTD	Long-term depression
LTP	Long-term potentiation
MAO	Monoamine oxidase
MDMA	3, 4-methylenedioxy-N-methylamphetamine
METH	Methamphetamine
NAPE	N-arachidonoyl phosphatidylethanolamine
NE	Norepinephrine
PKA	Protein kinases A
PKCΔ	Protein kinases C delta
ROSs	Reactive oxygen species
TAAR1	Trace amine associated receptor 1
THC	Δ9-tetrahydrocannabinol
TRPV1	Transient receptor potential cation channel subfamily V member 1
VMAT2	Vesicular monoamine transporter 2
VTA	Ventral tegmental area
β-PEA	β-phenylethylamine

References

1. Edeleano L. Ueber einige Derivate der Phenylmethacrylsäure und der Phenylisobuttersäure. Ber Dtsch Chem Ges. 1887;20:616–22.
2. Weisheit L, White RW. Methamphetamine: its history, pharmacology and treatment. Center City: Hazelden; 2009.
3. Abuse NIoD. NIDA Research Report Series: methamphetamine abuse and addiction. In: Rockville MD, editors. Dept. of Health and Human Services NIoH, National Institute on Drug Abuse. Research Report Series; 2002.

4. Anglin MD, Burke C, Perrochet B, Stamper E, Dawud-Noursi S. History of the methamphetamine problem. J Psychoactive Drugs. 2000;32(2):137–41.
5. D. S. Western Canadian Summit on methamphetamine: bringing together practitioners, policy makers and researchers: consensus panel report. Vancouver: Vancouver Coastal Health; 2005. p. 1–48.
6. Russell K, Dryden DM, Liang Y, Friesen C, O'Gorman K, Durec T, et al. Risk factors for methamphetamine use in youth: a systematic review. BMC Pediatr. 2008;8:48.
7. Iritani BJ, Hallfors DD, Bauer DJ. Crystal methamphetamine use among young adults in the USA. Addiction. 2007;102(7):1102–13.
8. Maxwell JC, Rutkowski BA. The prevalence of methamphetamine and amphetamine abuse in North America: a review of the indicators, 1992–2007. Drug Alcohol Rev. 2008;27(3):229–35.
9. McKetin R, Kozel N, Douglas J, Ali R, Vicknasingam B, Lund J, et al. The rise of methamphetamine in Southeast and East Asia. Drug Alcohol Rev. 2008;27(3):220–8.
10. United Nations, Publications. United Nations Office on Drugs and Crime: World Drug Report 2010. 2010.
11. Hollister L. Interactions of cannabis with other drugs in man. In: Ginzburg MCBHM, editor. Strategies for research on the interactions of drugs of abuse. Rockville: National Institute on Drug Abuse;1986.
12. Kalechstein AD, Newton TF, Longshore D, Anglin MD, van Gorp WG, Gawin FH. Psychiatric comorbidity of methamphetamine dependence in a forensic sample. J Neuropsychiatry Clin Neurosci. 2000;12(4):480–4.
13. Simon SL, Domier CP, Sim T, Richardson K, Rawson RA, Ling W. Cognitive performance of current methamphetamine and cocaine abusers. J Addict Dis. 2002;21(1):61–74.
14. Cottencin O, Rolland B, Karila L. New designer drugs (synthetic cannabinoids and synthetic cathinones): review of literature. Curr Pharm Des. 2014;20(25):4106–11.
15. Fisar Z. Phytocannabinoids and endocannabinoids. Curr Drug Abuse Rev. 2009;2(1):51–75.
16. Uchiyama N, Kawamura M, Kikura-Hanajiri R, Goda Y. URB-754: a new class of designer drug and 12 synthetic cannabinoids detected in illegal products. Forensic Sci Int. 2013;227(1–3):21–32.
17. Berry MD. Mammalian central nervous system trace amines. Pharmacologic amphetamines, physiologic neuromodulators. J Neurochem. 2004;90(2):257–71.
18. Bortolato M, Chen K, Shih JC. Monoamine oxidase inactivation: from pathophysiology to therapeutics. Adv Drug Deliv Rev. 2008;60(13–14):1527–33.
19. Paterson IA, Juorio AV, Boulton AA. 2-Phenylethylamine: a modulator of catecholamine transmission in the mammalian central nervous system? J Neurochem. 1990;55(6):1827–37.
20. Ishida K, Murata M, Katagiri N, Ishikawa M, Abe K, Kato M, et al. Effects of beta-phenylethylamine on dopaminergic neurons of the ventral tegmental area in the rat: a combined electrophysiological and microdialysis study. J Pharmacol Exp Ther. 2005;314(2):916–22.
21. Kuroki T, Tsutsumi T, Hirano M, Matsumoto T, Tatebayashi Y, Nishiyama K, et al. Behavioral sensitization to beta-phenylethylamine (PEA): enduring modifications of specific dopaminergic neuron systems in the rat. Psychopharmacology. 1990;102(1):5–10.
22. Sotnikova TD, Budygin EA, Jones SR, Dykstra LA, Caron MG, Gainetdinov RR. Dopamine transporter-dependent and -independent actions of trace amine beta-phenylethylamine. J Neurochem. 2004;91(2):362–73.
23. Ledonne A, Federici M, Giustizieri M, Pessia M, Imbrici P, Millan MJ, et al. Trace amines depress D(2)-autoreceptor-mediated responses on midbrain dopaminergic cells. Br J Pharmacol. 2010;160(6):1509–20.
24. Borowsky B, Adham N, Jones KA, Raddatz R, Artymyshyn R, Ogozalek KL, et al. Trace amines: identification of a family of mammalian G protein-coupled receptors. Proc Nat Acad Sci U S A. 2001;98(16):8966–71.
25. Miller GM, Verrico CD, Jassen A, Konar M, Yang H, Panas H, et al. Primate trace amine receptor 1 modulation by the dopamine transporter. J Pharmacol Exp Ther. 2005;313(3):983–94.

26. Xie Z, Westmoreland SV, Miller GM. Modulation of monoamine transporters by common biogenic amines via trace amine-associated receptor 1 and monoamine autoreceptors in human embryonic kidney 293 cells and brain synaptosomes. J Pharmacol Exp Ther. 2008;325(2):629–40.
27. Xie Z, Miller GM. Trace amine-associated receptor 1 is a modulator of the dopamine transporter. J Pharmacol Exp Ther. 2007;321(1):128–36.
28. Rothman RB, Baumann MH. Monoamine transporters and psychostimulant drugs. Eur J Pharmacol. 2003;479(1–3):23–40.
29. Han DD, Gu HH. Comparison of the monoamine transporters from human and mouse in their sensitivities to psychostimulant drugs. BMC Pharmacol. 2006;6:6.
30. Eshleman AJ, Henningsen RA, Neve KA, Janowsky A. Release of dopamine via the human transporter. Mol Pharmacol. 1994;45(2):312–6.
31. Sitte HH, Huck S, Reither H, Boehm S, Singer EA, Pifl C. Carrier-mediated release, transport rates, and charge transfer induced by amphetamine, tyramine, and dopamine in mammalian cells transfected with the human dopamine transporter. J Neurochem. 1998;71(3):1289–97.
32. Jones SR, Gainetdinov RR, Wightman RM, Caron MG. Mechanisms of amphetamine action revealed in mice lacking the dopamine transporter. J Neurosci. 1998;18(6):1979–86.
33. Bunzow JR, Sonders MS, Arttamangkul S, Harrison LM, Zhang G, Quigley DI, et al. Amphetamine, 3,4-methylenedioxymethamphetamine, lysergic acid diethylamide, and metabolites of the catecholamine neurotransmitters are agonists of a rat trace amine receptor. Mol Pharmacol. 2001;60(6):1181–8.
34. Fleckenstein AE, Metzger RR, Gibb JW, Hanson GR. A rapid and reversible change in dopamine transporters induced by methamphetamine. Eur J Pharmacol. 1997;323(2–3):R9–10.
35. Xie Z, Westmoreland SV, Bahn ME, Chen GL, Yang H, Vallender EJ, et al. Rhesus monkey trace amine-associated receptor 1 signaling: enhancement by monoamine transporters and attenuation by the D2 autoreceptor in vitro. J Pharmacol Exp Ther. 2007;321(1):116–27.
36. Shin EJ, Duong CX, Nguyen XK, Li Z, Bing G, Bach JH, et al. Role of oxidative stress in methamphetamine-induced dopaminergic toxicity mediated by protein kinase Cdelta. Behav Brain Res. 2012;232(1):98–113.
37. Sandoval V, Riddle EL, Ugarte YV, Hanson GR, Fleckenstein AE. Methamphetamine-induced rapid and reversible changes in dopamine transporter function: an in vitro model. J Neurosci. 2001;21(4):1413–9.
38. Lindemann L, Meyer CA, Jeanneau K, Bradaia A, Ozmen L, Bluethmann H, et al. Trace amine-associated receptor 1 modulates dopaminergic activity. J Pharmacol Exp Ther. 2008;324(3):948–56.
39. Erickson JD, Schafer MK, Bonner TI, Eiden LE, Weihe E. Distinct pharmacological properties and distribution in neurons and endocrine cells of two isoforms of the human vesicular monoamine transporter. Proc Natl Acad Sci U S A. 1996;93(10):5166–71.
40. Nickell JR, Krishnamurthy S, Norrholm S, Deaciuc G, Siripurapu KB, Zheng G, et al. Lobelane inhibits methamphetamine-evoked dopamine release via inhibition of the vesicular monoamine transporter-2. J Pharmacol Exp Ther. 2010;332(2):612–21.
41. Horton DB, Siripurapu KB, Norrholm SD, Culver JP, Hojahmat M, Beckmann JS, et al. meso-Transdiene analogs inhibit vesicular monoamine transporter-2 function and methamphetamine-evoked dopamine release. J Pharmacol Exp Ther. 2011;336(3):940–51.
42. Alvers KM, Beckmann JS, Zheng G, Crooks PA, Dwoskin LP, Bardo MT. The effect of VMAT2 inhibitor GZ-793 A on the reinstatement of methamphetamine-seeking in rats. Psychopharmacology. 2012;224(2):255–62.
43. Fumagalli F, Gainetdinov RR, Wang YM, Valenzano KJ, Miller GW, Caron MG. Increased methamphetamine neurotoxicity in heterozygous vesicular monoamine transporter 2 knockout mice. J Neurosci. 1999;19(7):2424–31.
44. Takahashi N, Miner LL, Sora I, Ujike H, Revay RS, Kostic V, et al. VMAT2 knockout mice: heterozygotes display reduced amphetamine-conditioned reward, enhanced amphetamine locomotion, and enhanced MPTP toxicity. Proc Natl Acad Sci U S A. 1997;94(18):9938–43.

45. Wang YM, Gainetdinov RR, Fumagalli F, Xu F, Jones SR, Bock CB, et al. Knockout of the vesicular monoamine transporter 2 gene results in neonatal death and supersensitivity to cocaine and amphetamine. Neuron. 1997;19(6):1285–96.
46. Miller HH, Shore PA, Clarke DE. In vivo monoamine oxidase inhibition by d-amphetamine. Biochemical Pharmacol. 1980;29(10):1347–54.
47. Kita T, Philbert MA, Wagner GC, Huang J, Lowndes HE. Methamphetamine-induced modification of dopamine metabolism in cultured striatal astrocytes. Pharmacol Toxicol. 1998;83(1):36–9.
48. Descarries L, Watkins KC, Garcia S, Bosler O, Doucet G. Dual character, asynaptic and synaptic, of the dopamine innervation in adult rat neostriatum: a quantitative autoradiographic and immunocytochemical analysis. J Comp Neurol. 1996;375(2):167–86.
49. Descarries L, Mechawar N. Ultrastructural evidence for diffuse transmission by monoamine and acetylcholine neurons of the central nervous system. Prog Brain Res. 2000;125:27–47.
50. Fuxe K, Dahlstrom A, Hoistad M, Marcellino D, Jansson A, Rivera A, et al. From the Golgi-Cajal mapping to the transmitter-based characterization of the neuronal networks leading to two modes of brain communication: wiring and volume transmission. Brain Res Rev. 2007;55(1):17–54.
51. Zoli M, Agnati LF. Wiring and volume transmission in the central nervous system: the concept of closed and open synapses. Prog Neurobiol. 1996;49(4):363–80.
52. Murray RM, Lappin J, Di Forti M. Schizophrenia: from developmental deviance to dopamine dysregulation. Eur Neuropsychopharmacol. 2008;18 Suppl 3:129–34.
53. Braff DL, Geyer MA. Sensorimotor gating and schizophrenia. Human and animal model studies. Arch Gen Psychiatry. 1990;47(2):181–8.
54. Swerdlow NR, Stephany N, Wasserman LC, Talledo J, Shoemaker J, Auerbach PP. Amphetamine effects on prepulse inhibition across-species: replication and parametric extension. Neuropsychopharmacology. 2003;28(4):640–50.
55. Chou YH, Karlsson P, Halldin C, Olsson H, Farde L. A PET study of D(1)-like dopamine receptor ligand binding during altered endogenous dopamine levels in the primate brain. Psychopharmacology. 1999;146(2):220–7.
56. Bergson C, Mrzljak L, Smiley JF, Pappy M, Levenson R, Goldman-Rakic PS. Regional, cellular, and subcellular variations in the distribution of D1 and D5 dopamine receptors in primate brain. J Neurosci. 1995;15(12):7821–36.
57. Yung KK, Bolam JP, Smith AD, Hersch SM, Ciliax BJ, Levey AI. Immunocytochemical localization of D1 and D2 dopamine receptors in the basal ganglia of the rat: light and electron microscopy. Neuroscience. 1995;65(3):709–30.
58. Sesack SR, Aoki C, Pickel VM. Ultrastructural localization of D2 receptor-like immunoreactivity in midbrain dopamine neurons and their striatal targets. J Neurosci. 1994;14(1):88–106.
59. Tirotta E DMC, Iitaka C, Ramos M, Holmes D, Borrelli E. U. Unraveling the role of dopamine receptors in vivo: lessons from knockout mice. In: KA N. The dopamine receptors. 2 ed. New York: Humana; 2010. p. 303–22.
60. Marcellino D, Kehr J, Agnati LF, Fuxe K. Increased affinity of dopamine for D(2)-like versus D(1) -like receptors. Relevance for volume transmission in interpreting PET findings. Synapse. 2012;66(3):196–203.
61. Rolinski Z, Scheel-Kruger J. The effect of dopamine and noradrenaline antagonists on amphetamine induced locomotor activity in mice and rats. Acta Pharmacol Toxicol. 1973;33(5):385–99.
62. Paulus MP, Geyer MA. A scaling approach to find order parameters quantifying the effects of dopaminergic agents on unconditioned motor activity in rats. Prog Neuropsychopharmacol Biol Psychiatry.1991;15(6):903–19.
63. Lapin IP, Rogawski MA. Effects of D1 and D2 dopamine receptor antagonists and catecholamine depleting agents on the locomotor stimulation induced by dizocilpine in mice. Behav Brain Res. 1995;70(2):145–51.

64. O'Neill MF, Shaw G. Comparison of dopamine receptor antagonists on hyperlocomotion induced by cocaine, amphetamine, MK-801 and the dopamine D1 agonist C-APB in mice. Psychopharmacology. 1999;145(3):237–50.
65. Ralph RJ, Varty GB, Kelly MA, Wang YM, Caron MG, Rubinstein M, et al. The dopamine D2, but not D3 or D4, receptor subtype is essential for the disruption of prepulse inhibition produced by amphetamine in mice. J Neurosci. 1999;19(11):4627–33.
66. Kim SE, Han SM. Nicotine- and methamphetamine-induced dopamine release evaluated with in-vivo binding of radiolabelled raclopride to dopamine D2 receptors: comparison with in-vivo microdialysis data. Int J Neuropsychopharmacol. 2009;12(6):833–41.
67. Gonon F. Prolonged and extrasynaptic excitatory action of dopamine mediated by D1 receptors in the rat striatum in vivo. J Neurosci. 1997;17(15):5972–8.
68. Centonze D, Picconi B, Baunez C, Borrelli E, Pisani A, Bernardi G, et al. Cocaine and amphetamine depress striatal GABAergic synaptic transmission through D2 dopamine receptors. Neuropsychopharmacology. 2002;26(2):164–75.
69. Cepeda C, Hurst RS, Altemus KL, Flores-Hernandez J, Calvert CR, Jokel ES, et al. Facilitated glutamatergic transmission in the striatum of D2 dopamine receptor-deficient mice. J Neurophysiol. 2001;85(2):659–70.
70. Chesselet MF, Plotkin JL, Wu N, Levine MS. Development of striatal fast-spiking GABAergic interneurons. Prog Brain Res. 2007;160:261–72.
71. Bamford NS, Zhang H, Schmitz Y, Wu NP, Cepeda C, Levine MS, et al. Heterosynaptic dopamine neurotransmission selects sets of corticostriatal terminals. Neuron. 2004;42(4):653–63.
72. Jones S, Kauer JA. Amphetamine depresses excitatory synaptic transmission via serotonin receptors in the ventral tegmental area. J Neurosci. 1999;19(22):9780–7.
73. Rothman RB, Baumann MH, Dersch CM, Romero DV, Rice KC, Carroll FI, et al. Amphetamine-type central nervous system stimulants release norepinephrine more potently than they release dopamine and serotonin. Synapse. 2001;39(1):32–41.
74. Ventura R, Cabib S, Alcaro A, Orsini C, Puglisi-Allegra S. Norepinephrine in the prefrontal cortex is critical for amphetamine-induced reward and mesoaccumbens dopamine release. J Neurosci. 2003;23(5):1879–85.
75. Del Arco A Martinez R Mora F. Amphetamine increases extracellular concentrations of glutamate in the prefrontal cortex of the awake rat: a microdialysis study. Neurochem Res. 1998;23(9):1153–8.
76. Miele M, Mura MA, Enrico P, Esposito G, Serra PA, Migheli R, et al. On the mechanism of d-amphetamine-induced changes in glutamate, ascorbic acid and uric acid release in the striatum of freely moving rats. Br J Pharmacol. 2000;129(3):582–8.
77. Paladini CA, Fiorillo CD, Morikawa H, Williams JT. Amphetamine selectively blocks inhibitory glutamate transmission in dopamine neurons. Nat Neurosci. 2001;4(3):275–81.
78. Ernst T, Chang L, Leonido-Yee M, Speck O. Evidence for long-term neurotoxicity associated with methamphetamine abuse: a 1H MRS study. Neurology. 2000;54(6):1344–9.
79. Volkow ND, Chang L, Wang GJ, Fowler JS, Ding YS, Sedler M, et al. Low level of brain dopamine D2 receptors in methamphetamine abusers: association with metabolism in the orbitofrontal cortex. Am J Psychiatry. 2001;158(12):2015–21.
80. Cubells JF, Rayport S, Rajendran G, Sulzer D. Methamphetamine neurotoxicity involves vacuolation of endocytic organelles and dopamine-dependent intracellular oxidative stress. J Neurosci. 1994;14(4):2260–71.
81. Larsen KE, Fon EA, Hastings TG, Edwards RH, Sulzer D. Methamphetamine-induced degeneration of dopaminergic neurons involves autophagy and upregulation of dopamine synthesis. J Neurosci. 2002;22(20):8951–60.
82. Cadet JL, Sheng P, Ali S, Rothman R, Carlson E, Epstein C. Attenuation of methamphetamine-induced neurotoxicity in copper/zinc superoxide dismutase transgenic mice. J Neurochem. 1994;62(1):380–3.
83. Hirata H, Ladenheim B, Rothman RB, Epstein C, Cadet JL. Methamphetamine-induced serotonin neurotoxicity is mediated by superoxide radicals. Brain Res. 1995;677(2):345–7.

84. Yamamoto BK, Zhu W. The effects of methamphetamine on the production of free radicals and oxidative stress. J Pharmacol Exp Ther. 1998;287(1):107–14.
85. Perfeito R, Cunha-Oliveira T, Rego AC. Reprint of: revisiting oxidative stress and mitochondrial dysfunction in the pathogenesis of Parkinson disease-resemblance to the effect of amphetamine drugs of abuse. Free Radic Biol Med. 2013;62:186–201.
86. Kiyatkin EA, Sharma HS. Acute methamphetamine intoxication: brain hyperthermia, blood-brain barrier, brain edema, and morphological cell abnormalities. Int Rev Neurobiol. 2009;88:65–100.
87. Bowyer JF, Davies DL, Schmued L, Broening HW, Newport GD, Slikker W, Jr., et al. Further studies of the role of hyperthermia in methamphetamine neurotoxicity. J Pharmacol Exp Ther. 1994;268(3):1571–80.
88. Davidson C, Gow AJ, Lee TH, Ellinwood EH. Methamphetamine neurotoxicity: necrotic and apoptotic mechanisms and relevance to human abuse and treatment. Brain Res Brain Res Rev. 2001;36(1):1–22.
89. Vergo S, Johansen JL, Leist M, Lotharius J. Vesicular monoamine transporter 2 regulates the sensitivity of rat dopaminergic neurons to disturbed cytosolic dopamine levels. Brain Res. 2007;1185:18–32.
90. Guillot TS, Shepherd KR, Richardson JR, Wang MZ, Li Y, Emson PC, et al. Reduced vesicular storage of dopamine exacerbates methamphetamine-induced neurodegeneration and astrogliosis. J Neurochem. 2008;106(5):2205–17.
91. Hadlock GC, Chu PW, Walters ET, Hanson GR, Fleckenstein AE. Methamphetamine-induced dopamine transporter complex formation and dopaminergic deficits: the role of D2 receptor activation. J Pharmacol Exp Ther. 2010;335(1):207–12.
92. Mark KA, Soghomonian JJ, Yamamoto BK. High-dose methamphetamine acutely activates the striatonigral pathway to increase striatal glutamate and mediate long-term dopamine toxicity. J Neurosci. 2004;24(50):11449–56.
93. Sonsalla PK, Nicklas WJ, Heikkila RE. Role for excitatory amino acids in methamphetamine-induced nigrostriatal dopaminergic toxicity. Science. 1989;243(4889):398–400.
94. Deng X, Cadet JL. Methamphetamine-induced apoptosis is attenuated in the striata of copper-zinc superoxide dismutase transgenic mice. Brain Res Mol Brain Res. 2000;83(1–2):121–4.
95. Deng X, Cai NS, McCoy MT, Chen W, Trush MA, Cadet JL. Methamphetamine induces apoptosis in an immortalized rat striatal cell line by activating the mitochondrial cell death pathway. Neuropharmacology. 2002;42(6):837–45.
96. Cadet JL, Krasnova IN, Jayanthi S, Lyles J. Neurotoxicity of substituted amphetamines: molecular and cellular mechanisms. Neurotox Res. 2007;11(3–4):183–202.
97. Hall W, Solowij N. Adverse effects of cannabis. Lancet. 1998;352(9140):1611–6.
98. Salo R, Nordahl TE, Natsuaki Y, Leamon MH, Galloway GP, Waters C, et al. Attentional control and brain metabolite levels in methamphetamine abusers. Biol Psychiatry. 2007;61(11):1272–80.
99. Degenhardt L, Hall W. Is cannabis use a contributory cause of psychosis? Can J Psychiatry. 2006;51(9):556–65.
100. Callaghan RC, Cunningham JK, Allebeck P, Arenovich T, Sajeev G, Remington G, et al. Methamphetamine use and schizophrenia: a population-based cohort study in California. Am J Psychiatry. 2012;169(4):389–96.
101. Arseneault L, Moffit TE, Caspi A, Taylor A. The targets of violence committed by young offenders with alcohol dependence, marijuana dependence and schizophrenia-spectrum disorders: findings from a birth cohort. Crim Behav Ment Health. 2002;12(2):155–68.
102. Henquet C, Krabbendam L, Spauwen J, Kaplan C, Lieb R, Wittchen HU, et al. Prospective cohort study of cannabis use, predisposition for psychosis, and psychotic symptoms in young people. BMJ. 2005;330(7481):11.
103. Evans MA, Martz R, Rodda BE, Lemberger L, Forney RB. Effects of marihuana-dextroamphetamine combination. Clin Pharmacol Ther. 1976;20(3):350–8.
104. Foltin RW, Fischman MW, Pippen PA, Kelly TH. Behavioral effects of cocaine alone and in combination with ethanol or marijuana in humans. Drug Alcohol Depend. 1993;32(2):93–106.

105. Hayase T, Yamamoto Y, Yamamoto K. Persistent anxiogenic effects of a single or repeated doses of cocaine and methamphetamine: interactions with endogenous cannabinoid receptor ligands. Behav Pharmacol. 2005;16(5–6):395–404.
106. Tambaro S, Bortolato M. Cannabinoid-related agents in the treatment of anxiety disorders: current knowledge and future perspectives. Recent Pat CNS Drug Discov. 2012;7(1):25–40.
107. Fowler JS, Volkow ND, Logan J, Alexoff D, Telang F, Wang GJ, et al. Fast uptake and long-lasting binding of methamphetamine in the human brain: comparison with cocaine. Neuroimage. 2008;43(4):756–63.
108. De Petrocellis L Cascio MG Di Marzo V. The endocannabinoid system: a general view and latest additions. Br J Pharmacol. 2004;141(5):765–74.
109. Mechoulam R, Parker LA. The endocannabinoid system and the brain. Annu Rev Psychol. 2013;64:21–47.
110. Howlett AC, Qualy JM, Khachatrian LL. Involvement of Gi in the inhibition of adenylate cyclase by cannabimimetic drugs. Mol Pharmacol. 1986;29(3):307–13.
111. Matsuda LA, Lolait SJ, Brownstein MJ, Young AC, Bonner TI. Structure of a cannabinoid receptor and functional expression of the cloned cDNA. Nature. 1990;346(6284):561–4.
112. Munro S, Thomas KL, Abu-Shaar M. Molecular characterization of a peripheral receptor for cannabinoids. Nature. 1993;365(6441):61–5.
113. Freund TF, Katona I, Piomelli D. Role of endogenous cannabinoids in synaptic signaling. Physiol Rev. 2003;83(3):1017–66.
114. Mackie K. Distribution of cannabinoid receptors in the central and peripheral nervous system. Handb Exp Pharmacol. 2005;168:299–325.
115. Morishita W, Alger BE. Evidence for endogenous excitatory amino acids as mediators in DSI of GABA(A)ergic transmission in hippocampal CA1. J Neurophysiol. 1999;82(5):2556–64.
116. Wilson RI, Nicoll RA. Endogenous cannabinoids mediate retrograde signalling at hippocampal synapses. Nature. 2001;410(6828):588–92.
117. Ohno-Shosaku T, Maejima T, Kano M. Endogenous cannabinoids mediate retrograde signals from depolarized postsynaptic neurons to presynaptic terminals. Neuron. 2001;29(3):729–38.
118. Varma N, Carlson GC, Ledent C, Alger BE. Metabotropic glutamate receptors drive the endocannabinoid system in hippocampus. J Neurosci. 2001;21(24):RC188.
119. Lovinger DM. Presynaptic modulation by endocannabinoids. Handb Exp Pharmacol. 2008;184:435–77.
120. Pertwee RG, Howlett AC, Abood ME, Alexander SP, Di Marzo V, Elphick MR, et al. International Union of Basic and Clinical Pharmacology. LXXIX. Cannabinoid receptors and their ligands: beyond CB(1) and CB(2). Pharmacol Rev. 2010;62(4):588–631.
121. Hudson BD, Hebert TE, Kelly ME. Ligand- and heterodimer-directed signaling of the CB(1) cannabinoid receptor. Mol Pharmacol. 2010;77(1):1–9.
122. Ferre S, Goldberg SR, Lluis C, Franco R. Looking for the role of cannabinoid receptor heteromers in striatal function. Neuropharmacology. 2009;56 Suppl 1:226–34.
123. Walter L, Stella N. Cannabinoids and neuroinflammation. Br J Pharmacol. 2004;141(5):775–85.
124. Van Sickle MD, Duncan M, Kingsley PJ, Mouihate A, Urbani P, Mackie K, et al. Identification and functional characterization of brainstem cannabinoid CB2 receptors. Science. 2005;310(5746):329–32.
125. Gong JP, Onaivi ES, Ishiguro H, Liu QR, Tagliaferro PA, Brusco A, et al. Cannabinoid CB2 receptors: immunohistochemical localization in rat brain. Brain Res. 2006;1071(1):10–23.
126. Mechoulam R, Ben-Shabat S, Hanus L, Ligumsky M, Kaminski NE, Schatz AR, et al. Identification of an endogenous 2-monoglyceride, present in canine gut, that binds to cannabinoid receptors. Biochem Pharmacol. 1995;50(1):83–90.
127. Sugiura T, Kondo S, Sukagawa A, Nakane S, Shinoda A, Itoh K, et al. 2-Arachidonoylglycerol: a possible endogenous cannabinoid receptor ligand in brain. Biochem Biophys Res Commun. 1995;215(1):89–97.

128. Devane WA, Hanus L, Breuer A, Pertwee RG, Stevenson LA, Griffin G, et al. Isolation and structure of a brain constituent that binds to the cannabinoid receptor. Science. 1992;258(5090):1946–9.
129. Okamoto Y, Morishita J, Tsuboi K, Tonai T, Ueda N. Molecular characterization of a phospholipase D generating anandamide and its congeners. J Biol Chem. 2004;279(7):5298–305.
130. Sun YX, Tsuboi K, Okamoto Y, Tonai T, Murakami M, Kudo I, et al. Biosynthesis of anandamide and N-palmitoylethanolamine by sequential actions of phospholipase A2 and lysophospholipase D. Biochem J. 2004;380(Pt 3):749–56.
131. Di Marzo V, Fontana A, Cadas H, Schinelli S, Cimino G, Schwartz JC, et al. Formation and inactivation of endogenous cannabinoid anandamide in central neurons. Nature. 1994;372(6507):686–91.
132. Beltramo M, Stella N, Calignano A, Lin SY, Makriyannis A, Piomelli D. Functional role of high-affinity anandamide transport, as revealed by selective inhibition. Science. 1997;277(5329):1094–7.
133. Hillard CJ, Campbell WB. Biochemistry and pharmacology of arachidonylethanolamide, a putative endogenous cannabinoid. J Lipid Res. 1997;38(12):2383–98.
134. Fegley D, Kathuria S, Mercier R, Li C, Goutopoulos A, Makriyannis A, et al. Anandamide transport is independent of fatty-acid amide hydrolase activity and is blocked by the hydrolysis-resistant inhibitor AM1172. Proc Natl Acad Sci U S A. 2004;101(23):8756–61.
135. Hillard CJ, Wilkison DM, Edgemond WS, Campbell WB. Characterization of the kinetics and distribution of N-arachidonylethanolamine (anandamide) hydrolysis by rat brain. Biochim Biophys Acta. 1995;1257(3):249–56.
136. Ueda N, Kurahashi Y, Yamamoto S, Tokunaga T. Partial purification and characterization of the porcine brain enzyme hydrolyzing and synthesizing anandamide. J Biol Chem. 1995;270(40):23823–7.
137. Cravatt BF, Giang DK, Mayfield SP, Boger DL, Lerner RA, Gilula NB. Molecular characterization of an enzyme that degrades neuromodulatory fatty-acid amides. Nature. 1996;384(6604):83–7.
138. Ade KK, Lovinger DM. Anandamide regulates postnatal development of long-term synaptic plasticity in the rat dorsolateral striatum. J Neurosci. 2007;27(9):2403–9.
139. Maccarrone M, Rossi S, Bari M, De Chiara V, Fezza F, Musella A, et al. Anandamide inhibits metabolism and physiological actions of 2-arachidonoylglycerol in the striatum. Nat Neurosci. 2008;11(2):152–9.
140. Maccarrone M, De Chiara V, Gasperi V, Viscomi MT, Rossi S, Oddi S, et al. Lipid rafts regulate 2-arachidonoylglycerol metabolism and physiological activity in the striatum. J Neurochem. 2009;109(2):371–81.
141. Rossi S, De Chiara V, Musella A, Sacchetti L, Cantarella C, Castelli M, et al. Preservation of striatal cannabinoid CB1 receptor function correlates with the antianxiety effects of fatty acid amide hydrolase inhibition. Mol Pharmacol. 2010;78(2):260–8.
142. Power BD, Stefanis NC, Dragovic M, Jablensky A, Castle D, Morgan V. Age at initiation of amphetamine use and age at onset of psychosis: the Australian Survey of High Impact Psychosis. Schizophr Res. 2014;152(1):300–2.
143. Muller-Vahl KR, Emrich HM. Cannabis and schizophrenia: towards a cannabinoid hypothesis of schizophrenia. Expert Rev Neurother. 2008;8(7):1037–48.
144. Giuffrida A, Leweke FM, Gerth CW, Schreiber D, Koethe D, Faulhaber J, et al. Cerebrospinal anandamide levels are elevated in acute schizophrenia and are inversely correlated with psychotic symptoms. Neuropsychopharmacology. 2004;29(11):2108–14.
145. Parolaro D, Realini N, Vigano D, Guidali C, Rubino T. The endocannabinoid system and psychiatric disorders. Exp Neurol. 2010;224(1):3–14.
146. De Marchi N De Petrocellis L Orlando P Daniele F Fezza F Di Marzo V. Endocannabinoid signalling in the blood of patients with schizophrenia. Lipids Health Dis. 2003;2:5.
147. Dean B, Sundram S, Bradbury R, Scarr E, Copolov D. Studies on [3H]CP-55940 binding in the human central nervous system: regional specific changes in density of cannabinoid-1

receptors associated with schizophrenia and cannabis use. Neuroscience. 2001;103(1):9–15.

148. Zavitsanou K, Garrick T, Huang XF. Selective antagonist [3H]SR141716 A binding to cannabinoid CB1 receptors is increased in the anterior cingulate cortex in schizophrenia. Prog Neuropsychopharmacol Biol Psychiatry. 2004;28(2):355–60.
149. Newell KA, Deng C, Huang XF. Increased cannabinoid receptor density in the posterior cingulate cortex in schizophrenia. Exp Brain Res. 2006;172(4):556–60.
150. Uriguen L, Garcia-Fuster MJ, Callado LF, Morentin B, La Harpe R, Casado V, et al. Immunodensity and mRNA expression of A2 A adenosine, D2 dopamine, and CB1 cannabinoid receptors in postmortem frontal cortex of subjects with schizophrenia: effect of antipsychotic treatment. Psychopharmacology. 2009;206(2):313–24.
151. Broome MR, Woolley JB, Tabraham P, Johns LC, Bramon E, Murray GK, et al. What causes the onset of psychosis? Schizophr Res. 2005 ;79(1):23–34.
152. Iyo M, Sekine Y, Mori N. Neuromechanism of developing methamphetamine psychosis: a neuroimaging study. Ann N Y Acad Sci. 2004;1025:288–95.
153. Ujike H, Katsu T, Okahisa Y, Takaki M, Kodama M, Inada T, et al. Genetic variants of D2 but not D3 or D4 dopamine receptor gene are associated with rapid onset and poor prognosis of methamphetamine psychosis. Prog Neuropsychopharmacol Biol Psychiatry. 2009;33(4):625–9.
154. Voruganti LN, Slomka P, Zabel P, Mattar A, Awad AG. Cannabis induced dopamine release: an in-vivo SPECT study. Psychiatry Res. 2001;107(3):173–7.
155. Bossong MG, van Berckel BN, Boellaard R, Zuurman L, Schuit RC, Windhorst AD, et al. Delta9-tetrahydrocannabinol induces dopamine release in the human striatum. Neuropsychopharmacology. 2009;34(3):759–66.
156. Stokes PR, Mehta MA, Curran HV, Breen G, Grasby PM. Can recreational doses of THC produce significant dopamine release in the human striatum? NeuroImage. 2009;48(1):186–90.
157 D'Souza DC, Sewell RA, Ranganathan M. Cannabis and psychosis/schizophrenia: human studies. Eur Arch Psychiatry Clin Neurosci. 2009;259(7):413–31.
158. Okahisa Y, Kodama M, Takaki M, Inada T, Uchimura N, Yamada M, et al. Association Study of Two Cannabinoid Receptor Genes, CNR1 and CNR2, with Methamphetamine Dependence. Curr Neuropharmacol. 2011;9(1):183–9.
159. Ujike H, Takaki M, Nakata K, Tanaka Y, Takeda T, Kodama M, et al. CNR1, central cannabinoid receptor gene, associated with susceptibility to hebephrenic schizophrenia. Molecular Psychiatry. 2002;7(5):515–8.
160. Martinez-Gras I, Hoenicka J, Ponce G, Rodriguez-Jimenez R, Jimenez-Arriero MA, Perez-Hernandez E, et al. (AAT)n repeat in the cannabinoid receptor gene, CNR1: association with schizophrenia in a Spanish population. Eur Arch Psychiatry Clin Neurosci. 2006;256(7):437–41.
161. Ballmaier M, Bortolato M, Rizzetti C, Zoli M, Gessa G, Heinz A, et al. Cannabinoid receptor antagonists counteract sensorimotor gating deficits in the phencyclidine model of psychosis. Neuropsychopharmacology. 2007;32(10):2098–107.
162. Tzavara ET, Degroot A, Wade MR, Davis RJ, Nomikos GG. CB1 receptor knockout mice are hyporesponsive to the behavior-stimulating actions of d-amphetamine: role of mGlu5 receptors. Eur Neuropsychopharmacol. 2009;19(3):196–204.
163. Meltzer HY, Arvanitis L, Bauer D, Rein W, Meta-Trial Study G. Placebo-controlled evaluation of four novel compounds for the treatment of schizophrenia and schizoaffective disorder. Am J Psychiatry. 2004;161(6):975–84.
164. Fergusson DM, Horwood LJ. Does cannabis use encourage other forms of illicit drug use? Addiction. 2000 Apr;95(4):505–20.
165. Morral AR, McCaffrey DF, Paddock SM. Reassessing the marijuana gateway effect. Addiction. 2002;97(12):1493–504.
166. Hall WD, Lynskey M. Is cannabis a gateway drug? Testing hypotheses about the relationship between cannabis use and the use of other illicit drugs. Drug Alcohol Rev. 2005;24(1):39–48.

167. Molitor F, Ruiz JD, Flynn N, Mikanda JN, Sun RK, Anderson R. Methamphetamine use and sexual and injection risk behaviors among out-of-treatment injection drug users. Am J Drug Alcohol Abuse. 1999;25(3):475–93.
168. Baskin-Sommers A, Sommers I. The co-occurrence of substance use and high-risk behaviors. J Adolesc Health. 2006;38(5):609–11.
169. Scott LA, Roxburgh A, Bruno R, Matthews A, Burns L. The impact of comorbid cannabis and methamphetamine use on mental health among regular ecstasy users. Addict Behav. 2012;37(9):1058–62.
170. Churchwell JC, Carey PD, Ferrett HL, Stein DJ, Yurgelun-Todd DA. Abnormal striatal circuitry and intensified novelty seeking among adolescents who abuse methamphetamine and cannabis. Dev Neurosci. 2012;34(4):310–7.
171. Oliere S, Joliette-Riopel A, Potvin S, Jutras-Aswad D. Modulation of the endocannabinoid system: vulnerability factor and new treatment target for stimulant addiction. Front Psychiatry. 2013;4:109.
172. Dlugos AM, Hamidovic A, Hodgkinson CA, Goldman D, Palmer AA, de Wit H. More aroused, less fatigued: fatty acid amide hydrolase gene polymorphisms influence acute response to amphetamine. Neuropsychopharmacology. 2010;35(3):613–22.
173. Sipe JC, Chiang K, Gerber AL, Beutler E, Cravatt BF. A missense mutation in human fatty acid amide hydrolase associated with problem drug use. Proc Natl Acad Sci U S A. 2002;99(12):8394–9.
174. Morita Y, Ujike H, Tanaka Y, Uchida N, Nomura A, Ohtani K, et al. A nonsynonymous polymorphism in the human fatty acid amide hydrolase gene did not associate with either methamphetamine dependence or schizophrenia. Neurosci Lett. 2005;376(3):182–7.
175. Comings DE, Muhleman D, Gade R, Johnson P, Verde R, Saucier G, et al. Cannabinoid receptor gene (CNR1): association with i.v. drug use. Mol Psychiatry. 1997;2(2):161–8.
176. Zhang PW, Ishiguro H, Ohtsuki T, Hess J, Carillo F, Walther D, et al. Human cannabinoid receptor 1: 5' exons, candidate regulatory regions, polymorphisms, haplotypes and association with polysubstance abuse. Mol Psychiatry. 2004;9(10):916–31.
177. Madsen MV, Peacock L, Werge T, Andersen MB. Effects of the cannabinoid CB1 receptor agonist CP55,940 and antagonist SR141716 A on d-amphetamine-induced behaviours in Cebus monkeys. J Psychopharmacol. 2006;20(5):622–8.
178. Cortright JJ, Lorrain DS, Beeler JA, Tang WJ, Vezina P. Previous exposure to delta9-tetrahydrocannibinol enhances locomotor responding to but not self-administration of amphetamine. J Pharmacol Exp Ther. 2011;337(3):724–33.
179. Vinklerova J, Novakova J, Sulcova A. Inhibition of methamphetamine self-administration in rats by cannabinoid receptor antagonist AM 251. J Psychopharmacol. 2002;16(2):139–43.
180. Schindler CW, Panlilio LV, Gilman JP, Justinova Z, Vemuri VK, Makriyannis A, et al. Effects of cannabinoid receptor antagonists on maintenance and reinstatement of methamphetamine self-administration in rhesus monkeys. Eur J Pharmacol. 2010;633(1–3):44–9.
181. Yu LL, Wang XY, Zhao M, Liu Y, Li YQ, Li FQ, et al. Effects of cannabinoid CB1 receptor antagonist rimonabant in consolidation and reconsolidation of methamphetamine reward memory in mice. Psychopharmacology. 2009;204(2):203–11.
182. Yu LL, Zhou SJ, Wang XY, Liu JF, Xue YX, Jiang W, et al. Effects of cannabinoid CB(1) receptor antagonist rimonabant on acquisition and reinstatement of psychostimulant reward memory in mice. Behav Brain Res. 2011;217(1):111–6.
183. Hiranita T, Nawata Y, Sakimura K, Yamamoto T. Methamphetamine-seeking behavior is due to inhibition of nicotinic cholinergic transmission by activation of cannabinoid CB1 receptors. Neuropharmacology. 2008;55(8):1300–6.
184. Rodriguez JS, Boctor SY, Flores LC, Phelix CF, Martinez JL, Jr. Local pretreatment with the cannabinoid CB1 receptor antagonist AM251 attenuates methamphetamine intra-accumbens self-administration. Neurosci Lett. 2011;489(3):187–91.
185. Anggadiredja K, Nakamichi M, Hiranita T, Tanaka H, Shoyama Y, Watanabe S, et al. Endocannabinoid system modulates relapse to methamphetamine seeking: possible mediation by the arachidonic acid cascade. Neuropsychopharmacology. 2004;29(8):1470–8.

186. Boctor SY, Martinez JL, Jr., Koek W, France CP. The cannabinoid CB1 receptor antagonist AM251 does not modify methamphetamine reinstatement of responding. Eur J Pharmacol. 2007;571(1):39–43.
187. Loewinger GC, Beckert MV, Tejeda HA, Cheer JF. Methamphetamine-induced dopamine terminal deficits in the nucleus accumbens are exacerbated by reward-associated cues and attenuated by CB1 receptor antagonism. Neuropharmacology. 2012;62(7):2192–201.
188. Wiskerke J, Stoop N, Schetters D, Schoffelmeer AN, Pattij T. Cannabinoid CB1 receptor activation mediates the opposing effects of amphetamine on impulsive action and impulsive choice. PloS one. 2011;6(10):e25856.
189. Parker LA, Burton P, Sorge RE, Yakiwchuk C, Mechoulam R. Effect of low doses of delta9-tetrahydrocannabinol and cannabidiol on the extinction of cocaine-induced and amphetamine-induced conditioned place preference learning in rats. Psychopharmacology. 2004;175(3):360–6.
190. Poncelet M, Barnouin MC, Breliere JC, Le Fur G, Soubrie P. Blockade of cannabinoid (CB1) receptors by 141716 selectively antagonizes drug-induced reinstatement of exploratory behaviour in gerbils. Psychopharmacology. 1999;144(2):144–50.
191. Tzavara ET, Davis RJ, Perry KW, Li X, Salhoff C, Bymaster FP, et al. The CB1 receptor antagonist SR141716 A selectively increases monoaminergic neurotransmission in the medial prefrontal cortex: implications for therapeutic actions. Br J Pharmacol. 2003;138(4):544–53.
192. Morra JT, Glick SD, Cheer JF. Neural encoding of psychomotor activation in the nucleus accumbens core, but not the shell, requires cannabinoid receptor signaling. J Neurosci. 2010;30(14):5102–7.
193. Morra JT, Glick SD, Cheer JF. Cannabinoid receptors mediate methamphetamine induction of high frequency gamma oscillations in the nucleus accumbens. Neuropharmacology. 2012;63(4):565–74.
194. Ferrer B, Gorriti MA, Palomino A, Gornemann I, de Diego Y, Bermudez-Silva FJ, et al. Cannabinoid CB1 receptor antagonism markedly increases dopamine receptor-mediated stereotypies. Eur J Pharmacol. 2007;559(2–3):180–3.
195. Marcellino D, Carriba P, Filip M, Borgkvist A, Frankowska M, Bellido I, et al. Antagonistic cannabinoid CB1/dopamine D2 receptor interactions in striatal CB1/D2 heteromers. A combined neurochemical and behavioral analysis. Neuropharmacology. 2008;54(5):815–23.
196. Houchi H, Babovic D, Pierrefiche O, Ledent C, Daoust M, Naassila M. CB1 receptor knockout mice display reduced ethanol-induced conditioned place preference and increased striatal dopamine D2 receptors. Neuropsychopharmacology. 2005;30(2):339–49.
197. Miller DK, Rodvelt KR, Constales C, Putnam WC. Analogs of SR-141716 A (Rimonabant) alter d-amphetamine-evoked [3H] dopamine overflow from preloaded striatal slices and amphetamine-induced hyperactivity. Life Sci. 2007;81(1):63–71.
198. Li X, Hoffman AF, Peng XQ, Lupica CR, Gardner EL, Xi ZX. Attenuation of basal and cocaine-enhanced locomotion and nucleus accumbens dopamine in cannabinoid CB1-receptor-knockout mice. Psychopharmacology. 2009;204(1):1–11.
199. Gorriti MA, Rodriguez de Fonseca F, Navarro M, Palomo T. Chronic (-)-delta9-tetrahydrocannabinol treatment induces sensitization to the psychomotor effects of amphetamine in rats. Eur J Pharmacol. 1999;365(2–3):133–42.
200. Landa L, Sulcova A, Slais K. Involvement of cannabinoid CB1 and CB2 receptor activity in the development of behavioural sensitization to methamphetamine effects in mice. Neuro Endocrinol Lett. 2006;27(1–2):63–9.
201. Corbille AG, Valjent E, Marsicano G, Ledent C, Lutz B, Herve D, et al. Role of cannabinoid type 1 receptors in locomotor activity and striatal signaling in response to psychostimulants. J Neurosci. 2007;27(26):6937–47.
202. Thiemann G, Di Marzo V, Molleman A, Hasenohrl RU. The CB(1) cannabinoid receptor antagonist AM251 attenuates amphetamine-induced behavioural sensitization while causing monoamine changes in nucleus accumbens and hippocampus. Pharmacol Biochem Behav. 2008;89(3):384–91.

203. Burchett SA, Hicks TP. The mysterious trace amines: protean neuromodulators of synaptic transmission in mammalian brain. Prog Neurobiol. 2006;79(5–6):223–46.
204. Xie Z, Miller GM. A receptor mechanism for methamphetamine action in dopamine transporter regulation in brain. J Pharmacol Exp Ther. 2009;330(1):316–25.
205. Tanda G, Pontieri FE, Di Chiara G. Cannabinoid and heroin activation of mesolimbic dopamine transmission by a common mu1 opioid receptor mechanism. Science. 1997;276(5321):2048–50.
206. Beltramo M, de Fonseca FR, Navarro M, Calignano A, Gorriti MA, Grammatikopoulos G, et al. Reversal of dopamine D(2) receptor responses by an anandamide transport inhibitor. J Neurosci. 2000;20(9):3401–7.
207. Gerdeman GL F-RJ. The endocannabinoid system in the physiology and pathophysiology of the basal ganglia. In: A K, editor. Cannabinoids and the brain. New York: Springer-Verlag; 2008. pp. 423–83.
208. Solinas M, Goldberg SR, Piomelli D. The endocannabinoid system in brain reward processes. Br J Pharmacol. 2008;154(2):369–83.
209. Sotnikova TD, Zorina OI, Ghisi V, Caron MG, Gainetdinov RR. Trace amine associated receptor 1 and movement control. Parkinsonism Relat Disord. 2008;14(Suppl 2):99–102.
210. Bradaia A, Trube G, Stalder H, Norcross RD, Ozmen L, Wettstein JG, et al. The selective antagonist EPPTB reveals TAAR1-mediated regulatory mechanisms in dopaminergic neurons of the mesolimbic system. Proc Natl Acad Sci U S A. 2009;106(47):20081–6.
211. Hernandez M, Berrendero F, Suarez I, Garcia-Gil L, Cebeira M, Mackie K, et al. Cannabinoid CB(1) receptors colocalize with tyrosine hydroxylase in cultured fetal mesencephalic neurons and their activation increases the levels of this enzyme. Brain Res. 2000;857(1–2):56–65.
212. Wenger T, Moldrich G, Furst S. Neuromorphological background of cannabis addiction. Brain Res Bull. 2003;61(2):125–8.
213. Lau T, Schloss P. The cannabinoid CB1 receptor is expressed on serotonergic and dopaminergic neurons. Eur J Pharmacol. 2008;578(2–3):137–41.
214. Laviolette SR, Grace AA. The roles of cannabinoid and dopamine receptor systems in neural emotional learning circuits: implications for schizophrenia and addiction. Cell Mol Life Sci: CMLS. 2006;63(14):1597–613.
215. Brown JA, Horvath S, Garbett K, Schmidt MJ, Everheart M, Gellert L, et al. The role of cannabinoid 1 receptor expressing interneurons in behavior. Neurobiol Dis. 2014;63:210–21.
216. Bortolato M, Frau R, Bini V, Luesu W, Loriga R, Collu M, et al. Methamphetamine neurotoxicity increases brain expression and alters behavioral functions of CB(1) cannabinoid receptors. J Psychiatr Res. 2010;44(14):944–55.
217. Piazza PV, Deminiere JM, Le Moal M, Simon H. Factors that predict individual vulnerability to amphetamine self-administration. Science. 1989;245(4925):1511–3.
218. Kelly TH, Robbins G, Martin CA, Fillmore MT, Lane SD, Harrington NG, et al. Individual differences in drug abuse vulnerability: d-amphetamine and sensation-seeking status. Psychopharmacology. 2006;189(1):17–25.
219. Giuffrida A, Parsons LH, Kerr TM, Rodriguez de Fonseca F, Navarro M, Piomelli D. Dopamine activation of endogenous cannabinoid signaling in dorsal striatum. Nat Neurosci. 1999;2(4):358–63.
220. Tzavara ET, Li DL, Moutsimilli L, Bisogno T, Di Marzo V, Phebus LA, et al. Endocannabinoids activate transient receptor potential vanilloid 1 receptors to reduce hyperdopaminergia-related hyperactivity: therapeutic implications. Biol Psychiatry. 2006;59(6):508–15.
221. Gardner EL. Endocannabinoid signaling system and brain reward: emphasis on dopamine. Pharmacol Biochem Behav. 2005;81(2):263–84.
222. Kleijn J, Wiskerke J, Cremers TI, Schoffelmeer AN, Westerink BH, Pattij T. Effects of amphetamine on dopamine release in the rat nucleus accumbens shell region depend on cannabinoid CB1 receptor activation. Neurochem Int. 2012;60(8):791–8.

223. Melis M, Pistis M, Perra S, Muntoni AL, Pillolla G, Gessa GL. Endocannabinoids mediate presynaptic inhibition of glutamatergic transmission in rat ventral tegmental area dopamine neurons through activation of CB1 receptors. J Neurosci. 2004;24(1):53–62.
224. Kreitzer AC, Malenka RC. Dopamine modulation of state-dependent endocannabinoid release and long-term depression in the striatum. J Neurosci. 2005;25(45):10537–45.
225. Yin HH, Lovinger DM. Frequency-specific and D2 receptor-mediated inhibition of glutamate release by retrograde endocannabinoid signaling. Proc Natl Acad Sci U S A. 2006;103(21):8251–6.
226. Shen W, Flajolet M, Greengard P, Surmeier DJ. Dichotomous dopaminergic control of striatal synaptic plasticity. Science. 2008;321(5890):848–51.
227. Chiu CQ, Puente N, Grandes P, Castillo PE. Dopaminergic modulation of endocannabinoid-mediated plasticity at GABAergic synapses in the prefrontal cortex. J Neurosci. 2010;30(21):7236–48.
228. Gubellini P, Picconi B, Bari M, Battista N, Calabresi P, Centonze D, et al. Experimental parkinsonism alters endocannabinoid degradation: implications for striatal glutamatergic transmission. J Neurosci. 2002;22(16):6900–7.
229. Lastres-Becker I, Cebeira M, de Ceballos ML, Zeng BY, Jenner P, Ramos JA, et al. Increased cannabinoid CB1 receptor binding and activation of GTP-binding proteins in the basal ganglia of patients with Parkinson's syndrome and of MPTP-treated marmosets. Eur J Neurosci. 2001;14(11):1827–32.
230. Di Marzo V Hill MP Bisogno T Crossman AR Brotchie JM. Enhanced levels of endogenous cannabinoids in the globus pallidus are associated with a reduction in movement in an animal model of Parkinson's disease. FASEB J. 2000;14(10):1432–8.
231. Kerr JN, Wickens JR. Dopamine D-1/D-5 receptor activation is required for long-term potentiation in the rat neostriatum in vitro. J Neurophysiol. 2001;85(1):117–24.
232 Calabresi P, Saiardi A, Pisani A, Baik JH, Centonze D, Mercuri NB, et al. Abnormal synaptic plasticity in the striatum of mice lacking dopamine D2 receptors. J Neurosci. 1997;17(12):4536–44.
233. Yamamoto Y, Nakanishi H, Takai N, Shimazoe T, Watanabe S, Kita H. Expression of N-methyl-D-aspartate receptor-dependent long-term potentiation in the neostriatal neurons in an in vitro slice after ethanol withdrawal of the rat. Neuroscience. 1999;91(1):59–68.
234. Kalivas PW, Sorg BA, Hooks MS. The pharmacology and neural circuitry of sensitization to psychostimulants. Behav Pharmacol. 1993;4(4):315–34.
235. Canales JJ, Capper-Loup C, Hu D, Choe ES, Upadhyay U, Graybiel AM. Shifts in striatal responsivity evoked by chronic stimulation of dopamine and glutamate systems. Brain. 2002;125(Pt 10):2353–63.
236. Capper-Loup C, Canales JJ, Kadaba N, Graybiel AM. Concurrent activation of dopamine D1 and D2 receptors is required to evoke neural and behavioral phenotypes of cocaine sensitization. J Neurosci. 2002;22(14):6218–27.
237. Szabo B, Muller T, Koch H. Effects of cannabinoids on dopamine release in the corpus striatum and the nucleus accumbens in vitro. J Neurochem. 1999;73(3):1084–9.
238. Szabo B, Siemes S, Wallmichrath I. Inhibition of GABAergic neurotransmission in the ventral tegmental area by cannabinoids. Eur J Neurosci. 2002;15(12):2057–61.
239. Robbe D, Alonso G, Duchamp F, Bockaert J, Manzoni OJ. Localization and mechanisms of action of cannabinoid receptors at the glutamatergic synapses of the mouse nucleus accumbens. J Neurosci. 2001;21(1):109–16.
240. Pistis M, Muntoni AL, Pillolla G, Gessa GL. Cannabinoids inhibit excitatory inputs to neurons in the shell of the nucleus accumbens: an in vivo electrophysiological study. Eur J Neurosci. 2002;15(11):1795–802.
241. Lupica CR, Riegel AC. Endocannabinoid release from midbrain dopamine neurons: a potential substrate for cannabinoid receptor antagonist treatment of addiction. Neuropharmacology. 2005;48(8):1105–16.
242. Wolf ME. The role of excitatory amino acids in behavioral sensitization to psychomotor stimulants. Prog Neurobiol. 1998;54(6):679–720.

243. Wise RA, Bozarth MA. A psychomotor stimulant theory of addiction. Psychol Rev. 1987;94(4):469–92.
244. Kalivas PW, Stewart J. Dopamine transmission in the initiation and expression of drug- and stress-induced sensitization of motor activity. Brain Res Brain Res Rev. 1991;16(3):223–44.
245. White FJ, Kalivas PW. Neuroadaptations involved in amphetamine and cocaine addiction. Drug Alcohol Depend. 1998;51(1–2):141–53.
246. Kalivas PW, Weber B. Amphetamine injection into the ventral mesencephalon sensitizes rats to peripheral amphetamine and cocaine. J Pharmacol Exp Ther. 1988;245(3):1095–102.
247. Vezina P, Stewart J. Amphetamine administered to the ventral tegmental area but not to the nucleus accumbens sensitizes rats to systemic morphine: lack of conditioned effects. Brain Res. 1990;516(1):99–106.
248. Jones S, Kornblum JL, Kauer JA. Amphetamine blocks long-term synaptic depression in the ventral tegmental area. J Neurosci. 2000;20(15):5575–80.
249. Tong ZY, Overton PG, Clark D. Chronic administration of (+)-amphetamine alters the reactivity of midbrain dopaminergic neurons to prefrontal cortex stimulation in the rat. Brain Res. 1995;674(1):63–74.
250. White FJ, Hu XT, Zhang XF, Wolf ME. Repeated administration of cocaine or amphetamine alters neuronal responses to glutamate in the mesoaccumbens dopamine system. J Pharmacol Exp Ther. 1995;273(1):445–54.
251. Fitzgerald LW, Ortiz J, Hamedani AG, Nestler EJ. Drugs of abuse and stress increase the expression of GluR1 and NMDAR1 glutamate receptor subunits in the rat ventral tegmental area: common adaptations among cross-sensitizing agents. J Neurosci. 1996;16(1):274–82.
252. Zhang XF, Hu XT, White FJ, Wolf ME. Increased responsiveness of ventral tegmental area dopamine neurons to glutamate after repeated administration of cocaine or amphetamine is transient and selectively involves AMPA receptors. J Pharmacol Exp Ther. 1997;281(2):699–706.
253. Cadogan AK, Alexander SP, Boyd EA, Kendall DA. Influence of cannabinoids on electrically evoked dopamine release and cyclic AMP generation in the rat striatum. J Neurochem. 1997;69(3):1131–7.
254. Glass M, Felder CC. Concurrent stimulation of cannabinoid CB1 and dopamine D2 receptors augments cAMP accumulation in striatal neurons: evidence for a Gs linkage to the CB1 receptor. J Neurosci. 1997;17(14):5327–33.
255. Gerdeman G, Lovinger DM. CB1 cannabinoid receptor inhibits synaptic release of glutamate in rat dorsolateral striatum. J Neurophysiol. 2001;85(1):468–71.
256. Herkenham M, Lynn AB, Johnson MR, Melvin LS, de Costa BR, Rice KC. Characterization and localization of cannabinoid receptors in rat brain: a quantitative in vitro autoradiographic study. J Neurosci. 1991;11(2):563–83.
257. Tsou K, Brown S, Sanudo-Pena MC, Mackie K, Walker JM. Immunohistochemical distribution of cannabinoid CB1 receptors in the rat central nervous system. Neuroscience. 1998;83(2):393–411.
258. Hermann H, Marsicano G, Lutz B. Coexpression of the cannabinoid receptor type 1 with dopamine and serotonin receptors in distinct neuronal subpopulations of the adult mouse forebrain. Neuroscience. 2002;109(3):451–60.
259. Julian MD, Martin AB, Cuellar B, Rodriguez De Fonseca F, Navarro M, Moratalla R, et al. Neuroanatomical relationship between type 1 cannabinoid receptors and dopaminergic systems in the rat basal ganglia. Neuroscience. 2003;119(1):309–18.
260. Martin AB, Fernandez-Espejo E, Ferrer B, Gorriti MA, Bilbao A, Navarro M, et al. Expression and function of CB1 receptor in the rat striatum: localization and effects on D1 and D2 dopamine receptor-mediated motor behaviors. Neuropsychopharmacology. 2008;33(7):1667–79.
261. Gonzalez B, Paz F, Floran L, Aceves J, Erlij D, Floran B. Cannabinoid agonists stimulate [3H]GABA release in the globus pallidus of the rat when G(i) protein-receptor coupling is restricted: role of dopamine D2 receptors. J Pharmacol Exp Ther. 2009;328(3):822–8.

262. Landa L, Jurajda M, Sulcova A. Altered dopamine D1 and D2 receptor mRNA expression in mesencephalon from mice exposed to repeated treatments with methamphetamine and cannabinoid CB1 agonist methanandamide. Neuro Endocrinol Lett. 2012;33(4):446–52.
263. Maneuf YP, Crossman AR, Brotchie JM. The cannabinoid receptor agonist WIN 55,212–2 reduces D2, but not D1, dopamine receptor-mediated alleviation of akinesia in the reserpine-treated rat model of Parkinson's disease. Exp Neurol. 1997;148(1):265–70.
264. Rodriguez de Fonseca F Del Arco I Martin-Calderon JL Gorriti MA Navarro M. Role of the endogenous cannabinoid system in the regulation of motor activity. Neurobiol Dis. 1998;5(6 Pt B):483–501.
265. Andersson M, Usiello A, Borgkvist A, Pozzi L, Dominguez C, Fienberg AA, et al. Cannabinoid action depends on phosphorylation of dopamine- and cAMP-regulated phosphoprotein of 32 kDa at the protein kinase A site in striatal projection neurons. J Neurosci. 2005;25(37):8432–8.
266. Lane DA, Chan J, Fitzgerald ML, Kearn CS, Mackie K, Pickel VM. Quinpirole elicits differential in vivo changes in the pre- and postsynaptic distributions of dopamine D(2) receptors in mouse striatum: relation to cannabinoid-1 (CB(1)) receptor targeting. Psychopharmacology. 2012;221(1):101–13.
267. Calabresi P, Maj R, Pisani A, Mercuri NB, Bernardi G. Long-term synaptic depression in the striatum: physiological and pharmacological characterization. J Neurosci. 1992;12(11):4224–33.
268. Choi S, Lovinger DM. Decreased probability of neurotransmitter release underlies striatal long-term depression and postnatal development of corticostriatal synapses. Proc Natl Acad Sci U S A. 1997;94(6):2665–70.
269. Centonze D, Gubellini P, Picconi B, Calabresi P, Giacomini P, Bernardi G. Unilateral dopamine denervation blocks corticostriatal LTP. J Neurophysiol. 1999;82(6):3575–9.
270. Kung VW, Hassam R, Morton AJ, Jones S. Dopamine-dependent long term potentiation in the dorsal striatum is reduced in the R6/2 mouse model of Huntington's disease. Neuroscience. 2007;146(4):1571–80.
271. Stern EA, Jaeger D, Wilson CJ. Membrane potential synchrony of simultaneously recorded striatal spiny neurons in vivo. Nature. 1998;394(6692):475–8
272. Chiang YC, Chen JC. The role of the cannabinoid type 1 receptor and down-stream cAMP/DARPP-32 signal in the nucleus accumbens of methamphetamine-sensitized rats. J Neurochem. 2007;103(6):2505–17.
273. Calabresi P, Gubellini P, Centonze D, Picconi B, Bernardi G, Chergui K, et al. Dopamine and cAMP-regulated phosphoprotein 32 kDa controls both striatal long-term depression and long-term potentiation, opposing forms of synaptic plasticity. J Neurosci. 2000;20(22):8443–51.
274. Kearn CS, Blake-Palmer K, Daniel E, Mackie K, Glass M. Concurrent stimulation of cannabinoid CB1 and dopamine D2 receptors enhances heterodimer formation: a mechanism for receptor cross-talk? Mol Pharmacol. 2005;67(5):1697–704.
275. Huang YC, Wang SJ, Chiou LC, Gean PW. Mediation of amphetamine-induced long-term depression of synaptic transmission by CB1 cannabinoid receptors in the rat amygdala. J Neurosci. 2003;23(32):10311–20.
276. Chavez AE, Chiu CQ, Castillo PE. TRPV1 activation by endogenous anandamide triggers postsynaptic long-term depression in dentate gyrus. Nat Neurosci. 2010;13(12):1511–8.
277. Mezey E, Toth ZE, Cortright DN, Arzubi MK, Krause JE, Elde R, et al. Distribution of mRNA for vanilloid receptor subtype 1 (VR1), and VR1-like immunoreactivity, in the central nervous system of the rat and human. Proc Natl Acad Sci U S A. 2000;97(7):3655–60.
278. Bisogno T, Hanus L, De Petrocellis L, Tchilibon S, Ponde DE, Brandi I, et al. Molecular targets for cannabidiol and its synthetic analogues: effect on vanilloid VR1 receptors and on the cellular uptake and enzymatic hydrolysis of anandamide. Br J Pharmacol. 2001;134(4):845–52.
279. Costa B, Giagnoni G, Franke C, Trovato AE, Colleoni M. Vanilloid TRPV1 receptor mediates the antihyperalgesic effect of the nonpsychoactive cannabinoid, cannabidiol, in a rat model of acute inflammation. Br J Pharmacol. 2004;143(2):247–50.

280. De Petrocellis L, Ligresti A, Moriello AS, Allara M, Bisogno T, Petrosino S, et al. Effects of cannabinoids and cannabinoid-enriched Cannabis extracts on TRP channels and endocannabinoid metabolic enzymes. Br J Pharmacol. 2011;163(7):1479–94.
281. Moreira FA, Guimaraes FS. Cannabidiol inhibits the hyperlocomotion induced by psychotomimetic drugs in mice. Eur J Pharmacol. 2005;512(2–3):199–205.
282. Starowicz K, Nigam S, Di Marzo V. Biochemistry and pharmacology of endovanilloids. Pharmacol Ther. 2007;114(1):13–33.
283. Hu SS, Bradshaw HB, Benton VM, Chen JS, Huang SM, Minassi A, et al. The biosynthesis of N-arachidonoyl dopamine (NADA), a putative endocannabinoid and endovanilloid, via conjugation of arachidonic acid with dopamine. Prostaglandins Leukot Essent Fatty Acids. 2009;81(4):291–301.
284. Bobrov MY, Lizhin AA, Andrianova EL, Gretskaya NM, Frumkina LE, Khaspekov LG, et al. Antioxidant and neuroprotective properties of N-arachidonoyldopamine. Neurosci Lett. 2008;431(1):6–11.
285. Oz M, Jaligam V, Galadari S, Petroianu G, Shuba YM, Shippenberg TS. The endogenous cannabinoid, anandamide, inhibits dopamine transporter function by a receptor-independent mechanism. J Neurochem. 2010;112(6):1454–64.
286. Onaivi ES, Green MR, Martin BR. Pharmacological characterization of cannabinoids in the elevated plus maze. J Pharmacol Exp Ther. 1990;253(3):1002–9.
287. Ortega-Alvaro A, Aracil-Fernandez A, Garcia-Gutierrez MS, Navarrete F, Manzanares J. Deletion of CB2 cannabinoid receptor induces schizophrenia-related behaviors in mice. Neuropsychopharmacology. 2011;36(7):1489–504.
288. Nossoll M, Teuchert-Noodt G, Dawirs RR. A single dose of methamphetamine in neonatal gerbils affects adult prefrontal gamma-aminobutyric acid innervation. Eur J Pharmacol. 1997;340(2–3):R3–5.
289. Gonzalez R, Rippeth JD, Carey CL, Heaton RK, Moore DJ, Schweinsburg BC, et al. Neurocognitive performance of methamphetamine users discordant for history of marijuana exposure. Drug Alcohol Depend. 2004;76(2):181–90.
290. Jin KL, Mao XO, Goldsmith PC, Greenberg DA. CB1 cannabinoid receptor induction in experimental stroke. Ann Neurol. 2000;48(2):257–61.
291. Fernandez-Lopez D, Martinez-Orgado J, Nunez E, Romero J, Lorenzo P, Moro MA, et al. Characterization of the neuroprotective effect of the cannabinoid agonist WIN-55212 in an in vitro model of hypoxic-ischemic brain damage in newborn rats. Pediatr Res. 2006;60(2):169–73.
292. Nagayama T, Sinor AD, Simon RP, Chen J, Graham SH, Jin K, et al. Cannabinoids and neuroprotection in global and focal cerebral ischemia and in neuronal cultures. J Neurosci. 1999;19(8):2987–95.
293. Marsicano G, Goodenough S, Monory K, Hermann H, Eder M, Cannich A, et al. CB1 cannabinoid receptors and on-demand defense against excitotoxicity. Science. 2003;302(5642):84–8.
294. Marchalant Y, Rosi S, Wenk GL. Anti-inflammatory property of the cannabinoid agonist WIN-55212-2 in a rodent model of chronic brain inflammation. Neuroscience. 2007;144(4):1516–22.
295. Solbrig MV, Hermanowicz N. Cannabinoid rescue of striatal progenitor cells in chronic Borna disease viral encephalitis in rats. J Neurovirol. 2008;14(3):252–60.
296. van der Stelt M, Di Marzo V. Cannabinoid receptors and their role in neuroprotection. Neuromolecular Med. 2005;7(1–2):37–50.
297. Romero J, Berrendero F, Perez-Rosado A, Manzanares J, Rojo A, Fernandez-Ruiz JJ, et al. Unilateral 6-hydroxydopamine lesions of nigrostriatal dopaminergic neurons increased CB1 receptor mRNA levels in the caudate-putamen. Life Sci. 2000;66(6):485–94.
298. Shen M, Piser TM, Seybold VS, Thayer SA. Cannabinoid receptor agonists inhibit glutamatergic synaptic transmission in rat hippocampal cultures. J Neurosci. 16(14):4322–34.
299. Shen M, Thayer SA. Cannabinoid receptor agonists protect cultured rat hippocampal neurons from excitotoxicity. Mol Pharmacol. 1998;54(3):459–62.

300. Nadler V, Mechoulam R, Sokolovsky M. Blockade of 45Ca2+ influx through the N-methyl-D-aspartate receptor ion channel by the non-psychoactive cannabinoid HU-211. Brain Res. 1993;622(1–2):79–85.
301. Hampson AJ, Grimaldi M, Axelrod J, Wink D. Cannabidiol and (-)Delta9-tetrahydrocannabinol are neuroprotective antioxidants. Proc Natl Acad Sci U S A. 1998;95(14):8268–73.
302. Borges RS, Batista J, Jr., Viana RB, Baetas AC, Orestes E, Andrade MA, et al. Understanding the molecular aspects of tetrahydrocannabinol and cannabidiol as antioxidants. Molecules. 2013;18(10):12663–74.
303. Lavie G, Teichner A, Shohami E, Ovadia H, Leker RR. Long term cerebroprotective effects of dexanabinol in a model of focal cerebral ischemia. Brain Res. 2001;901(1–2):195–201.
304. Leker RR, Shohami E, Abramsky O, Ovadia H. Dexanabinol; a novel neuroprotective drug in experimental focal cerebral ischemia. J Neurol Sci. 1999;162(2):114–9.
305. Louw DF, Yang FW, Sutherland GR. The effect of delta-9-tetrahydrocannabinol on forebrain ischemia in rat. Brain Res. 2000;857(1–2):183–7.
306. Panikashvili D, Simeonidou C, Ben-Shabat S, Hanus L, Breuer A, Mechoulam R, et al. An endogenous cannabinoid (2-AG) is neuroprotective after brain injury. Nature. 2001;413(6855):527–31.
307. Sinor AD, Irvin SM, Greenberg DA. Endocannabinoids protect cerebral cortical neurons from in vitro ischemia in rats. Neurosci Lett. 2000;278(3):157–60.
308. van der Stelt M, Veldhuis WB, van Haaften GW, Fezza F, Bisogno T, Bar PR, et al. Exogenous anandamide protects rat brain against acute neuronal injury in vivo. J Neurosci. 2001;21(22):8765–71.
309. Knoller N, Levi L, Shoshan I, Reichenthal E, Razon N, Rappaport ZH, et al. Dexanabinol (HU-211) in the treatment of severe closed head injury: a randomized, placebo-controlled, phase II clinical trial. Crit Care Med. 2002;30(3):548–54.
310. Bortolato M, Piomelli D. The endocannabinoid system and anxiety responses. In: Blanchard RJ BD, Griebel G, Nutt D, editors. Hand book of Anxiety and Fear. Amsterdam: Elsevier; 2008. pp. 303–25.

Chapter 17
Cannabinoid-Dopamine Interactions: Modulation of Midbrain DA Neurons by Endocannabinoids

François Georges and Miriam Melis

Abstract A key feature of human and animal behavior is to learn from environmental stimuli to adapt efficiently. Under physiological conditions, dopaminergic (DA) neurons are used to evaluate and learn new sensory information and adjust behavior to maximize reward and minimize aversive consequences. The two main DA pathways in the mesencephalon originate from the substantia nigra pars compacta (SNpc) and the ventral tegmental area (VTA). Both *in vivo* and *in vitro* studies have established that DA neurons exhibit spontaneous spike firing that is driven by intrinsic electrophysiological properties, with their activity modulated by afferent inputs and a number of neuromodulators, including endocannabinoids. In the VTA and SNpc, cannabinoid type 1-(CB_1) and ionotropic transient receptor potential vanilloid type 1 (TRPV1) receptors are abundantly expressed as well as their endogenous ligands, mainly anandamide and 2-arachidonoylglycerol. This chapter attempts to summarize some of the major research findings demonstrating that SNpc and VTA DA neurons vary significantly in their molecular and physiological properties according to target location, and that endocannabinoids act on GABAergic, glutamatergic and cholinergic terminals to participate in discrete mechanisms aimed at DA cell homeostatic regulation. As a result, given the role of the endocannabinoid system in modulating DA neuronal function of the SNpc and the VTA, they might take part in associative learning, reward signaling, goal directed behavior, motor skill learning and action-habit transformation. These considerations help explaining the correlation between an unbalanced endocannabinoid signal and altered DA-dependent processes underpinning diverse pathological conditions of both nigrostriatal and mesocorticolimbic systems.

Keywords Dopamine neurons · Ventral tegmental area · Substantia nigra pars compacta · Endocannabinoids · Peroxisome-proliferator-activated receptors · Reward · Rodent · Synaptic plasticity · Vanilloid receptors

F. Georges (✉)
IINS-UMR5297, CNRS, Bordeaux, France
e-mail: francois.georges@u-bordeaux.fr

M. Melis
Department of Biomedical Sciences, University of Cagliari, Cagliari, Italy

P. Campolongo, L. Fattore (eds.), *Cannabinoid Modulation of Emotion, Memory, and Motivation,* DOI 10.1007/978-1-4939-2294-9_17

Introduction

The brain constantly receives and evaluates sensory information, and adjusts behavioral output in a flexible fashion in order to maximize reward, and minimize aversive consequences. Midbrain dopamine (DA) pathways control key physiological functions related to locomotor activity, short-term and working memory, associative learning, attention, and novelty encoding. Consequently, abnormal DA system function has long been implicated in both neurological and psychiatric disorders. The endocannabinoid system controls most of the above cited physiological functions, and participates in modulation of neuronal excitability and various forms of synaptic plasticity of midbrain DA neurons [1, 2].

In the present chapter, we will focus on the two main DA pathways originating in the mesencephalon: the substantia nigra pars compacta (SNpc) and the ventral tegmental area (VTA), whose modulation by endogenous cannabinoids has been extensively reviewed [2–5].

The endogenous system comprises a metabolic apparatus to produce, release and eventually transport the endogenous ligands (i.e. endocannabinoids), and their receptors (i.e. cannabinoid receptors, CB). In the VTA and SNpc, cannabinoid type 1-(CB_1) receptors are abundantly expressed [6–8] as well as their endogenous ligands, mainly anandamide and 2-arachidonoylglycerol (2-AG) [9, 10]. The canonical mechanism of action proposed for endocannabinoids posits an *on demand* release of these lipid molecules by the postsynaptic neuron [11], which retrogradely travel across the synaptic cleft to bind to and activate CB_1 receptors located on the presynaptic terminals [12, 13]. The result of such an activation is a reduction of neurotransmitter release in a short- or a long-term manner [12, 14]. Through this mechanism of action, endocannabinoid signaling contributes to different forms of short- and long-term synaptic plasticity at both excitatory and inhibitory synapses [12, 15, 16].

Long-term forms of synaptic plasticity are widely considered indicative of long lasting adaptations of individual synapses, circuits or neural networks, underlying changes in behavior. By mediating and/or regulating long-term forms of synaptic plasticity, endocannabinoids can, therefore, influence the repertoire of enduring modifications of individual synapses, circuits or neural networks and the consequent behavior [15].

Short-term forms of synaptic plasticity, which are rapid means for a bidirectional and reversible modulation of synaptic strength, serve as important mechanisms to modify synaptic functions during computation [17]. Particularly, short-term forms of synaptic plasticity are often viewed as dynamic filters used for transmission of specific input frequencies or patterns, and are key in mechanisms regulating a proper scaling of synaptic inputs [17–21]. By modulating this dynamic gain control mechanism, endocannabinoids can guarantee a proper accuracy of input information [22]. These general rules apply to the VTA [2, 3] as well to the SNpc [23, 24].

Early work demonstrated that DA neurons located in the SNpc and the VTA provide a prediction-error signal for reward-mediated learning [25, 26], and play a central role in encoding positive and negative motivational signals [27, 28]. In striking contrast, negative outcomes are supposed to generate a negative (inhibitory) prediction-error signal in the same neuronal population. Although DA neurons exhibit considerable heterogeneity regarding their inputs, their projection targets and their basic pharmacological properties [29–31], it was widely assumed that they exhibit homogeneous appetitive and aversive coding across the entire population. The interpretation of these results led researchers to the oversimplified assumption that midbrain DA neurons may operate as a single functional unit. However, this hypothesis is not consistent with recent findings [32]. Particularly, it has been reported that a sub-population of DA neurons exhibit excitatory responses to novel aversive sensory stimuli [27, 33], and maintain this excitatory pattern as the animals learn about negative outcome associated with these aversive stimuli [27]. In addition, aversive stimuli evoke DA release at projection targets [34–36]. Finally, yet importantly, DA plays a critical role during learning of aversive association in both immature and mature brain [37, 38].

As any group of neurons defined by their location and neurotransmitter content, DA cell function is determined by the circuit in which they are embedded, and the behavioral output for this latter. The notion that sub-populations of midbrain DA neurons are integrated in distinct neuronal circuits is also substantiated by studies showing different patterns of DA release in discrete VTA projection targets. As an example, infusion of the AMPA/kainate receptor antagonist LY293558 into the VTA increases the DA levels in the nucleus accumbens (NAc), but decreases them in the prefrontal cortex (PFC) [39]. Furthermore, Lammel and colleagues [40, 41] demonstrated that an appetitive or an aversive behavioral state has an opposite effect on the plastic properties of VTA DA excitatory synapses in a projection specific manner. Notably, although it is still controversial, the scientific community has accepted that specific SNpc or VTA DA sub-populations, each with specific projection target areas and differential afferent inputs, might have different integrative electrophysiological properties and behavioral functions [42].

Anatomical Organization and Molecular Diversity of the Midbrain DA Neurons

Morphology of Midbrain DA Neurons

Within the VTA and the SNpc, DA neurons expressing tyrosine hydroxylase (TH, i.e. the rate-limiting enzyme in DA synthesis) are interspersed with GABAergic neurons [43, 44]. Importantly, a population of glutamatergic neurons in the VTA has also been identified [45, 46]. DA neurons are medium to large size cells with a body diameter ranging from 12 to 25 μm [47, 48]. They display morphological variations in their shape (i.e. ovoid, medium sized or fusiform) and in the orientation of their dentritic arborization [48], which suggest a degree of diversity at least among SNpc DA neurons. DA neurons emit sparse and relatively unbranched dendritic arborization. Most of their dendrites are smooth and/or occasionally and sparsely invested with "spine-like appendages" [48]. Anatomical evidence suggests that SNpc DA neurons projecting to the striatum could be under modulation of the VTA, since they extend their dendritic arborization in this neighboring area [48]. These findings support the notion that SNpc and VTA are part of a single continuous cell group, rather than two separate nuclei [49, 50].

Molecular Diversity of Midbrain DA Neurons

The VTA and the SNpc are two highly heterogeneous brain structures in terms of their neuronal phenotype. In the VTA and SNpc, the DA neurons represent 65 and 70 % of the total number of neurons, respectively [43]. Moreover, *in situ* hybridization and RT-qPCR coupled to laser-capture approaches revealed distinct populations of DA neurons spread across the VTA and SNpc [30, 43, 46].

Key markers of DA neurons include mRNA for TH, DA transporter (DAT), vesicular monoamine transporter 2 (VMAT2), G-protein-coupled inwardly rectifying potassium channel subunit 2 (GIRK2), DA D2 receptor (D2) and vesicular glutamatergic transporter type 2 (vGluT2). The abundance of these biochemical markers co-vary according to the localization of the DA neuron within the VTA or the SNpc, and to their projection targets.

The identification of the molecular diversity of DA neurons is primarily important for the unraveling of DA system function. In addition, the identifying of this variety is key for a proper interpretation of findings obtained with optogenetics, which accesses and controls network activity at a single cell level by using cell-type *specific promoter* elements [42]. For instance, 80 % of the TH immunolabeled neurons in the medial part of the VTA were negative for the transcript encoding DAT [46], thus suggesting that optogenenic approaches targeting the DAT promoter only manipulate a subset of DA neurons. Additionally, 82 % of the TH immunolabeled neurons in the anterior part of the VTA were negative for the transcript encoding DAT, and did not release DA in their terminal region (i.e. lateral habenula [51]). Consequently, it appears that neither targeting the TH promoter with optogenenics always and necessarily results in modifications in DA release in the target region. Remarkably, current evidence suggests that molecular diversity is more evident among DA neurons in the VTA rather than in SNpc [42]. Hence, functional diversity has been often reported between posterior versus anterior, and lateral versus medial, DA neurons *in vivo* [29, 30, 41, 52].

Physiology of Midbrain DA Neurons

A better understanding of the functions of midbrain DA neurons derives from the pioneering electrophysiological studies carried out by performing single unit intra- and extra- cellular recordings *in vivo* [53–58]. At first, DA neurons were identified and characterized within the SNpc [53]. In rodent and awake nonhuman primate, VTA and SNpc DA neurons recorded *in vivo* display a characteristic electrophysiological signature that allows them to be distinguished from non-DA neurons in surrounding regions of the VTA or in the pars reticulata of the substantia nigra (SNpr) [55, 56, 59–61]. Subsequently, these electrophysiological features have been reexamined and reviewed [62]. The current criteria for their identification include: (i) an action potential with biphasic or triphasic waveform, with a duration longer than 2.2 ms; (ii) a slow spontaneous firing rate (2–9 Hz); (iii) a regular and/or burst spontaneous firing pattern, this latter being characterized by a spike-amplitude decrement; (iv) the inhibition of spontaneous activity following administration of DA D2 receptor agonists, and subsequent reversal by DA D2 receptors antagonists.

Midbrain DA neurons fire bursts of action potentials in response to sensory stimuli leading to transient increases in extracellular DA levels in target areas [63]. *In vivo* DA neurons fire bursts of action potentials upon depolarization of their membrane and display pauses after GABA conductance activation [64]. Bursts are supposed to mediate reward information coding, whereas pauses signal the absence of an expected reward. Accumulating evidence shows that an increased glutamatergic or cholinergic drive produces the characteristic bursting pattern observed *in vivo* [65–67], and that NMDARs in DA neurons modulate burst firing and DA release in postsynaptic brain regions [68]. Furthermore, bursting activity of DA neurons is sufficient to mediate behavioral conditioning in freely-behaving rats [69]. Conversely, phasic changes in tonic activity of GABAergic afferents are a potential extrinsic mechanism able to trigger bursts and pauses in DA neurons [64]. Consequently, DA neurons are regulated by an integrated network of inputs [70], and their firing pattern is sculpted through the balance of activation of GABAergic inputs and NMDA receptor-mediated excitatory inputs.

Connectivity of SNpc and VTA DA Systems

Midbrain DA ascending pathways are organized in three major tracts (Fig. 17.1). The mesolimbic and mesocortical pathways project from the VTA to the NAcc, limbic areas, i.e. amygdala, septal area, bed nucleus of the stria terminalis (BNST), and the PFC, and are mainly implicated in associative learning, reward signaling and goal directed behavior [26, 28, 71–73]. The nigrostriatal pathway projects from the SNpc to the dorsal striatum, and is primarily involved in the regulation of motor activity, exploration and action selection [42]. Most of the knowledge regarding the mapping of inputs to VTA and SNpc DA neurons derives from classical tract-tracing studies [74] and viral-based tracing approaches [70] (Fig. 17.2).

Only a small proportion of cortical excitatory inputs innervate VTA and SNpc DA neurons [70]. Notably, VTA DA neurons receive fewer cortical inputs when compared with SNpc DA neurons. Cortical inputs to SNpc DA neurons are mostly coming from the primary and secondary motor cortices, whereas the only major cortical projection to the VTA originates from the PFC (mainly prelimbic and infralimbic cortices) [70, 75]. The BNST sends most of its projections to VTA DA neurons [76–78], and less to SNpc DA neurons [70]. From the midbrain and hindbrain, VTA and SNpc DA neurons receive the largest input from the superior col-

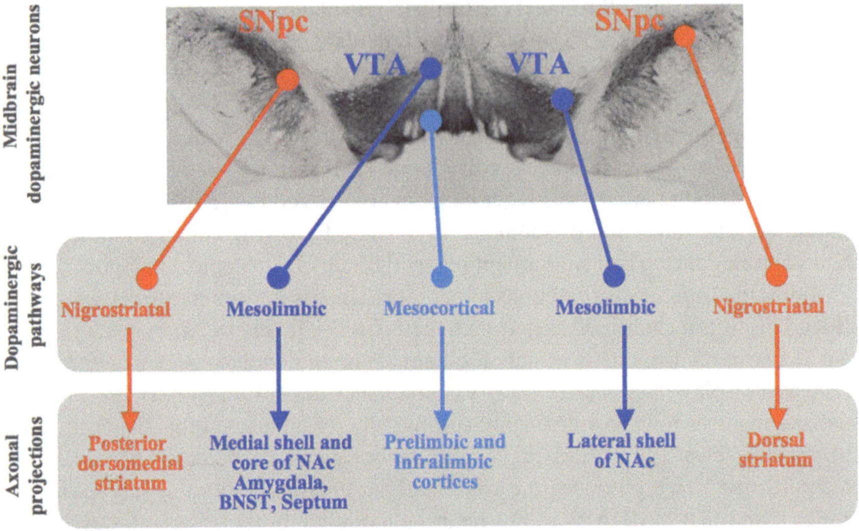

Fig. 17.1 Schematic organization of the three major midbrain ascending pathways. The mesolimbic and mesocortical pathways project from the ventral tegmental area (*VTA*) to the nucleus accumbens (*NAc*), limbic areas such as amygdala, septum, bed nucleus of the stria terminalis (*BNST*) and the prelimbic and infralimbic cortices. The nigrostriatal pathway projects from the pars compacta of the substantia nigra (*SNpc*) to the dorsal striatum

liculus, the periaqueductal gray (PAG) and the dorsal raphe (DR). The pedunculotegmental nucleus (PPTg) preferentially projects to the SNpc DA neurons, whereas laterodorsal tegmental nucleus (LDTg) preferentially projects to VTA DA neurons. The dorsal striatum (DS), the external globus pallidus (GPe) and the SNpr are major inhibitory GABAergic inputs controlling SNpc DA cell activity. Within the midbrain, the tail of the ventral tegmental area (tVTA) [79], also named the rostromedial tegmental nucleus (RMTg) [80], appears as a GABAergic inhibitory structure heavily projecting to the VTA and SNpc. Hence, the tVTA/RMTg regulates the activity of midbrain DA systems [81, 82]. In summary, although the input-output connectome of midbrain DA neurons has been partially elucidated at the structural level, the connectivity at the level of individual DA neuron requires to be further explored [83].

Integrative Properties of SNpc and VTA DA Neurons

Change in Excitability: Integration and Balance Between Inhibitory and Excitatory Inputs

Similar to other neuronal types, DA cell dendrites receive information from tens of thousands of synaptic inputs. This information are coordinated and stored via highly complex processes of dendritic integration of both inhibitory and excitatory

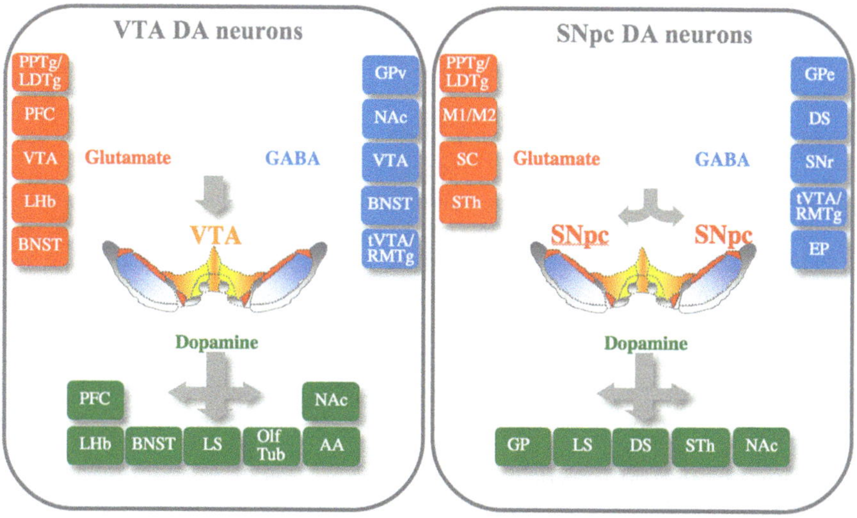

Fig. 17.2 Schematic organization of the principal afferents and efferents to the VTA (*left*) and SNpc (*right*) dopamine neurons. For clarity, only some of the projection are shown. *Red* indicate excitatory glutamatergic structure, *blue*, inhibitory GABAergic structures and *green*, dopaminergic target areas. Glutamatergic and GABAergic control the excitability of VTA and SNpc dopaminergic neurons. The pedunculotegmental nucleus (*PPTg*) preferentially projects to SNpc dopaminergic neurons, whereas laterodorsal tegmental nucleus (*LDTg*) preferentially projects to VTA dopaminergic neurons. The VTA dopaminergic neurons receive fewer cortical inputs than SNpc dopaminergic neurons. *SC* superior colliculus, *PAG* the periaqueductal *gray*, *DR* dorsal raphe, *DS* The dorsal striatum, *GPe* external globus pallidus tVTA/RMTg tail of the ventral tegmental area/rostromedial tegmental nucleus. There are only a few overlaps between dopaminergic efferents from the VTA and the SNpc. *PFC* prefrontal cortex, *LHb* lateral habenula, *BNST* bed nucleus of the stria terminalis, *LS* lateral septum, *OlfTub* olfactory tubercle, *AA* amygdala, *NAc* nucleus accumbens, *STh* subthalamic nucleus, *GP* globus pallidus, *DS* dorsal striatum

synaptic inputs. This integrative computation can influence sub-threshold membrane potentials, and play a role in the switch between tonic and phasic DA signal. Endocannabinoids, by serving as retrograde messengers and key modulators of synaptic functions, participate in this switch [2]. Indeed, they can finely tune firing activity and pattern of VTA DA neurons [84, 85] and, consequently, their phasic DA release in the NAc [86]. Hence, the emerging picture is that the endocannabinoid system acts as a local device for DA neurons to switch their firing pattern and activity in response to stimuli not only in the VTA [84, 85] but also in the SNpc [23, 24]. Given that physiological significance of endocannabinoid signaling at synapses onto DA neurons is reflected in the activity of these cells in response to input stimulation, these lipid molecules ultimately finely tune phasic versus tonic, and *vice versa*, DA release in the terminal regions [86, 87]. These considerations highlight how powerfully the endocannabinoid system might regulate not only DA volume transmission, but also DA modulation of cortical and subcortical information processing. In addition, they help explaining the correlation between an unbalanced endocannabinoid signal and altered DA-dependent processes underpinning diverse pathological conditions of both nigrostriatal and mesocorticolimbic systems [4].

Role of Endocannabinoids on GABA Afferents

As already mentioned, a tight regulation of DA levels in terminal regions is crucial as DA regulates key features of motivated behaviors to provide the adaptability of behavioral outputs required for species survival, such as for instance approach towards and withdrawal from rewarding and aversive stimuli, respectively [71, 88, 89].

DA neuronal activity results from the finest regulation of intrinsic and extrinsic mechanisms. Both VTA and SNpc DA neurons are subject to major background GABA inputs, whose activation results in either inhibition of their spontaneous activity (in terms of both firing activity and number of active cells) and/or in triggering bursts and pauses in DA cells [64, 90]. As a result, the dissection of which synapse is equipped with molecular architecture of a given endocannabinoid is extremely relevant from a functional point of view.

From an anatomical point of view, CB_1 receptors are located on GABAergic synapses onto VTA and SNpc DA cells [6, 8]. In the VTA, 2-AG biosynthetic enzyme diacylglycerol (DAG) lipase is found in DA cells at the level of the plasma membrane, and the main degrading enzyme monoacylglycerol (MAG) lipase is localized at a presynaptic level [6]. In the SNpc, instead, the molecular determinants for 2-AG have not been identified yet. In addition, anatomical evidence indicates the presence of the enzyme N-acyl phosphatidylethanolamine phospholipase D (NAPE-PLD) within the midbrain [91], thus supporting the notion that *N*-acylethanolamines such as anandamide and endogenous ligands to peroxisome proliferator-activated receptor-α (PPARα) (i.e. oleoylethanolamide, palmitoylethanolamide) are abundant under basal conditions in midbrain slices [1, 10].

Notably, the endocannabinoid/vanilloid N-arachidonoyl-dopamine (NADA) can be also detected within the SNpc, but only upon K^+-induced depolarization or activation of postsynaptic metabotropic glutamate receptor type-1 (mGluR1) on SNpc DA cells [8, 24]. Hence, in the SNpc, electrophysiological evidence points to a role of tonic NADA released by DA cells upon mGluR1 activation resulting from glutamate spillover from nearby synapses likely arising from subthalamic nucleus [24]. Thus, concomitant activation of excitatory inputs might be associated to endocannabinoid-mediated inhibition of GABAergic afferents, thereby enhancing DA cell responsiveness to excitatory stimuli and resulting in burst firing [24]. In this scenario (i.e. SNpc), the endocannabinoid produced by DA cells *on demand* (i.e. cell membrane depolarization, activation of postsynaptic muscarinic receptors) [23, 24] is most likely 2-AG, which mediates depolarization-induced suppression of inhibition (DSI), a short-term form of synaptic plasticity. Particularly, 2-AG would activate CB_1 receptors on striatonigral terminals, which in turn decrease GABA release and lead to intrinsic inhibition of DA cells [23]. Notably, Yanovsky et al. [23] questioned the canonical mode of action, and described for the first time a tonic endocannabinoid tone at these synapses. Hence, they suggested that endocannabinoids and DA are co-released within the SNpc to modulate striatonigral GABAergic fibers through the activation of CB_1 and D1 receptors, respectively. Remarkably, both DA- and endocannabinoid- induced changes in inhibition occur independently from each other. Indeed, activation of D1 receptors by somato-dendritically released DA would emphasize extrinsic inhibition to the expenses of a decreased intrinsic inhibition due to the counteraction of DSI. On the other hand, CB_1 receptor activation would promote intrinsic inhibition [23]. Given that both D1 and CB_1 receptors are located on striatonigral fibers, whereas D1 receptors are only on pallidonigral fibers but not on intrinsic fibers, co-release of DA and endocannabinoids might most likely occur upon different scenarios, such as discrete types of discharge rates and patterns of DA neurons. This would ultimately allow for a differential tuning of inhibition of DA cells within SNpc, and would be consistent with the spatio-temporal definition of endocannabinoid actions at synapses [23] (Fig. 17.3).

In the VTA, electrophysiological evidence converges to a role of 2-AG in modulating GABA inputs, whereas there is no indication supporting a role for either anandamide or NADA in regulating plasticity at these synapses [85, 92, 93]. Indeed, either intracellular loading of DAG lipase inhibitors or G-protein inhibitors into DA neurons proved to block endocannabinoid- mediated actions on discrete GABA receptors [85, 92, 93]. These observations are supported by the localization of 2-AG-synthesizing enzyme DAG lipase in the DA cell [6], which would release 2-AG following group I mGluR activation [92]. Noteworthy, release of 2-AG by VTA DA cells would occur under conditions similar to those of SNpc DA cells. Particularly, it has been shown that under conditions favoring bursting 2-AG decreases either $GABA_A$- and/or $GABA_B$- mediated responses of VTA DA cells [85, 92]. In fact, activation of group I mGluRs or increased intracellular Ca^{2+}concentrations, and block of small conductance calcium-activated potassium channels (SK channels) [85, 94] can lead to 2-AG production and its dependent effects [92]. Thus, DA neuron responsiveness to excitatory input would increase, and in the presence of a diminished inhibition, the balance of activity might most likely shift toward excitation and bursting [95, 96] (Fig. 17.4). Notably, sex differences in 2-AG effects have been revealed in the rat VTA [93]. In particular, electrophysiological evidence suggests that female rats displayed a tonic 2-AG signaling acting upon CB_1 receptors onto discrete GABA inputs onto VTA DA cells. Noteworthy, no sex dichotomy is found in CB_1 receptor expression and function in the VTA, whereas activational effects of sex hormones regulate the density of these receptors [97]. In particular, CB_1 receptor density decreases in both the PFC and amygdala of female rats. Noteworthy, both of these regions are critically involved in decision-making and learning processes underlying goal-directed behaviours [98, 99], in which important gender differences have been described [100]. Thus, such an enhanced 2-AG tone onto inhibitory inputs onto VTA DA cells might contribute not only to disinhibition of DA neurons, but also to sex-dependent differences in their function under normal and pathological conditions even before gonadal function fully matures [101–103]. This is in agreement with the increasing number of studies [101, 104–106] highlighting that sex/gender differences of brain structure and network function are not simply restricted to structures primarily involved in sexual behavior (see also Chap. 12).

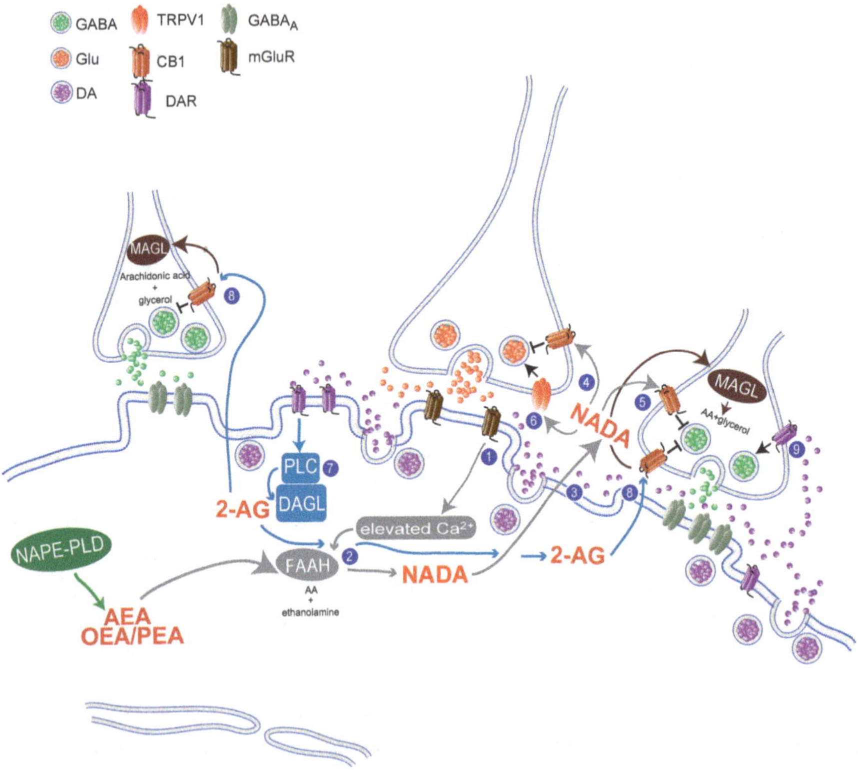

Fig. 17.3 Schematic organization of endocannabinoid system onto dopamine neurons of the Substiantia Nigra pars compacta. Proposed mechanism by which endocannabinoids, namely 2-arachidonoylglycerol (*2-AG*) and N-arachidonoyldopamine (*NADA*), regulate synaptic transmission onto dopamine (*DA*) neurons of the pars compact of Substantia Nigra (*SNpc*) are illustrated. Glutamate spillover activates postsynaptic type 1 metabotropic glutamate receptors (*mGluR*) (*1*) [24] with consequent increase in intracellular Ca^{2+} (*2*) [24], thus leading to generation of NADA. Arachidonic acid for NADA synthesis is provided by the FAAH, which appears to be directly involved in NADA synthesis [186]. Subsequently, NADA binds to presynaptic CB_1 receptors expressed on glutamate (*4*) [8] and GABA terminals (*5*) [8, 24] to reduce neurotransmitter release. NADA can also bind TRPV1 receptors located on glutamatergic terminals (*6*) [8] and finely tune SNpc DA neuron activity. Activation of postsynaptic dopamine D2 (*DAR*) cooperates to enhance 2-AG synthesis. 2-AG is produced *on demand* via a Ca^{2+}-independent mechanism involving activation of phospholipase C (PLC) (*7*) [24], which in turn cleaves phosphatidylinositol 4,5-bisphosphate (PIP2) into inositol trisphosphate (IP3) and DAG. DAG is then hydrolyzed by diacylglycerol lipase (*DAGL*). 2-AG moves retrogradely across the synaptic cleft to activate CB_1 receptors located on GABAergic terminals (*8*) [23]. A subset of presynaptic GABAergic terminals (i.e. striatonigral) co-express DAD1 receptors (*DAR*), whose activation by somatodendritically released DA counteracts 2-AG effects (*9*) [23]. The strengthening of extrinsic inhibition in the network reduces local (i.e. intranigral) inhibition. The two discrete retrograde signaling mechanisms (i.e. 2-AG and DA) operate independently, and their functional need remains elusive. However, the co-expression of DAR and CB_1 receptors on striatonigral fibers, while DAR are exclusively present on pallidonigral fibers and absent in intranigral fibers, allows the two retrograde signals for changes the influence of intrinsic versus extrinsic inhibition [23]. Whether these terminals are equipped with the monoacylglycerol (MAG) lipase (*MAGL*) has to be established yet.

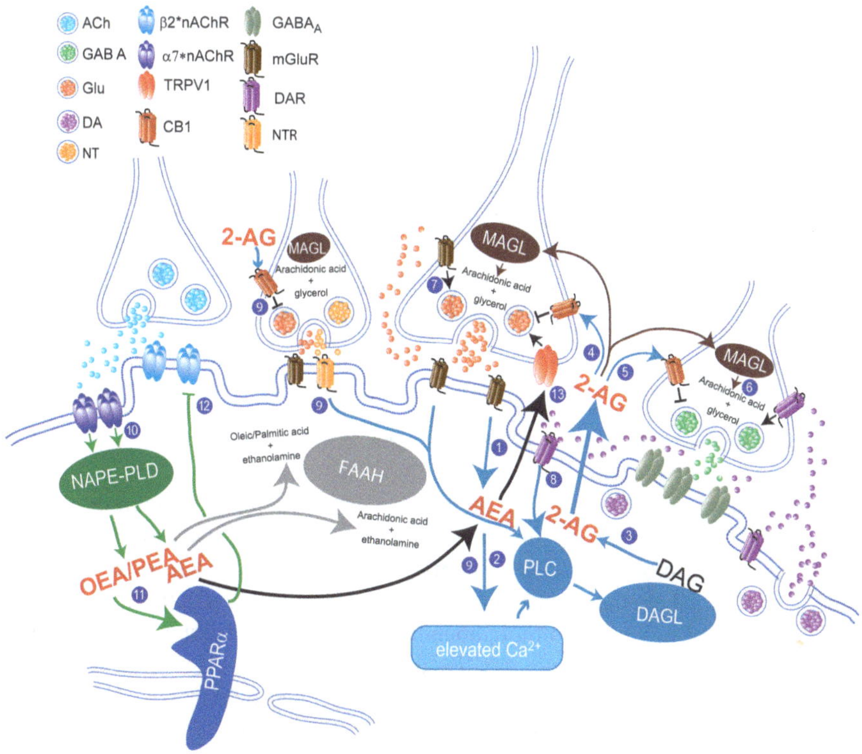

Fig. 17.4 Schematic organization of endocannabinoid system onto dopamine neurons of the ventral tegmental area. Proposed mechanisms by which endocannabinoids (i.e. 2-arachidonoylglycerol and anandamide) and N-acylethanolamines (*OEA* and *PEA*) regulate synaptic transmission onto ventral tegmental area (*VTA*) dopamine (*DA*) neurons are illustrated. Anandamide (*AEA*) and 2-arachidonoylglycerol (*2-AG*) are produced *on demand* by two Ca^{2+}-dependent enzymes, which are N-acylphosphatidylethanolamine hydrolyzing phospholipase D (*NAPE-PLD*) and diacylglycerol lipase (*DAGL*), respectively. Activation of metabotropic glutamate receptors (*mGluR*) (*1*) [84] increases intracellular Ca^{2+} (*2*) [84, 85] that activates phospholipase C (*PLC*), which in turn cleaves phosphatidylinositol 4,5-bisphosphate (PIP2) into inositol trisphosphate (IP3) and DAG [84, 154]. DAG is then hydrolyzed by DAGL to generate 2-AG (*3*) [84]. 2-AG binds to presynaptic CB_1 receptors expressed on glutamate (*4*) [84, 154] and GABA terminals (*5*) [85, 93, 187] to depress neurotransmitter release [84, 154]. 2-AG is mainly degraded by the enzyme MAG lipase (*MAGL*) expressed in both glutamatergic and GABAergic terminals (*6*) [6, 93]. Presynaptic activation of mGluRs on glutamatergic terminals concurs with 2-AG to dampen further excitability of DA cells by decreasing probability of glutamate release (*7*) [85]. Activation of postsynaptic dopamine D2 (*DAR*) cooperates to enhance 2-AG synthesis and release (*8*) [154, 187]. Activation of post-synaptic neurotensin receptors (*NTR*) leads to G-protein dependent activation of PLC, and consequent generation of PIP2, IP3 and DAG (*9*) [115]. Subsequent hydrolysis of DAG by DAGL generates 2-AG, which then moves retrogradely across the synaptic cleft to activate CB_1 receptors located on a subset of glutamatergic terminals co-releasing NT, thus resulting in inhibition of glutamate and NT release [115]. Whether these terminals are equipped with MAGL has to be established yet. AEA is synthesized, along with OEA and PEA, by NAPE-PLD (*10*). AEA activates TRPV1 receptors located on presynaptic glutamatergic terminals (*13*) [154], whose activation leads to increased spontaneous activity of DA cells [87, 113]. OEA and PEA are instead endogenous ligands of type α Peroxisome proliferator-activated receptors (*PPARα*), whose activation

Although immunocytochemical investigation of CB_1 receptors failed to precisely identify the origin of GABAergic afferents [6], electrophysiological evidence suggests CB_1 receptors to be localized onto those inputs arising from ventral pallidum [85, 93, 107], RMTg nucleus [93, 108] and local GABA neurons [109]. Given that these discrete inputs play important roles in controlling the number of spontaneously active DA neurons [90] and their own discharge rate [108, 110, 111], it is paramount to examine whether these synapses are differently equipped/enriched with discrete players of 2-AG signalling machinery. Hence, since GABA removal induces disinhibition bursts [64], the endocannabinoids NADA and 2-AG might be likely involved in transiently silencing inhibitory synapses onto midbrain DA cells, thereby contributing to their phasic excitation in the framework of multiple signaling modalities.

Role of Endocannabinoids on Glutamatergic Afferents

Endocannabinoid modulation of excitatory synapses impinging upon midbrain DA neurons has been extensively investigated within the past decade. CB_1 receptors have been found to be located on glutamatergic/asymmetric synapses in the SNpc and VTA [6, 8], where they serve as targets for endocannabinoids. Particularly, CB_1 receptors have been identified more abundantly on vesicular glutamate transporter 1 (VGLUT1)-positive terminals, predicted to be of cortical origin, rather than on VGLUT2- expressing terminals, expected to be of subcortical origin [74], in close proximity to VTA DA neuron dendrites [7].

To date, SNpc and VTA DA cells appear to synthesize and release diverse endocannabinoids acting on different targets, in agreement with their spatial definition. For instance, in the SNpc, 2-AG has not been reported to modulate excitatory inputs onto DA neurons. In contrast, anandamide was first shown to be released upon K^+-induced depolarization and to enhance glutamatergic inputs via activation of ionotropic transient receptor potential vanilloid type 1 (TRPV1) receptors [112]. Subsequently, NADA was also shown to modulate DA cell excitability through activation of either TRPV1 or CB_1 receptor [8]. Importantly, under basal conditions NADA concentrations are barely detectable within the SNpc [8, 24], thus supporting the hypothesis that it could be released only upon particular functional states of DA cells. Since NADA synthesis depends upon DA synthesis itself, a plausible scenario might be that under basal conditions low levels of NADA are released to preferentially activate TRPV1 receptors and enhance glutamate release. Instead,

(*11*) [113, 124] leads to phosphorylation and, thereby, negative modulation of somatodendritic nicotinic acetylcholine receptors containing the β2 subunit (*β2*nAChRs*) (*12*) [10, 113, 124]. Synthesis of OEA and PEA depends upon activation of somatodendritic nAChRs containing the α7 subunit (α7nAChRs) as well as increases in intracellular Ca^{2+} (*10*) [10]. The increased levels of OEA and PEA acting via PPARα serve to prevent aberrant hypercholinergic-driven excitation of DA neurons (*12*) [10]. AEA, OEA and PEA are mainly degraded by the fatty acid amide hydrolase (*FAAH*) located on the postsynaptic compartment [6] upon either prolonged or excessive depolarization, NADA binds to CB1 receptors to dampen glutamatergic transmission [8] (Fig. 17.3). Whether a similar scenario applies to the neighboring cells of the VTA is still not determined, although anandamide excites these cells via TRPV1 [113], whose activation results in increased DA levels in the NAc [87].

In the VTA, CB_1 receptors are localized on asymmetric synapses at the opposite site of the main synthesizing enzyme for 2-AG, that is the DAG lipase [6], within the DA cells. Several lines of evidence indicate 2-AG as the key endocannabinoid released *on demand* by VTA DA neurons, which mediates both short- and long-term forms of synaptic plasticity. Particularly, 2-AG mediates depolarization-induced suppression of excitation (DSE) [9], a form of short-term plasticity that most likely serves to limit pathological excitation of DA neurons, such as that observed under ischemic-reperfusion injury [9]. Additionally, 2-AG is released by DA neurons during behaviourally relevant patterns of synaptic activity, such as a brief burst of excitatory synaptic activity [84]. Under these conditions, mGluR1 activation and enhanced intracellular Ca^{2+} levels contribute to its synthesis and release, ultimately leading to transient and selective silencing of excitatory inputs onto the neuron itself. Thus, 2-AG ensures a fine modulation of both spike and burst probability of VTA DA cells [84] (Fig. 17.4).

2-AG also plays a role in diverse forms of long-term synaptic plasticity. Particularly, it mediates long-term depression (LTD) [114, 115], and negatively gates long-term potentiation (LTP) at these synapses [7]. Indeed, low frequency stimulation (LFS)-induced LTD requires 2-AG, since pharmacological inhibition of either phospholipase C (PLC) or DAG lipase, both critical for 2-AG biosynthesis, abolished LFS–LTD, whose induction also necessitates an increased postsynaptic intracellular Ca^{2+} through L-type Ca^{2+} channels [114]. Another form of 2-AG-mediated LTD is expressed by VTA DA cells, and this is the insulin-dependent LTD [116]. Indeed, acute activation of insulin receptors is able to induce a Ca^{2+}- independent release of 2-AG, which binds to and activate presynaptic CB_1 receptors located selectively on glutamatergic terminals, and induces LTD [116]. Notably, 2-AG also mediates LTD in response to activation of neurotensin (NT) receptors [115]. Hence, it appears that under conditions resulting in a significant release of NT in the VTA, NT co-released with glutamate from VGLUT-positive axon terminals could negatively regulate excitatory inputs onto DA neurons and, therefore, LTP induction. Notably, NT-induced long lasting depression of glutamate release via 2-AG requires Gq-protein-mediated activation of PLC and subsequent DAG-lipase activity, but not raises in Ca^{2+} levels. In turn, 2-AG via CB_1 receptor activation reduces glutamate release by inhibiting voltage-dependent Ca^{2+} channels (VDCC) [115] (Fig. 17.4). This finding is particularly relevant from a pathophysiological perspective. Indeed, the observations that NT CSF levels are dramatically reduced in drug-free schizophrenic patients, and that an altered NT neurotransmission within the VTA can be associated with mesolimbic DA hyperactivity characteristic of schizophrenia [117], highlight the relevance of such a tight and long lasting regulation of DA cell excitability by NT via CB_1 receptors. Remarkably, 2-AG, released by DA neurons and through activation of CB_1 receptors on VGLUT1-positive terminals, also negatively regulates spike time-dependent LTP induction [7]. Thus, it appears that under circumstances of strengthened excitatory plasticity, DA cells would release 2-AG, which mediates LTD and impairs LTP at the same synapses to protect DA cells from aberrant excitation, while simultaneously silencing inhibitory afferents [92]. Altogether, these findings might help elucidating the paradox, and explain the controversy, that exposure to *Cannabis* might be either a self-medication for psychotic patients and psycothomimetic [118, 119]. Hence, one has always to take into account that differences exist between the effects of CB_1 receptor activation by endogenous ligands and CB_1 receptor agonists (e.g. *Cannabis*, spice drugs), and, finally yet importantly, that endocannabinoid system states may differ among and within individuals.

Role of Endocannabinoids on Cholinergic Afferents

Midbrain DA cell firing and pattern are also powerfully controlled by extrinsic cholinergic inputs arising from the LDTg and the PPTg nuclei [120] through activation of nicotinic receptors (nAChRs) [121, 122]. Midbrain DA cells express two major forms of nAChRs, high-affinity β2*-nAChRs and low-affinity α7-nAChRs [123], where β2*-nAChRs enable the transition from tonic to phasic activity [122]. Remarkably, β2*-nAChRs can be negatively modulated by endogenous ligands of PPARα, such as *N*-acylethanolamines and fatty acids [1, 10, 124]. Notably, the *N*-acylethanolamines oleoyl-ethanolamide (OEA) and palmitoyl-ethanolamide (PEA) are regarded as belonging to the "extended family" of endocannabinoids [125]. In fact, both the enzyme FAAH and NAPE-PLD, key in degradation and synthesis of the endocannabinoid anandamide, tightly regulate levels of other *N*-acylethanolamines along with anandamide [126, 127]. Thus, endogenous PPARα ligands, such as the anorectic OEA [128] and the anti-inflammatory PEA [129], by sharing with anandamide both the anabolic and degradative pathway [130] can produce an indirect activation of other receptors and the so-called 'entourage effect' [131–134].

N-acylethanolamines, by activating PPARα, decrease spontaneous activity of VTA DA cells and the number of spontaneously active DA neurons through a rapid non-genomic mechanism [10, 124]. These effects, rapid in onset and blocked by the tyrosine kinase inhibitor genistein [113], are indicative of phosphorylation of β2*-nAChRs as the underlying mechanism of PPARα-mediated actions [10, 124]. *N*-acylethanolamines are found in all mammalian tissues [135] and are abundantly present within the midbrain [1, 10], where they enable VTA DA cells to switch between tonic/phasic modes of activity that are tightly regulated by β2*-nAChRs [10, 122]. Additionally, their synthesis and/or release occurs *on demand* upon α7-nAChRs activation [10]. Hence, in VTA DA neurons, acetylcholine and *N*-acylethanolamines appear to control each other in a negative feedback mechanism, where high acetylcholine enhance OEA and PEA levels to negatively modulate β2*-nAChRs downstream to PPARα activation in order to prevent aberrant DA cell excitation [10] (Fig. 17.4). Thus, by modulating VTA DA cell excitability, PPARα may have consequences for a number of behavioral responses known to be sensitive to the function of DA circuits. Since physiological activation of these neurons occurs across three dimensions that affect firing rate, firing pattern and the proportion of spontaneously active neurons, and that PPARα activation has been shown to affect at least two of these [124], it is very important to further examine the role of these nuclear receptors within the midbrain. Particularly, given the prominent categorical difference between SNpc and VTA DA neurons with respect to energy metabolism [136], one could speculate that some might be less susceptible to metabolic distress. This is particularly important because mitochondrial dysfunction appears to be critical to the pathogenesis of sporadic Parkinson's disease, which is due to degeneration of DA neurons within the SNpc [137] whereas those within the VTA are spared. Therefore, whether or not DA cells within the SNpc are under PPARα regulation similar to the VTA is a critical issue.

Physiological Events Triggering the Release of Endocannabinoids in Midbrain DA Regions

The aforementioned molecular machinery for 2-AG negative feedback pathway is remarkably conserved at both glutamatergic and GABAergic synapses in the VTA [6], and most likely in the SNpc. However, a prominent role for NADA has been described exclusively within the SNpc [8, 24]. Thus, it appears that endocannabinoids, by regulating DA cell activity either in an homo- or hetero- synaptic fashion, may not only contribute to the rewarding/teaching signal encoded by these neurons, but may also regulate the start/stop of sequence learning [138]. Additionally, endocannabinoids may participate in discrete mechanisms aimed at DA cell homeostatic regulation. As a result, given the role of the endocannabinoid system in modulating DA neuronal function, they might take part in motor skill learning and in the action-habit transformation. Hence, several and diametrically opposite events in life are able to trigger endocannabinoid synthesis and release from midbrain DA cells, such as for instance the pursuit of natural rewards, physical exercise, stress and noxious stimuli.

Endocannabinoids have long been involved in appetitive-motivational aspects of reward-directed behaviors [139–141]. The first demonstration that 2-AG is the endocannabinoid enhancing neural mechanisms of cue-motivated reward seeking, thereby supporting its key role for multiple forms of synaptic plasticity within the VTA [2], has been elegantly provided by Cheer's group [142]. In particular, the Authors showed the existence of a single neural signaling mechanism through which CB_1 receptor antagonists can effectively reduce the influence that environmental cues exert over motivated behavior. Indeed, it is well established that the motivational state of the individual regulates those appetitive behaviors involving the pursuit of reward [143, 144]. Thus, Oleson and colleagues demonstrated that the disruption of 2-AG signaling in the VTA simultaneously decreases cue-evoked reward seeking as well as DA levels in the NAc shell [142]. Conversely, pharmacological enhancement of 2-AG signaling (i.e. by inhibition of its main degrading enzyme) within the VTA produces the opposite behavioral output. Therefore, the Authors indicate 2-AG in the VTA as the main endocannabinoid involved in mediating cue-motivated reward directed behaviors [142]. Notably, several lines of evidence postulate that cue-encoding VTA DA cells form discrete neural assemblies with GABAergic synapses, which will consequently allow for a fine-tuning of DA neural activity itself during reward seeking. Accordingly, activation of CB_1 receptors located on discrete GABAergic terminals have been shown to decrease GABA release onto VTA DA neurons [85, 93, 107–109], thereby resulting in their disinhibition.

Similarly, it has been proven that physical exercise enhances the endocannabinoid system in both humans [145] and rodents [32, 146–148], and that its involvement may take part in beneficial effects of exercise. Notably, the reinforcing properties of wheel running are ascribed to activation of the endocannabinoid system on wheel-based seeking which, similarly to the pursuit of reward, is a form of appetitive behavior [148]. Remarkably, activation of CB_1 receptors on GABAergic terminals onto VTA DA cells exerts a tonic stimulatory influence on voluntary running performance [147]. Particularly, an endocannabinoid-dependent stimulation of CB_1 receptors located on VTA GABAergic terminals can define running performance as endocannabinoids regulate DA-dependent reward-directed processes. In turn, both acute and repeated voluntary exercise considerably affect VTA DA cell activity [147]. Particularly, following acute and repeated voluntary wheel running GABA-$CB_1^{-/-}$ mice display a marked decline in the number of spontaneously active DA neurons, which also show a reduction in both firing rate and bursting activity. Although the nature of the endocannabinoid exerting this tonic control on inhibitory transmission via CB_1 receptor activation has to be elucidated yet, this elegant study highlights and remarks the role of this endogenous system in the regulation of VTA DA neuron activity after both acute and repeated voluntary running. In this scenario, shortly after running the absence of CB_1 receptors at inhibitory synapses shifts excitation towards inhibition of DA neuronal activity, and their behavioral output [147]. Consistent with these findings are the observations that CB_1 deficient mice exhibit dramatic motor deficits [149], and lack endocannabinoid-dependent LTD at the indirect pathway within the basal ganglia [150].

Excitatory synaptic transmission onto VTA DA cells is also potentiated by feeding-related peptides [151, 152], and this might underlie the motivation to obtain food [151, 153]. On the other hand, insulin in the VTA might reduce the salience of cues and/or contexts associated with food by reducing the strength of excitatory synapses onto DA neurons via 2-AG [116]. In fact, electrophysiological observations suggest that when plasmatic levels of insulin are increased, such as immediately following a caloric meal consumption (i.e. sweet high fat meal), 2-AG levels increase within the VTA, where its effects are unmasked by pharmacological blockade of CB_1 receptors [116]. Thus, it appears that insulin receptor activation triggers 2-AG synthesis and release from VTA DA neurons to activate CB_1 receptors on a subset of excitatory terminals. This results in a decreased glutamate release from presynaptic terminals without affecting GABA transmission [116], in according to the spatial definition of endocannabinoid actions. Notably, these findings do not reflect the effort exerted by the animal to obtain a palatable food, but they show that enhanced circulating levels of insulin via CB_1 receptor activation may reduce simple appetitive behaviors displayed routinely before food consumption as well as the salience of food-related cues [116].

As often mentioned in this chapter, midbrain DA neurons are potential indicators of salient, pleasant, or noxious stimuli, and they are tightly regulated by both excitatory and inhibitory inputs equipped with both CB_1 and/or $TRPV_1$ receptors. As a result, these neurons via endocannabinoids can integrate signals from the periphery, such as those induced by nociceptive and stressful stimuli. In particular, in freely moving rats a peripheral noxious stimulation of the tail determined a $TRPV_1$-dependent increase in extracellular DA levels in the NAc [87]. Notably, no identification of the endogenous mediator(s) has been provided yet, and no direct measurements of endocannabinoids/endovanilloids have been performed under those conditions. However, the observations that activation of $TRPV_1$ in acute midbrain slice preparations increases the firing rate of VTA DA cells [87, 113], a mechanism dependent upon presynaptic facilitation of glutamatergic transmission, suggest the involvement of an endovanilloid/endocannabinoid molecule (e.g. anandamide, NADA) [87, 154]. A similar scenario occurs within the SNpc in DA cells [8, 112]. Remarkably, to date, different studies propose discrete lipid molecules as mediators of presynaptic facilitation of glutamatergic transmission in the VTA and SNpc. Hence, while in the VTA anandamide is in the spotlight for such a fine tuning of DA cell excitability via presynaptic inhibition or facilitation of glutamate release downstream of activation of CB_1 and $TRPV_1$ receptors [154], respectively, in the SNpc NADA appears to be acting in such a position [8].

Noteworthy, as mentioned above in this chapter, anatomical and functional heterogeneity of midbrain DA neurons has been well established [30, 41, 42, 155]. Accordingly, a subset of VTA DA neurons can decrease their impulse activity in response to a noxious stimulus, such as the tail-pinch, as a result from tVTA/RMTg activation [108, 111]. Consequently, given that the tVTA/RMTg nucleus strongly projects to the SNpc, and that tVTA/RMTg afferents onto VTA DA cells are under negative control of CB_1 receptors [93, 108, 156], one would expect these latter to be also located on tVTA/RMTg terminals onto SNpc cells. However, since different subsets of midbrain DA neurons receive topographic inputs from different subregions of the tVTA/RMTg [80], this prediction might not be correct.

Midbrain DA neurons also confer the individual with the capability to update and adapt to formerly learned behavioral responses in a changing environment, and this is essential for coping with adverse events. Notably, mesocortical DA transmission has long been involved in cognitive flexibility processes contributing to behavioral flexibility [157], which takes account of both set-shifting and reversal learning. Importantly, all of these phenomena are markedly impaired by stress [158–161]. In addition, converging evidence suggests the involvement of the endocannabinoid system in stress-induced responses [162–168]. Hence, genetic deletion of CB_1 receptors induces hypersensitivity to stress [162, 169], being rodents more vulnerable to stress-induced, depression-like changes in behavior and gene expression [162, 164, 168, 170–174]. Conversely, FAAH deficient mice display reduced anxiety-like behaviors in both the elevated plus maze and the light-dark box, both effects requiring CB_1 receptors activation [175, 176]. Remarkably, while endocannabinoid levels are influenced by acute and chronic stressors in diverse brain regions [168, 177, 178], direct measurements within the VTA and SNpc have never been performed. However, in socially isolated rats changes in mRNA expression of DAGL isozyme β were found in both the VTA and SNpc [179]. Noteworthy, important excitatory inputs onto VTA DA cells undergo aberrant plasticity in response to stress in CB_1 receptor deficient mice [180], and this might ultimately influence VTA DA neuronal activity. In particular, the anterolateral portion of the BNST shows a widespread connectivity with those systems that are phylogenetically conserved to process stress signals [181]. Hence, activation of the PFC (in particular of the infralimbic portion) enhances excitatory afferents to the BNST, which in turn excites BNST neurons projecting to the VTA [76, 77], thus leading to their phasic activation [182]. In this scenario, CB_1 receptor activation within the BNST dampens VTA DA cell excitation induced by PFC stimulation, thus unveiling the role of endocannabinoids in this neural circuit [182, 183]. Therefore, the observation that in CB_1 receptor deficient mice acute stress shifts synaptic plasticity at excitatory afferents onto BNST neurons from LTD to LTP [180] was somehow predictable. Accordingly, pharmacological blockade of CB_1 receptors within the PFC is not able to reduce stress-evoked increases of extracellular DA levels in the PFC [184]. Nonetheless, caution should be used when comparing the aforementioned studies given that the procedures to induce acute stress are different, being acute restraint stress [180] and acute tail-pinch stress [184], respectively. Conversely, while one study reported that rat prefrontal glucocorticoid receptors play a role in stress-induced enhanced DA levels within the PFC [184], the other ruled out the involvement of mouse glucocorticoid receptor activation in the switch of CB_1 receptor-dependent synaptic plasticity at excitatory synapses onto BNST [180]. Irrespective of the diverse species and stressors used, however, it is worth to mention that acute restraint stress does increase both the firing rate and bursting pattern of VTA DA neurons in awake rats [185]. Whether or not the endocannabinoid system is involved in such an increased frequency and pattern of discharge of VTA DA neurons remains to be elucidated.

Concluding Remarks

The endocannabinoid system plays a fundamental role in making short- and long-term modifications to DA neural circuits and related behaviours. Nevertheless, many facets of this interplay are still unclear. For instance, it remains to elucidate whether or not there is a bias for one of the diverse endocannabinoids to be mobilized by DA cells *on demand*, or to be tonically produced depending on the state the synapse resides. Since CB_1 receptors are abundantly expressed in the midbrain, and both anandamide and NADA serve two masters (i.e. CB_1 and TRPV1 receptors) located on terminals impinging upon midbrain DA cells, it is tempting to speculate that neuromodulatory functions of the "extended family" of endocannabinoids might help keeping synaptic efficacy within those dynamic ranges that guarantee a proper integration of interoceptive stimuli and sensory information. Importantly, this balance is processed in order to facilitate motor control and promote learning.

The number of exciting discoveries brought up to the scientific community almost on a daily basis highlights the importance of such a fine regulation. Hence, given that deficits in DA neuromodulation contribute to the pathophysiology of several neuropsychiatric disorders, and the emerging and prominent role of the endocannabinoid system in modulating DA neuronal activity and transmission, pharmacotherapies aimed at precisely regulating the endogenous levels represent a promising treatment for diverse psychiatric and neurological disorders. Equally important is that potential therapeutic benefits of cannabis and cannabinoids are currently under heavy analysis in many countries worldwide.

Reference

1. Melis M, Carta G, Pistis M, Banni S. Physiological role of peroxisome proliferator-activated receptors type alpha on dopamine systems. CNS Neurol Disord Drug Targets. 2013;12:70–7.
2. Melis M, Pistis M. Hub and switches: endocannabinoid signalling in midbrain dopamine neurons. Philos Trans R Soc Lond B Biol Sci. 2012;367:3276–85.
3. Melis M, Muntoni AL, Pistis M. Endocannabinoids and the processing of value-related signals. Front Pharmacol. 2012;3:7.
4. Fernandez-Ruiz J, Hernandez M, Ramos JA. Cannabinoid-dopamine interaction in the pathophysiology and treatment of CNS disorders. CNS Neurosci Ther. 2010;16:e72–e91.
5. Melis M, Pistis P. Endocannabinoid signaling in midbrain dopamine neurons: more than physiology? Curr Neuropharmacol. 2007;5:268–77.
6. Matyas F, Urban GM, Watanabe M, Mackie K, Zimmer A, Freund TF, Katona I. Identification of the sites of 2-arachidonoylglycerol synthesis and action imply retrograde endocannabinoid signaling at both GABAergic and glutamatergic synapses in the ventral tegmental area. Neuropharmacology. 2008;54:95–107.
7. Kortleven C, Fasano C, Thibault D, Lacaille JC, Trudeau LE. The endocannabinoid 2-arachidonoylglycerol inhibits long-term potentiation of glutamatergic synapses onto ventral tegmental area dopamine neurons in mice. Eur J Neurosci. 2011;33:1751–60.
8. Marinelli S, Di Marzo V, Florenzano F, Fezza F, Viscomi MT, van der Stelt M, Bernardi G, Molinari M, Maccarrone M, Mercuri NB. N-arachidonoyl-dopamine tunes synaptic transmission onto dopaminergic neurons by activating both cannabinoid and vanilloid receptors. Neuropsychopharmacology. 2007;32:298–308.
9. Melis M, Pillolla G, Bisogno T, Minassi A, Petrosino S, Perra S, Muntoni AL, Lutz B, Gessa GL, Marsicano G, Di Marzo V, Pistis M. Protective activation of the endocannabinoid system during ischemia in dopamine neurons. Neurobiol Dis. 2006;24:15–27.
10. Melis M, Scheggi S, Carta G, Madeddu C, Lecca S, Luchicchi A, Cadeddu F, Frau R, Fattore L, Fadda P, Ennas MG, Castelli MP, Fratta W, Schilstrom B, Banni S, De Montis MG, Pistis M. PPARalpha regulates cholinergic-driven activity of midbrain dopamine neurons via a novel mechanism involving alpha7 nicotinic acetylcholine receptors. J Neurosci. 2013;33:6203–11.
11. Di Marzo V, Fontana A, Cadas H, Schinelli S, Cimino G, Schwartz JC, Piomelli D. Formation and inactivation of endogenous cannabinoid anandamide in central neurons. Nature. 1994;372:686–91.
12. Alger BE. Endocannabinoids at the synapse a decade after the dies mirabilis (29 March 2001): what we still do not know. J Physiol. 2012;590:2203–12.
13. Kano M, Ohno-Shosaku T, Hashimotodani Y, Uchigashima M, Watanabe M. Endocannabinoid-mediated control of synaptic transmission. Physiol Rev. 2009;89:309–80.
14. Katona I, Freund TF. Multiple functions of endocannabinoid signaling in the brain. Annu Rev Neurosci. 2012;35:529–58.

15. Castillo PE, Younts TJ, Chavez AE, Hashimotodani Y. Endocannabinoid signaling and synaptic function. Neuron. 2012;76:70–81.
16. Heifets BD, Castillo PE. Endocannabinoid signaling and long-term synaptic plasticity. Annu Rev Physiol. 2009;71:283–306.
17. Abbott LF, Regehr WG. Synaptic computation. Nature. 2004;431:796–803.
18. Dittman JS, Kreitzer AC, Regehr WG. Interplay between facilitation, depression, and residual calcium at three presynaptic terminals. J Neurosci. 2000;20:1374–85.
19. Klyachko VA, Stevens CF. Excitatory and feed-forward inhibitory hippocampal synapses work synergistically as an adaptive filter of natural spike trains. PLoS Biol. 2006;4:e207.
20. Fortune ES, Rose GJ. Short-term synaptic plasticity as a temporal filter. Trends Neurosci. 2001;24:381–5.
21. Lewis JE, Maler L. Dynamics of electrosensory feedback: short-term plasticity and inhibition in a parallel fiber pathway. J Neurophysiol. 2002;88:1695–706.
22. Abbott LF, Varela JA, Sen K, Nelson SB. Synaptic depression and cortical gain control. Science. 1997;275:220–4.
23. Yanovsky Y, Mades S, Misgeld U. Retrograde signaling changes the venue of postsynaptic inhibition in rat substantia nigra. Neuroscience. 2003;122:317–28.
24. Freestone PS, Guatteo E, Piscitelli F, di Marzo V, Lipski J, Mercuri NB. Glutamate spillover drives endocannabinoid production and inhibits GABAergic transmission in the Substantia Nigra pars compacta. Neuropharmacology. 2013;79C:467–75.
25. Schultz W. Predictive reward signal of dopamine neurons. J Neurophysiol. 1998;80:1–27.
26. Wise RA. Dopamine, learning and motivation. Nat Rev Neurosci. 2004;5:483–94.
27. Matsumoto M, Hikosaka O. Two types of dopamine neuron distinctly convey positive and negative motivational signals. Nature. 2009;459:837–41.
28. Schultz W. Behavioral dopamine signals. Trends Neurosci. 2007;30:203–10.
29. Ford CP, Mark GP, Williams JT. Properties and opioid inhibition of mesolimbic dopamine neurons vary according to target location. J Neurosci. 2006;26:2788–97.
30. Lammel S, Hetzel A, Hackel O, Jones I, Liss B, Roeper J. Unique properties of mesoprefrontal neurons within a dual mesocorticolimbic dopamine system. Neuron. 2008;57:760–73.
31. Margolis EB, Lock H, Hjelmstad GO, Fields HL. The ventral tegmental area revisited: is there an electrophysiological marker for dopaminergic neurons? J Physiol. 2006;577:907–24.
32. Ilango A, Kesner AJ, Keller KL, Stuber GD, Bonci A, Ikemoto S. Similar roles of substantia nigra and ventral tegmental dopamine neurons in reward and aversion. J Neurosci. 2014;34:817–22.
33. Brischoux F, Chakraborty S, Brierley DI, Ungless MA. Phasic excitation of dopamine neurons in ventral VTA by noxious stimuli. Proc Natl Acad Sci U S A. 2009;106:4894–9.
34. Abercrombie ED, Keefe KA, DiFrischia DS, Zigmond MJ. Differential effect of stress on in vivo dopamine release in striatum, nucleus accumbens, and medial frontal cortex. J Neurochem. 1989;52:1655–8.
35. Kalivas PW, Churchill L, Romanides A. Involvement of the pallidal-thalamocortical circuit in adaptive behavior. Ann N Y Acad Sci. 1999;877:64–70.
36. Kalivas PW, Nakamura M. Neural systems for behavioral activation and reward. Curr Opin Neurobiol. 1999;9:223–7.
37. Barr GA, Moriceau S, Shionoya K, Muzny K, Gao P, Wang S, Sullivan RM. Transitions in infant learning are modulated by dopamine in the amygdala. Nat Neurosci. 2009;12:1367–9.
38. Pezze MA, Feldon J. Mesolimbic dopaminergic pathways in fear conditioning. Prog Neurobiol. 2004;74:301–20.
39. Takahata R, Moghaddam B. Target-specific glutamatergic regulation of dopamine neurons in the ventral tegmental area. J Neurochem. 2000;75:1775–8.
40. Lammel S, Ion DI, Roeper J, Malenka RC. Projection-specific modulation of dopamine neuron synapses by aversive and rewarding stimuli. Neuron. 2011;70:855–62.
41. Lammel S, Lim BK, Ran C, Huang KW, Betley MJ, Tye KM, Deisseroth K, Malenka RC. Input-specific control of reward and aversion in the ventral tegmental area. Nature. 2012;491:212–7.

42. Roeper J. Dissecting the diversity of midbrain dopamine neurons. Trends Neurosci. 2013;36:336–42.
43. Nair-Roberts RG, Chatelain-Badie SD, Benson E, White-Cooper H, Bolam JP, Ungless MA. Stereological estimates of dopaminergic, GABAergic and glutamatergic neurons in the ventral tegmental area, substantia nigra and retrorubral field in the rat. Neuroscience. 2008;152:1024–31.
44. Henny P, Brown MT, Northrop A, Faunes M, Ungless MA, Magill PJ, Bolam JP. Structural correlates of heterogeneous in vivo activity of midbrain dopaminergic neurons. Nat Neurosci. 2012;15:613–9.
45. Dobi A, Margolis EB, Wang HL, Harvey BK, Morales M. Glutamatergic and nonglutamatergic neurons of the ventral tegmental area establish local synaptic contacts with dopaminergic and nondopaminergic neurons. J Neurosci. 2010;30:218–29.
46. Li SX, Wei YM, Shi HS, Luo YX, Ding ZB, Xue YX, Lu L, Yu CX. Glycogen synthase kinase-3beta in the ventral tegmental area mediates diurnal variations in cocaine-induced conditioned place preference in rats. Addict Biol. 2014;19(6):996–1005.
47. Spiga S, Serra GP, Puddu MC, Foddai M, Diana M. Morphine withdrawal-induced abnormalities in the VTA: confocal laser scanning microscopy. Eur J Neurosci. 2003;17:605–12.
48. Tepper JM, Sawyer SF, Groves PM. Electrophysiologically identified nigral dopaminergic neurons intracellularly labeled with HRP: light-microscopic analysis. J Neurosci. 1987;7:2794–806.
49. Fallon JH. Collateralization of monoamine neurons: mesotelencephalic dopamine projections to caudate, septum, and frontal cortex. J Neurosci. 1981;1:1361–8.
50. Fallon JH, Riley JN, Moore RY. Substantia nigra dopamine neurons: separate populations project to neostriatum and allocortex. Neurosci Lett. 1978;7:157–62.
51. Stamatakis AM, Jennings JH, Ung RL, Blair GA, Weinberg RJ, Neve RL, Boyce F, Mattis J, Ramakrishnan C, Deisseroth K, Stuber GD. A unique population of ventral tegmental area neurons inhibits the lateral habenula to promote reward. Neuron. 2013;80:1039–53.
52. Brown MT, Henny P, Bolam JP, Magill PJ. Activity of neurochemically heterogeneous dopaminergic neurons in the substantia nigra during spontaneous and driven changes in brain state. J Neurosci. 2009;29:2915–25.
53. Bunney BS, Walters JR, Roth RH, Aghajanian GK. Dopaminergic neurons: effect of antipsychotic drugs and amphetamine on single cell activity. J Pharmacol Exp Ther. 1973;185:560–71.
54. Guyenet PG, Aghajanian GK. Antidromic identification of dopaminergic and other output neurons of the rat substantia nigra. Brain Res. 1978;150:69–84.
55. Grace AA, Bunney BS. The control of firing pattern in nigral dopamine neurons: burst firing. J Neurosci. 1984;4:2877–90.
56. Grace AA, Bunney BS. The control of firing pattern in nigral dopamine neurons: single spike firing. J Neurosci. 1984;4:2866–76.
57. Tepper JM, Nakamura S, Young SJ, Groves PM. Autoreceptor-mediated changes in dopaminergic terminal excitability: effects of striatal drug infusions. Brain Res. 1984;309:317–33.
58. Tepper JM, Young SJ, Groves PM. Autoreceptor-mediated changes in dopaminergic terminal excitability: effects of increases in impulse flow. Brain Res. 1984;309:309–16.
59. Gariano RF, Tepper JM, Sawyer SF, Young SJ, Groves PM. Mesocortical dopaminergic neurons. 1. Electrophysiological properties and evidence for soma-dendritic autoreceptors. Brain Res Bull. 1989;22:511–6.
60. Grace AA, Bunney BS. Intracellular and extracellular electrophysiology of nigral dopaminergic neurons—1. Identification and characterization. Neuroscience. 1983;10:301–15.
61. Romo R, Schultz W. Dopamine neurons of the monkey midbrain: contingencies of responses to active touch during self-initiated arm movements. J Neurophysiol. 1990;63:592–606.
62. Ungless MA, Grace AA. Are you or aren't you? Challenges associated with physiologically identifying dopamine neurons. Trends Neurosci. 2012;35:422–30.
63. Gonon FG. Nonlinear relationship between impulse flow and dopamine released by rat midbrain dopaminergic neurons as studied by in vivo electrochemistry. Neuroscience. 1988;24:19–28.

64. Lobb CJ, Wilson CJ, Paladini CA. A dynamic role for GABA receptors on the firing pattern of midbrain dopaminergic neurons. J Neurophysiol. 2010;104:403–13.
65. Blythe SN, Atherton JF, Bevan MD. Synaptic activation of dendritic AMPA and NMDA receptors generates transient high-frequency firing in substantia nigra dopamine neurons in vitro. J Neurophysiol. 2007;97:2837–50.
66. Blythe SN, Wokosin D, Atherton JF, Bevan MD. Cellular mechanisms underlying burst firing in substantia nigra dopamine neurons. J Neurosci. 2009;29:15531–41.
67. Kitai ST, Shepard PD, Callaway JC, Scroggs R. Afferent modulation of dopamine neuron firing patterns. Curr Opin Neurobiol. 1999;9:690–7.
68. Zweifel LS, Parker JG, Lobb CJ, Rainwater A, Wall VZ, Fadok JP, Darvas M, Kim MJ, Mizumori SJ, Paladini CA, Phillips PE, Palmiter RD. Disruption of NMDAR-dependent burst firing by dopamine neurons provides selective assessment of phasic dopamine-dependent behavior. Proc Natl Acad Sci U S A. 2009;106:7281–8.
69. Tsai HC, Zhang F, Adamantidis A, Stuber GD, Bonci A, de Lecea L, Deisseroth K. Phasic firing in dopaminergic neurons is sufficient for behavioral conditioning. Science. 2009;324:1080–4.
70. Watabe-Uchida M, Zhu L, Ogawa SK, Vamanrao A, Uchida N. Whole-brain mapping of direct inputs to midbrain dopamine neurons. Neuron. 2012;74:858–73.
71. Fields HL, Hjelmstad GO, Margolis EB, Nicola SM. Ventral tegmental area neurons in learned appetitive behavior and positive reinforcement. Annu Rev Neurosci. 2007;30:289–316.
72. Ikemoto S. Dopamine reward circuitry: two projection systems from the ventral midbrain to the nucleus accumbens-olfactory tubercle complex. Brain Res Rev. 2007;56:27–78.
73. Redgrave P, Gurney K, Reynolds J. What is reinforced by phasic dopamine signals? Brain Res Rev. 2008;58:322–39.
74. Geisler S, Derst C, Veh RW, Zahm DS. Glutamatergic afferents of the ventral tegmental area in the rat. J Neurosci. 2007;27:5730–43.
75. Carr DB, Sesack SR. Projections from the rat prefrontal cortex to the ventral tegmental area: target specificity in the synaptic associations with mesoaccumbens and mesocortical neurons. J Neurosci. 2000;20:3864–73.
76. Georges F, Aston-Jones G. Potent regulation of midbrain dopamine neurons by the bed nucleus of the stria terminalis. J Neurosci. 2001;21:RC160.
77. Georges F, Aston-Jones G. Activation of ventral tegmental area cells by the bed nucleus of the stria terminalis: a novel excitatory amino acid input to midbrain dopamine neurons. J Neurosci. 2002;22:5173–87.
78. Jennings JH, Sparta DR, Stamatakis AM, Ung RL, Pleil KE, Kash TL, Stuber GD. Distinct extended amygdala circuits for divergent motivational states. Nature. 2013;496:224–8.
79. Kaufling J, Veinante P, Pawlowski SA, Freund-Mercier MJ, Barrot M. Afferents to the GABAergic tail of the ventral tegmental area in the rat. J Comp Neurol. 2009;513:597–621.
80. Jhou TC, Geisler S, Marinelli M, Degarmo BA, Zahm DS. The mesopontine rostromedial tegmental nucleus: a structure targeted by the lateral habenula that projects to the ventral tegmental area of Tsai and substantia nigra compacta. J Comp Neurol. 2009;513:566–96.
81. Barrot M, Sesack SR, Georges F, Pistis M, Hong S, Jhou TC. Braking dopamine systems: a new GABA master structure for mesolimbic and nigrostriatal functions. J Neurosci. 2012;32:14094–101.
82. Bourdy R, Barrot M. A new control center for dopaminergic systems: pulling the VTA by the tail. Trends Neurosci. 2012;35:681–90.
83. Matsuda W, Furuta T, Nakamura KC, Hioki H, Fujiyama F, Arai R, Kaneko T. Single nigrostriatal dopaminergic neurons form widely spread and highly dense axonal arborizations in the neostriatum. J Neurosci. 2009;29:444–53.
84. Melis M, Perra S, Muntoni AL, Pillolla G, Lutz B, Marsicano G, Di Marzo V, Gessa GL, Pistis M. Prefrontal cortex stimulation induces 2-arachidonoyl-glycerol-mediated suppression of excitation in dopamine neurons. J Neurosci. 2004;24:10707–15.

85. Riegel AC, Lupica CR. Independent presynaptic and postsynaptic mechanisms regulate endocannabinoid signaling at multiple synapses in the ventral tegmental area. J Neurosci. 2005;24:11070–8.
86. Cheer JF, Wassum KM, Sombers LA, Heien ML, Ariansen JL, Aragona BJ, Phillips PE, Wightman RM. Phasic dopamine release evoked by abused substances requires cannabinoid receptor activation. J Neurosci. 2007;27:791–5.
87. Marinelli S, Pascucci T, Bernardi G, Puglisi-Allegra S, Mercuri NB. Activation of TRPV1 in the VTA excites dopaminergic neurons and increases chemical- and noxious-induced dopamine release in the nucleus accumbens. Neuropsychopharmacology. 2005;30:864–70.
88. Ungless MA, Magill PJ, Bolam JP. Uniform inhibition of dopamine neurons in the ventral tegmental area by aversive stimuli. Science. 2004;303:2040–2.
89. Oleson EB, Cachope R, Fitoussi A, Cheer JF. Tales from the dark side: do neuromodulators of drug withdrawal require changes in endocannabinoid tone? Prog Neuropsychopharmacol Biol Psychiatry. 2013.
90. Valenti O, Grace AA. Antipsychotic drug-induced increases in ventral tegmental area dopamine neuron population activity via activation of the nucleus accumbens-ventral pallidum pathway. Int J Neuropsychopharmacol. 2010;13:845–60.
91. Egertova M, Simon GM, Cravatt BF, Elphick MR. Localization of N-acyl phosphatidylethanolamine phospholipase D (NAPE-PLD) expression in mouse brain: a new perspective on N-acylethanolamines as neural signaling molecules. J Comp Neurol. 2008;506:604–15.
92. Pan B, Hillard CJ, Liu QS. Endocannabinoid signaling mediates cocaine-induced inhibitory synaptic plasticity in midbrain dopamine neurons. J Neurosci. 2008;28:1385–97.
93. Melis M, De Felice M, Lecca S, Fattore L, Pistis M. Sex-specific tonic 2-arachidonoylglycerol signaling at inhibitory inputs onto dopamine neurons of Lister Hooded rats. Front Integr Neurosci. 2013;7(93):1–13.
94. Fiorillo CD, Williams JT. Glutamate mediates an inhibitory postsynaptic potential in dopamine neurons. Nature. 1998;394:78–82.
95. Overton PG, Clark D. Burst firing in midbrain dopaminergic neurons. Brain Res Brain Res Rev. 1997;25:312–34.
96. Cooper DC. The significance of action potential bursting in the brain reward circuit. Neurochem Int. 2002;41:333–40.
97. Castelli MP, Fadda P, Casu A, Spano MS, Casti A, Fratta W, Fattore L. Male and female rats differ in brain cannabinoid CB1 receptor density and function and in behavioural traits predisposing to drug addiction: effect of ovarian hormones. Curr Pharm Des. 2013.
98. Schoenbaum G, Chiba AA, Gallagher M. Orbitofrontal cortex and basolateral amygdala encode expected outcomes during learning. Nat Neurosci. 1998;1:155–9.
99. Bechara A, Damasio H, Damasio AR, Lee GP. Different contributions of the human amygdala and ventromedial prefrontal cortex to decision-making. J Neurosci. 1999;19:5473–81.
100. van den Bos R, Homberg J, de Visser L. A critical review of sex differences in decision-making tasks: focus on the Iowa Gambling Task. Behav Brain Res. 2013;238:95–108.
101. Fattore L, Melis M, Fadda P, Fratta W. Sex differences in addictive disorders. Front Neuroendocrinol. in press.
102. Anker JJ, Carroll ME. Females are more vulnerable to drug abuse than males: evidence from preclinical studies and the role of ovarian hormones. Curr Top Behav Neurosci. 2011;8:73–96.
103. Harrison PJ, Tunbridge EM. Catechol-O-methyltransferase (COMT): a gene contributing to sex differences in brain function, and to sexual dimorphism in the predisposition to psychiatric disorders. Neuropsychopharmacology. 2008;33:3037–45.
104. Huang GZ, Woolley CS. Estradiol acutely suppresses inhibition in the hippocampus through a sex-specific endocannabinoid and mGluR-dependent mechanism. Neuron. 2012;74:801–8.
105. Vrticka P, Neely M, Walter Shelly E, Black JM, Reiss AL. Sex differences during humor appreciation in child-sibling pairs. Soc Neurosci. 2013;8:291–304.

106. Chung WC, Auger AP. Gender differences in neurodevelopment and epigenetics. Pflugers Arch. 2013;465:573–84.
107. Melis M, Pillolla G, Perra S, Colombo G, Muntoni AL, Pistis M. Electrophysiological properties of dopamine neurons in the ventral tegmental area of Sardinian alcohol-preferring rats. Psychopharmacology (Berl). 2009;201:471–81.
108. Lecca S, Melis M, Luchicchi A, Muntoni AL, Pistis M. Inhibitory inputs from rostromedial tegmental neurons regulate spontaneous activity of midbrain dopamine cells and their responses to drugs of abuse. Neuropsychopharmacology. 2012;37:1164–76.
109. Szabo B, Siemes S, Wallmichrath I. Inhibition of GABAergic neurotransmission in the ventral tegmental area by cannabinoids. Eur J Neurosci. 2002;15:2057–61.
110. Hong S, Jhou TC, Smith M, Saleem KS, Hikosaka O. Negative reward signals from the lateral habenula to dopamine neurons are mediated by rostromedial tegmental nucleus in primates. J Neurosci. 2011;31:11457–71.
111. Jhou TC, Fields HL, Baxter MG, Saper CB, Holland PC. The rostromedial tegmental nucleus (RMTg), a GABAergic afferent to midbrain dopamine neurons, encodes aversive stimuli and inhibits motor responses. Neuron. 2009;61:786–800.
112. Marinelli S, Di Marzo V, Berretta N, Matias I, Maccarrone M, Bernardi G, Mercuri NB. Presynaptic facilitation of glutamatergic synapses to dopaminergic neurons of the rat substantia nigra by endogenous stimulation of vanilloid receptors. J Neurosci. 2003;23:3136–44.
113. Melis M, Pillolla G, Luchicchi A, Muntoni AL, Yasar S, Goldberg SR, Pistis M. Endogenous fatty acid ethanolamides suppress nicotine-induced activation of mesolimbic dopamine neurons through nuclear receptors. J Neurosci. 2008;28:13985–94.
114. Haj-Dahmane S, Shen RY. Regulation of plasticity of glutamate synapses by endocannabinoids and the cyclic-AMP/protein kinase A pathway in midbrain dopamine neurons. J Physiol. 2010;588:2589–604.
115. Kortleven C, Bruneau LC, Trudeau LE. Neurotensin inhibits glutamate-mediated synaptic inputs onto ventral tegmental area dopamine neurons through the release of the endocannabinoid 2-AG. Neuropharmacology. 2012;63:983–91.
116. Labouebe G, Liu S, Dias C, Zou H, Wong JC, Karunakaran S, Clee SM, Phillips AG, Boutrel B, Borgland SL. Insulin induces long-term depression of ventral tegmental area dopamine neurons via endocannabinoids. Nat Neurosci. 2013;16:300–8.
117. Binder EB, Kinkead B, Owens MJ, Nemeroff CB. The role of neurotensin in the pathophysiology of schizophrenia and the mechanism of action of antipsychotic drugs. Biol Psychiatry. 2001;50:856–72.
118. D'Souza DC, Perry E, MacDougall L, Ammerman Y, Cooper T, Wu YT, Braley G, Gueorguieva R, Krystal JH. The psychotomimetic effects of intravenous delta-9-tetrahydrocannabinol in healthy individuals: implications for psychosis. Neuropsychopharmacology. 2004;29:1558–72.
119. Kolliakou A, Joseph C, Ismail K, Atakan Z, Murray RM. Why do patients with psychosis use cannabis and are they ready to change their use? Int J Dev Neurosci. 2011;29:335–46.
120. Lodge DJ, Grace AA. The laterodorsal tegmentum is essential for burst firing of ventral tegmental area dopamine neurons. Proc Natl Acad Sci U S A. 2006;103:5167–72.
121. Schilstrom B, Rawal N, Mameli-Engvall M, Nomikos GG, Svensson TH. Dual effects of nicotine on dopamine neurons mediated by different nicotinic receptor subtypes. Int J Neuropsychopharmacol. 2003;6:1–11.
122. Mameli-Engvall M, Evrard A, Pons S, Maskos U, Svensson TH, Changeux JP, Faure P. Hierarchical control of dopamine neuron-firing patterns by nicotinic receptors. Neuron. 2006;50:911–21.
123. Clarke PB, Schwartz RD, Paul SM, Pert CB, Pert A. Nicotinic binding in rat brain: autoradiographic comparison of [3H]acetylcholine, [3H]nicotine, and [125I]-alpha-bungarotoxin. J Neurosci. 1985;5:1307–15.
124. Melis M, Carta S, Fattore L, Tolu S, Yasar S, Goldberg SR, Fratta W, Maskos U, Pistis M. Peroxisome proliferator-activated receptors-alpha modulate dopamine cell activity through nicotinic receptors. Biol Psychiatry. 2010;68:256–64.

125. Pistis M, Melis M. From surface to nuclear receptors: the endocannabinoid family extends its assets. Curr Med Chem. 2010;17:1450–67.
126. Ueda N, Puffenbarger RA, Yamamoto S, Deutsch DG. The fatty acid amide hydrolase (FAAH). Chem Phys Lipids. 2000;108:107–21.
127. Tsuboi K, Okamoto Y, Ikematsu N, Inoue M, Shimizu Y, Uyama T, Wang J, Deutsch DG, Burns MP, Ulloa NM, Tokumura A, Ueda N. Enzymatic formation of N-acylethanolamines from N-acylethanolamine plasmalogen through N-acylphosphatidylethanolamine-hydrolyzing phospholipase D-dependent and -independent pathways. Biochim Biophys Acta. 2011;1811:565–77.
128. Fu J, Gaetani S, Oveisi F, Lo Verme J, Serrano A, Rodríguez De Fonseca F, Rosengarth A, Luecke H, Di Giacomo B, Tarzia G, Piomelli D. Oleylethanolamide regulates feeding and body weight through activation of the nuclear receptor PPAR-alpha. Nature. 2003;425:90–3.
129. LoVerme J, Russo R, La Rana G, Fu J, Farthing J, Mattace-Raso G, Meli R, Hohmann A, Calignano A, Piomelli D. Rapid broad-spectrum analgesia through activation of peroxisome proliferator-activated receptor-alpha. J Pharmacol Exp Ther. 2006;319:1051–61.
130. Lambert DM, Di Marzo V. The palmitoylethanolamide and oleamide enigmas: are these two fatty acid amides cannabimimetic? Curr Med Chem. 1999;6:757–73.
131. Smart D, Jonsson KO, Vandevoorde S, Lambert DM, Fowler CJ. 'Entourage' effects of N-acyl ethanolamines at human vanilloid receptors. Comparison of effects upon anandamide-induced vanilloid receptor activation and upon anandamide metabolism. Br J Pharmacol. 2002;136:452–8.
132. Di Marzo V, Melck D, Orlando P, Bisogno T, Zagoory O, Bifulco M, Vogel Z, De Petrocellis L. Palmitoylethanolamide inhibits the expression of fatty acid amide hydrolase and enhances the anti-proliferative effect of anandamide in human breast cancer cells. Biochem J. 2001;358:249–55.
133. De Petrocellis L, Davis JB, Di Marzo V. Palmitoylethanolamide enhances anandamide stimulation of human vanilloid VR1 receptors. FEBS Lett. 2001;506:253–6.
134. Jonsson KO, Vandevoorde S, Lambert DM, Tiger G, Fowler CJ. Effects of homologues and analogues of palmitoylethanolamide upon the inactivation of the endocannabinoid anandamide. Br J Pharmacol. 2001;133:1263–75.
135. Hansen HH, Ikonomidou C, Bittigau P, Hansen SH, Hansen HS. Accumulation of the anandamide precursor and other N-acylethanolamine phospholipids in infant rat models of in vivo necrotic and apoptotic neuronal death. J Neurochem. 2001;76:39–46.
136. Greene JG. Gene expression profiles of brain dopamine neurons and relevance to neuropsychiatric disease. J Physiol. 2006;575:411–6.
137. Hauser MA, Li YJ, Xu H, Noureddine MA, Shao YS, Gullans SR, Scherzer CR, Jensen RV, McLaurin AC, Gibson JR, Scott BL, Jewett RM, Stenger JE, Schmechel DE, Hulette CM, Vance JM. Expression profiling of substantia nigra in Parkinson disease, progressive supranuclear palsy, and frontotemporal dementia with parkinsonism. Arch Neurol. 2005;62:917–21.
138. Jin X, Costa RM. Start/stop signals emerge in nigrostriatal circuits during sequence learning. Nature. 2010;466:457–62.
139. Ward SJ, Dykstra LA. The role of CB1 receptors in sweet versus fat reinforcement: effect of CB1 receptor deletion, CB1 receptor antagonism (SR141716A) and CB1 receptor agonism (CP-55940). Behav Pharmacol. 2005;16:381–8.
140. Cota D, Marsicano G, Lutz B, Vicennati V, Stalla GK, Pasquali R, Pagotto U. Endogenous cannabinoid system as a modulator of food intake. Int J Obes Relat Metab Disord. 2003;27:289–301.
141. Mahler SV, Smith KS, Berridge KC. Endocannabinoid hedonic hotspot for sensory pleasure: anandamide in nucleus accumbens shell enhances 'liking' of a sweet reward. Neuropsychopharmacology. 2007;32:2267–78.
142. Oleson EB, Beckert MV, Morra JT, Lansink CS, Cachope R, Abdullah RA, Loriaux AL, Schetters D, Pattij T, Roitman MF, Lichtman AH, Cheer JF. Endocannabinoids shape ac-

cumbal encoding of cue-motivated behavior via CB1 receptor activation in the ventral tegmentum. Neuron. 2012;73:360–73.
143. Bindra D. A unfied interpretation of emotion and motivation. Ann N Y Acad Sci. 1969;159:1071–83.
144. Kelley SP, Mittleman G. Effects of hippocampal damage on reward threshold and response rate during self-stimulation of the ventral tegmental area in the rat. Behav Brain Res. 1999;99:133–41.
145. Sparling PB, Giuffrida A, Piomelli D, Rosskopf L, Dietrich A. Exercise activates the endocannabinoid system. Neuroreport. 2003;14:2209–11.
146. De Chiara V, Errico F, Musella A, Rossi S, Mataluni G, Sacchetti L, Siracusano A, Castelli M, Cavasinni F, Bernardi G, Usiello A, Centonze D. Voluntary exercise and sucrose consumption enhance cannabinoid CB1 receptor sensitivity in the striatum. Neuropsychopharmacology. 2010;35:374–87.
147. Dubreucq S, Durand A, Matias I, Bénard G, Richard E, Soria-Gomez E, Glangetas C, Groc L, Wadleigh A, Massa F, Bartsch D, Marsicano G, Georges F, Chaouloff F. Ventral tegmental area cannabinoid type-1 receptors control voluntary exercise performance. Biol Psychiatry. 2013;73:895–903.
148. Rasmussen EB, Hillman C. Naloxone and rimonabant reduce the reinforcing properties of exercise in rats. Exp Clin Psychopharmacol. 2011;19:389–400.
149. Zimmer A, Zimmer AM, Hohmann AG, Herkenham M, Bonner TI. Increased mortality, hypoactivity, and hypoalgesia in cannabinoid CB1 receptor knockout mice. Proc Natl Acad Sci U S A. 1999;96:5780–5.
150. Kreitzer AC, Malenka RC. Endocannabinoid-mediated rescue of striatal LTD and motor deficits in Parkinson's disease models. Nature. 2007;445:643–7.
151. Borgland SL, Chang SJ, Bowers MS, Thompson JL, Vittoz N, Floresco SB, Chou J, Chen BT, Bonci A. Orexin A/hypocretin-1 selectively promotes motivation for positive reinforcers. J Neurosci. 2009;29:11215–25.
152. Abizaid A, Liu ZW, Andrews ZB, Shanabrough M, Borok E, Elsworth JD, Roth RH, Sleeman MW, Picciotto MR, Tschöp MH, Gao XB, Horvath TL. Ghrelin modulates the activity and synaptic input organization of midbrain dopamine neurons while promoting appetite. J Clin Invest. 2006;116:3229–39.
153. Thompson JL, Borgland SL. A role for hypocretin/orexin in motivation. Behav Brain Res. 2011;217:446–53.
154. Melis M, Pistis M, Perra S, Muntoni AL, Pillolla G, Gessa GL. Endocannabinoids mediate presynaptic inhibition of glutamatergic transmission in rat ventral tegmental area dopamine neurons through activation of CB1 receptors. J Neurosci. 2004;24:53–62.
155. Neuhoff H, Neu A, Liss B, Roeper J. I(h) channels contribute to the different functional properties of identified dopaminergic subpopulations in the midbrain. J Neurosci. 2002;22:1290–302.
156. Lecca S, Melis M, Luchicchi A, Ennas MG, Castelli MP, Muntoni AL, Pistis M. Effects of drugs of abuse on putative rostromedial tegmental neurons, inhibitory afferents to midbrain dopamine cells. Neuropsychopharmacology. 2011;36:589–602.
157. Floresco SB, Magyar O. Mesocortical dopamine modulation of executive functions: beyond working memory. Psychopharmacology (Berl). 2006;188:567–85.
158. Liston C, Miller MM, Goldwater DS, Radley JJ, Rocher AB, Hof PR, Morrison JH, McEwen BS. Stress-induced alterations in prefrontal cortical dendritic morphology predict selective impairments in perceptual attentional set-shifting. J Neurosci. 2006;26:7870–4.
159. Cerqueira JJ, Mailliet F, Almeida OF, Jay TM, Sousa N. The prefrontal cortex as a key target of the maladaptive response to stress. J Neurosci. 2007;27:2781–7.
160. Butts KA, Weinberg J, Young AH, Phillips AG. Glucocorticoid receptors in the prefrontal cortex regulate stress-evoked dopamine efflux and aspects of executive function. Proc Natl Acad Sci U S A. 2011;108:18459–64.
161. Butts KA, Floresco SB, Phillips AG. Acute stress impairs set-shifting but not reversal learning. Behav Brain Res. 2013;252:222–9.

162. Haller J, Varga B, Ledent C, Barna I, Freund TF. Context-dependent effects of CB1 cannabinoid gene disruption on anxiety-like and social behaviour in mice. Eur J Neurosci. 2004;19:1906–12.
163. Patel S, Cravatt BF, Hillard CJ. Synergistic interactions between cannabinoids and environmental stress in the activation of the central amygdala. Neuropsychopharmacology. 2005;30:497–507.
164. Aso E, Ozaita A, Valdizan EM, Ledent C, Pazos A, Maldonado R, Valverde O. BDNF impairment in the hippocampus is related to enhanced despair behavior in CB1 knockout mice. J Neurochem. 2008;105:565–72.
165. Hill MN, McLaughlin RJ, Bingham B, Shrestha L, Lee TT, Gray JM, Hillard CJ, Gorzalka BB, Viau V. Endogenous cannabinoid signaling is essential for stress adaptation. Proc Natl Acad Sci U S A. 2010;107:9406–11.
166. Hill MN, Patel S, Campolongo P, Tasker JG, Wotjak CT, Bains JS. Functional interactions between stress and the endocannabinoid system: from synaptic signaling to behavioral output. J Neurosci. 2010;30:14980–6.
167. Riebe CJ, Wotjak CT. Endocannabinoids and stress. Stress. 2011;14:384–97.
168. Dubreucq S, Matias I, Cardinal P, et al. Genetic dissection of the role of cannabinoid type-1 receptors in the emotional consequences of repeated social stress in mice. Neuropsychopharmacology. 2012;37:1885–900.
169. Hill MN, Hillard CJ, McEwen BS. Alterations in corticolimbic dendritic morphology and emotional behavior in cannabinoid CB1 receptor-deficient mice parallel the effects of chronic stress. Cereb Cortex. 2011;21:2056–64.
170. Haller J, Bakos N. Stress-induced social avoidance: a new model of stress-induced anxiety? Physiol Behav. 2002;77:327–32.
171. Haller J, Bakos N, Rodriguiz RM, Caron MG, Wetsel WC, Liposits Z. Behavioral responses to social stress in noradrenaline transporter knockout mice: effects on social behavior and depression. Brain Res Bull. 2002;58:279–84.
172. Uriguen L, Perez-Rial S, Ledent C, Palomo T, Manzanares J. Impaired action of anxiolytic drugs in mice deficient in cannabinoid CB1 receptors. Neuropharmacology. 2004;46:966–73.
173. Steiner MA, Wanisch K, Monory K, Marsicano G, Borroni E, Bächli H, Holsboer F, Lutz B, Wotjak CT. Impaired cannabinoid receptor type 1 signaling interferes with stress-coping behavior in mice. Pharmacogenomics J. 2008;8:196–208.
174. Aso E, Ozaita A, Serra MA, Maldonado R. Genes differentially expressed in CB1 knockout mice: involvement in the depressive-like phenotype. Eur Neuropsychopharmacol. 2011;21:11–22.
175. Kathuria S, Gaetani S, Fegley D, Valiño F, Duranti A, Tontini A, Mor M, Tarzia G, La Rana G, Calignano A, Giustino A, Tattoli M, Palmery M, Cuomo V, Piomelli D. Modulation of anxiety through blockade of anandamide hydrolysis. Nat Med. 2003;9:76–81.
176. Moreira FA, Kaiser N, Monory K, Lutz B. Reduced anxiety-like behaviour induced by genetic and pharmacological inhibition of the endocannabinoid-degrading enzyme fatty acid amide hydrolase (FAAH) is mediated by CB1 receptors. Neuropharmacology. 2008;54:141–50.
177. Patel S, Roelke CT, Rademacher DJ, Hillard CJ. Inhibition of restraint stress-induced neural and behavioural activation by endogenous cannabinoid signalling. Eur J Neurosci. 2005;21:1057–69.
178. Rademacher DJ, Meier SE, Shi L, Ho WS, Jarrahian A, Hillard CJ. Effects of acute and repeated restraint stress on endocannabinoid content in the amygdala, ventral striatum, and medial prefrontal cortex in mice. Neuropharmacology. 2008;54:108–16.
179. Robinson SA, Loiacono RE, Christopoulos A, Sexton PM, Malone DT. The effect of social isolation on rat brain expression of genes associated with endocannabinoid signaling. Brain Res. 2010;1343:153–67.
180. Glangetas C, Girard D, Groc L, Marsicano G, Chaouloff F, Georges F. Stress switches cannabinoid type-1 (CB1) receptor-dependent plasticity from LTD to LTP in the bed nucleus of the stria terminalis. J Neurosci. 2013;33:19657–63.

181. Davis M, Walker DL, Miles L, Grillon C. Phasic vs sustained fear in rats and humans: role of the extended amygdala in fear vs anxiety. Neuropsychopharmacology. 2010;35:105–35.
182. Massi L, Elezgarai I, Puente N, Reguero L, Grandes P, Manzoni OJ, Georges F. Cannabinoid receptors in the bed nucleus of the stria terminalis control cortical excitation of midbrain dopamine cells in vivo. J Neurosci. 2008;28:10496–508.
183. Puente N, Cui Y, Lassalle O, Lafourcade M, Georges F, Venance L, Grandes P, Manzoni OJ. Polymodal activation of the endocannabinoid system in the extended amygdala. Nat Neurosci. 2011;14:1542–7.
184. Butts KA, Phillips AG. Glucocorticoid receptors in the prefrontal cortex regulate dopamine efflux to stress via descending glutamatergic feedback to the ventral tegmental area. Int J Neuropsychopharmacol. 2013;16:1799–807.
185. Anstrom KK, Woodward DJ. Restraint increases dopaminergic burst firing in awake rats. Neuropsychopharmacology. 2005;30:1832–40.
186. Hu SS, Bradshaw HB, Benton VM, Chen JS, Huang SM, Minassi A, Bisogno T, Masuda K, Tan B, Roskoski R Jr, Cravatt BF, Di Marzo V, Walker JM. The biosynthesis of N-arachidonoyl dopamine (NADA), a putative endocannabinoid and endovanilloid, via conjugation of arachidonic acid with dopamine. Prostaglandins Leukot Essent Fatty Acids. 2009;81:291–301.
187. Pan B, Hillard CJ, Liu QS. D2 dopamine receptor activation facilitates endocannabinoid-mediated long-term synaptic depression of GABAergic synaptic transmission in midbrain dopamine neurons via cAMP-protein kinase A signaling. J Neurosci. 2008;28:14018–30.

Erratum to: Cannabinoid-Alcohol Interactions

Luis A. Natividad, Paola Maccioni, Loren H. Parsons and Giancarlo Colombo

Erratum to:
Chapter 14 in P. Campolongo, L. Fattore (eds.), ***Cannabinoid Modulation of Emotion, Memory, and Motivation,***
DOI 10.1007/978-1-4939-2294-9_14

The publisher regrets the incorrect affiliations for Giancarlo Colombo and Paola Maccioni in Chapter 14 opening page and in the Contributors List of the print and online versions of this book. Their correct affiliations have been updated in the original chapter and are mentioned below:

Giancarlo Colombo, Institute of Neuroscience-Cagliari, National Research Council (CNR), Monserrato, Italy.

Paola Maccioni, Institute of Neuroscience-Cagliari, National Research Council (CNR), Monserrato, Italy.

The online version of the original chapter can be found under
DOI 10.1007/978-1-4939-2294-9_14

L. H. Parsons (✉)
Committee on the Neurobiology of Addictive Disorders
The Scripps Research Institute, La Jolla, CA, USA
e-mail: lparsons@scripps.edu

L. A. Natividad
Research Associate, Committee on the Neurobiology of Addictive Disorders
The Scripps Research Institute, La Jolla, CA, USA

P. Maccioni
Institute of Neuroscience-Cagliari, National Research Council (CNR), Monserrato, Italy

G. Colombo
Institute of Neuroscience-Cagliari, National Research Council (CNR), Monserrato, Italy

P. Campolongo, L. Fattore (eds.), *Cannabinoid Modulation of Emotion, Memory, and Motivation,* DOI 10.1007/978-1-4939-2294-9_18

FSC
www.fsc.org
MIX
Papier | Fördert gute Waldnutzung
FSC® C083411

Kolibri 360 GmbH
Ferdinand-Jühlke-Straße 7
99095 Erfurt, Deutschland
produktsicherheit@kolibri360.de